# Probability and Its Applications

*Published in association with the Applied Probability Trust*

*Editors:* S. Asmussen, J. Gani, P. Jagers, T.G. Kurtz

Photo of Charles Stein, in front, with, from left to right in the rear, Qi-Man Shao, Louis Chen and Larry Goldstein, taken at a conference at Stanford University held in honor of Charles Stein's 90th birthday on March 22nd, 2010

For further titles published in this series, go to www.springer.com/series/1560

Louis H.Y. Chen • Larry Goldstein • Qi-Man Shao

# Normal Approximation by Stein's Method

Louis H.Y. Chen
Department of Mathematics
National University of Singapore
10 Lower Kent Ridge Road
Singapore 119076
Republic of Singapore
matchyl@nus.edu.sg

Qi-Man Shao
Department of Mathematics
Hong Kong University of Science and Technology
Clear Water Bay, Kowloon
Hong Kong
China
maqmshao@ust.hk

Larry Goldstein
Department of Mathematics KAP 108
University of Southern California
Los Angeles, CA 90089-2532
USA
larry@usc.edu



ISSN 1431-7028
ISBN 978-3-642-26565-5
ISBN 978-3-642-15007-4 (eBook)
DOI 10.1007/978-3-642-15007-4
Springer Heidelberg Dordrecht London New York

Mathematics Subject Classification (2010): 60F05, 60B12, 62E17



*Cover design*: VTEX, Vilnius

Printed on acid-free paper

Springer is part of Springer Science+Business Media (www.springer.com)

*This book is dedicated to Charles Stein.*
*We also dedicate this book to our families.*

*Annabelle, Yitian, Yipei*
*Nancy*
*Jiena and Wenqi*

# Preface

Stein's method has developed considerably since its first appearance in 1972, and presently shows every sign that its range in theory and applications will continue to expand. Nevertheless, there must be some point along this continuing path when the method reaches a certain level of maturity that a thorough, self contained treatment, highlighted with a sampling of its many successes, is warranted. The authors of this book believe that now is this time.

In the years since Stein's method for the normal was introduced, the recognition of its power has only slowly begun to percolate throughout the probability community, helped along, no doubt, by the main references in the field over the last many years, first, the monograph of Stein (1986), the compilation of Diaconis and Holmes (2004), and the series of Barbour and Chen (2005a, 2005b). Nevertheless, to use one barometer, to date there exist only a small number of books or monographs, targeted generally and accessible at the graduate or undergraduate level, that make any mention of Stein's method for the normal at all, in particular, the texts of Stroock (2000) and Ross and Peköz (2007). With a thorough building up of the fundamentals necessary to cover the many forms that Stein's method for the normal can take to date, and the inclusion of a large number of recent developments in both theory and applications, we hope this book on normal approximation will continue to accelerate the appreciation, understanding, and use of Stein's method. Indeed, as interest in the method has steadily grown, this book was partly written to add to the list we can give in response to the many queries we have received over the years, regarding sources where one can go to learn more about the method, and, moreover, to get a sense of whether it can be applied to new situations.

We have many to thank for this book's existence. The first author would like to thank Charles Stein for his ideas which the former learned from him as a student and which has been a rich source of inspiration to him over the years. He would also like to thank his co-authors, Andrew Barbour, Kwok-Pui Choi, Xiao Fang, Yu-Kiang Leong, Qi-Man Shao and Aihua Xia, from whom he has benefited substantially through many stimulating discussions.

The second author first heard about Stein's method, for the Poisson case, in a lecture by Persi Diaconis, and he thanks his first teachers in that area, Richard Arratia and Louis Gordon, for conveying a real sense of the use of the Stein equation, and

Michael Waterman for providing a fountain of wonderful applications. He learned the most about the normal approximation version of the method, and about its applications, from his work with Yosi Rinott, to whom he is most grateful. He has also benefited greatly through all his other collaborations where Stein's method played a role, most notably those with Gesine Reinert, as well as with Aihua Xia, Mathew Penrose, and Haimeng Zhang.

The third author would like to thank Louis Chen for introducing him to Stein's method, and for the inspiration and insight he has provided.

All the authors would like to thank the Institute for Mathematical Sciences, at the National University of Singapore, for their support of the many Singapore conferences, which served as a nexus for the dissemination of the most recent discoveries by the participants, and for the creation of a perfect environment for the invention of new ideas.

For comments and suggestions regarding the preparation of this work the authors would particularly like to thank Jason Fulman, Ivan Nourdin and Giovanni Peccati for their guidance on the material in Chap. 14. Additionally, we thank Subhankar Ghosh and Wenxin Zhou for their help in various stages of the preparation of this book, and proofreading, and Xiao Fang for his assistance and help in writing parts of Chap. 7, on Discretized Normal Approximation. The first author was partially supported by the Tan Chin Tuan Centennial Professorship Grant C-389-000-010-101 at the National University of Singapore during the time this manuscript was prepared, the second author acknowledges the grant support of NSA-AMS 091026, and the third author acknowledges grant support from Hong Kong Research Grants Council (CERG-602608 and 603710).

For updates and further information on this book, please visit: http://mizar.usc.edu/~larry/nabsm.html.

# Contents

# Chapter 1
# Introduction

## 1.1 The Central Limit Theorem

The Central Limit Theorem is one of the most striking and useful results in probability and statistics, and explains why the normal distribution appears in areas as diverse as gambling, measurement error, sampling, and statistical mechanics. In essence, the Central Limit Theorem in its classical form states that a normal approximation applies to the distribution of quantities that can be modeled as the sum of many independent contributions, all of which are roughly the same size.

Thus mathematically justified, at least asymptotically, in practice the normal law may be used to approximate quantities ranging from a $p$-value of a hypothesis tests, the probability that a manufacturing process will remain in control or the chance of observing an unusual conductance reading in a laboratory experiment. However, even though in practice sample sizes may be large, or may appear to be sufficient for the purposes at hand, depending on that and other factors, the normal approximation may or may not be accurate. It is here the need for the evaluation of the quality of the normal approximation arises, which is the topic of this book.

The seeds of the Central Limit Theorem, or CLT, lie in the work of Abraham de Moivre, who, around the year 1733, not being able to secure himself an academic appointment, supported himself consulting on problems of probability and gambling. He approximated the limiting probabilities of the binomial distribution, the one which governs the behavior of the number

$$S_n = X_1 + \cdots + X_n \tag{1.1}$$

of successes in an experiment which consists of $n$ independent Bernoulli trials, each one having the same probability $p \in (0, 1)$ of success. de Moivre realized that even though the sum

$$P(S_n \le m) = \sum_{k \le m} \binom{n}{k} p^k (1-p)^{n-k}$$

that yields the cumulative probability of $m$ or fewer successes becomes unwieldy for even moderate values of $n$, there exists an easily computable, normal approximation to such probabilities that can be quite accurate even for moderate values of $n$.

L.H.Y. Chen et al., *Normal Approximation by Stein's Method*,
Probability and Its Applications,
DOI 10.1007/978-3-642-15007-4_1, © Springer-Verlag Berlin Heidelberg 2011

Only many years later with the work of Laplace around 1820 did it begin to be systematically realized that the normal limit holds in much greater generality. The result was the classical Central Limit Theorem, which states that $W_n \to_d Z$, that is, $W_n$ converges in distribution to $Z$, whenever

$$W_n = (S_n - n\mu)/\sqrt{n\sigma^2} \tag{1.2}$$

is the standardization of a sum $S_n$, as in (1.1), of independent and identically distributed random variables each with mean $\mu$ and variance $\sigma^2$. Here, $Z$ denotes a standard normal variable, that is, one with distribution function $P(Z \le x) = \Phi(x)$ given by

$$\Phi(x) = \int_{-\infty}^{x} \varphi(u)\,du \quad \text{where } \varphi(u) = \frac{1}{\sqrt{2\pi}} \exp\left(-\frac{1}{2}u^2\right),$$

and we say a sequence of random variables $Y_n$ is said to converge in distribution to $Y$, written $Y_n \to_d Y$, if

$$\lim_{n\to\infty} P(Y_n \le x) = P(Y \le x) \quad \text{for all continuity points } x \text{ of } P(Y \le x). \tag{1.3}$$

Generalizing further, but still keeping the variables independent, the question of when a sum of independent but not necessarily identically distributed random variables is asymptotically normal is essentially completely answered by the Lindeberg–Feller–Lévy Theorem (see Feller 1968b), which shows that the Lindeberg condition is sufficient, and nearly necessary, for the normal limit to hold. For a more detailed, and delightful account of the history of the CLT, we refer the reader to LeCam (1986).

When the quantity $W_n$ given by (1.2) is a normalized sum of i.i.d. variables $X_1, \ldots, X_n$ with finite third moment, the works of Berry (1941) and Esseen (1942) were the first to give a bound on the normal approximation error, in terms of some universal constant $C$, of the form

$$\sup_{z\in\mathbb{R}} \left| P(W_n \le z) - P(Z \le z) \right| \le \frac{CE|X_1|^3}{\sqrt{n}}.$$

This prototype bound has since been well studied, generalized and applied in practice, and it appears in many guises in the pages that follows. Esseen's original upper bound on $C$ of magnitude 7.59 has been markedly decreased over the years, the record currently now held by Tyurin (2010) who proved $C \le 0.4785$.

With the independent case tending toward resolution, attention can now turn to situations where the variables exhibit dependence. However, as there are countless ways variables can fail to be independent, no single technique can be used to address all situations, and no theorem parallel to the Lindeberg–Feller–Lévy theorem is ever to be expected in this greater generality. Consequently, the literature for validating the normal approximation in the presence of dependence now fragments somewhat into various techniques which can handle certain specific structures, or assumptions, two notable examples being central limit theorems proved under mixing conditions, and those results that can be applied to martingales.

Characteristic function methods have proved essential in making progress in the analysis of dependence, and though they are quite powerful, they rely on handling distributions through their transforms. In doing so it is doubtless that some probabilistic intuition is lost. In essence, the Stein method replaces the complex valued characteristic function with a real characterizing equation through which the random variable, in its original domain, may be manipulated, and in particular, coupled.

## 1.2 A Brief History of Stein's Method

Stein's method for normal approximation made its first appearance in the ground breaking work of Stein (1972), and it was here that the characterization of the normal distribution on which this book is based was first presented. That is, the fact that $Z \sim \mathcal{N}(0, \sigma^2)$ if and only if

$$E\big[Zf(Z)\big] = \sigma^2 E\big[f'(Z)\big], \tag{1.4}$$

for all absolutely continuous functions $f$ for which the above expectations exist. Very soon thereafter the work of Chen (1975) followed, applying the characterizing equation method to the Poisson distribution based on the parallel fact that $X \sim \mathcal{P}(\lambda)$, a Poisson variable with parameter $\lambda$, if and only if

$$E\big[Zf(Z)\big] = \lambda E\big[f(Z+1)\big],$$

for all functions $f$ for which the expectations above exist. From this point it seemed to take a number of years for the power of the method in both the normal and Poisson cases to become fully recognized; for Poisson approximation using Stein's method, see, for instance, the work of Arratia et al. (1989), and Barbour et al. (1992). The key identity (1.4) for the normal was, however, put to good use in the meantime.

In another landmark paper, Stein (1981) applied the characterization that he had proved earlier for the purpose of normal approximation to derive minimax estimates for the mean of a multivariate normal distribution in dimensions three or larger. In particular, he shows, using the multivariate version of (1.4), that when $\mathbf{X}$ has the normal distribution with mean $\boldsymbol{\theta}$ and identity covariance matrix, then the mean squared error risk of the estimate $\mathbf{X} + g(\mathbf{X})$, for an almost everywhere differentiable function $g : \mathbb{R}^p \to \mathbb{R}^p$, is unbiasedly estimated by $p + \|g(\mathbf{X})\|^2 + 2\nabla \cdot g(\mathbf{X})$. This 1981 work builds on the earlier and rather remarkable and surprising result of Stein (1956), that shows that the usual sample mean estimate $\overline{\mathbf{X}}$ for the true mean $\boldsymbol{\theta}$ of a multivariate normal distribution $\mathcal{N}_p(\boldsymbol{\theta}, \mathbf{I})$ is not admissible in dimensions three and greater; the multivariate normal characterization given in Stein (1981) provides a rather streamlined proof of this very counterintuitive fact.

Returning to normal approximation, by 1986 Stein's method was sufficiently cohesive that its foundations and some illustrative examples could be laid out in the manuscript of Stein (1986), with the exchangeable pair approach being one notable cornerstone. This manuscript also considers approximations using the binomial and the Poisson, and other probability estimates related to but not directly concerning the normal. In the realm of normal approximation, this work rather convincingly demonstrated the potential of the method under dependence by showing how it could be used to assess the quality of approximations for the distribution of the number of empty cells in an allocation model, and the number of isolated trees in the Erdös–Rényi random graph. For a personal history up to this time from the view point of Charles Stein, see his recollections in DeGroot (1986).

The period following the publication of Stein's 1986 manuscript saw a veritable explosion in the number of ideas and applications in the area, a fact well illustrated by the wide range of topics covered here, as well as in the two volumes of Barbour and Chen (2005b, 2005c), and those referred to in the bibliographies thereof. Including up to the present day, powerful extensions and applications of the method continue to be discovered that were, at the time of its invention, completely unanticipated.

## 1.3 The Basic Idea of Stein's Method

To show a random variable $W$ has a distribution close to that of a target distribution, say that of the random variable $Z$, one can compare the values of the expectations of the two distributions on some class of functions. For instance, one can compare the characteristic function $\phi(u) = Ee^{iuW}$ of $W$ to that of $Z$, thus encapsulating all expectations of the family of functions $e^{iuz}$ for $u \in \mathbb{R}$. And indeed, as this family of functions is rich enough, closeness of the characteristic functions implies closeness of the distributions. When studying the sum of random variables, and independent random variables in particular, the characteristic function is a natural choice, as convolution in the space of measures become products in the realm of characteristic functions. Powerful as they may be, one may lose contact with probabilistic intuition when handling complex functions in the transform domain. Stein's method, based instead on a direct, random variable characterization of a distribution, allows the manipulation of the distribution through constructions involving the basic random quantities of which $W$ is composed, and coupling can begin to play a large role.

Consider, then, testing for the closeness of the distributions of $W$ and $Z$ by evaluating the difference between the expectations $Eh(W)$ and $Eh(Z)$ over some collection of functions $h$. At first there appears to be no handle that we can apply, the task as stated being perhaps overly general. Nevertheless, it seems clear that if the distribution of $W$ is close to the distribution of $Z$ then the difference $Eh(W) - Eh(Z)$ should be small for many functions $h$. Specializing the problem, for a specific distribution, we may evaluate the difference by relying on a characterization of $Z$. For instance, by (1.4), the distribution of a random variable $Z$ is $\mathcal{N}(0, 1)$ if and only if

$$E\big(f'(Z) - Zf(Z)\big) = 0 \tag{1.5}$$

for all absolutely continuous functions $f$ for which the expectation above exists. Again, if the distribution of $W$ is close to that of $Z$, then evaluating the left hand side of (1.5) when $Z$ is replaced by $W$ should result in something small. Putting these two differences together, from the Stein characterization (1.5) we arrive at the Stein equation

$$f'(w) - wf(w) = h(w) - Eh(Z). \tag{1.6}$$

Now, given $h$, one solves (1.6) for $f$, evaluates the left hand side of (1.6) at $W$ and takes the expectation, obtaining $Eh(W) - Eh(Z)$.

Perhaps at first glance the problem has not been made any easier, as the evaluation of $Eh(W) - Eh(Z)$ has been replaced by the need to compute $E(f'(W) - Wf(W))$. Yet the form of what is required to evaluate is based on the normal characterization, and, somehow, for this reason, the expectation lends itself to calculation for $W$ for which approximation by the normal is appropriate. Borrowing, essentially, the following 'leave one out' idea from Stein's original 1972 paper, let $\xi_1, \ldots, \xi_n$ be independent mean zero random variables with variances $\sigma_1^2, \ldots, \sigma_n^2$ summing to one, and set

$$W = \sum_{i=1}^{n} \xi_i.$$

Then, with $W^{(i)} = W - \xi_i$, for some given $f$, we have

$$E\big(Wf(W)\big) = E\sum_{i=1}^{n} \xi_i f(W) = E\sum_{i=1}^{n} \xi_i f\big(W^{(i)} + \xi_i\big).$$

If $f$ is differentiable, then the summand may be expanded as

$$\xi_i f\big(W^{(i)} + \xi_i\big) = \xi_i f\big(W^{(i)}\big) + \xi_i^2 \int_0^1 f'\big(W^{(i)} + u\xi_i\big)\,du,$$

and, since $W^{(i)}$ and $\xi_i$ are independent, the first term on the right hand side vanishes when taking expectation, yielding

$$E\big(Wf(W)\big) = E\sum_{i=1}^{n} \xi_i^2 \int_0^1 f'\big(W^{(i)} + u\xi_i\big)\,du.$$

On the other hand, again with reference to the left hand side of (1.6), since $\sigma_1^2, \ldots, \sigma_n^2$ sum to 1, and $\xi_i$ and $W^{(i)}$ are independent, we may write

$$\begin{aligned}
Ef'(W) &= E\sum_{i=1}^{n} \sigma_i^2 f'(W) \\
&= E\sum_{i=1}^{n} \sigma_i^2 f'\big(W^{(i)}\big) + E\sum_{i=1}^{n} \sigma_i^2\big(f'(W) - f'\big(W^{(i)}\big)\big) \\
&= E\sum_{i=1}^{n} \xi_i^2 f'\big(W^{(i)}\big) + E\sum_{i=1}^{n} \sigma_i^2\big(f'(W) - f'\big(W^{(i)}\big)\big).
\end{aligned}$$

Taking the difference we obtain the expectation of the left hand side of (1.6) at $W$,

$$\begin{aligned}
E\big(f'(W) - Wf(W)\big) &= E\sum_{i=1}^{n} \xi_i^2 \int_0^1 \big(f'\big(W^{(i)}\big) - f'\big(W^{(i)} + u\xi_i\big)\big)\,du \\
&\quad + E\sum_{i=1}^{n} \sigma_i^2\big(f'(W) - f'\big(W^{(i)}\big)\big). \qquad (1.7)
\end{aligned}$$

When $n$ is large, as $\xi_1, \ldots, \xi_n$ are random variables of comparable size, it now becomes apparent why this expectation is small, no matter the distribution of the summands. Indeed, $W$ and $W^{(i)}$ only differ by the single variable $\xi_i$, accounting for roughly $1/\sqrt{n}$ of the total variance, so the differences in both terms above are small.

To make the case more convincingly, when $f$ has a bounded second derivative, then for all $u \in [0, 1]$, with $\|g\|$ denoting the supremum norm of a function $g$, the mean value theorem yields

$$\left|f'\big(W^{(i)}\big) - f'\big(W^{(i)} + u\xi_i\big)\right| \le |\xi_i| \|f''\|.$$

As this bound applies as well to the second term in (1.7), it being the case $u = 1$, when $\xi_i$ has third moments we obtain

$$\begin{aligned} \left|E\big(f'(W) - Wf(W)\big)\right| &\le \|f''\| \sum_{i=1}^{n} \big(E\left|\xi_i^3\right| + \sigma_i^2 E|\xi_i|\big) \\ &\le 2\|f''\| \sum_{i=1}^{n} E\left|\xi_i^3\right|, \end{aligned} \tag{1.8}$$

by Hölder's inequality.

The calculation reveals the need for the understanding of the smoothness relation between the solution $f$ and the given function $h$. For starters, we see directly from (1.6) that $f$ always has one more degree of smoothness than $h$, which, naturally, helps. However, as the original question was regarding the evaluation of the difference of expectations $Eh(W) - Eh(Z)$ expressed in terms of $h$, we see that in order to answer using (1.8) that bounds on quantities such as $\|f''\|$ must be provided in terms of some corresponding bound involving $h$. It is also worth noting that this illustration, and therefore also the original paper of Stein, contains the germ of several of the couplings which we will develop and apply later on, the present one bearing the most similarity to the analysis of local dependence.

The resemblance between Stein's 'leave one out' approach and the method of Lindeberg (see, for instance, Section 8.6 of Breiman 1986) is worth some exploration. Let $X_1, X_2, \ldots$ be i.i.d. mean zero random variables with variance 1, and for each $n$ let

$$\xi_{i,n} = \frac{X_i}{\sqrt{n}}, \quad i = 1, \ldots, n, \tag{1.9}$$

the elements of a triangular array. The basic idea of Lindeberg is to compare the sum

$$W_n = \xi_{1,n} + \cdots + \xi_{n,n}$$

to the sum

$$Z_n = Z_{1,n} + \cdots + Z_{n,n}$$

of mean zero, i.i.d. normals $Z_{1,n}, \ldots, Z_{n,n}$ with $\mathrm{Var}(Z_n) = 1$. Let $h$ be a twice differentiable bounded function on $\mathbb{R}$ such that $h''$ is uniformly continuous and

$$M = \sup_{x\in\mathbb{R}} |h''(x)| < \infty. \tag{1.10}$$

For such an $h$, the quantity

$$\delta(\epsilon) = \sup_{|x-y|\le\epsilon} |h''(x) - h''(y)|$$

is bounded over $\epsilon \in \mathbb{R}$ and satisfies $\lim_{\epsilon\downarrow 0} \delta(\epsilon) = 0$.

Write the difference $Eh(W_n) - Eh(Z_n)$ as the telescoping sum

$$Eh(W_n) - Eh(Z_n) = E\sum_{i=1}^{n} h(V_{i,n}) - h(V_{i-1,n}), \tag{1.11}$$

where

$$V_{i,n} = \sum_{j=1}^{i} \xi_{j,n} + \sum_{j=i+1}^{n} Z_{j,n},$$

with the usual convention that an empty sum is zero. In this way, the variables interpolate between $W_n = V_{n,n}$ and $Z_n = V_{0,n}$. Writing

$$U_{i,n} = \sum_{j=1}^{i-1} \xi_{j,n} + \sum_{j=i+1}^{n} Z_{j,n},$$

a Taylor expansion on the summands in (1.11) yields

$$\begin{aligned} h(V_{i,n}) - h(V_{i-1,n}) &= h(U_{i,n} + \xi_{i,n}) - h(U_{i,n} + Z_{i,n}) \\ &= (\xi_{i,n} - Z_{i,n})h'(U_{i,n}) \\ &\quad + \frac{1}{2}\xi_{i,n}^2 h''(U_{i,n} + u\xi_{i,n}) - \frac{1}{2}Z_{i,n}^2 h''(U_{i,n} + vZ_{i,n}), \end{aligned}$$

for some $u, v \in [0, 1]$. Since $h'$ can grow at most linearly the expectation of the first term exists, and, as $\xi_{i,n}$ and $Z_{i,n}$ are independent of $U_{i,n}$, equals zero.

Considering the expectation of the remaining second order terms, write

$$E\xi_{i,n}^2 h''(U_{i,n} + u\xi_{i,n}) = E\big(\xi_{i,n}^2 h''(U_{i,n})\big) + \alpha E\big(\xi_{i,n}^2 \delta(|\xi_{i,n}|)\big),$$

for some $\alpha \in [-1, 1]$, with a similar equality holding for the expectation of the last term. As $E\xi_{i,n}^2 = EZ_{i,n}^2$, taking the difference of the second order terms, using independence, and that $\xi_{i,n}$ and $Z_{i,n}$ are identically distributed, respectively, for $i = 1, \ldots, n$, yields

$$E|h(V_{i,n}) - h(V_{i-1,n})| \le \frac{1}{2}\Big(E\big(\xi_{1,n}^2 \delta(|\xi_{1,n}|)\big) + E\big(Z_{1,n}^2 \delta(|Z_{1,n}|)\big)\Big). \tag{1.12}$$

Recalling (1.9), we have

$$E\big(\xi_{1,n}^2 \delta(|\xi_{1,n}|)\big) = \frac{1}{n} E\big(X_1^2 \delta(n^{-1/2}|X_1|)\big),$$

with a similar equality holding for the second term of (1.12). Hence, by (1.11), with $Z$ now denoting a standard normal variable, summing yields

$$\left|Eh(W_n) - Eh(Z)\right| \le \frac{1}{2}\big(E\big(X_1^2\delta\big(n^{-1/2}|X_1|\big)\big) + E\big(Z^2\delta\big(n^{-1/2}|Z|\big)\big)\big).$$

By (1.10), $\delta(\epsilon) \le 2M$ for all $\epsilon \in \mathbb{R}$, so $X_1^2\delta(n^{-1/2}|X_1|) \le 2MX_1^2$. As $X_1^2\delta(n^{-1/2}|X_1|) \to 0$ almost surely as $n \to \infty$, the dominated convergence theorem implies the first term above tends to zero. Applying the same reasoning to the second term we obtain

$$\lim_{n\to\infty}\left|Eh(W_n) - Eh(Z)\right| = 0. \tag{1.13}$$

As the class of functions $h$ for which we have obtained $Eh(W_n) \to Eh(Z)$ is rich enough, we have shown $W_n \to_d Z$.

Both the Stein and Lindeberg approaches proceed through calculations that 'leave one out.' However, the Stein approach seems more finely tuned to the target distribution, using the solution of a differential equation tailored to the normal. Moreover, use of the Stein differential equation provides that the functions $f$ being evaluated on the variables of interest have one degree of smoothness over that of the basic test functions $h$ which are used to gauge the distance between $W$ and $Z$. However, the main practical difference between Stein's method and that of Lindeberg, as far as outcome, is the former's additional benefit of providing a bound on the distance to the target, and not only convergence in distribution; witness the difference between conclusions (1.8) and (1.13). Furthermore, Stein's method allows for a variety of ways in which variables can be handled in the Stein equation, the 'leave one out' approach being just the beginning.

## 1.4 Outline and Summary

We begin in Chap. 2 by introducing and working with the fundamentals of Stein's method. First we prove the Stein characterization (1.4) for the normal, and develop bounds on the Stein equation (1.6) that will be required throughout our treatment; the multivariate Stein equation for the normal, and its solution by the generator method, is also introduced here.

The 'leave one out' coupling considered in Sect. 1.3 is but one variation on the many ways in which variables close to the one of interest can enter the Stein equation, and is in particular related to some of the couplings we consider later on to handle locally dependent variables. Four additional, and somewhat overlapping, basic methods for handling variables in the Stein equation are introduced in Chap. 2: the $K$-function approach, the original exchangeable pair method of Stein, and the zero bias and size bias transformations. Illustrations of how these methods allow for various manipulations in the Stein equation are provided, as well as a number of examples, some of which will continue as themes and illustrations for the remainder of the book. The independent case, of course, serves as one important testing ground throughout. A framework that includes some of our approaches is considered in Sect. 2.4. Some technical calculations for bounds to the Stein equation appear in the Appendix to Chap. 2, as do other such calculations in subsequent chapters.

Chapter 3 focuses on the independent case. The goal is to demonstrate a version of the classical Berry–Esseen theorem using Stein's method. Along the way techniques are developed for obtaining $L^1$ bounds, and the Lindeberg central limit theorem is shown as well. The Berry–Esseen theorem is first demonstrated for the case where the random variables are bounded. The boundedness condition is then relaxed in two ways, first by concentration inequalities, then by induction. This chapter concludes with a lower bound for the Berry–Esseen inequality. As seen in the chapter dependency diagram that follows, Chaps. 2 and 3 form much of the basis of this work.

Chapter 4 develops a theory for obtaining $L^1$ bounds using the zero bias coupling, and a main result is obtained which can be applied in non-independent settings. A number of examples are presented for illustration. The case of independence is considered first, with an $L^1$ Berry–Esseen bound followed by the demonstration of a type of contraction principle satisfied by sums of independent variables which implies, or even in a way explains, normal convergence. Bounds in $L^1$ are then proved for hierarchical structures, that is, self similar, fractal type objects whose scale at small levels is replicated on the larger. Then, making our first departure from independence we prove $L^1$ bounds for the projections of random vectors having distribution concentrated on regular convex sets in Euclidean space. Next, illustrating a different coupling, $L^1$ bounds to the normal for the combinatorial central limit theorem are given. Though the combinatorial central limit theorem contains simple random sampling as a particular case, somewhat better bounds may be obtained by applying specifics in the special case; hence, an $L^1$ bound is given for the case of simple random sampling alone. Next we present Chatterjee's $L^1$ theorem for functions of independent random variables, and apply it to the approximation of the distribution of the volume covered by randomly placed spheres in the Euclidean torus. Results are then given for sums of locally dependent random variables, with applications including the number of local maxima on a graph. Chapter 4 concludes with a consideration of a class of smooth functions, contained in the one which may be used to determine the $L^1$ distance, for which convergence to the normal is at the accelerated rate of $1/n$, subject to a vanishing third moment assumption.

The theme of Chap. 5 is to provide upper bounds in the $L^\infty$, or Kolmogorov distance, that can be applied when certain bounded couplings can be constructed. Various bounds to the normal for a random variable $W$ are formed by constructing an auxiliary random variable, say $\widetilde{W}$, on the same space as $W$. We have in mind here, in particular, the cases where $\widetilde{W}$ has the same distribution as $W$, or the zero bias or size bias distribution of $W$. The resulting bound is often interpretable, sometimes directly, as a distance between $W$ and $\widetilde{W}$, a small bound being a reflection of a small distance. Heuristically, being able to make a close coupling to $W$, shows, in a sense, that perturbing $W$ has only a weak effect. Being able to make a close coupling shows the dependence making up $W$ is weak, and, as a random variable has an approximate normal distribution when it depends on many small weakly dependent factors, such a $W$ should be approximately normal. The bounded couplings studied in this chapter, ones where $|W - \widetilde{W}| \le \delta$ with probability one for some $\delta$, and are often much easier to manage than unbounded ones. Chapter 5 provides results when bounded zero bias, exchangeable pair, or size bias couplings can be constructed. The chapter concludes with the use of smoothing inequalities to obtain distances between $W$ and the normal over general function classes, one special case being the derivation of Kolmogorov distance bounds when bounded size bias couplings exist.

Chapter 6 applies the $L^\infty$ results of Chap. 5 to a number of applications, all of which involve dependence. Dependence can loosely be classified into two types, first, the local type, such as when each variable has a small neighborhood outside of which the remaining variables are independent, and second, dependence with a global nature. Chapter 6 deals mainly with global dependence but begins to also touch upon local dependence, a topic more thoroughly explored in Chap. 9. Regarding global dependence, the analysis of the combinatorial central limit theorem, studied in $L^1$ in Chap. 4, is continued here with the goal of obtaining $L^\infty$ results. Results for the classical case are given, where the permutation is uniformly chosen over the symmetric group, as well as for the case where the permutation is chosen with distribution constant over some conjugacy class, such as the class of involutions. Two approaches are considered, one using the zero bias coupling and one using induction. Normal approximation bounds for the so called lightbulb process are also given in this chapter, again an example of handling global dependence, this time using the size bias coupling. The anti-voter model is also studied, handled by the exchangeable pair technique, as is the binary expansion of a random integer. Results for the occurrences of patterns in graphs and permutations, an example of local dependence, are handled using the size bias method.

Returning to the independent case, and inspired by use of the continuity correction for the normal approximation of the binomial, in Chap. 7 we consider the approximation of independent sums of integer valued random variables by the discretized normal distribution, in the total variation metric. The main result is shown by obtaining bounds between the zero biased distribution of the sum and the normal, and then treating the coupled zero biased variable as a type of perturbation.

Continuing our consideration of the independent case, in Chap. 8 we derive non-uniform bounds for sums of independent random variables. In particular, by use of non-uniform concentration inequalities and the Bennett–Hoeffding inequality we provide bounds for the absolute difference between the distribution function $F(z)$ of a sum of independent variables and the normal $\Phi(z)$, which may depend on $z \in \mathbb{R}$. Non-uniform bounds serve as a counterpoint to the earlier derived supremum norm bounds that are not allowed to vary with $z$, and give information on how the quality of the normal approximation varies over $\mathbb{R}$.

In Chap. 9 we consider local dependence using the $K$-function approach, and obtain both uniform and non-uniform Berry–Esseen bounds. The results are applied to certain scan statistics, and yield a general theorem when the local dependence can be expressed in terms of a dependency graph whose vertices are the underlying variables, and where two non-intersecting subsets of variables are independent anytime there is no edge in the graph connecting a element of one subset with the other.

In Chap. 10 we develop uniform and non-uniform bounds for non-linear functions $T(X_1, \ldots, X_n)$, of independent random variables $X_1, \ldots, X_n$, that can be well approximated by a linear term plus a non-linear remainder. Applications include $U$-statistics, $L$-statistics and random sums. Randomized concentration inequalities are established in order to develop the theory necessary to cover these examples.

In previous chapters we have measured the accuracy of approximations using differences between two distributions. For the most part, the resulting measures are sensitive to the variations between distributions in their bulk, that is, measures like the $L^1$ or $L^\infty$ norm typically compare two distributions in the region where most of their mass is concentrated. In contrast, in Chap. 11, we consider moderate deviations of distributions, and rather than consider a difference, compare the ratio of the distribution function of the variable $W$ of interest to that of the normal. Information on small probabilities in the tail become available in this way. Applications of the results of this chapter include the combinatorial central limit theorem, the anti-voter model, the binary expansion of a random integer, and the Curie–Weiss model.

In Chap. 12 we consider multivariate normal approximation, extending both the size bias and exchangeable pair methods to this setting. In the latter case we show how in some cases the exchangeable pair 'linearity condition' can be achieved by embedding the problem in a higher dimension. Applications of both methods are applied to problems in random graphs.

We momentarily depart from normal approximation in Chap. 13. We confine ourselves to approximations by continuous distributions for which the methods of the previous chapters may be extended. As one application, we approximate the distribution of the total spin of the Curie–Weiss model from statistical physics, at the critical inverse temperature, by a distribution with density proportional to $\exp(-x^4/12)$ using the method of exchangeable pairs. We also develop bounds for approximation by the exponential distribution, and apply it to the spectrum of the Bernoulli Laplace Markov chain, and first passage times for Markov chains.

In Chap. 14 we consider two applications of Stein's method, each of which go well beyond the confines of the method's originally intended uses; the approximation of the distribution of characters of elements chosen uniformly from compact Lie groups, and of random variables in a fixed Wiener chaos of Brownian motion, using the tools of Malliavin calculus. Regarding the first topic, the study of random characters is in some sense a generalization to abstract groups of the study of traces of random matrices, a framework into which the combinatorial central limit theorem can be made to fit. As for the second, joining Stein's method to Malliavin calculus shows that the underlying fundamentals of Stein's method, in particular the basic characterization of the normal which can be shown by integration by parts, can be extended, with great benefit, to abstract Wiener spaces.

As for what this book fails to include, narrowing in as it does on what can be shown in the realm of normal approximation by Stein's method, we do not consider, most notably, transform methods, mixing, or martingales. For these topics, having more history than the one presently considered, sources already abound.

We stress to the reader that this book need not at all be read in a linear fashion, especially if one is interested in applications and is willing to forgo the proofs of the theorems on which the applications are based. The following diagram reflects the dependence of each chapter on the others.

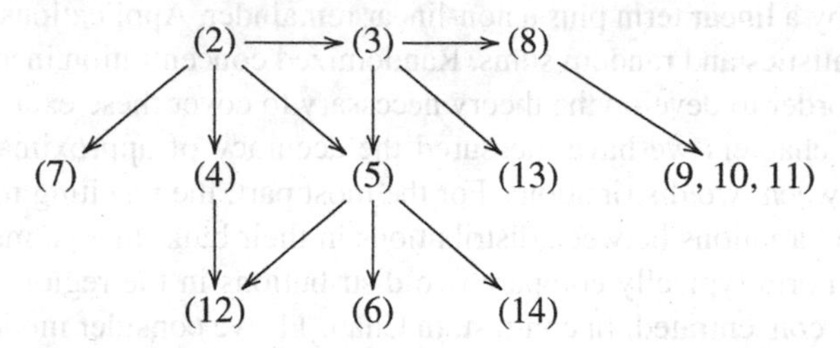

# Chapter 2
# Fundamentals of Stein's Method

We begin by giving a detailed account of the fundamentals of Stein's method, starting with Stein's characterization of the normal distribution and the basic properties of the solution to the Stein equation. Then we provide an outline of the basic Stein identities and distributional transformations which play a large role in coupling constructions, introducing first the construction of the $K$ function for independent random variables, the exchangeable pair approach due to Stein, the zero bias transformation for random variable with mean zero and variance one, and lastly the size bias transformation for non-negative random variables with finite mean. We conclude the chapter with a framework under which a number of Stein identities can be placed, and a proposition for normal approximation using Lipschitz functions. Some of the more technical results on bounds to the Stein equation can be found in the Appendix to this chapter.

## 2.1 Stein's Equation

Stein's method rests on the following characterization of the distribution of a standard normal variable $Z$, given in Stein (1972).

**Lemma 2.1** *If $W$ has a standard normal distribution, then*

$$Ef'(W) = E\big[Wf(W)\big], \tag{2.1}$$

*for all absolutely continuous functions $f:\mathbb{R}\to\mathbb{R}$ with $E|f'(Z)|<\infty$. Conversely, if* (2.1) *holds for all bounded, continuous and piecewise continuously differentiable functions $f$ with $E|f'(Z)|<\infty$, then $W$ has a standard normal distribution.*

Though there is no known definitive method for the construction of a characterizing identity, of the type given in Lemma 2.1, for the distribution of a random variable $Y$ in general, two main contenders emerge. The first one we might call the 'density approach.' If $W$ has density $p(w)$ then in many cases one can replace the coefficient $W$ on the right hand side of (2.1) by $-p'(W)/p(W)$; this approach is pursued in

L.H.Y. Chen et al., *Normal Approximation by Stein's Method*,
Probability and Its Applications,
DOI 10.1007/978-3-642-15007-4_2, © Springer-Verlag Berlin Heidelberg 2011

Chap. 13 to study approximations by non-normal distributions. In another avenue, one which we might call the 'generator approach', we seek a Markov process that has as its stationary distribution the one of interest. In this case, the generator, or some variation thereof, of such a process has expectation zero when applied to sufficiently smooth functions, giving the difference between the two sides of (2.1). In Sect. 2.3.2 we discuss the relation between the generator method and exchangeable pairs, and in Sect. 2.2 its relation to the solution of the Stein equation, the differential equation motivated by the characterization (2.1). In fact, we now prove one direction of Lemma 2.1 using the Stein equation (2.2).

**Lemma 2.2** *For fixed $z \in \mathbb{R}$ and $\Phi(z) = P(Z \le z)$, the cumulative distribution function of $Z$, the unique bounded solution $f(w) := f_z(w)$ of the equation*

$$f'(w) - wf(w) = \mathbf{1}_{\{w \le z\}} - \Phi(z) \tag{2.2}$$

*is given by*

$$f_z(w) = \begin{cases} \sqrt{2\pi} e^{w^2/2} \Phi(w)[1 - \Phi(z)] & \text{if } w \le z, \\ \sqrt{2\pi} e^{w^2/2} \Phi(z)[1 - \Phi(w)] & \text{if } w > z. \end{cases} \tag{2.3}$$

*Proof* Multiplying both sides of (2.2) by the integrating factor $e^{-w^2/2}$ yields

$$\left(e^{-w^2/2} f(w)\right)' = e^{-w^2/2}\left(\mathbf{1}_{\{w \le z\}} - \Phi(z)\right).$$

Integration now yields

$$\begin{aligned} f_z(w) &= e^{w^2/2} \int_{-\infty}^{w} \left[\mathbf{1}_{\{x \le z\}} - \Phi(z)\right] e^{-x^2/2} \, dx \\ &= -e^{w^2/2} \int_{w}^{\infty} \left[\mathbf{1}_{\{x \le z\}} - \Phi(z)\right] e^{-x^2/2} \, dx, \end{aligned}$$

which is equivalent to (2.3). Lemma 2.3 below shows $f_z(w)$ is bounded.

The general solution to (2.2) is given by $f_z(w)$ plus some constant multiple, say $ce^{w^2/2}$, of the solution to the homogeneous equation. Hence the only bounded solution is obtained by taking $c = 0$. □

*Proof of Lemma 2.1 Necessity.* Let $f$ be an absolutely continuous function satisfying $E|f'(Z)| < \infty$. If $W$ has a standard normal distribution then

$$\begin{aligned} Ef'(W) &= \frac{1}{\sqrt{2\pi}} \int_{-\infty}^{\infty} f'(w) e^{-w^2/2} \, dw \\ &= \frac{1}{\sqrt{2\pi}} \int_{-\infty}^{0} f'(w) \left(\int_{-\infty}^{w} -xe^{-x^2/2} \, dx\right) dw \\ &\quad + \frac{1}{\sqrt{2\pi}} \int_{0}^{\infty} f'(w) \left(\int_{w}^{\infty} xe^{-x^2/2} \, dx\right) dw. \end{aligned}$$

By Fubini's theorem, it thus follows that

$$\begin{aligned}
Ef'(W) &= \frac{1}{\sqrt{2\pi}} \int_{-\infty}^{0} \left( \int_{x}^{0} f'(w)\,dw \right) (-x) e^{-x^2/2}\,dx \\
&\quad + \frac{1}{\sqrt{2\pi}} \int_{0}^{\infty} \left( \int_{0}^{x} f'(w)\,dw \right) x e^{-x^2/2}\,dx \\
&= \frac{1}{\sqrt{2\pi}} \int_{-\infty}^{\infty} \left[ f(x) - f(0) \right] x e^{-x^2/2}\,dx \\
&= E\left[ W f(W) \right].
\end{aligned}$$

*Sufficiency.* The function $f_z$ as given in (2.3) is clearly continuous and piecewise continuously differentiable; Lemma 2.3 below shows $f_z$ is bounded as well. Hence, if (2.1) holds for all bounded, continuous and continuously differentiable functions, then by (2.2)

$$0 = E\left[ f_z'(W) - W f_z(W) \right] = E\left[ \mathbf{1}_{\{W \le z\}} - \Phi(z) \right] = P(W \le z) - \Phi(z).$$

Thus $W$ has a standard normal distribution. □

When $f$ is an absolutely continuous and bounded function, one can prove (2.1) holds for a standard normal $W$ using integration by parts, as in this case

$$\begin{aligned}
E\left[ W f(W) \right] &= \frac{1}{\sqrt{2\pi}} \int_{-\infty}^{\infty} w f(w) e^{-w^2/2}\,dw \\
&= -\frac{1}{\sqrt{2\pi}} \int_{-\infty}^{\infty} f(w)\, d\left( e^{-w^2/2} \right) \\
&= \frac{1}{\sqrt{2\pi}} \int_{-\infty}^{\infty} f'(w) e^{-w^2/2}\,dw \\
&= Ef'(W).
\end{aligned}$$

For a given real valued measurable function $h$ with $E|h(Z)| < \infty$ we denote $Eh(Z)$ by $Nh$ and call

$$f'(w) - w f(w) = h(w) - Nh \tag{2.4}$$

the Stein equation for $h$, or simply the Stein equation. Note that (2.2) is the special case of (2.4) for $h(w) = \mathbf{1}_{\{w \le z\}}$. By the same method of integrating factors that produced (2.3) one may show that the unique bounded solution of (2.4) is given by

$$\begin{aligned}
f_h(w) &= e^{w^2/2} \int_{-\infty}^{w} \left( h(x) - Nh \right) e^{-x^2/2}\,dx \\
&= -e^{w^2/2} \int_{w}^{\infty} \left( h(x) - Nh \right) e^{-x^2/2}\,dx.
\end{aligned} \tag{2.5}$$

## 2.2 Properties of the Solutions

We now list some properties of the solutions (2.3) and (2.5) to the Stein equations (2.2) and (2.4), respectively, that are required to determine error bounds in our various approximations to come. We defer the detailed proofs of Lemmas 2.3 and 2.4

to an Appendix since they are somewhat technical. As the arguments used to prove these bounds do not themselves figure in the methods themselves, the reader may skip them if they so choose. We begin with the solution $f_z$ to (2.2).

**Lemma 2.3** *Let $z \in \mathbb{R}$ and let $f_z$ be given by* (2.3). *Then*

$$wf_z(w) \text{ is an increasing function of } w. \tag{2.6}$$

*Moreover, for all real $w$, $u$ and $v$,*

$$\left|wf_z(w)\right| \le 1, \qquad \left|wf_z(w) - uf_z(u)\right| \le 1 \tag{2.7}$$

$$\left|f_z'(w)\right| \le 1, \qquad \left|f_z'(w) - f_z'(u)\right| \le 1 \tag{2.8}$$

$$0 < f_z(w) \le \min\left(\sqrt{2\pi}/4, 1/|z|\right) \tag{2.9}$$

*and*

$$\left|(w+u)f_z(w+u) - (w+v)f_z(w+v)\right| \le \left(|w| + \sqrt{2\pi}/4\right)\left(|u| + |v|\right). \tag{2.10}$$

We mostly use (2.8) and (2.9) for our approximations. If one does not care much about constants, the bounds

$$\left|f_z'(w)\right| \le 2 \quad \text{and} \quad 0 < f_z(w) \le \sqrt{\pi/2}$$

may be easily obtained by using the well-known inequality

$$1 - \Phi(w) \le \min\left(\frac{1}{2}, \frac{1}{w\sqrt{2\pi}}\right)e^{-w^2/2}, \qquad w > 0. \tag{2.11}$$

Next, we consider (2.5), the solution $f_h$ to the Stein equation (2.4). For any real valued function $h$ on $\mathbb{R}^p$ let

$$\|h\| = \sup_{x \in \mathbb{R}^p} \left|h(x)\right|.$$

**Lemma 2.4** *For a given function $h : \mathbb{R} \to \mathbb{R}$, let $f_h$ be the solution* (2.5) *to the Stein equation* (2.4). *If $h$ is bounded, then*

$$\|f_h\| \le \sqrt{\pi/2}\left\|h(\cdot) - Nh\right\| \quad \text{and} \quad \|f_h'\| \le 2\left\|h(\cdot) - Nh\right\|. \tag{2.12}$$

*If $h$ is absolutely continuous, then*

$$\|f_h\| \le 2\|h'\|, \quad \|f_h'\| \le \sqrt{2/\pi}\,\|h'\| \quad \text{and} \quad \|f_h''\| \le 2\|h'\|. \tag{2.13}$$

Some of the results that follow are shown by letting $h(w)$ be the indicator of $(-\infty, z]$ with a linear decay to zero over an interval of length $\alpha > 0$, that is, the function

$$h(w) = \begin{cases} 1 & w \le z, \\ 1 + (z - w)/\alpha & z < w \le z + \alpha, \\ 0 & w > z + \alpha. \end{cases} \tag{2.14}$$

The following bounds for the solution to the Stein equation for the smoothed indicator appear in Chen and Shao (2004).

**Lemma 2.5** *For $z \in \mathbb{R}$ and $\alpha > 0$, let $f$ be the solution* (2.5) *to the Stein equation* (2.4) *for the smoothed indicator function* (2.14). *Then, for all $w, v \in \mathbb{R}$,*

$$0 \le f(w) \le 1, \quad |f'(w)| \le 1, \quad |f'(w) - f'(v)| \le 1 \tag{2.15}$$

*and*

$$|f'(w+v) - f'(w)| \le |v| \left(1 + |w| + \frac{1}{\alpha} \int_0^1 \mathbf{1}_{[z,z+\alpha]}(w + rv)dr\right). \tag{2.16}$$

For multivariate approximations we consider an extension of the Stein equation (2.4) to $\mathbb{R}^p$. For a twice differentiable function $g : \mathbb{R}^p \to \mathbb{R}$ let $\nabla g$ and $D^2 g$ denote the gradient and second derivative, or Hessian matrix, of $g$ respectively and let $\mathrm{Tr}(A)$ be the trace of a matrix $A$. Let $\mathbf{Z}$ be multivariate normal vector in $\mathbb{R}^p$ with mean zero and identity covariance matrix. For a test function $h : \mathbb{R}^p \to \mathbb{R}$ and for $u \ge 0$ define

$$(T_u h)(\mathbf{w}) = E\{h(\mathbf{w}e^{-u} + \sqrt{1 - e^{-2u}}\mathbf{Z})\}. \tag{2.17}$$

Letting $Nh = Eh(\mathbf{Z})$, the following lemma provides bounds on the solution of the 'multivariate generator' method for solutions to the Stein equation

$$(\mathcal{A}g)(\mathbf{w}) = h(\mathbf{w}) - Nh \quad \text{where } (\mathcal{A}g)(\mathbf{w}) = \mathrm{Tr}\, D^2 g(\mathbf{w}) - \mathbf{w} \cdot \nabla g(\mathbf{w}). \tag{2.18}$$

We note that in one dimension (2.18) reduces to (2.4) with one extra derivative, that is, to

$$g''(w) - wg'(w) = h(w) - Nh. \tag{2.19}$$

For a vector $\mathbf{k} = (k_1, \ldots, k_p)$ of nonnegative integers and a function $h : \mathbb{R}^p \to \mathbb{R}$, let

$$h^{(\mathbf{k})}(\mathbf{w}) = \frac{\partial^{|\mathbf{k}|}}{\prod_{j=1}^p \partial w_{k_j}} h(\mathbf{w}) \quad \text{where } |\mathbf{k}| = \sum_{j=1}^p k_j,$$

and for a matrix $A \in \mathbb{R}^{p \times p}$, let

$$\|A\| = \max_{1 \le i,j \le p} |a_{ij}|.$$

**Lemma 2.6** *If $h : \mathbb{R}^p \to \mathbb{R}$ has three bounded derivatives then*

$$g(\mathbf{w}) = -\int_0^\infty [T_u h(\mathbf{w}) - Nh]\, du \tag{2.20}$$

*solves* (2.18), *and if the* **k***th partial derivative of $h$ exists then*

$$\|g^{(\mathbf{k})}\| \le \frac{1}{k} \|h^{(\mathbf{k})}\|.$$

*Further, for any $\boldsymbol{\mu} \in R^p$ and positive definite $p \times p$ matrix $\Sigma$, $f$ defined by the change of variable*

$$f(\mathbf{w}) = g(\Sigma^{-1/2}(\mathbf{w} - \boldsymbol{\mu})) \tag{2.21}$$

*solves*

$$\operatorname{Tr} \Sigma D^2 f(\mathbf{w}) - (\mathbf{w} - \boldsymbol{\mu}) \cdot \nabla f(\mathbf{w}) = h\big(\Sigma^{-1/2}(\mathbf{w} - \boldsymbol{\mu})\big) - Nh, \tag{2.22}$$

*and satisfies*

$$\|f^{(\mathbf{k})}\| \le \frac{p^k}{k} \|\Sigma^{-1/2}\|^k \|h^{(\mathbf{k})}\|. \tag{2.23}$$

The operator $\mathcal{A}$ in (2.18) is the generator of the Ornstein–Uhlenbeck process in $\mathbb{R}^p$, whose stationary distribution is the standard normal. The operator $(T_u h)(\mathbf{w})$ in (2.17) is the expected value of $h$ evaluated at the position of the Ornstein–Uhlenbeck process at time $u$, when it has initial position $\mathbf{w}$ at time 0. Equations of the form $\mathcal{A}g = h - Eh(\mathbf{Z})$ may be solved more generally by (2.20) when $\mathcal{A}$ is the generator of a Markov process with stationary distribution $\mathbf{Z}$, see Ethier and Kurtz (1986). Indeed, the generator method may be employed to solve the Stein equation for distributions other than the normal, see, for instance, Barbour et al. (1992) for the Poisson, and Luk (1994) for the Gamma distribution.

For the specific case at hand, the proof of Lemma 2.6 can be found in Barbour (1990), see equations (2.23) and (2.5), and also in Götze (1991). Essentially, following Barbour (1990) one shows that $g$ is a solution, and that under the assumptions above, differentiating (2.20) and applying the dominated convergence yields

$$g^{(\mathbf{k})}(\mathbf{w}) = -\int_0^\infty e^{-ku} E\big\{h^{(\mathbf{k})}\big(\mathbf{w}e^{-u} + \sqrt{1 - e^{-2u}}\mathbf{Z}\big)\big\}\, du.$$

The bounds then follow by straightforward calculations.

## 2.3 Construction of Stein Identities

Stein's equation (2.4) is the starting point of Stein's method. To prove that a mean zero, variance one random variable $W$ can be approximated by a standard normal distribution, that is, to show that $Eh(W) - Eh(Z)$ is small for some large class of functions $h$, rather than estimating this difference directly, we solve (2.4) for a given $h$ and show that $E[f'(W) - Wf(W)]$ is small instead. As we shall see, this latter quantity is often much easier to deal with than the former, as various identities and couplings may be applied to handle it.

In essence Stein's method shows that the distribution of two random variables are close by using the fact that they satisfy similar identities. For example, in Sect. 2.3.1, we demonstrate that when $W$ is the sum of independent mean zero random variables $\xi_1, \ldots, \xi_n$ whose variances sum to 1, then

$$E\big[Wf(W)\big] = Ef'\big(W^{(I)} + \xi_I^*\big)$$

where $W^{(I)}$ is the sum $W$ with a random summand $\xi_I$ removed, and $\xi_I^*$ is a random variable independent of $W^{(I)}$. Hence $W$ satisfies an identity very much like the characterization (2.1) for the normal.

We present four different approaches, or variations, for handling the Stein equation. Sect. 2.3.1 introduces the $K$ function method when $W$ is a sum of independent random variables. In Sect. 2.3.2 we present the exchangeable pair approach of Stein, which works well when $W$ has a certain dependency structure. We then discuss the zero bias distribution and the associated transformation, which, in principle, may be applied for arbitrary mean zero random variables having finite variance. We note that the $K$ function method of Sect. 2.3.1 and the zero bias method of Sect. 2.3.3 are essentially identical in the simple context of sums of independent random variables, but these approaches will later diverge. Size bias transformations, and some associated couplings, presented in Sect. 2.3.3 are closely related to those for zero biasing; the size bias method is most naturally applied to non-negative variables such as counts.

### *2.3.1 Sums of Independent Random Variables*

In this subsection we consider the most elementary case and apply Stein's method to justify the normal approximation of the sum $W$ of independent random variables $\xi_1, \xi_2, \ldots, \xi_n$ satisfying

$$E\xi_i = 0, \quad 1 \le i \le n \quad \text{and} \quad \sum_{i=1}^{n} E\xi_i^2 = 1.$$

Set

$$W = \sum_{i=1}^{n} \xi_i \quad \text{and} \quad W^{(i)} = W - \xi_i,$$

and define

$$K_i(t) = E\{\xi_i(\mathbf{1}_{\{0 \le t \le \xi_i\}} - \mathbf{1}_{\{\xi_i \le t < 0\}})\}. \tag{2.24}$$

It is easy to check that $K_i(t) \ge 0$ for all real $t$, and that

$$\int_{-\infty}^{\infty} K_i(t)\,dt = E\xi_i^2 \quad \text{and} \quad \int_{-\infty}^{\infty} |t| K_i(t)\,dt = \frac{1}{2} E|\xi_i|^3. \tag{2.25}$$

Let $h$ be a measurable function with $E|h(Z)| < \infty$, and let $f = f_h$ be the corresponding solution of the Stein equation (2.4). Our goal is to estimate

$$Eh(W) - Nh = E\{f'(W) - Wf(W)\}. \tag{2.26}$$

The argument below is fundamental to the $K$ function approach, with many of the following tricks reappearing repeatedly in the sequel.

Since $\xi_i$ and $W^{(i)}$ are independent for each $1 \le i \le n$, we have

$$E[Wf(W)] = \sum_{i=1}^{n} E[\xi_i f(W)]$$
$$= \sum_{i=1}^{n} E\{\xi_i [f(W) - f(W^{(i)})]\},$$

where the last equality follows because $E\xi_i = 0$. Writing the final difference in integral form, we thus have

$$E[Wf(W)] = \sum_{i=1}^{n} E\left\{\xi_i \int_0^{\xi_i} f'(W^{(i)} + t)\,dt\right\}$$
$$= \sum_{i=1}^{n} E\left\{\int_{-\infty}^{\infty} f'(W^{(i)} + t)\xi_i(\mathbf{1}_{\{0 \le t \le \xi_i\}} - \mathbf{1}_{\{\xi_i \le t < 0\}})\,dt\right\}$$
$$= \sum_{i=1}^{n} \int_{-\infty}^{\infty} E\{f'(W^{(i)} + t)\} K_i(t)\,dt, \tag{2.27}$$

from the definition of $K_i$ and again using independence. However, from

$$\sum_{i=1}^{n} \int_{-\infty}^{\infty} K_i(t)\,dt = \sum_{i=1}^{n} E\xi_i^2 = 1, \tag{2.28}$$

it follows that

$$Ef'(W) = \sum_{i=1}^{n} \int_{-\infty}^{\infty} E\{f'(W)\} K_i(t)\,dt. \tag{2.29}$$

Thus, by (2.27) and (2.29),

$$E\{f'(W) - Wf(W)\} = \sum_{i=1}^{n} \int_{-\infty}^{\infty} E\{f'(W) - f'(W^{(i)} + t)\} K_i(t)\,dt. \tag{2.30}$$

Since $K_i(t)$ is non-negative and $\int_{-\infty}^{\infty} K_i(t)\,dt = E\xi_i^2$, the ratio $K_i(t)/E\xi_i^2$ can be regarded as a probability density function. Let $\xi_i^*$, $i = 1, \ldots, n$ be independent random variables, independent of $\xi_j$ for $j \ne i$, having density function $K_i(t)/E\xi_i^2$ for each $i$. Let $I$ be a random index, independent of $\{\xi_i,\ \xi_i^*,\ i = 1, \ldots, n\}$ with distribution

$$P(I = i) = E\xi_i^2.$$

Then we may rewrite (2.27) as

$$E[Wf(W)] = Ef'(W^{(I)} + \xi_I^*) \tag{2.31}$$

and (2.30) as

$$E\{f'(W) - Wf(W)\} = E\{f'(W) - f'(W^{(I)} + \xi_I^*)\}.$$

Equations (2.27), and (2.30) play a key role in proving good normal approximations. Note in particular that (2.30) is an *equality*, and that (2.27) and (2.30) hold for all bounded absolutely continuous functions $f$. It is easy to see that bounds on the solution $f$ such as those furnished by Lemma 2.4 can now come into play to bound the expected difference in (2.30), and therefore the left hand side of (2.26).

### 2.3.2 Exchangeable Pairs

Suppose now that $W$ is an arbitrary random variable, in particular, not necessarily a sum. A number of variations of Stein's method introduce an auxiliary random variable coupled to $W$ possessing certain properties. In the exchangeable pair approach, (see Stein 1986) one constructs $W'$ on the same probability space as $W$ in such a way that $(W, W')$ is an exchangeable pair, that is, such that $(W, W') =_d (W', W)$, where $=_d$ signifies equality in distribution. The exchangeable pair approach makes essential use of the elementary fact that, if $(W, W')$ is an exchangeable pair, then

$$Eg(W, W') = 0 \tag{2.32}$$

for all antisymmetric measurable functions $g(x, y)$ such that the expected value above exists.

The key identities applied in the exchangeable pair approach are given in Lemma 2.7, for which we require the following definition.

**Definition 2.1** If the pair $(W, W')$ is exchangeable and satisfies the 'linear regression condition'

$$E(W'|W) = (1 - \lambda)W \tag{2.33}$$

with $\lambda \in (0, 1)$, then we call $(W, W')$ a $\lambda$-Stein pair, or more simply, a Stein pair.

One heuristic explanation of why property (2.33) should be of any importance in normal approximation is that it is parallel to the conditional expectation property enjoyed by the bivariate normal distribution. That is, if $Z, Z'$ have the bivariate normal distribution then the conditional expectation of $Z'$ given $Z$ is linear, specifically

$$E(Z'|Z) = \mu_1 + \sigma_1 \rho \left( \frac{Z - \mu_2}{\sigma_2} \right),$$

where $\sigma_1^2$ and $\sigma_2^2$ are the variances of $Z'$ and $Z$, respectively, and $\rho$ is the correlation coefficient. Hence, when $Z$ and $Z'$ have mean zero and equal variance, we obtain (2.33),

$$E(Z'|Z) = (1 - \lambda)Z,$$

with $\lambda = 1 - \rho$.

**Lemma 2.7** *Let $(W, W')$ be a Stein pair and $\Delta = W - W'$. Then*

$$EW = 0 \quad \text{and} \quad E\Delta^2 = 2\lambda EW^2 \quad \text{if } EW^2 < \infty. \tag{2.34}$$

*Furthermore, when $EW^2 < \infty$, for every absolutely continuous function $f$ satisfying $|f(w)| \le C(1 + |w|)$, we have*

$$E\big[Wf(W)\big] = \frac{1}{2\lambda} E\big\{(W - W')\big(f(W) - f(W')\big)\big\}, \tag{2.35}$$

$$E\big[Wf(W)\big] = E\left\{\int_{-\infty}^{\infty} f'(W + t)\hat{K}(t)\,dt\right\}, \tag{2.36}$$

*and*

$$\begin{aligned} &E\big[f'(W) - EWf(W)\big] \\ &\quad = Ef'(W)\left(1 - \frac{\Delta^2}{2\lambda}\right) + E\int_{-\infty}^{\infty} \big(f'(W) - f'(W + t)\big)\hat{K}(t)\,dt, \end{aligned} \tag{2.37}$$

*where*

$$\hat{K}(t) = \frac{\Delta}{2\lambda}(\mathbf{1}_{\{-\Delta \le t \le 0\}} - \mathbf{1}_{\{0 < t \le -\Delta\}}) \tag{2.38}$$

*satisfies*

$$\int_{-\infty}^{\infty} \hat{K}(t)\,dt = \frac{\Delta^2}{2\lambda}. \tag{2.39}$$

*Proof* Taking expectation in (2.33) yields, by exchangeability, $EW = EW' = (1 - \lambda)EW$ so $EW = 0$. Furthermore, as

$$EW'W = E\big(E(W'W|W)\big) = E\big(WE(W'|W)\big) = (1 - \lambda)EW^2,$$

we have

$$E(W' - W)^2 = 2EW^2 - 2EW'W = 2\lambda EW^2.$$

Next we exploit (2.32) with the antisymmetric function $g(x, y) = (x - y)(f(y) + f(x))$, for which $Eg(W, W')$ exists, because of the growth assumption on $f$. Identity (2.32) yields

$$\begin{aligned} 0 &= E\big\{(W - W')\big(f(W') + f(W)\big)\big\} \\ &= E\big\{(W - W')\big(f(W') - f(W)\big)\big\} + 2E\big\{f(W)(W - W')\big\} \\ &= E\big\{(W - W')\big(f(W') - f(W)\big)\big\} + 2E\big\{f(W)E(W - W'|W)\big\} \\ &= E\big\{(W - W')\big(f(W') - f(W)\big)\big\} + 2\lambda E\big\{Wf(W)\big\}, \end{aligned}$$

this last by (2.33). Rearranging this equality yields (2.35), and now

$$
\begin{aligned}
E\big[Wf(W)\big] &= \frac{1}{2\lambda}E\big\{(W-W')\big(f(W)-f(W')\big)\big\}\\
&= \frac{1}{2\lambda}E\big\{\Delta\big(f(W)-f(W-\Delta)\big)\big\}\\
&= \frac{1}{2\lambda}E\int_{-\Delta}^{0}\Delta f'(W+t)\,dt\\
&= E\int_{-\infty}^{\infty} f'(W+t)\hat{K}(t)\,dt. \qquad (2.40)
\end{aligned}
$$

This proves (2.36).

Now note that integrating (2.38) yields (2.39) and so to prove (2.37), we need only observe that

$$
Ef'(W) = E\left\{f'(W)\left(1-\frac{\Delta^2}{2\lambda}\right)\right\} + E\left\{\int_{-\infty}^{\infty} f'(W)\hat{K}(t)\,dt\right\}
$$

and subtract using (2.40). □

As the linear regression condition (2.33) may at times be too restrictive, it can be replaced by

$$
E(W-W' \mid W) = \lambda(W-R), \qquad (2.41)
$$

where $R$ is a random variable of small order. Following the proof of (2.36), if $W'$ and $W$ are mean zero exchangeable random variables with finite second moments, and (2.41) holds for some $\lambda \in (0,1)$ and random variable $R$, then

$$
E\big[Wf(W)\big] = E\int_{-\infty}^{\infty} f'(W+t)\hat{K}(t)\,dt + E\big[Rf(W)\big], \qquad (2.42)
$$

with $\hat{K}(t)$ given by (2.38).

We present three examples that give the flavor of the construction of exchangeable pairs; sometimes we will denote the pair by $(W', W'')$, instead of by $(W, W')$.

*Example 2.1* (Independent random variables) Let $\{\xi_i, 1 \le i \le n\}$ be independent random variables with zero means and $\sum_{i=1}^n E\xi_i^2 = 1$, and put $W = \sum_{i=1}^n \xi_i$. Let $\{\xi_i',\ i = 1,\dots,n\}$ be an independent copy of $\{\xi_i,\ i=1,\dots,n\}$, and let $I$ have uniform distribution on $\{1,2,\dots,n\}$, independent of $\{\xi_i, \xi_i',\ i=1,\dots,n\}$. Define $W' = W - \xi_I + \xi_I'$. Then $(W, W')$ is an exchangeable pair, and it is easy to verify

$$
E(W' \mid W) = \left(1-\frac{1}{n}\right)W,
$$

so that (2.33) is satisfied with $\lambda = 1/n$.

The exchangeable pair above is a special case of the following general construction.

*Example 2.2* (Exchangeable pair by substitution) Let $W = g(\xi_1, \dots, \xi_n)$, and $\xi_i'$ have the conditional distribution of $\xi_i$ given $\xi_j$, $1 \le j \ne i \le n$. Let $I$ be a random index uniformly distribution over $\{1, \dots, n\}$, independent of $\{\xi_i, \xi_i', \ i = 1, \dots, n\}$. Define $W' = g(\xi_1, \dots, \xi_{I-1}, \xi_I', \xi_{I+1}, \dots, \xi_n)$. That is, in the definition of $W$, the $\xi_I$ is replaced by $\xi_I'$ while the other variables remain the same. Then $(W, W')$ is an exchangeable pair. We note that unlike Example 2.1 the linearly condition (2.33) is not automatically satisfied.

*Example 2.3* (Combinatorial Central Limit Theorem) For a given array $\{a_{ij}\}_{1 \le i,j \le n}$ of real numbers and $\pi = \pi'$ a random permutation, let

$$Y' = \sum_{i=1}^{n} a_{i,\pi'(i)}. \tag{2.43}$$

Classically, $\pi'$ is taken to be uniformly distributed over the symmetric group $\mathcal{S}_n$; we specialize to that case here, and study it in Sects. 4.4 and 6.1.1, but also consider alternative permutation distributions in Sect. 6.1.2.

Let

$$a_{\bullet\bullet} = \frac{1}{n^2} \sum_{i,j=1}^{n} a_{ij}, \quad a_{i\bullet} = \frac{1}{n} \sum_{j=1}^{n} a_{ij} \quad \text{and} \quad a_{\bullet j} = \frac{1}{n} \sum_{i=1}^{n} a_{ij}. \tag{2.44}$$

Using that $\pi'$ is uniform one easily obtains $EY' = na_{\bullet\bullet} = \sum_i a_{i\bullet} = \sum_i a_{\bullet\pi(i)}$, and therefore

$$Y' - EY' = \sum_{i=1}^{n} (a_{i,\pi(i)} - a_{\bullet\bullet}) = \sum_{i=1}^{n} (a_{i,\pi(i)} - a_{i\bullet} - a_{\bullet\pi(i)} + a_{\bullet\bullet}). \tag{2.45}$$

As our goal is to derive bounds to the normal for the standardized variable $(Y' - EY')/\sqrt{\mathrm{Var}(Y')}$, without loss of generality we may replace $a_{ij}$ by $a_{ij} - a_{i\bullet} - a_{\bullet j} + a_{\bullet\bullet}$, and assume

$$a_{i\bullet} = a_{\bullet j} = a_{\bullet\bullet} = 0. \tag{2.46}$$

Let $\tau_{ij}$ be the permutation that transposes $i$ and $j$, $\pi'' = \pi'\tau_{ij}$ and $Y''$ be given by (2.43) with $\pi'$ replaced by $\pi''$. Since $\pi''(k) = \pi'(k)$ for $k \notin \{i, j\}$ while $\pi''(i) = \pi'(j)$ and $\pi''(j) = \pi'(i)$, we have

$$Y'' - Y' = b\big(i, j, \pi(i), \pi(j)\big), \tag{2.47}$$

where $b(i, j, k, l) = a_{il} + a_{jk} - (a_{ik} + a_{jl})$. Taking $(I, J)$ to be independent of $\pi'$, with the uniform distribution over all pairs satisfying $1 \le I \ne J \le n$, the permutations $\pi'$ and $\pi'' = \tau_{IJ}\pi'$ are exchangeable, and hence so are $Y'$ and $Y''$.

To prove that the linear regression property (2.33) is satisfied, write

$$Y'' - Y' = (a_{I,\pi'(J)} + a_{J,\pi'(I)}) - (a_{I,\pi'(I)} + a_{J,\pi'(J)}). \tag{2.48}$$

Taking the conditional expectation given $\pi'$, using (2.46), we obtain

$$\begin{aligned}E(Y''-Y'|\pi') &= 2\left(-\frac{1}{n}\sum_{i=1}^n a_{i,\pi'(i)}+\frac{1}{n(n-1)}\sum_{i\neq j}a_{i,\pi'(j)}\right)\\&= -2\left(\frac{1}{n}\sum_{i=1}^n a_{i,\pi'(i)}+\frac{1}{n(n-1)}\sum_{i=1}^n a_{i,\pi'(i)}\right)=-\frac{2}{n-1}Y'.\end{aligned}$$

As the right hand side is measurable with respect to $Y'$, we conclude that

$$E(Y''|Y')=\left(1-\frac{2}{n-1}\right)Y',$$

demonstrating that $Y', Y''$ is a $2/(n-1)$-Stein pair.

One particular special case of note is when $a_{ij}=b_i c_j$ where $b_1,\ldots,b_n$ are any real numbers and the values $c_j\in\{0,1\}$, $j=1,\ldots,n$ satisfy

$$\sum_{i=1}^n c_j=m.$$

In this case, as any set of $m$ values from $\{b_1,\ldots,b_n\}$ are as likely to be summed to yield $Y'$ as any other set of that same size, $Y'$ is the sum of a simple random sample of size $m$ from a population whose numerical characteristics are given by $\{b_i,\ i=1,\ldots,n\}$.

It is worth mentioning a connection between the exchangeable pair and the generator approach which gave the solutions and bounds to the Stein equation in Lemma 2.6. To see the connection, let $(W,W')$ be a $\lambda$-Stein pair and rewrite

$$E(W'|W)=(1-\lambda)W \quad \text{as } E(W'-W|W)=-\lambda W.$$

If one can construct a sequence $W_1, W_2,\ldots$ such that

$$(W_t, W_{t+1}) =_d (W,W'), \quad \text{for } t=1,2,\ldots,$$

then $E(W_{t+1}-W_t|W_t)=-\lambda W_t$, and so, with $\Delta W_t=W_{t+1}-W_t$ we have

$$\Delta W_t=-\lambda W_t+\epsilon_t \quad \text{where } E[\epsilon_t|W_t]=0,$$

a recursion reminiscent of the stochastic differential equation for the Ornstein–Uhlenbeck process,

$$dW_t=-\lambda W_t+\sigma dB_t$$

where $B_t$ is a Brownian motion.

It is sometimes possible to produce the sequence $W_1, W_2,\ldots$ as the successive states of a reversible Markov chain in stationarity. Or, looking at this construction in another way, for a given $W$ of interest, one may be able to create a Stein pair by constructing a reversible Markov chain with stationary distribution $W$. As an illustration, consider the sum $Y$ of a simple random sample $S=\{X_1,\ldots,X_n\}$ of size $n$ of $N$ population characteristics $\mathcal{A}=\{a_1,\ldots,a_N\}$ which have been centered to satisfy

$$\sum_{i=1}^{N} a_i = 0. \tag{2.49}$$

Given a simple random sample $S_0$, one may construct a Markov chain $S_0, S_1, \ldots$, whose state space consists of all size $n$ subsets of $\mathcal{A}$, by interchanging at time step $n$ a randomly chosen element of $S_n$ with one from the complement of $S_n$ to form $S_{n+1}$. The chain is in equilibrium and is reversible, hence the sets $S_n, n = 0, 1, \ldots$ are identically distributed, and $(S_n, S_{n+1})$ is exchangeable. In particular, the sums $Y_n$ and $Y_{n+1}$ of $S_n$ and $S_{n+1}$ respectively, are exchangeable and have the same distribution as $Y$. This construction is, essentially, the one used in Theorem 4.10, and it is shown there that the linearity condition (2.33) holds under the centering (2.49). This method for the construction of exchangeable pairs features prominently in the analysis of the anti-voter model in Sect. 6.4.

### 2.3.3 *Zero Bias*

Stein's characterization (2.1) of the standard normal $Z$ can be easily extended to the mean zero normal family in general. In particular, a simple change of variable in Lemma 2.1 shows that $X$ is $\mathcal{N}(0, \sigma^2)$ if and only if

$$\sigma^2 Ef'(X) = E\big[Xf(X)\big] \tag{2.50}$$

for all absolutely continuous functions for which these expectations exist. Though the left and right hand sides of (2.50) will only be equal at the normal, one can create an identity in the same spirit that holds more generally. In particular, as introduced in Goldstein and Reinert (1997), given $X$ with mean zero and variance $\sigma^2$, we say that $X^*$ has the $X$-zero bias distribution if

$$\sigma^2 Ef'\big(X^*\big) = E\big[Xf(X)\big] \tag{2.51}$$

for all absolutely continuous functions $f$ for which these expectations exist. It is convenient to regard (2.51) as giving rise to a transformation mapping the distribution of $X$ to that of $X^*$. Indeed, the characterization in Lemma 2.1 can be restated as saying that the normal distribution is the unique fixed point of the zero bias transformation.

It is the uniqueness of the fixed point of the zero bias transformation, that is, the fact that $X^*$ has the same distribution as $X$ only when $X$ is normal, that provides a probabilistic reason for a normal approximation to hold. If the distribution of a random variable $X$ gets mapped to an $X^*$ which is close in distribution to $X$, then $X$ is close to the zero bias transformation's unique fixed point, that is, close to the normal.

This same reasoning indicates that not only should a normal approximation be justified whenever the distribution of $X$ is close to that of $X^*$, but that the quality of the approximation can be measured in terms of their distance. Though this claim will later be made precise in a number of ways, for now one can see how it might be formalized by observing that a coupling of a mean zero, variance one $W$ to such a $W^*$ can be used in the Stein equation (2.4) as

$$Eh(W) - Nh = E\big[f'(W) - Wf(W)\big] = E\big[f'(W) - f'(W^*)\big].$$

Hence, when $W$ and $W^*$ are close, the right hand side, and so also the left hand side, will be small.

While the zero bias transformation fixes the mean zero normal, for non-normal distributions, in some sense, the transformation moves them closer to normality. For example, let $\xi \in \{0, 1\}$ be a Bernoulli random variable with success probability $p \in (0, 1)$. Centering $\xi$ to form the mean zero discrete random variable $X = \xi - p$ having variance $\sigma^2 = p(1 - p)$, substitution into the right hand side of (2.51) yields

$$\begin{aligned} E\big[Xf(X)\big] &= E\big[(\xi - p)f(\xi - p)\big] \\ &= p(1-p)f(1-p) - (1-p)pf(-p) \\ &= \sigma^2\big[f(1-p) - f(-p)\big] \\ &= \sigma^2 \int_{-p}^{1-p} f'(u)\,du \\ &= \sigma^2 Ef'(U), \end{aligned}$$

for $U$ uniformly distributed over $[-p, 1-p]$. Hence, with $=_d$ indicating the equality of two random variables in distribution, and $\mathcal{U}[a, b]$ denoting the uniform distribution on the finite interval $[a, b]$,

$$(\xi - p)^* =_d U \quad \text{where } U \sim \mathcal{U}[-p, 1-p]. \tag{2.52}$$

As hinted at by the Bernoulli example, the following lemma shows that the zero bias distribution exists and is absolutely continuous for every $X$ having mean zero and some finite, positive variance.

**Proposition 2.1** *Let $X$ be a random variable with mean zero and finite positive variance $\sigma^2$. Then there exists a unique distribution for $X^*$ such that*

$$Ef'(X^*) = \sigma^2 E\big[Xf(X)\big] \tag{2.53}$$

*for every absolutely continuous function $f$ for which $E|Xf(X)| < \infty$.*

*Moreover, the distribution of $X^*$ is absolutely continuous with density*

$$p^*(x) = E\big[X\mathbf{1}(X > x)\big]/\sigma^2 = -E\big[X\mathbf{1}(X \le x)\big]/\sigma^2 \tag{2.54}$$

*and distribution function*

$$G^*(x) = E\big[X(X - x)\mathbf{1}(X \le x)\big]/\sigma^2. \tag{2.55}$$

*Proof* We prove the claims assuming $\sigma^2 = 1$, the extension to the general case being straightforward. First, regarding (2.54), we note that the second equality holds since $EX = 0$. It follows that $p^*(x)$ is nonnegative, using the first form for $x \ge 0$, and the second for $x < 0$.

To prove that we may write $E[Xf(X)]$ as the expectation on the left hand side of (2.53), in terms of an absolutely continuous variable $X^*$ with density $p^*(x)$, let $f(x) = \int_0^x g$ with $g$ a nonnegative function which is integrable on compact domains. Then by Fubini's theorem,

$$\begin{aligned}\int_0^\infty f'(u)E\big[X\mathbf{1}(X>u)\big]\,du &= \int_0^\infty g(u)E\big[X\mathbf{1}(X>u)\big]\,du \\ &= E\left(X\int_0^\infty g(u)\mathbf{1}(X>u)\,du\right) \\ &= E\left(X\int_0^{X\vee 0} g(u)\,du\right) \\ &= E\big[Xf(X)\mathbf{1}(X\ge 0)\big].\end{aligned}$$

A similar argument over $(-\infty, 0]$ yields

$$\int_{-\infty}^\infty f'(u)E\big[X\mathbf{1}(X>u)\big]\,du = E\big[Xf(X)\big], \tag{2.56}$$

where both sides may be $+\infty$. If $f(x) = \int_0^x g$ with $E|Xf(X)| < \infty$, then taking the difference of the contributions from the positive and negative parts of $g$ shows that (2.56) continues to hold over this larger class of functions, as it does for $f$ satisfying the conditions of the theorem by writing $f(x) = \int_0^x g + f(0)$ and using that the mean of $X$ is zero. Taking $f(x) = x$ shows that $p^*(x)$ integrates to one and is therefore a density, whence the left hand side of (2.56) may be written as $Ef'(X^*)$ for $X^*$ with density $p^*(x)$. The distribution of $X^*$ is clearly unique, as $Ef'(X^*) = Ef'(Y^*)$ for all, say, continuously differentiable functions $f$ with compact support, implies $X^* =_d Y^*$.

Integrating the density $p^*$ to obtain the distribution function $G^*$, we have

$$\begin{aligned}G^*(x) &= -E\left(X\int_{-\infty}^x \mathbf{1}(X\le u)\,du\right) \\ &= -E\left(X\int_X^x du\,\mathbf{1}(X\le x)\right) \\ &= E\big[X(X-x)\mathbf{1}(X\le x)\big].\end{aligned}$$

□

The characterization (2.51) also specifies a relationship between the moments of $X$ and $X^*$. One of the most useful of these relations is the one which results from applying (2.51) with $f(x) = (1/2)x^2\,\mathrm{sgn}(x)$, for which $f'(x) = |x|$, yielding

$$\sigma^2 E|X^*| = \frac{1}{2}E|X|^3 \quad \text{where } \sigma^2 = \mathrm{Var}(X). \tag{2.57}$$

In particular, we see that $E|X|^3 < \infty$ if and only if $E|X^*| < \infty$.

We have observed that the zero bias distribution of a mean zero Bernoulli variable with support $\{-p, 1-p\}$ is uniform on $[-p, 1-p]$, and it is easy to see from (2.54) that, more generally, if $x$ is such that $P(X > x) = 0$, then the same holds for all $y > x$, and $p^*(y) = 0$ for all such $y$, while if $x$ is such that $P(X > x) > 0$ then

$p^*(x) > 0$. As similar statements hold when considering $x$ for which $P(X \le x) = 0$, letting support$(X)$ be the support of the distribution of $X$, if

$$a = \inf \text{support}(X) \quad \text{and} \quad b = \sup \text{support}(X)$$

are finite then support$(X^*) = [a, b]$. One can verify that the support continues to be given by this relation, with any closed endpoint replaced by the corresponding open one, when any of the values of $a$ or $b$ are infinite. One consequence of this fact is that if $X$ is bounded by some constant then $X^*$ is also bounded by the same constant, that is,

$$|X| \le C \quad \text{implies} \quad |X^*| \le C. \tag{2.58}$$

The zero bias transformation enjoys the following scaling, or linearity property. If $X$ is a mean zero random variable with finite variance, and $X^*$ has the $X$-zero biased distribution, then for all $a \neq 0$

$$(aX)^* =_d aX^*. \tag{2.59}$$

The verification of this claim follows directly from (2.51), as letting $\sigma^2 = \text{Var}(X)$ and $g(x) = f(ax)$, we find

$$\begin{aligned}
(a\sigma)^2 E f'(aX^*) &= a\sigma^2 E g'(X^*) \\
&= aE\big[Xg(X)\big] \\
&= E\big[(aX) f(aX)\big] \\
&= (a\sigma)^2 E f'\big((aX)^*\big).
\end{aligned}$$

But by far the most important properties of the zero bias transformation are those like the ones given in the following lemma.

**Lemma 2.8** *Let $\xi_i$, $i = 1, \ldots, n$ be independent mean zero random variables with* $\text{Var}(\xi_i) = \sigma_i^2$ *summing to* 1. *Let $\xi_i^*$ have the $\xi_i$-zero bias distribution with $\xi_i^*$, $i = 1, \ldots, n$ mutually independent, and $\xi_i^*$ independent of $\xi_i$ for all $j \neq i$. Further, let $I$ be a random index, independent of $\xi_i, \xi_i^*, i = 1, \ldots, n$ with distribution*

$$P(I = i) = \sigma_i^2. \tag{2.60}$$

*Then*

$$W^* =_d W - \xi_I + \xi_I^*, \tag{2.61}$$

*where $W^*$ has the $W$-zero bias distribution.*

In other words, upon replacing the variable $\xi_I$ by $\xi_I^*$ in the sum $W = \sum_{i=1}^n \xi_i$ we obtain a variable with the $W$-zero bias distribution. The distributional identity (2.61) indicates that a normal approximation is justified when the difference $\xi_I - \xi_I^*$ is small, since then the distribution of $W$ will be close to that of $W^*$. To prepare for the proof, note that we may write the variables $\xi_I$ and $\xi_I^*$ selected by $I$ using indicators as follows

$$\xi_I = \sum_{i=1}^n \mathbf{1}\{I=i\}\xi_i \quad \text{and} \quad \xi_I^* = \sum_{i=1}^n \mathbf{1}\{I=i\}\xi_i^*,$$

from which it is clear, writing $\mathcal{L}$ for the distribution, or law of a random variable, that the distributions of $\xi_I$ and $\xi_I^*$ are the mixtures

$$\mathcal{L}(\xi_I) = \sum_{i=1}^n \mathcal{L}(\xi_i)\sigma_i^2 \quad \text{and} \quad \mathcal{L}(\xi_I^*) = \sum_{i=1}^n \mathcal{L}(\xi_i^*)\sigma_i^2.$$

*Proof* Let $W^*$ have the $W$-zero bias distribution. Then for all absolutely continuous functions $f$ for which the following expectations exist,

$$\begin{aligned}
E\big[f'(W^*)\big] &= E\big[Wf(W)\big] \\
&= E\left[\sum_{i=1}^n \xi_i f(W)\right] \\
&= \sum_{i=1}^n E\big[\xi_i f(W - \xi_i + \xi_i)\big] \\
&= \sum_{i=1}^n E\big[\sigma_i^2 f'(W - \xi_i + \xi_i^*)\big] \\
&= E\left[\sum_{i=1}^n f'(W - \xi_i + \xi_i^*)\mathbf{1}(I=i)\right] \\
&= E\big[f'(W - \xi_I + \xi_I^*)\big],
\end{aligned}$$

where independence is used in the fourth and fifth equalities. The equality of the expectations of $W^*$ and $W - \xi_I + \xi_I^*$ over this class of functions is sufficient to guarantee (2.61), that is, that these two random variables have the same distribution, as in the proof of Proposition 2.1. □

When handling the sum of independent random variables, the zero bias method and the $K$ function approach of Sect. 2.3.1 are essentially equivalent, with the former providing a probabilistic formulation of the latter. To begin to see the connection, note that by (2.54) and (2.24) the zero bias density $p^*(t)$ and the $K(t)$ function are almost sure multiplies,

$$p^*(t) = K(t)/\sigma^2.$$

In particular, by Lemma 2.8, with $K_i(t)$ the function (2.24) corresponding to $\xi_i$, integrating against the density of $\xi_i^*$ yields

$$Ef(W^*) = \sum_{i=1}^n Ef(W^{(i)} + \xi_i^*)\sigma_i^2 = \sum_{i=1}^n \int_{-\infty}^{\infty} Ef(W^{(i)} + t)K_i(t)\,dt. \quad (2.62)$$

Likewise, that $p^*(x)$ is a density function, and the moment identity (2.57), are probabilistic interpretations of the two equalities in (2.25), respectively, in terms of random variables. In addition, we note the correspondence between Lemma 2.8 and identity (2.31). To later explore the relationship between the zero bias method and the general Stein identity in Sect. 2.4, note now that if $W$ and $W^*$ are defined on the same space then trivially from the defining zero bias identity (2.51) we have

$$E\big[Wf(W)\big] = Ef'(W+\Delta) \quad \text{where } \Delta = W^* - W.$$

Though the $K$ function approach and zero biasing are essentially completely parallel when dealing with sums of independent variables, these two views each give rise to useful, and separate, ways of handling different classes of examples. In addition to its ties to the $K$ function approach, we will see in Proposition 4.6 that zero biasing is also connected to the exchangeable pair.

### *2.3.4 Size Bias*

The size bias and zero bias transformations are close relatives, and as such, size bias and zero bias couplings can be used in the Stein equation in somewhat similar manners. The size bias transformation is defined on the class of non-negative random variables $X$ with finite non-zero means. For such an $X$ with mean $EX = \mu$, we say $X^s$ has the $X$-size biased distribution if for all functions $f$ for which $E[Xf(X)]$ exists,

$$E\big[Xf(X)\big] = \mu Ef\big(X^s\big). \tag{2.63}$$

We note that this characterization for size biasing is of the same form as (2.51) for zero biasing, but with the mean replacing the variance, and $f$ replacing $f'$ for the evaluation of the biased variable.

To place size biasing in the framework of Sect. 2.4 to follow, we note that when $\mathrm{Var}(X) = \sigma^2$ and $W = (X-\mu)/\sigma$, and, with a slight abuse of notation, $W^s = (X^s-\mu)/\sigma$, if $X$ and $X^s$ are defined on the same space, identity (2.63) can be written

$$E\big[Wf(W)\big] = \frac{\mu}{\sigma}E\big[f\big(W^s\big) - f(W)\big] = E\int_{-\infty}^{\infty} f'(W+t)\hat{K}(t)\,dt, \tag{2.64}$$

where

$$\hat{K}(t) = \frac{\mu}{\sigma}\big(\mathbf{1}_{\{0\le t\le W^s-W\}} - \mathbf{1}_{\{W^s-W\le t<0\}}\big). \tag{2.65}$$

The characterization (2.63) is easily seen to be the same as the more common specification of the size bias distribution $F^s(x)$ as the one which is absolutely continuous with respect to the distribution $F(x)$ of $X$ with Radon Nikodym derivative

$$\frac{dF^s(x)}{dF(x)} = \frac{x}{\mu}. \tag{2.66}$$

Hence, parallel to property (2.58) for zero bias, here we have

$$0 \le X \le C \quad \text{implies} \quad 0 \le X^s \le C. \tag{2.67}$$

Moreover, if $X$ is absolutely continuous with density $p(x)$, then $X^s$ is also absolutely continuous, and has density $xp(x)/\mu$. Size biasing also enjoys a scaling property. If $X^s$ has the $X$-size bias distribution, then for $a > 0$

$$(aX)^s = aX^s$$

by an argument nearly identical to the one that proves (2.59).

Size biasing can occur, possibly unwanted, when applying various sampling designs where items associated with larger outcomes are more likely to be chosen. For instance, when sampling an individual in a population at random, their report of the number of siblings in their family is size biased. Size biasing is also responsible for the well known waiting time paradox (see Feller 1968b), but can also be used to advantage, in particular, to form unbiased ratio estimates (Midzuno 1951).

Lemma 2.8 carries over with only minor changes when replacing zero biasing by size biasing, though the variable replaced is now selected proportional to its mean, rather its variance. Moreover, the size bias construction generalizes easily to the case where the sum is of dependent random variables. In particular, let $\mathbf{X} = \{X_\alpha, \alpha \in \mathcal{A}\}$ be a collection of nonnegative random variables with finite, nonzero means $\mu_\alpha = EX_\alpha$. For $\alpha \in \mathcal{A}$, we say that $\mathbf{X}^\alpha$ has the $\mathbf{X}$ distribution biased in direction, or coordinate, $\alpha$ if

$$EX_\alpha f(\mathbf{X}) = \mu_\alpha Ef\left(\mathbf{X}^\alpha\right) \tag{2.68}$$

for all real valued functions $f$ for which the expectation of the left hand side exists.

Parallel to (2.66), if $F(\mathbf{x})$ is the distribution of $\mathbf{X}$, then the distribution $F^\alpha(\mathbf{x})$ of $\mathbf{X}^\alpha$ satisfies

$$\frac{dF^\alpha(\mathbf{x})}{dF(\mathbf{x})} = \frac{x_\alpha}{\mu_\alpha}. \tag{2.69}$$

By considering functions $f$ which depend only on $x_\alpha$, it is easy to verify that $X_\alpha^\alpha =_d X_\alpha^s$, that is, that $X_\alpha^\alpha$ has the $X_\alpha$-size biased distribution.

A consequence of the following proposition is a method for size biasing sums of dependent variables.

**Proposition 2.2** *Let $\mathcal{A}$ be an arbitrary index set, and let $\mathbf{X} = \{X_\alpha, \alpha \in \mathcal{A}\}$ be a collection of nonnegative random variables with finite means. For any subset $B \subset \mathcal{A}$, set*

$$X_B = \sum_{\beta \in B} X_\beta \quad \text{and} \quad \mu_B = EX_B.$$

*Suppose $B \subset \mathcal{A}$ with $0 < \mu_B < \infty$, and for $\beta \in B$ let $\mathbf{X}^\beta$ have the $\mathbf{X}$-size biased distribution in coordinate $\beta$ as in Definition 2.68. Let $I$ be a random index, independent of $\mathbf{X}$, with distribution*

$$P(I = \beta) = \frac{\mu_\beta}{\mu_B}.$$

*Then* $\mathbf{X}^B = \mathbf{X}^I$, *that is, the collection* $\mathbf{X}^B$ *which is equal to* $\mathbf{X}^\beta$ *with probability* $\mu_\beta/\mu_B$, *satisfies*

$$E\big[X_B f(\mathbf{X})\big] = \mu_B E f\big(\mathbf{X}^B\big) \tag{2.70}$$

*for all real valued functions* $f$ *for which these expectations exist.*

*If* $f$ *is a function of* $X_A = \sum_{\alpha\in A} X_\alpha$ *only, then*

$$E\big[X_B f(X_A)\big] = \mu_B E f\big(X_A^B\big) \quad \text{where } X_A^B = \sum_{\alpha\in A} X_\alpha^B,$$

*and when* $A = B$ *we have* $EX_A f(X_A) = \mu_A E f(X_A^A)$, *and that* $X_A^A$ *has the* $X_A$*-size biased distribution.*

*Proof* Without loss of generality, assume $\mu_\beta > 0$ for all $\beta \in \mathcal{A}$. By (2.68) we have

$$E\big[X_\beta f(\mathbf{X})\big]/\mu_\beta = E f\big(\mathbf{X}^\beta\big).$$

Multiplying by $\mu_\beta/\mu_B$, summing over $\beta \in B$ and recalling $\mathbf{X}^B$ is a mixture yields (2.70). The remainder of the lemma now follows as special cases. □

By the last claim of the lemma, to achieve the size bias distribution of the sum $X_\mathcal{A} = \sum_{\alpha\in\mathcal{A}} X_\alpha$ of all the variables in the collection, one mixes over the distributions of $X_\mathcal{A}^\beta = \sum_{\alpha\in\mathcal{A}} X_\alpha^\beta$ using the random index with distribution

$$P(I = \beta) = \frac{\mu_\beta}{\sum_{\alpha\in\mathcal{A}} \mu_\alpha}. \tag{2.71}$$

Hence, by randomization over $\mathcal{A}$, a construction of $\mathbf{X}^\beta$ for every coordinate $\beta$ leads to a construction of $X^s_\mathcal{A}$.

We may size bias in coordinates by applying the following procedure. Let $\mathcal{A} = \{1,\dots,n\}$ for notational ease. For given $i \in \{1,\dots,n\}$, write the joint distribution of $\mathbf{X}$ as a product of the marginal distribution of $X_i$ times the conditional distribution of the remaining variables given $X_i$,

$$dF(\mathbf{x}) = dF_i(x_i)dF(x_1,\dots,x_{i-1},x_{i+1},\dots,x_n|x_i), \tag{2.72}$$

which gives a factorization of (2.69) as

$$dF^i(\mathbf{x}) = dF_i^i(x_i)dF(x_1,\dots,x_{i-1},x_{i+1},\dots,x_n|x_i), \tag{2.73}$$

where $dF_i^i(x_i) = (x_i/\mu_i)dF_i(x_i)$.

The representation (2.73) says that one may form $\mathbf{X}^i$ by first generating $X_i^i$ having the $X_i$-sized biased distribution, and then the remaining variables from their original distribution, conditioned on $x_i$ taking on its newly chosen sized biased value. For $\mathbf{X}$ already given, a coupling between the sum of $Y = X_1 + \cdots + X_n$ and $Y^s$ can be generated by first constructing, for every $i$, the biased variable $X_i^i$ and then 'adjusting' the remaining variables $X_j, j \ne i$ as necessary so that they have the correct conditional distribution. Mixing then yields $Y^s$. Typically the goal is to adjust the variables as little as possible in order to have the resulting bounds to normality small.

The following important corollary of Proposition 2.2 handles the case where the variables $X_1, \ldots, X_n$ are independent, so that (2.72) reduces to

$$dF(\mathbf{x}) = dF_i(x_i)dF(x_1, \ldots, x_{i-1}, x_{i+1}, \ldots, x_n).$$

The following result is parallel to Lemma 2.8.

**Corollary 2.1** *Let $Y = \sum_{i=1}^n X_i$, where $X_1, \ldots, X_n$ are independent, nonnegative random variables with means $EX_i = \mu_i$, $i = 1, \ldots, n$. Let $I$ be a random index with distribution given by* (2.71), *independent of all other variables. Then, upon replacing the summand $X_I$ selected by $I$ with a variable $X_I^s$ having its size biased distribution, independent of $X_j$ for $j \neq I$, we obtain*

$$Y^I =_d Y - X_I + X_I^s,$$

*a variable having the $Y$-size bias distribution.*

*Proof* Letting $\mathbf{X} = (X_1, \ldots, X_n)$, the vector

$$\mathbf{X}^i = \left(X_1, \ldots, X_{i-1}, X_i^s, X_{i+1}, \ldots, X_n\right)$$

has the $\mathbf{X}$-size biased distribution in coordinate $i$, as the conditional distribution in (2.73) is the same as the unconditional one. Now apply Proposition 2.2. □

In other words, when the variables are independent and $X_i$ is replaced by its size biased version, here is no need to change any of the remaining variables $X_j$, $j \neq i$ in order for them to have their original conditional distribution given the new value $X_i^s$.

As shown in Goldstein and Reinert (2005), size biasing and zero biasing are both special cases of a general form of distributional biasing, where given a 'biasing function' $P(x)$ with $m \in \{0, 1, \ldots\}$ sign changes, and a distribution $X$ which satisfies the $m - 1$ orthogonality relations $EX^i P(X) = 0$, $i = 0, \ldots, m-1$, there exists a distribution $X^{(P)}$ satisfying

$$E\left[P(X)f(X)\right] = \alpha E f^{(m)}\left(X^{(P)}\right) \tag{2.74}$$

when $\alpha = EP(X)X^m/m! > 0$.

For example, for zero biasing the function $P(x) = x$ has $m = 1$ sign change, so the identity involves the first derivative $f'$, and we require that the distribution of $X$ satisfies the single orthogonality relation $E(1 \cdot X) = EX = 0$, and set $\alpha = EX^2 = \sigma^2$. For size biasing $P(x) = \max\{x, 0\}$, which has $m = 0$ sign changes, so no derivatives of $f$ are involved, and neither are there any orthogonality relations to be satisfied, and $\alpha = EX$. Letting $X^\square$ be characterized by (2.74) with $P(x) = x^2$, since $P(x)$ has no sign changes the distribution $X^\square$ exists for any distribution $X$ with finite second moment, and $\alpha = EX^2$. In this particular case, where

$$E\left[X^2 f(X)\right] = EX^2 E f\left(X^\square\right) \tag{2.75}$$

for all functions $f$ for which $E|Xf(X)| < \infty$, we say that $X^\square$ has the $X$-square bias distribution. As in (2.69) for the size biased distribution, the distribution of $X^\square$ can also be characterized by its Radon–Nikodym derivative with respect to the distribution of $X$, as we do in Proposition 2.3, below.

By comparing Lemma 2.8 with Corollary 2.1 one can already see that zero and size biasing are closely related. Another relation between the two is given by the following proposition.

**Proposition 2.3** *Let $X$ be a symmetric random variable with finite, non-zero variance $\sigma^2$, and let $X^\square$ have the $X$-square bias distribution, that is,*

$$dF^\square(x) = \frac{x^2 dF(x)}{\sigma^2}.$$

*Then, with $U \sim \mathcal{U}[-1,1]$ independent of $X^\square$, the variable*

$$X^* \stackrel{d}{=} UX^\square$$

*has the $X$-zero bias distribution.*

*Proof* Since $X$ is symmetric with finite second moment, $EX = 0$ and $EX^2 = \sigma^2$. For an absolutely continuous function $f$ with derivative $g \in C_c$, the collection of continuous functions having compact support, using the characterization (2.75) for the fourth equality below, we have

$$\begin{aligned}
\sigma^2 Eg(UX^\square) &= \sigma^2 Ef'(UX^\square) \\
&= \frac{\sigma^2}{2} E \int_{-1}^{1} f'(uX^\square)du \\
&= \frac{\sigma^2}{2} E\left(\frac{f(X^\square) - f(-X^\square)}{X^\square}\right) \\
&= \frac{1}{2} E\left(X^2 \frac{f(X) - f(-X)}{X}\right) \\
&= \frac{1}{2} E\big(X\big(f(X) - f(-X)\big)\big) \\
&= \frac{1}{2}\big(EXf(X) + E(-X)f(-X)\big) \\
&= E\big[Xf(X)\big].
\end{aligned}$$

Hence, if $X^*$ has the $X$-zero bias distribution,

$$\sigma^2 Eg(UX^\square) = E\big[Xf(X)\big] = \sigma^2 E\big[f'(X^*)\big] = \sigma^2 Eg(X^*).$$

As the expectation of $g(UX^\square)$ and $g(X^*)$ agree for all $g \in C_c$, the random variables $UX^\square$ and $X^*$ must be equal in distribution. □

## 2.4 A General Framework for Stein Identities and Normal Approximation for Lipschitz Functions

Identity (2.42)

$$E\big[Wf(W)\big] = E\int_{-\infty}^{\infty} f'(W+t)\hat{K}(t)dt + E\big[Rf(W)\big], \tag{2.76}$$

arose when allowing for the possibility that a given exchangeable pair may not satisfy the linearity condition (2.33) exactly. The function $\hat{K}(t)$ may be random, and, to obtain a good bound, $R$ should be a random variable so that the second term $E[Rf(W)]$ is of smaller order than the first. The exchangeable pair and size bias identities, (2.36) and (2.64), respectively, are both the special case of (2.76) when $R=0$. For the first case, the function $\hat{K}(t)$ is given by (2.38), and by (2.65) in the second.

Though the zero bias identity (2.51) with $\sigma^2=1$ does not fit the mold of (2.76) precisely, in somewhat the same spirit, with $\Delta = W^* - W$ we have

$$EWf(W) = Ef'(W+\Delta), \tag{2.77}$$

holding for all absolutely continuous functions $f$ for which the expectations above exist. The following proposition provides a general bound for normal approximation for smooth functions when (2.76) or (2.77) holds.

**Proposition 2.4** *Let $h$ be an absolutely continuous function with $\|h'\| < \infty$ and $\mathcal{F}$ any $\sigma$-algebra containing $\sigma\{W\}$.*

(i) *If (2.76) holds, then*

$$\big|Eh(W) - Nh\big| \le \|h'\|\left(\sqrt{\frac{2}{\pi}}E|1-\hat{K}_1| + 2E\hat{K}_2 + 2E|R|\right), \tag{2.78}$$

*where*

$$\hat{K}_1 = E\left\{\int_{-\infty}^{\infty} \hat{K}(t)\,dt \,|\mathcal{F}\right\} \quad \text{and} \quad \hat{K}_2 = \int_{-\infty}^{\infty} \big|t\hat{K}(t)\big|\,dt.$$

(ii) *If (2.77) holds, then*

$$\big|Eh(W) - Nh\big| \le 2\|h'\|E|\Delta|. \tag{2.79}$$

*Proof* Let $f_h$ be the solution (2.5) to the Stein equation (2.4). We note that by (2.13), both $f_h$ and $f_h'$ are bounded. We may assume the expectations on the right hand side of (2.78) are finite, as otherwise the result is trivial. By (2.4) and (2.76),

$$\begin{aligned}
Eh(W) - Nh &= E\big[f_h'(W) - Wf_h(W)\big] \\
&= Ef_h'(W) - E\int_{-\infty}^{\infty} f_h'(W+t)\hat{K}(t)\,dt - E\big[Rf_h(W)\big] \\
&= Ef_h'(W)(1-\hat{K}_1) + E\int_{-\infty}^{\infty}\big\{f_h'(W) - f_h'(W+t)\big\}\hat{K}(t)\,dt \\
&\quad - E\big[Rf_h(W)\big].
\end{aligned}$$

By the properties of the Stein solution $f_h$ given in (2.13) and the mean value theorem, we have

$$\left|Ef_h'(W)(1-\hat{K}_1)\right| \le \|h'\|\sqrt{\frac{2}{\pi}}E|1-\hat{K}_1|,$$

$$\left|E\int_{-\infty}^{\infty}\{f_h'(W)-f_h'(W+t)\}\hat{K}(t)\,dt\right| \le E\int_{-\infty}^{\infty}2\|h'\|\left|t\hat{K}(t)\right|dt = 2\|h'\|E\hat{K}_2$$

and

$$\left|E\left[Rf_h(W)\right]\right| \le 2\|h'\|E|R|.$$

This proves (2.78).

Next, (2.79) follows from (2.13) and

$$\begin{aligned}\left|Eh(W)-Nh\right| &= \left|E\left(f_h'(W)-Wf_h(W)\right)\right| \\ &= \left|E\left(f_h'(W)-f_h'(W+\Delta)\right)\right| \\ &\le \|f_h''\|E|\Delta|.\end{aligned}$$

□

We will explore smooth function bounds extensively in Chap. 4.

## Appendix

Here we prove Lemmas 2.3 and 2.4, giving the basic properties of the solutions to the Stein equations (2.2) and (2.4). The proof of Lemma 2.3, and part of Lemma 2.4, follow Stein (1986), while parts of the proof of Lemma 2.4 are due to Stroock (2000) and Raič (2004) (see also Chatterjee 2008).

Before beginning, note that from (2.2) and (2.3) it follows that

$$\begin{aligned} f_z'(w) &= wf_z(w) + \mathbf{1}_{\{w\le z\}} - \Phi(z) \\ &= \begin{cases} wf_z(w)+1-\Phi(z) & \text{for } w<z, \\ wf_z(w)-\Phi(z) & \text{for } w>z, \end{cases} \\ &= \begin{cases} (\sqrt{2\pi}we^{w^2/2}\Phi(w)+1)(1-\Phi(z)) & \text{for } w<z, \\ (\sqrt{2\pi}we^{w^2/2}(1-\Phi(w))-1)\Phi(z) & \text{for } w>z, \end{cases} \end{aligned} \tag{2.80}$$

and

$$\left(wf_z(w)\right)' = \begin{cases} \sqrt{2\pi}(1-\Phi(z))((1+w^2)e^{w^2/2}\Phi(w)+\frac{w}{\sqrt{2\pi}}) & \text{if } w<z, \\ \sqrt{2\pi}\Phi(z)((1+w^2)e^{w^2/2}(1-\Phi(w))-\frac{w}{\sqrt{2\pi}}) & \text{if } w>z. \end{cases} \tag{2.81}$$

*Proof of Lemma 2.3* Since $f_z(w) = f_{-z}(-w)$, we need only consider the case $z \ge 0$. Note that for $w > 0$

$$\int_w^{\infty} e^{-x^2/2}\,dx \le \int_w^{\infty}\frac{x}{w}e^{-x^2/2}\,dx = \frac{e^{-w^2/2}}{w},$$

and that

$$\int_w^\infty e^{-x^2/2}\,dx \geq \frac{we^{-w^2/2}}{1+w^2},$$

by comparing the derivatives of the two functions and their values at $w=0$. Thus

$$\frac{we^{-w^2/2}}{(1+w^2)\sqrt{2\pi}} \leq 1-\Phi(w) \leq \frac{e^{-w^2/2}}{w\sqrt{2\pi}}. \tag{2.82}$$

Applying the lower bound in inequality (2.82) to the form $(wf_z(w))'$ for $w>z$ in (2.81), we see that this derivative is nonnegative, thus yielding (2.6). Now, in view of (2.82) and the fact that $wf_z(w)$ is increasing, taking limits using (2.3) we have,

$$\lim_{w\to-\infty} wf_z(w)=\Phi(z)-1 \quad\text{and}\quad \lim_{w\to\infty} wf_z(w)=\Phi(z), \tag{2.83}$$

and (2.7) follows.

Now, using that $wf_z(w)$ is an increasing function of $w$, (2.83) and (2.80),

$$0<f_z'(w)\leq zf_z(z)+1-\Phi(z)<1 \quad \text{for } w<z \tag{2.84}$$

and

$$-1<zf_z(z)-\Phi(z)\leq f_z'(w)<0 \quad \text{for } w>z, \tag{2.85}$$

proving the first inequality of (2.8). For the second, note that for any $w$ and $u$ we therefore have

$$\left|f_z'(w)-f_z'(u)\right| \leq zf_z(z)+1-\Phi(z)-\big(zf_z(z)-\Phi(z)\big)=1.$$

Next, observe that by (2.84) and (2.85), $f_z(w)$ attains its maximum at $z$. Thus

$$0<f_z(w)\leq f_z(z)=\sqrt{2\pi}e^{z^2/2}\Phi(z)\big(1-\Phi(z)\big).$$

By (2.82), $f_z(z)\leq 1/z$. To finish the proof of (2.9), let

$$g(z)=\Phi(z)\big(1-\Phi(z)\big)-\frac{e^{-z^2/2}}{4} \quad\text{and}\quad g_1(z)=\frac{1}{\sqrt{2\pi}}+\frac{z}{4}-\frac{2\Phi(z)}{\sqrt{2\pi}}.$$

Observe that $g'(z)=e^{-z^2/2}g_1(z)$ and that

$$g_1(0)=0,\ g_1'(0)<0,\ g_1''(z)=\frac{z}{\pi}e^{-z^2/2} \quad\text{and}\quad \lim_{z\to\infty} g_1(z)=\infty.$$

Hence $g_1$ is convex on $[0,\infty)$, and there exists $z_1>0$ such that $g_1(z)<0$ for $z<z_1$ and $g_1(z)>0$ for $z>z_1$. In particular, on $[0,\infty)$ the function $g(z)$ decreases for $z<z_1$ and increases for $z>z_1$, so its supremum must be attained at either $z=0$ or $z=\infty$, that is,

$$g(z)\leq \max\big(g(0),g(\infty)\big)=0 \quad \text{for all } z\in[0,\infty),$$

which is equivalent to $f_z(z)\leq\sqrt{2\pi}/4$. This completes the proof of (2.9).

To verify the last inequality (2.10), write

$$\begin{aligned}&(w+u)f_z(w+u)-(w+v)f_z(w+v)\\&\quad=w\bigl(f_z(w+u)-f_z(w+v)\bigr)+uf_z(w+u)-vf_z(w+v)\end{aligned}$$

and apply the mean value theorem and (2.8) on the first term, and (2.9) on the second. □

*Proof of Lemma 2.4* Let $\tilde{h}(w)=h(w)-Nh$ and put $c_0=\|\tilde{h}\|$ and let $c_1=\|h'\|$ if $h$ is absolutely continuous, and $c_1=\infty$ otherwise. Since $\tilde{h}$ and $f_h$ are unchanged when $h$ is replaced by $h-h(0)$, we may assume that $h(0)=0$. Therefore $|h(t)|\le c_1|t|$ and $|Nh|\le c_1E|Z|=c_1\sqrt{2/\pi}$.

We first prove the two bounds on $f_h$ itself. From the expression (2.5) for $f_h$ it follows that

$$\begin{aligned}\bigl|f_h(w)\bigr|&\le\begin{cases}e^{w^2/2}\int_{-\infty}^{w}|\tilde{h}(x)|e^{-x^2/2}\,dx & \text{if } w\le 0,\\ e^{w^2/2}\int_{w}^{\infty}|\tilde{h}(x)|e^{-x^2/2}\,dx & \text{if } w\ge 0\end{cases}\\&\le e^{w^2/2}\min\biggl(c_0\int_{|w|}^{\infty}e^{-x^2/2}\,dx,\ c_1\int_{|w|}^{\infty}\bigl(|x|+\sqrt{2/\pi}\,\bigr)e^{-x^2/2}\,dx\biggr)\\&\le\min\bigl(\sqrt{\pi/2}c_0,2c_1\bigr),\end{aligned}$$

where in the last inequality we obtain

$$e^{w^2/2}\int_{|w|}^{\infty}e^{-x^2/2}\,dx\le\sqrt{\pi/2}$$

by applying (2.82) to show that the function on the left hand side above has a negative derivative for $w\ge 0$, and therefore that its maximum is achieved at $w=0$. We note that the first bound in the minimum applies if $h$ is only bounded, thus yielding the first claim in (2.12), while if $h$ is only absolutely continuous the second bound holds, yielding the first claim in (2.13).

Moving to bounds on $f_h'$, by (2.4) for $w\ge 0$,

$$\begin{aligned}\bigl|f_h'(w)\bigr|&\le\bigl|h(w)-Nh\bigr|+we^{w^2/2}\int_{w}^{\infty}\bigl|h(x)-Nh\bigr|e^{-x^2/2}\,dx\\&\le c_0+c_0we^{w^2/2}\int_{w}^{\infty}e^{-x^2/2}\,dx\le 2c_0,\end{aligned}$$

using (2.82). A similar argument may be applied for $w<0$, proving the remaining claim in (2.12).

To prove the second claim in (2.13), when $h$ is absolutely continuous write

$$\begin{aligned}h(x)-Nh&=\frac{1}{\sqrt{2\pi}}\int_{-\infty}^{\infty}\bigl[h(x)-h(u)\bigr]e^{-u^2/2}\,du\\&=\frac{1}{\sqrt{2\pi}}\int_{-\infty}^{x}\int_{u}^{x}h'(t)e^{-u^2/2}\,dt\,du-\frac{1}{\sqrt{2\pi}}\int_{x}^{\infty}\int_{x}^{u}h'(t)e^{-u^2/2}\,dt\,du\\&=\int_{-\infty}^{x}h'(t)\Phi(t)\,dt-\int_{x}^{\infty}h'(t)\bigl(1-\Phi(t)\bigr)\,dt,\end{aligned}\tag{2.86}$$

from which it follows that

$$\begin{aligned}
f_h(w) &= e^{w^2/2} \int_{-\infty}^{w} \big[h(x) - Nh\big] e^{-x^2/2}\, dx \\
&= e^{w^2/2} \int_{-\infty}^{w} \left( \int_{-\infty}^{x} h'(t)\Phi(t)\, dt - \int_{x}^{\infty} h'(t)\big(1 - \Phi(t)\big)\, dt \right) e^{-x^2/2}\, dx \\
&= -\sqrt{2\pi} e^{w^2/2} \big(1 - \Phi(w)\big) \int_{-\infty}^{w} h'(t)\Phi(t)\, dt \\
&\quad - \sqrt{2\pi} e^{w^2/2} \Phi(w) \int_{w}^{\infty} h'(t)\big[1 - \Phi(t)\big]\, dt. \qquad (2.87)
\end{aligned}$$

Now, from (2.4), (2.87) and (2.86),

$$\begin{aligned}
f_h'(w) &= w f_h(w) + h(w) - Nh \\
&= \big(1 - \sqrt{2\pi} w e^{w^2/2} \big(1 - \Phi(w)\big)\big) \int_{-\infty}^{w} h'(t)\Phi(t)\, dt \\
&\quad - \big(1 + \sqrt{2\pi} w e^{w^2/2} \Phi(w)\big) \int_{w}^{\infty} h'(t)\big(1 - \Phi(t)\big)\, dt.
\end{aligned}$$

Hence

$$\begin{aligned}
\|f_h'\| \le \|h'\| \sup_{w\in\mathbb{R}} \bigg( &\big|1 - \sqrt{2\pi} w e^{w^2/2} \big(1 - \Phi(w)\big)\big| \int_{-\infty}^{w} \Phi(t)\, dt \\
&+ \big|1 + \sqrt{2\pi} w e^{w^2/2} \Phi(w)\big| \int_{w}^{\infty} \big(1 - \Phi(t)\big)\, dt \bigg).
\end{aligned}$$

By integration by parts,

$$\begin{aligned}
&\int_{-\infty}^{w} \Phi(t)\, dt = w\Phi(w) + \frac{e^{-w^2/2}}{\sqrt{2\pi}} \quad \text{and} \\
&\int_{w}^{\infty} \big(1 - \Phi(t)\big)\, dt = -w\big(1 - \Phi(w)\big) + \frac{e^{-w^2/2}}{\sqrt{2\pi}}.
\end{aligned} \qquad (2.88)$$

Thus,

$$\begin{aligned}
\|f_h'\| \le \|h'\| \sup_{w\in\mathbb{R}} \bigg( &\big|1 - \sqrt{2\pi} w e^{w^2/2} \big(1 - \Phi(w)\big)\big| \left( w\Phi(w) + \frac{e^{-w^2/2}}{\sqrt{2\pi}} \right) \\
&+ \big|1 + \sqrt{2\pi} w e^{w^2/2} \Phi(w)\big| \left( -w\big(1 - \Phi(w)\big) + \frac{e^{-w^2/2}}{\sqrt{2\pi}} \right) \bigg).
\end{aligned}$$

One may now verify that the term inside the brackets attains its maximum value of $\sqrt{2/\pi}$ at $w = 0$.

Now we prove the final claim of (2.13). Differentiating (2.4) gives

$$\begin{aligned}
f_h''(w) &= w f_h'(w) + f_h(w) + h'(w) \\
&= \big(1 + w^2\big) f_h(w) + w\big(h(w) - Nh\big) + h'(w). \qquad (2.89)
\end{aligned}$$

From (2.89), (2.87), (2.86), (2.82) and (2.88) we obtain

$$
\begin{aligned}
|f_h''(w)| &\le |h'(w)| + \left|(1+w^2) f_h(w) + w\big(h(w) - Nh\big)\right| \\
&\le |h'(w)| + \left|\left(w - \sqrt{2\pi}(1+w^2)e^{w^2/2}\big(1-\Phi(w)\big)\right)\int_{-\infty}^{w} h'(t)\Phi(t)\,dt\right| \\
&\quad + \left|\left(-w - \sqrt{2\pi}(1+w^2)e^{w^2/2}\Phi(w)\right)\int_{w}^{\infty} h'(t)\big(1-\Phi(t)\big)\,dt\right| \\
&\le |h'(w)| + c_1\left(-w + \sqrt{2\pi}(1+w^2)e^{w^2/2}\big(1-\Phi(w)\big)\right)\int_{-\infty}^{w} \Phi(t)\,dt \\
&\quad + c_1\left(w + \sqrt{2\pi}(1+w^2)e^{w^2/2}\Phi(w)\right)\int_{w}^{\infty} \big(1-\Phi(t)\big)\,dt \\
&= |h'(w)| \\
&\quad + c_1\left(-w + \sqrt{2\pi}(1+w^2)e^{w^2/2}\big(1-\Phi(w)\big)\right)\left(w\Phi(w) + \frac{e^{-w^2/2}}{\sqrt{2\pi}}\right) \\
&\quad + c_1\left(w + \sqrt{2\pi}(1+w^2)e^{w^2/2}\Phi(w)\right)\left(-w\big(1-\Phi(w)\big) + \frac{e^{-w^2/2}}{\sqrt{2\pi}}\right) \\
&= |h'(w)| + c_1 \le 2c_1,
\end{aligned}
$$

as desired. □

We now present the proof of Lemma 2.5 for bounds on the solution $f(w)$ to the Stein equation for the linearly smoothed indicator function (2.14). For this case Bolthausen (1984) proved the inequalities $|f(w)| \le 1$, $|f'(w)| \le 2$, and, through use of the latter, the bound (2.16) with the factor of $|w|$ replaced by $2|w|$.

*Proof of Lemma 2.5* As in (2.87) in the proof of Lemma 2.3, letting

$$
\eta(w) = \sqrt{2\pi}e^{w^2/2}\Phi(w),
$$

we have

$$
f(w) = -\eta(-w)\int_{-\infty}^{w} h'(t)\Phi(t)\,dt - \eta(w)\int_{w}^{\infty} h'(t)\Phi(-t)\,dt. \tag{2.90}
$$

For $z \le w \le z+\alpha$, we therefore have

$$
\begin{aligned}
f(w) &= \eta(-w)\int_{z}^{w} \frac{\Phi(t)}{\alpha}dt + \eta(w)\int_{w}^{z+\alpha} \frac{\Phi(-t)}{\alpha}\,dt \\
&\le \frac{\eta(-w)\Phi(w)(w-z)}{\alpha} + \frac{\eta(w)\Phi(-w)(z+\alpha-w)}{\alpha} \\
&= \eta(w)\Phi(-w) \\
&= \sqrt{2\pi}e^{w^2/2}\Phi(w)\Phi(-w).
\end{aligned} \tag{2.91}
$$

By symmetry we may take $w \ge 0$ without loss of generality. Then, using the fact that $\Phi(w)/w$ is decreasing, and straightforward inequalities, we derive

$$\sqrt{2\pi}e^{w^2/2}\Phi(w)\Phi(-w) \le \min\left(\frac{\sqrt{2\pi}}{2}, \frac{1}{w}\right)\Phi(w) \le \frac{\sqrt{2\pi}}{2}\Phi\left(\sqrt{\frac{2}{\pi}}\right) < 1,$$

showing $f(w) \le 1$ for $w \in [z, z+\alpha]$.

Next, note that $\Phi(z) \le Nh \le \Phi(z+\alpha)$, and let $f_z(w)$ be the solution to the Stein equation for the function $\mathbf{1}_{\{w \le z\}}$. For $w < z$, since $e^{w^2/2}\Phi(w)$ is increasing, we obtain

$$\begin{aligned} f(w) &= \sqrt{2\pi}(1 - Nh)e^{w^2/2}\Phi(w) \\ &\le \sqrt{2\pi}\big(1 - \Phi(z)\big)e^{w^2/2}\Phi(w) \\ &\le \sqrt{2\pi}\big(1 - \Phi(z)\big)e^{z^2/2}\Phi(z) \\ &= f_z(z) \le \sqrt{2\pi}/4, \end{aligned} \tag{2.92}$$

using (2.9). Similarly, for $w > z + \alpha$,

$$\begin{aligned} f(w) &= \sqrt{2\pi}Nhe^{w^2/2}\big(1 - \Phi(w)\big) \\ &\le \sqrt{2\pi}\Phi(z+\alpha)e^{w^2/2}\big(1 - \Phi(w)\big) \\ &\le \sqrt{2\pi}\Phi(w)e^{w^2/2}\big(1 - \Phi(w)\big) \\ &= f_w(w) \le \sqrt{2\pi}/4, \end{aligned} \tag{2.93}$$

showing that $f(w) \le 1$ for all $w \in \mathbb{R}$. The proof of the first claim of (2.15) is completed by showing the lower bound, which follows from the three expressions (2.91), (2.92) and (2.93), proving that $f(w) \ge 0$ over the three intervals $(-\infty, z)$, $[z, z+\alpha]$ and $(\alpha, \infty)$, respectively.

For the second claim, starting again from (2.5), we have

$$\begin{aligned} e^{-w^2/2}f(w) &= \int_{-\infty}^{w} h(x)e^{-x^2/2}dx - \int_{-\infty}^{w} e^{-x^2/2}Nhdx \\ &= \int_{-\infty}^{w} h(x)e^{-x^2/2}dx - \frac{1}{\sqrt{2\pi}}\int_{-\infty}^{w} e^{-x^2/2}\int_{-\infty}^{\infty} h(t)e^{-t^2/2}\,dt\,dx \\ &= \int_{-\infty}^{w} h(x)e^{-x^2/2}dx - \Phi(w)\int_{-\infty}^{\infty} h(t)e^{-t^2/2}\,dt \\ &= \big(1 - \Phi(w)\big)\int_{-\infty}^{w} h(x)e^{-x^2/2}dx - \Phi(w)\int_{w}^{\infty} h(t)e^{-t^2/2}\,dt. \end{aligned}$$

Hence,

$$\begin{aligned} f(w) &= e^{w^2/2}\big(1 - \Phi(w)\big)\int_{-\infty}^{w} h(x)e^{-x^2/2}dx - e^{w^2/2}\Phi(w)\int_{w}^{\infty} h(t)e^{-t^2/2}\,dt \\ &= \frac{1}{\sqrt{2\pi}}\eta(-w)\int_{-\infty}^{w} h(x)e^{-x^2/2}dx - \frac{1}{\sqrt{2\pi}}\eta(w)\int_{w}^{\infty} h(t)e^{-t^2/2}\,dt, \end{aligned}$$

and taking the derivative, we obtain

$$\begin{aligned}
f'(w) &= -\frac{1}{\sqrt{2\pi}}\eta'(-w)\int_{-\infty}^{w} h(x)e^{-x^2/2}dx + \frac{1}{\sqrt{2\pi}}\eta(-w)e^{-w^2/2}h(w) \\
&\quad - \frac{1}{\sqrt{2\pi}}\eta'(w)\int_{w}^{\infty} h(t)e^{-t^2/2}dt + \frac{1}{\sqrt{2\pi}}\eta(w)e^{-w^2/2}h(w) \\
&= h(w)\big(\Phi(w)+\Phi(-w)\big) \\
&\quad - \frac{1}{\sqrt{2\pi}}\left(\eta'(-w)\int_{-\infty}^{w} h(x)e^{-x^2/2}dx + \eta'(w)\int_{w}^{\infty} h(t)e^{-t^2/2}dt\right) \\
&= h(w) - g(w),
\end{aligned}$$

where we have set

$$g(w) = \frac{1}{\sqrt{2\pi}}\left(\eta'(-w)\int_{-\infty}^{w} h(x)e^{-x^2/2}dx + \eta'(w)\int_{w}^{\infty} h(x)e^{-x^2/2}dx\right).$$

Since $\eta'(w) \geq 0$, we have

$$\begin{aligned}
&\inf_x h(x)\big(\eta'(-w)\Phi(w) + \eta'(w)\Phi(-w)\big) \\
&\quad \leq g(w) \leq \sup_x h(x)\big(\eta'(-w)\Phi(w) + \eta'(w)\Phi(-w)\big).
\end{aligned}$$

However, noting

$$\eta'(-w)\Phi(w) + \eta'(w)\Phi(-w) = 1, \tag{2.94}$$

it follows that

$$\inf_x h(x) - \sup_x h(x) \leq f'(w) \leq \sup_x h(x) - \inf_x h(x),$$

that is, $|f'(w)| \leq 1$, proving the second claim in (2.15).

For the third claim in (2.15), differentiating (2.90) yields

$$f'(w) = \eta'(-w)\int_{-\infty}^{w} h'(t)\Phi(t)\,dt - \eta'(w)\int_{w}^{\infty} h'(t)\Phi(-t)\,dt.$$

For $w < z$ we have

$$f'(w) = \eta'(w)\int_{z}^{z+\alpha} \frac{\Phi(-t)}{\alpha}dt,$$

for $w \in [z, z+\alpha]$,

$$f'(w) = -\eta'(-w)\int_{z}^{w} \frac{\Phi(t)}{\alpha}dt + \eta'(w)\int_{w}^{z+\alpha} \frac{\Phi(-t)}{\alpha}dt,$$

and for $w > z+\alpha$

$$f'(w) = -\eta'(-w)\int_{z}^{z+\alpha} \frac{\Phi(t)}{\alpha}dt.$$

Hence, we may write

$$f'(w) = -\frac{1}{\alpha}\int_{z}^{z+\alpha} G(w,t)\,dt,$$

where

$$G(w,t)=\begin{cases}-\eta'(w)\Phi(-t) & \text{when } w\le t,\\ \eta'(-w)\Phi(t) & \text{when } w>t.\end{cases}$$

Now writing

$$\eta(w)=\sqrt{2\pi}\,e^{w^2/2}\Phi(w)=\int_{-\infty}^{0}e^{-s^2/2-sw}ds,$$

applying the dominated convergence theorem to differentiate under the integral, we obtain

$$\eta''(w)=\int_{-\infty}^{0}s^2e^{-s^2/2-sw}ds,$$

and therefore

$$\frac{\partial G(w,t)}{\partial w}=\begin{cases}-\eta''(w)\Phi(-t) & \text{when } w<t,\\ -\eta''(-w)\Phi(t) & \text{when } w>t.\end{cases}$$

Hence, for any fixed $t$, the function $G(w,t)$ is decreasing in $w$ for $w<t$ and $w>t$, and, moreover, satisfies

$$\lim_{w\to-\infty}G(w,t)=0,\qquad \lim_{w\to\infty}G(w,t)=0,$$

and

$$\lim_{w\uparrow t}G(w,t)=-\eta(t)\Phi(-t)<0\quad\text{and}\quad \lim_{w\downarrow t}G(w,t)=\eta'(-t)\Phi(t)>0.$$

Now, from (2.94), it follows that

$$\left|G(w,t)-G(v,t)\right|\le\eta'(t)\Phi(-t)+\eta'(-t)\Phi(t)=1,$$

and hence

$$\begin{aligned}\left|f'(w)-f'(v)\right|&=\left|\frac{1}{\alpha}\int_z^{z+\alpha}\left[G(w,t)-G(v,t)\right]dt\right|\\&\le\frac{1}{\alpha}\int_z^{z+\alpha}\left|G(w,t)-G(v,t)\right|dt\\&\le\frac{1}{\alpha}\int_z^{z+\alpha}1\,dt=1.\end{aligned}$$

Lastly, to demonstrate (2.16), we apply the mean value theorem and the first two bounds in (2.15) to write

$$\begin{aligned}&\left|f'(w+v)-f'(w)\right|\\&=\left|vf(w+v)+w\big(f(w+v)-f(w)\big)+h(w+v)-h(w)\right|\\&\le|v|\left(1+|w|+\frac{1}{\alpha}\int_0^1\mathbf{1}_{[z,z+\alpha]}(w+rv)dr\right).\end{aligned}$$

□

# Chapter 3
# Berry–Esseen Bounds for Independent Random Variables

In this chapter we illustrate some of the main ideas of the Stein method by proving the classical Lindeberg central limit theorem and the Berry–Esseen inequality for sums of independent random variables. We begin with Lipschitz functions, which suffice to prove the Lindeberg theorem. We then prove the Berry–Esseen inequality by developing a concentration inequality. Throughout this chapter we assume that $W = \xi_1 + \cdots + \xi_n$ where $\xi_1, \ldots, \xi_n$ are independent random variables satisfying

$$E\xi_i = 0, \quad 1 \le i \le n \quad \text{and} \quad \sum_{i=1}^{n} \mathrm{Var}(\xi_i) = 1. \tag{3.1}$$

Though we focus on the independent case, the ideas developed here provide a basis for handling more general situations, see, for instance, Theorem 3.5 and its consequence, Theorem 5.2.

Recall that the supremum, $L^\infty$, or Kolmogorov distance between two distribution functions $F$ and $G$ is given by

$$\|F - G\|_\infty = \sup_{z \in \mathbb{R}} \big|F(z) - G(z)\big|.$$

The main goal of this chapter is to prove the Berry–Esseen inequality, first shown by Berry (1941), and Esseen (1942), which gives a uniform bound between $F$, the distribution function of $W$, and $\Phi$, that of the standard normal $Z$, of the form

$$\|F - \Phi\|_\infty \le C \sum_{i=1}^{n} E|\xi_i|^3 \tag{3.2}$$

where $C$ is an absolute constant. The upper bound on the smallest possible value of $C$ has decreased from Esseen's original estimate of 7.59 to its current value of 0.4785 by Tyurin (2010). After proving the Lindeberg and Berry–Esseen theorems, the latter using both the concentration inequality and inductive approaches, we end the chapter with a lower bound on $\|F - \Phi\|_\infty$.

L.H.Y. Chen et al., *Normal Approximation by Stein's Method*,
Probability and Its Applications,
DOI 10.1007/978-3-642-15007-4_3, 

## 3.1 Normal Approximation with Lipschitz Functions

We recall that $h : \mathbb{R} \to \mathbb{R}$ is a Lipschitz continuous function if there exists a constant $K$ such that

$$\left|h(x) - h(y)\right| \le K|x - y| \quad \text{for all } x, y \in \mathbb{R}.$$

Equivalently, $h$ is Lipschitz continuous if and only if $h$ is absolutely continuous with $\|h'\| < \infty$.

**Theorem 3.1** *Let $W = \sum_{i=1}^n \xi_i$ be the sum of mean zero independent random variables $\xi_i, 1 \le i \le n$ with $\sum_{i=1}^n \mathrm{Var}(\xi_i) = 1$, and $h$ a Lipschitz continuous function. If $E|\xi_i|^3 < \infty$ for $i = 1, \ldots, n$, then*

$$\left|Eh(W) - Nh\right| \le 3\|h'\|\gamma, \tag{3.3}$$

*where*

$$\gamma = \sum_{i=1}^n E|\xi_i|^3. \tag{3.4}$$

*Proof* By Lemma 2.8, (2.77) holds with $\Delta = \xi_I^* - \xi_I$, where $\xi_i^*$ has the $\xi_i$-zero bias distribution and is independent of $\xi_j$, $j \ne i$, and $I$ is a random index with distribution (2.60), independent of all other variables. Invoking (2.79) of Proposition 2.4,

$$\begin{aligned}
\left|Eh(W) - Nh\right| &\le 2\|h'\|E\left|\xi_I^* - \xi_I\right| \\
&= 2\|h'\|\sum_{i=1}^n E\left|\xi_i^* - \xi_i\right|E\xi_i^2 \\
&\le 2\|h'\|\sum_{i=1}^n \left(E\left|\xi_i^*\right|E\xi_i^2 + E|\xi_i|E\xi_i^2\right) \\
&= 2\|h'\|\sum_{i=1}^n \left(\frac{1}{2}E|\xi_i|^3 + E|\xi_i|E\xi_i^2\right) \\
&\le 3\|h'\|\sum_{i=1}^n E|\xi_i|^3,
\end{aligned}$$

where we have invoked (2.57) to obtain the second equality, followed by Hölder's inequality. □

The constant 3 is improved to 1 in Corollary 4.2.

The following theorem shows that one can bound $|Eh(W) - Nh|$ in terms of sums of the truncated second and third moments

$$\beta_2 = \sum_{i=1}^n E\xi_i^2 \mathbf{1}_{\{|\xi_i|>1\}} \quad \text{and} \quad \beta_3 = \sum_{i=1}^n E|\xi_i|^3 \mathbf{1}_{\{|\xi_i|\le 1\}}, \tag{3.5}$$

without the need to assume the existence of third moments as in Theorem 3.1.

**Theorem 3.2** *If $W = \sum_{i=1}^n \xi_i$ is the sum of mean zero independent random variables $\xi_i, 1 \le i \le n$ with $\sum_{i=1}^n \mathrm{Var}(\xi_i) = 1$, then for any Lipschitz function $h$*

$$\left|Eh(W) - Nh\right| \le \|h'\|(4\beta_2 + 3\beta_3). \tag{3.6}$$

*Proof* We adopt the same notation as in the proof of Theorem 3.1. The key observation is that we can follow the proof of (2.79) in Proposition 2.4, but instead of applying $|f_h'(W) - f_h'(W + \Delta)| \le 2\|h'\||\Delta|$, we instead use

$$\left|f_h'(W) - f_h'(W + \Delta)\right| \le \min\bigl(2\|f_h'\|, \|f_h''\||\Delta|\bigr) \le 2\|h'\|\bigl(1 \wedge |\Delta|\bigr), \tag{3.7}$$

which holds by (2.13), where $a \wedge b$ denotes $\min(a, b)$. Hence

$$\begin{aligned} \left|Eh(W) - Nh\right| &\le 2\|h'\|E\bigl(1 \wedge |\xi_I^* - \xi_I|\bigr) \\ &\le 2\|h'\|E\bigl(1 \wedge \bigl(|\xi_I^*| + |\xi_I|\bigr)\bigr) \\ &\le 2\|h'\|\bigl(E\bigl(1 \wedge |\xi_I^*|\bigr) + E\bigl(1 \wedge |\xi_I|\bigr)\bigr). \end{aligned} \tag{3.8}$$

Letting $\mathrm{sign}(x)$ be $+1$ for $x > 0$, $-1$ for $x < 0$ and $0$ for $x = 0$, setting

$$f(x) = x\mathbf{1}_{|x|>1} + \frac{1}{2}x^2\,\mathrm{sign}(x)\mathbf{1}_{|x|\le 1} \quad \text{we have} \quad f'(x) = 1 \wedge |x|.$$

Hence, (2.60) and (2.51) now yield

$$\begin{aligned} E\bigl(1 \wedge |\xi_I^*|\bigr) &= \sum_{i=1}^n E\bigl(1 \wedge |\xi_i^*|\bigr)E\xi_i^2 \\ &= \sum_{i=1}^n E\left(\xi_i^2\mathbf{1}_{|\xi_i|>1} + \frac{1}{2}|\xi_i|^3\mathbf{1}_{|\xi_i|\le 1}\right) = \beta_2 + \frac{1}{2}\beta_3. \end{aligned} \tag{3.9}$$

We recall the fact that if $g$ and $h$ are increasing functions, then $Eg(\xi)Eh(\xi) \le Eg(\xi)h(\xi)$. Now, regarding the second term in (3.8), since both $1 \wedge |x|$ and $x^2$ are increasing functions of $|x|$, again applying (2.60),

$$\begin{aligned} E\bigl(1 \wedge |\xi_I|\bigr) &= \sum_{i=1}^n E\bigl(1 \wedge |\xi_i|\bigr)E\xi_i^2 \le E\sum_{i=1}^n \bigl(1 \wedge |\xi_i|\bigr)\xi_i^2 \\ &\le E\sum_{i=1}^n \xi_i^2\mathbf{1}_{\{|\xi_i|>1\}} + |\xi_i|^3\mathbf{1}_{\{|\xi_i|\le 1\}} = \beta_2 + \beta_3. \end{aligned} \tag{3.10}$$

Substituting the bounds (3.9) and (3.10) into (3.8) now gives the result. □

One cannot derive a sharp Berry–Esseen bound for $W$ using the smooth function bounds (3.3) or (3.6). Nevertheless, as noted by Erickson (1974), these smooth function bounds imply a weak $L^\infty$ bound, as highlighted in the following theorem.

**Theorem 3.3** *Assume that there exists a $\delta$ such that, for any Lipschitz function $h$,*

$$\left|Eh(W) - Nh\right| \le \delta\|h'\|. \tag{3.11}$$

*Then*

$$\sup_{z\in\mathbb{R}}\big|P(W\le z)-\Phi(z)\big|\le 2\delta^{1/2}. \tag{3.12}$$

Proposition 2.4 shows that (3.11) is satisfied under conditions (2.76) or (2.77). Though the resulting bound $\delta$ has the optimal rate in many applications, see, for example, (3.3) and (3.6), the rate of a Berry–Esseen bound of the type (3.12) may not be optimal.

*Proof* We can assume that $\delta\le 1/4$, since otherwise (3.12) is trivial. Let $\alpha=\delta^{1/2}(2\pi)^{1/4}$, and for some fixed $z\in\mathbb{R}$ define

$$h_\alpha(w)=\begin{cases}1 & \text{if } w\le z,\\ 0 & \text{if } w\ge z+\alpha,\\ \text{linear} & \text{if } z<w<z+\alpha.\end{cases}$$

Then $h$ is Lipschitz continuous with $\|h'\|=1/\alpha$, and hence, by (3.11),

$$\begin{aligned}P(W\le z)-\Phi(z)&\le Eh_\alpha(W)-Nh_\alpha+Nh_\alpha-\Phi(z)\\&\le\frac{\delta}{\alpha}+P(z\le Z\le z+\alpha)\\&\le\frac{\delta}{\alpha}+\frac{\alpha}{\sqrt{2\pi}}.\end{aligned}$$

Therefore

$$P(W\le z)-\Phi(z)\le 2(2\pi)^{-1/4}\delta^{1/2}\le 2\delta^{1/2}.$$

Similarly, we have

$$P(W\le z)-\Phi(z)\ge -2\delta^{1/2},$$

proving (3.12). □

## 3.2 The Lindeberg Central Limit Theorem

Let $\xi_1,\ldots,\xi_n$ be independent random variables satisfying $E\xi_i=0,\ 1\le i\le n$ and $\sum_{i=1}^n\operatorname{Var}(\xi_i)=1$, and let $W=\sum_{i=1}^n\xi_i$. The classical Lindeberg central limit theorem states that

$$\sup_{z\in\mathbb{R}}\big|P(W\le z)-\Phi(z)\big|\to 0\quad\text{as } n\to\infty$$

if the Lindeberg condition is satisfied, that is, if for all $\varepsilon>0$

$$\sum_{i=1}^n E\xi_i^2\mathbf{1}_{\{|\xi_i|>\varepsilon\}}\to 0\quad\text{as } n\to\infty. \tag{3.13}$$

With $\beta_2$ and $\beta_3$ as in (3.5), observe that for any $0 < \varepsilon < 1$,

$$\begin{aligned}\beta_2 + \beta_3 &= \sum_{i=1}^n E\xi_i^2 \mathbf{1}_{\{|\xi_i|>1\}} + \sum_{i=1}^n E|\xi_i|^3 \mathbf{1}_{\{|\xi_i|\le 1\}} \\ &\le \sum_{i=1}^n E\xi_i^2 \mathbf{1}_{\{|\xi_i|>1\}} + \sum_{i=1}^n E\xi_i^2 \mathbf{1}_{\{\varepsilon<|\xi_i|\le 1\}} + \sum_{i=1}^n \varepsilon E\xi_i^2 \mathbf{1}_{\{|\xi_i|\le \varepsilon\}} \\ &\le \sum_{i=1}^n E\xi_i^2 \mathbf{1}_{\{|\xi_i|>\varepsilon\}} + \varepsilon. \end{aligned} \tag{3.14}$$

Hence, if the Lindeberg condition (3.13) holds, then (3.14) implies $\beta_2 + \beta_3 \to 0$ as $n \to \infty$, since $\varepsilon$ is arbitrary. Therefore, by Theorems 3.2 and 3.3,

$$\sup_{z\in\mathbb{R}} \left| P(W \le z) - \Phi(z) \right| \le 4(\beta_2 + \beta_3)^{1/2} \to 0 \quad \text{as } n \to \infty, \tag{3.15}$$

thus proving the Lindeberg central limit theorem.

In Sect. 3.5 we prove the partial converse, that if $\max_{1\le i\le n} E\xi_i^2 \to 0$, then the Lindeberg condition (3.13) is necessary for normal convergence.

## 3.3 Berry–Esseen Inequality: The Bounded Case

In the previous section, the smooth function bounds in Theorem 3.3 (see also Proposition 2.4) are of order $O(\delta)$, while the $L^\infty$ bounds are only of the larger order $O(\delta^{1/2})$. Here, we turn to deriving $L^\infty$ bounds which are of comparable order to those of the smooth function bounds. We will use the notation introduced in Sect. 2.3.1,

$$\begin{aligned} W = \sum_{i=1}^n \xi_i, \quad W^{(i)} = W - \xi_i, \\ \text{and} \quad K_i(t) = E\xi_i (\mathbf{1}_{\{0\le t\le \xi_i\}} - \mathbf{1}_{\{\xi_i\le t<0\}}). \end{aligned} \tag{3.16}$$

For bounded $\xi_i$, we are ready to apply (2.27) to obtain the following Berry–Esseen bound.

**Theorem 3.4** *Let $\xi_1, \xi_2, \ldots, \xi_n$ be independent random variables with zero means satisfying $\sum_{i=1}^n \mathrm{Var}(\xi_i) = 1$, and $W^{(i)}$ and $K_i(t)$ as in* (3.16). *Then*

$$\left| \sum_{i=1}^n \int_{-\infty}^{\infty} P\big(W^{(i)} + t \le z\big) K_i(t)\, dt - \Phi(z) \right| \le 2.44\gamma \tag{3.17}$$

*where $\gamma$ is given in* (3.4).

*If in addition $|\xi_i| \le \delta_0$ for $1 \le i \le n$, then*

$$\sup_{z\in\mathbb{R}} \left| P(W \le z) - \Phi(z) \right| \le 3.3\delta_0. \tag{3.18}$$

Before starting the proof, we note that by (2.62), we may write (3.17) as

$$\left|P(W^* \le z) - \Phi(z)\right| \le 2.44\gamma. \tag{3.19}$$

*Proof* For $z \in \mathbb{R}$, let $f = f_z$ be the solution of the Stein equation (2.2). From (2.27) and (2.2),

$$\begin{aligned}
E\{Wf(W)\} &= \sum_{i=1}^n \int_{-\infty}^{\infty} E\{f'(W^{(i)}+t)\}K_i(t)\,dt \\
&= \sum_{i=1}^n \int_{-\infty}^{\infty} E\{(W^{(i)}+t)f(W^{(i)}+t) + \mathbf{1}_{\{W^{(i)}+t\le z\}} - \Phi(z)\}K_i(t)\,dt.
\end{aligned}$$

Reorganizing this equality, using $\sum_{i=1}^n \int_{-\infty}^{\infty} K_i(t)\,dt = 1$ from (2.28), and recalling $K_i(t)$ is real yields

$$\begin{aligned}
&\sum_{i=1}^n \int_{-\infty}^{\infty} P(W^{(i)}+t \le z)K_i(t)\,dt - \Phi(z) \\
&\quad = \sum_{i=1}^n \int_{-\infty}^{\infty} E\{Wf(W) - (W^{(i)}+t)f(W^{(i)}+t)\}K_i(t)\,dt. \qquad (3.20)
\end{aligned}$$

Now, by (2.10), we may bound the absolute value of (3.20) by

$$\begin{aligned}
&\sum_{i=1}^n E\int_{-\infty}^{\infty} \left|Wf(W) - (W^{(i)}+t)f(W^{(i)}+t)\right|K_i(t)\,dt \\
&= \sum_{i=1}^n E\int_{-\infty}^{\infty} \left|(W^{(i)}+\xi_i)f(W^{(i)}+\xi_i) - (W^{(i)}+t)f(W^{(i)}+t)\right|K_i(t)\,dt \\
&\le \sum_{i=1}^n \int_{-\infty}^{\infty} E\left(\left|W^{(i)}\right| + \sqrt{2\pi}/4\right)\left(|\xi_i| + |t|\right)K_i(t)\,dt \\
&\le (1+\sqrt{2\pi}/4)\sum_{i=1}^n \int_{-\infty}^{\infty} \left(E|\xi_i| + |t|\right)K_i(t)\,dt,
\end{aligned}$$

since $E(W^{(i)})^2 \le 1$ and $\xi_i$ and $W^{(i)}$ are independent. Hence, recalling (2.25), we have

$$\begin{aligned}
&\left|\sum_{i=1}^n \int_{-\infty}^{\infty} P(W^{(i)}+t \le z)K_i(t)\,dt - \Phi(z)\right| \\
&\quad \le (1+\sqrt{2\pi}/4)\sum_{i=1}^n \left\{E|\xi_i|E\xi_i^2 + \frac{1}{2}E|\xi_i|^3\right\} \\
&\quad \le \frac{3}{2}(1+\sqrt{2\pi}/4)\gamma \le 2.44\gamma
\end{aligned}$$

proving (3.17).

The proof would be finished if $P(W^{(i)}+t\le z)$ could be replaced by $P(W\le z)$, since $\sum_{i=1}^n\int_{-\infty}^{\infty}K_i(t)\,dt=1$. Note that $|\xi_i|\le\delta_0$ implies that $K_i(t)=0$ for $|t|>\delta_0$, and when both $|t|$ and $|\xi_i|$ are bounded by $\delta_0$ then

$$P\big(W^{(i)}+t\le z\big)=P(W-\xi_i+t\le z)\ge P(W\le z-2\delta_0). \tag{3.21}$$

Replacing $z$ by $z+2\delta_0$ in (3.17) and (3.21) we obtain

$$\begin{aligned}2.44\gamma&\ge\sum_{i=1}^n\int_{-\infty}^{\infty}P\big(W^{(i)}+t\le z+2\delta_0\big)K_i(t)\,dt-\Phi(z+2\delta_0)\\&\ge\sum_{i=1}^n\int_{-\infty}^{\infty}P(W\le z)K_i(t)\,dt-\Phi(z+2\delta_0)\\&\ge P(W\le z)-\Phi(z)-\frac{2\delta_0}{\sqrt{2\pi}},\end{aligned}$$

where we have applied (2.28) followed by an elementary inequality. Next, as $|\xi_i|\le\delta_0$ for all $i=1,\ldots,n$,

$$\gamma=\sum_{i=1}^n E|\xi_i|^3\le\delta_0\sum_{i=1}^n E|\xi_i|^2=\delta_0,$$

from which we now obtain

$$P(W\le z)-\Phi(z)\le 2.44\gamma+\frac{2\delta_0}{\sqrt{2\pi}}\le 3.3\delta_0. \tag{3.22}$$

The proof is completed by proving the corresponding lower bound using similar reasoning. □

The key ingredient in the proof of Theorem 3.4 is to rewrite $E[Wf(W)]$ in terms of a functional of $f'$. We now formulate a result along these same lines, taking as our basis the Stein identity (2.76).

**Theorem 3.5** *For $W$ any random variable, suppose that for every $z\in\mathbb{R}$ there exist a random variable $R_1$ and random function $\hat{K}(t)\ge 0$, $t\in\mathbb{R}$, and constants $\delta_0$ and $\delta_1$ not depending on $z$, such that $|ER_1|\le\delta_1$ and*

$$EWf_z(W)=E\int_{|t|\le\delta_0}f_z'(W+t)\hat{K}(t)\,dt+ER_1, \tag{3.23}$$

*where $f_z$ is the solution of the Stein equation (2.2). Then*

$$\sup_{z\in\mathbb{R}}\big|P(W\le z)-\Phi(z)\big|\le\delta_0\big(1.1+E\big[|W|\hat{K}_1\big]\big)+2.7E|1-\hat{K}_1|+\delta_1, \tag{3.24}$$

*where $\hat{K}_1=E(\int_{|t|\le\delta_0}\hat{K}(t)\,dt\mid W)$.*

*Proof* We can assume $\delta_0\le 1$ because (3.24) is trivial otherwise. Using that $f_z$ satisfies the Stein equation (2.2), and the nonnegativity of $\hat{K}(t)$, we have

$$
\begin{aligned}
&E\int_{|t|\le\delta_0} f_z'(W+t)\hat{K}(t)\,dt \\
&= E\int_{|t|\le\delta_0} \big(\mathbf{1}_{\{W+t\le z\}} - \Phi(z)\big)\hat{K}(t)\,dt + E\int_{|t|\le\delta_0} (W+t)f_z(W+t)\hat{K}(t)\,dt \\
&\le E\int_{|t|\le\delta_0} \big(\mathbf{1}_{\{W\le z+\delta_0\}} - \Phi(z)\big)\hat{K}(t)\,dt + E\int_{|t|\le\delta_0} (W+t)f_z(W+t)\hat{K}(t)\,dt \\
&= E\big(\mathbf{1}_{\{W\le z+\delta_0\}} - \Phi(z)\big)\hat{K}_1 + E\int_{|t|\le\delta_0} (W+t)f_z(W+t)\hat{K}(t)\,dt,
\end{aligned}
$$

where the inequality holds because $-t\le\delta_0$.

Now, writing $\hat{K}_1 = 1-(1-\hat{K}_1)$, we find that

$$
\begin{aligned}
&E\int_{|t|\le\delta_0} f_z'(W+t)\hat{K}(t)\,dt \\
&\le P(W\le z+\delta_0) - \Phi(z) + E|1-\hat{K}_1| + E\int_{|t|\le\delta_0} (W+t)f_z(W+t)\hat{K}(t)\,dt \\
&\le P(W\le z+\delta_0) - \Phi(z+\delta_0) + \frac{\delta_0}{\sqrt{2\pi}} \\
&\quad + E|1-\hat{K}_1| + E\int_{|t|\le\delta_0} (W+t)f_z(W+t)\hat{K}(t)\,dt.
\end{aligned}
$$

Thus, rearranging and using (3.23) to obtain the first equality,

$$
\begin{aligned}
&P(W\le z+\delta_0) - \Phi(z+\delta_0) \\
&\ge -\frac{\delta_0}{\sqrt{2\pi}} - E|1-\hat{K}_1| + E\int_{|t|\le\delta_0} f_z'(W+t)\hat{K}(t)\,dt \\
&\quad - E\int_{|t|\le\delta_0} (W+t)f_z(W+t)\hat{K}(t)\,dt \\
&= -\frac{\delta_0}{\sqrt{2\pi}} - E|1-\hat{K}_1| + EWf_z(W) - ER_1 \\
&\quad - E\int_{|t|\le\delta_0} (W+t)f_z(W+t)\hat{K}(t)\,dt \\
&= -\frac{\delta_0}{\sqrt{2\pi}} - E|1-\hat{K}_1| + E\big[Wf_z(W)(1-\hat{K}_1)\big] - ER_1 \\
&\quad + E\int_{|t|\le\delta_0} \big\{Wf_z(W) - (W+t)f_z(W+t)\big\}\hat{K}(t)\,dt \\
&\ge -\frac{\delta_0}{\sqrt{2\pi}} - 2E|1-\hat{K}_1| - \delta_1 - \int_{|t|\le\delta_0} E\big(|W| + \sqrt{2\pi}/4\big)|t|\hat{K}(t)\,dt,
\end{aligned}
$$

this last by (2.7), the hypotheses $|ER_1|\le\delta_1$, and (2.10). Hence,

$$
\begin{aligned}
&P(W\le z+\delta_0) - \Phi(z+\delta_0) \\
&\ge -\frac{\delta_0}{\sqrt{2\pi}} - 2E|1-\hat{K}_1| - \delta_1 - E\int_{|t|\le\delta_0} \big(|W| + 0.7\big)\delta_0\hat{K}(t)\,dt
\end{aligned}
$$

$$
\begin{aligned}
&= -\frac{\delta_0}{\sqrt{2\pi}} - 2E|1-\hat{K}_1| - \delta_1 - \delta_0 E\big(|W|+0.7\big)\hat{K}_1 \\
&\ge -\frac{\delta_0}{\sqrt{2\pi}} - 2E|1-\hat{K}_1| - \delta_1 - \delta_0\big\{E\big[|W|\hat{K}_1\big] + 0.7 + 0.7E|1-\hat{K}_1|\big\} \\
&\ge -\delta_0\big(1.1 + E\big[|W|\hat{K}_1\big]\big) - 2.7E|1-\hat{K}_1| - \delta_1, \qquad (3.25)
\end{aligned}
$$

recalling that $\delta_0 \le 1$. A similar argument gives

$$
\begin{aligned}
&P(W \le z - \delta_0) - \Phi(z-\delta_0) \\
&\quad \le \delta_0\big(1.1 + E\big[|W|\hat{K}_1\big]\big) + 2.7E|1-\hat{K}_1| + \delta_1, \qquad (3.26)
\end{aligned}
$$

completing the proof of (3.24). □

In Chap. 5 we illustrate how to use Theorem 3.5 to obtain Berry–Esseen bounds in various applications.

## 3.4 The Berry–Esseen Inequality for Unbounded Variables

Theorem 3.4 demonstrates the Berry–Esseen inequality when $W$ is a sum of uniformly bounded, mean zero, independent random variables $\xi_1, \ldots, \xi_n$ with variances summing to one. Here we drop the boundedness restriction and prove, using two different methods, that there exists a universal constant $C$ such that

$$
\sup_{z\in\mathbb{R}}\big|P(W \le z) - \Phi(z)\big| \le C\gamma \quad \text{where } \gamma = \sum_{i=1}^{n} E|\xi_i|^3. \qquad (3.27)
$$

Tyurin (2010) has shown that $C$ can be taken 0.4785. Both of our two approaches, using concentration inequalities in Sect. 3.4.1, and an inductive method in Sect. 3.4.2, lead to somewhat larger constants, but as the sequel shows, these approaches generalize to many cases where the independence condition can be dropped.

### *3.4.1 The Concentration Inequality Approach*

Noting that (3.17) in Theorem 3.4 holds without the uniform boundedness restriction, with $W^{(i)}$ as in (3.16) we see that one can prove the Berry–Esseen inequality more generally by showing that

$$
P\big(W^{(i)} + t \le z\big) \quad \text{is close to} \quad P(W \le z) = P\big(W^{(i)} + \xi_i \le z\big),
$$

which it suffices to have a good bound for $P(a \le W^{(i)} \le b)$. Intuitively, the distribution of $W^{(i)}$ is close to the standard normal, and hence we should be able to bound $P(a \le W^{(i)} \le b)$ using some multiple of $b-a$. This heuristic is made precise by the concentration inequality

**Lemma 3.1** *For all real $a < b$, and for every $1 \le i \le n$,*

$$P\big(a \le W^{(i)} \le b\big) \le \sqrt{2}(b-a) + 2(\sqrt{2}+1)\gamma \tag{3.28}$$

*where $\gamma$ is as in* (3.27).

We remark that Chen (1998) was the first to apply the concentration inequality approach to independent but non-identically distributed variables. Postponing the proof of (3.28) to the end of this section, we demonstrate the following Berry–Esseen bound with a constant of 9.4.

**Theorem 3.6** *Let $\xi_1, \xi_2, \ldots, \xi_n$ be independent random variables with zero means, satisfying $\sum_{i=1}^n \mathrm{Var}(\xi_i) = 1$. Then $W = \sum_{i=1}^n \xi_i$ satisfies*

$$\sup_{z \in \mathbb{R}} \big| P(W \le z) - \Phi(z) \big| \le 9.4\gamma \quad \text{where } \gamma = \sum_{i=1}^n E|\xi_i|^3. \tag{3.29}$$

*Proof* With $W^{(i)}$ and $K_i(t)$ as in (3.16), by (2.25) and (3.28) we have

$$\begin{aligned}
&\left| \sum_{i=1}^n \int_{-\infty}^{\infty} P\big(W^{(i)} + t \le z\big) K_i(t)\,dt - P(W \le z) \right| \\
&\quad = \left| \sum_{i=1}^n \int_{-\infty}^{\infty} \big(P\big(W^{(i)} + t \le z\big) - P(W \le z)\big) K_i(t)\,dt \right| \\
&\quad \le \sum_{i=1}^n \int_{-\infty}^{\infty} \big| P\big(W^{(i)} + t \le z\big) - P(W \le z) \big| K_i(t)\,dt \\
&\quad = \sum_{i=1}^n \int_{-\infty}^{\infty} \big| P\big(W^{(i)} + t \le z\big) - P\big(W^{(i)} + \xi_i \le z\big) \big| K_i(t)\,dt \\
&\quad = \sum_{i=1}^n \int_{-\infty}^{\infty} E\big\{ P\big(z - t \vee \xi_i \le W^{(i)} \le z - t \wedge \xi_i \mid \xi_i\big) \big\} K_i(t)\,dt \\
&\quad \le \sum_{i=1}^n \int_{-\infty}^{\infty} E\big\{ \sqrt{2}\big(|t| + |\xi_i|\big) + 2(\sqrt{2}+1)\gamma \big\} K_i(t)\,dt \\
&\quad = \sqrt{2} \sum_{i=1}^n \left( \frac{1}{2} E|\xi_i|^3 + E|\xi_i| E\xi_i^2 \right) + 2(\sqrt{2}+1)\gamma \\
&\quad \le (3.5\sqrt{2} + 2)\gamma \le 6.95\gamma,
\end{aligned} \tag{3.30}$$

where we have again applied (2.25). Invoking (3.17) now yields the claim. □

As in Theorem 3.2, one can dispense with the third moment assumption in Theorem 3.6 and replace $\gamma$ in (3.29) by $\beta_2 + \beta_3$, defined in (3.5); we leave the details to the reader. Additionally, with a more refined concentration inequality, the constant can be reduced further, resulting in

$$\sup_{z\in\mathbb{R}}\big|P(W\le z)-\Phi(z)\big|\le 4.1(\beta_2+\beta_3), \tag{3.31}$$

see Chen and Shao (2001).

We now prove the concentration inequality (3.28). The idea is to use the fact that if $f'$ equals the indicator $\mathbf{1}_{[a,b]}$ of some interval, then $Ef'(W)=P(a\le W\le b)$. This fixes $f$ up to a constant, and choosing $f((a+b)/2)=0$ the norm $\|f\|=(b-a)/2$ takes on its minimal value, yielding the smallest factor in the right hand side of the inequality

$$\big|EWf(W)\big|\le\frac{1}{2}(b-a)E|W|\le\frac{1}{2}(b-a),$$

which holds whenever $EW^2\le 1$.

*Proof of Lemma 3.1* Define $\delta=\gamma$ and take

$$f(w)=\begin{cases}-\frac{1}{2}(b-a)-\delta & \text{if } w<a-\delta,\\ w-\frac{1}{2}(b+a) & \text{if } a-\delta\le w\le b+\delta,\\ \frac{1}{2}(b-a)+\delta & \text{for } w>b+\delta,\end{cases} \tag{3.32}$$

so that $f'=\mathbf{1}_{[a-\delta,b+\delta]}$, and $\|f\|=\frac{1}{2}(b-a)+\delta$. Set

$$\hat{K}_j(t)=\xi_j(\mathbf{1}_{\{-\xi_j\le t\le 0\}}-\mathbf{1}_{\{0<t\le-\xi_j\}})\quad\text{and}\quad\hat{K}(t)=\sum_{j=1}^n\hat{K}_j(t). \tag{3.33}$$

Since $\xi_j$ and $W^{(i)}-\xi_j$ are independent for $j\ne i$, $\xi_i$ is independent of $W^{(i)}$, and $E\xi_j=0$ for all $j$, similarly to (2.27), we have

$$\begin{aligned}&EW^{(i)}f\big(W^{(i)}\big)-E\xi_i f\big(W^{(i)}-\xi_i\big)\\&=\sum_{j=1}^n E\xi_j\big[f\big(W^{(i)}\big)-f\big(W^{(i)}-\xi_j\big)\big]\\&=\sum_{j=1}^n E\xi_j\int_{-\xi_j}^0 f'\big(W^{(i)}+t\big)\,dt\\&=\sum_{j=1}^n E\int_{-\infty}^{\infty} f'\big(W^{(i)}+t\big)\hat{K}_j(t)\,dt\\&=E\int_{-\infty}^{\infty} f'\big(W^{(i)}+t\big)\hat{K}(t)\,dt.\end{aligned} \tag{3.34}$$

Noting that $f'(t)\ge 0$ and $\hat{K}(t)\ge 0$, we have by the definition of $f$

$$E\int_{-\infty}^{\infty} f'(W^{(i)}+t)\hat{K}(t)\,dt \geq E\int_{|t|\leq\delta} f'(W^{(i)}+t)\hat{K}(t)\,dt$$
$$\geq E\mathbf{1}_{\{a\leq W^{(i)}\leq b\}}\int_{|t|\leq\delta}\hat{K}(t)\,dt.$$

Letting $K(t)=E\hat{K}(t)$, we may write this last expression as

$$E\mathbf{1}_{\{a\leq W^{(i)}\leq b\}}\int_{|t|\leq\delta}\big[\hat{K}(t)-K(t)\big]\,dt + P\big(a\leq W^{(i)}\leq b\big)\int_{|t|\leq\delta}K(t)\,dt. \quad (3.35)$$

As in (2.28) and (2.25), respectively, the function $K(t)$ is a density and $E|T|=\gamma/2$ for $T$ so distributed. Hence, for the integral in the second term of (3.35), recalling $\delta=\gamma$,

$$\int_{|t|\leq\delta}K(t)\,dt = P\big(|T|\leq\delta\big) = 1-P\big(|T|>\delta\big)\geq 1-\frac{\gamma}{2\delta}=1/2.$$

For the first term of (3.35), applying the Cauchy–Schwarz inequality and integrating yields the bound

$$\mathrm{Var}\bigg(\int_{|t|\leq\delta}\hat{K}(t)\,dt\bigg)^{1/2} \leq \bigg(\mathrm{Var}\bigg\{\sum_{j=1}^{n}|\xi_j|\min\big(\delta,|\xi_j|\big)\bigg\}\bigg)^{1/2}$$
$$\leq\bigg(\sum_{j=1}^{n}E\xi_j^2\min\big(\delta,|\xi_j|\big)^2\bigg)^{1/2}$$
$$\leq\delta\bigg(\sum_{j=1}^{n}E\xi_j^2\bigg)^{1/2}=\delta.$$

Hence, from (3.34) and (3.35) we obtain

$$EW^{(i)}f\big(W^{(i)}\big)-E\xi_i f\big(W^{(i)}-\xi_i\big)\geq\frac{1}{2}P\big(a\leq W^{(i)}\leq b\big)-\delta. \quad (3.36)$$

On the other hand, recalling that $\|f\|\leq\frac{1}{2}(b-a)+\delta$, we have

$$EW^{(i)}f\big(W^{(i)}\big)-E\xi_i f\big(W^{(i)}-\xi_i\big)$$
$$\leq\bigg(\frac{1}{2}(b-a)+\delta\bigg)\big(E\big|W^{(i)}\big|+E|\xi_i|\big)$$
$$\leq\frac{1}{\sqrt{2}}\big(\big(E\big|W^{(i)}\big|\big)^2+\big(E|\xi_i|\big)^2\big)^{1/2}(b-a+2\delta)$$
$$\leq\frac{1}{\sqrt{2}}\big(E\big|W^{(i)}\big|^2+E|\xi_i|^2\big)^{1/2}(b-a+2\delta)$$
$$=\frac{1}{\sqrt{2}}(b-a+2\delta). \quad (3.37)$$

Combining (3.36) and (3.37) thus gives

$$P\big(a \le W^{(i)} \le b\big) \le \sqrt{2}(b-a) + (2\sqrt{2}+2)\delta = \sqrt{2}(b-a) + 2(\sqrt{2}+1)\gamma$$

as desired. □

By reasoning as above, and as in the proofs of Theorem 8.1 and Propositions 10.1 and 10.2, one can prove the following stronger concentration inequality.

**Proposition 3.1** *If $W$ is the sum of the independent mean zero random variables $\xi_1, \ldots, \xi_n$, then for all real $a < b$*

$$P(a \le W \le b) \le b - a + 2(\beta_2 + \beta_3) \tag{3.38}$$

*where $\beta_2$ and $\beta_3$ are defined in* (3.5). *In addition, if $W^{(i)} = W - \xi_i$, then*

$$P\big(a \le W^{(i)} \le b\big) \le \sqrt{2}(b-a) + (\sqrt{2}+1)(\beta_2 + \beta_3) \tag{3.39}$$

*for every $1 \le i \le n$.*

We leave the proof to the reader. Clearly, $\beta_2 + \beta_3 \le \gamma$, so Proposition 3.1 not only relaxes the moment assumption required by (3.28) but improves the constant as well.

### 3.4.2 An Inductive Approach

In this section we prove the following Berry–Esseen inequality by induction.

**Theorem 3.7** *Let $\xi_1, \xi_2, \ldots, \xi_n$ be independent random variables with zero means, satisfying $\sum_{i=1}^n \mathrm{Var}(\xi_i) = 1$. Then $W = \sum_{i=1}^n \xi_i$ satisfies*

$$\sup_{z\in\mathbb{R}} \big|P(W \le z) - \Phi(z)\big| \le 10\gamma \quad \textit{where } \gamma = \sum_{i=1}^n E|\xi_i|^3. \tag{3.40}$$

Though the constant produced is not optimal, the inductive approach is quite useful in more general settings when the removal of some variables leaves a structure similar to the original one; see Theorem 6.2 in Sect. 6.1.1 for one example involving dependence where the inductive method succeeds, and references to other such examples. Use of induction in the independent case appears in the text of Stroock (2000).

*Proof* Without loss of generality we may assume $E\xi_i^2 \ne 0$ for all $i = 1, \ldots, n$. Let

$$\tau_i^2 = E\big(W^{(i)}\big)^2 \quad \text{and} \quad \tau = \min_{1\le i\le n} \tau_i.$$

Since (3.40) is trivial if $\gamma \ge 1/10$, we can assume $\gamma < 1/10$. Since

$$1 = EW^2 = E\big(\big(W^{(i)}\big)^2 + \xi_i\big)^2 = E\big(W^{(i)}\big)^2 + E\xi_i^2 \le E\big(W^{(i)}\big)^2 + \big(E|\xi_i|^3\big)^{2/3},$$

we have

$$\tau^2 \geq 1 - \gamma^{2/3} \geq 0.7845. \tag{3.41}$$

When $n = 1$, since $\gamma = E|\xi_1|^3 \geq (E\xi_1^2)^{3/2} = 1$, inequality (3.40) is trivially true. Now take $n \geq 2$ and assume that (3.40) has been established for a sum composed of fewer than $n$ summands. Then for all $i = 1, \ldots, n$ and $a < b$, with $C = 10$ we have

$$\begin{aligned}
&P\big(a < W^{(i)} \leq b\big) \\
&\quad = \Phi(b/\tau_i) - \Phi(a/\tau_i) + P\big(W^{(i)} \leq b\big) - \Phi(b/\tau_i) \\
&\quad\quad - \big\{P\big(W^{(i)} \leq a\big) - \Phi(a/\tau_i)\big\} \\
&\quad \leq \frac{2C}{\tau_i^3} \sum_{j \neq i} E|\xi_j|^3 + \frac{b-a}{\sqrt{2\pi}\,\tau_i} \\
&\quad \leq 2.88C\gamma + (b-a)/2,
\end{aligned} \tag{3.42}$$

using (3.41) twice in the final inequality.

Let $\xi_i^*$ have the $\xi_i$-zero bias distribution and be independent of $\xi_j, j \neq i$, and let $I$ be a random index, independent of all other variables, with distribution (2.60). Then, by Lemma 2.8, letting $\delta = 2\gamma$, we have

$$\begin{aligned}
&P(W^* \leq z) - P(W \leq z - 2\delta) \\
&\quad = P\big(W^{(I)} + \xi_I^* \leq z\big) - P\big(W^{(I)} + \xi_I \leq z - 2\delta\big) \\
&\quad \geq -EP\big(z - \xi_I^* \leq W^{(I)} \leq z - \xi_I - 2\delta \,|\, \xi_I, \xi_I^*\big)\mathbf{1}\big(\xi_I^* \geq \xi_I + 2\delta\big) \\
&\quad \geq -E\big(2.88C\gamma + \big(\xi_I^* - \xi_I\big)/2 - \delta\big)\mathbf{1}\big(\xi_I^* \geq \xi_I + 2\delta\big) \\
&\quad \geq -2.88C\gamma P\big(\xi_I^* - \xi_I \geq 2\delta\big) - E\big(\xi_I^* - \xi_I\big)\mathbf{1}\big(\xi_I^* \geq \xi_I + 2\delta\big)/2 - \delta,
\end{aligned}$$

where we have invoked (3.42) to obtain the second inequality. By Theorem 4.3, $\xi_i$ and $\xi_i^*$ may be coupled so that

$$E\big|\xi_i^* - \xi_i\big| \leq \frac{E|\xi_i|^3}{2E\xi_i^2} \quad \text{so, by (2.60),} \quad E\big|\xi_I^* - \xi_I\big| \leq \gamma/2.$$

But now

$$P\big(\xi_I^* - \xi_I \geq 2\delta\big) \leq \gamma/(4\delta) \quad \text{and} \quad E\big(\xi_I^* - \xi_I\big)\mathbf{1}\big(\xi_I^* \geq \xi_I + 2\delta\big) \leq \gamma/2.$$

Hence, recalling $\delta = 2\gamma$,

$$P(W^* \leq z) - P(W \leq z - 2\delta) \geq -2.88C\gamma/8 - \gamma/4 - 2\gamma = -5.85\gamma.$$

Thus, by (3.19),

$$\begin{aligned}
P(W \leq z - 2\delta) - \Phi(z - 2\delta) &\leq P(W^* \leq z) - \Phi(z - 2\delta) + 5.85\gamma \\
&\leq 2.44\gamma + \frac{4\gamma}{\sqrt{2\pi}} + 5.85\gamma < 10\gamma.
\end{aligned}$$

Similarly, we may obtain

$$P(W \leq z + 2\delta) - \Phi(z + 2\delta) \geq -10\gamma,$$

thus completing the proof. □

## 3.5 A Lower Berry–Esseen Bound

Again, let $\xi_1, \ldots, \xi_n$ be independent random variables with zero means satisfying $\sum_{i=1}^n \mathrm{Var}(\xi_i) = 1$. Feller (1935) and Lévy (1935) proved independently (see LeCam 1986) that if the Feller–Lévy condition

$$\max_{1 \le i \le n} E\xi_i^2 \to 0, \tag{3.43}$$

is satisfied, then the Lindeberg condition (3.13) is necessary for the central limit theorem. The theorem below is due to Hall and Barbour (1984) who used Stein's method to provide not only a nice proof of the necessity, but also a lower bound for the $L^\infty$ distance between the distribution of $W$ and the normal.

**Theorem 3.8** *Let $\xi_1, \xi_2, \ldots, \xi_n$ be independent random variables with zero means and finite variances $E\xi_i^2 = \sigma_i^2$, $1 \le i \le n$, satisfying $\sum_{i=1}^n \sigma_i^2 = 1$, and let $W = \sum_{i=1}^n \xi_i$. Then there exists an absolute constant $C$ such that for all $\varepsilon > 0$,*

$$\begin{aligned}
&\left(1 - e^{-\varepsilon^2/4}\right) \sum_{i=1}^n E\xi_i^2 \mathbf{1}_{\{|\xi_i| > \varepsilon\}} \\
&\quad \le C\left(\sup_{z \in \mathbb{R}} \left|P(W \le z) - \Phi(z)\right| + \sum_{i=1}^n \sigma_i^4\right).
\end{aligned} \tag{3.44}$$

Clearly, the Feller–Lévy condition (3.43) implies that $\sum_{i=1}^n \sigma_i^4 \le \max_{1 \le i \le n} \sigma_i^2 \to 0$ as $n \to \infty$. Therefore, if $W$ is asymptotically normal,

$$\sum_{i=1}^n E\xi_i^2 \mathbf{1}_{\{|\xi_i| > \varepsilon\}} \to 0$$

as $n \to \infty$ for every $\varepsilon > 0$, that is, the Lindeberg condition is satisfied.

*Proof* Once again, the argument starts with the Stein equation

$$E\left\{f_h'(W) - W f_h(W)\right\} = Eh(W) - Nh, \tag{3.45}$$

for a function $h$ yet to be chosen. Taking $h$ absolutely continuous with $\int_{-\infty}^{\infty} |h'(w)| dw < \infty$, we may integrate by parts and obtain the bound

$$\begin{aligned}
\left|Eh(W) - Nh\right| &= \left|\int_{-\infty}^{\infty} h'(w)\left\{P(W \le w) - \Phi(w)\right\} dw\right| \\
&\le \delta \int_{-\infty}^{\infty} \left|h'(w)\right| dw,
\end{aligned} \tag{3.46}$$

where $\delta = \sup_{z \in \mathbb{R}} |P(W \le z) - \Phi(z)|$.

For the left hand side of (3.45), in the usual way, because $\xi_i$ and $W^{(i)} = W - \xi_i$ are independent, and $E\xi_i = 0$, we have

$$EWf_h(W) = \sum_{i=1}^n E\xi_i^2 f_h'(W^{(i)})$$
$$+ \sum_{i=1}^n E\{\xi_i(f_h(W^{(i)}+\xi_i) - f_h(W^{(i)}) - \xi_i f_h'(W^{(i)}))\},$$

and, because $\sum_{i=1}^n \sigma_i^2 = 1$,

$$Ef_h'(W) = \sum_{i=1}^n \sigma_i^2 Ef_h'(W^{(i)}) + \sum_{i=1}^n \sigma_i^2 E\{f_h'(W) - f_h'(W^{(i)})\},$$

with the last term easily bounded by $\frac{1}{2}\|f_h'''\|\sum_{i=1}^n \sigma_i^4$. Hence

$$\left| E\{f_h'(W) - Wf_h(W)\} - \sum_{i=1}^n E\xi_i^2 g(W^{(i)}, \xi_i)\right| \le \frac{1}{2}\|f_h'''\|\sum_{i=1}^n \sigma_i^4, \tag{3.47}$$

where

$$g(w,y) = g_h(w,y) = -y^{-1}\{f_h(w+y) - f_h(w) - yf_h'(w)\}.$$

Intuitively, if the distribution of $W$ is close to that of the standard normal $Z$, taken to be independent of the $\xi_i$'s, then

$$R_1 := \sum_{i=1}^n E\xi_i^2 g(W^{(i)}, \xi_i) \quad \text{and} \quad R := \sum_{i=1}^n E\xi_i^2 g(Z, \xi_i),$$

should be close to one another.

Taking (3.46) and (3.47) together, we will be able to compute a lower bound for $\delta$, if we can produce an absolutely continuous function $h$ satisfying $\int_{-\infty}^{\infty}|h'(w)|\,dw < \infty$ for which $Eg_h(Z,y)$ is of constant sign, provided also that $\|f_h'''\| < \infty$. In practice, it is easier to look for a suitable $f$, and then define $h(w) = f'(w) - wf(w)$. The function $g$ is zero for any linear function $f$, and when $f$ is an even function then $Eg(Z,y)$ is odd. Choosing $f$ to be the odd function $f(y) = y^3$ yields $Eg(Z,y) = -y^2$, of constant sign. Unfortunately, this $f$ fails to yield an $h$ satisfying $\int_{-\infty}^{\infty}|h'(w)|\,dw < \infty$.

A good choice is $f(w) = we^{-w^2/2}$, which behaves much like the sum of a linear and a cubic function for those values of $w$ where $Z$ puts most of its mass, yet decays to zero quickly when $|w|$ is large. Making the computations, we have

$$Eg(Z,y) = -\frac{y^{-1}}{\sqrt{2\pi}}\int_{-\infty}^{\infty}\{(w+y)e^{-(w+y)^2/2}$$
$$- we^{-w^2/2} - ye^{-w^2/2}(1-w^2)\}e^{-w^2/2}\,dw$$
$$= \frac{1}{2\sqrt{2}}\left(1 - e^{-y^2/4}\right), \tag{3.48}$$

a nonnegative function which satisfies

$$Eg(Z,y) \geq \frac{1}{2\sqrt{2}}\left(1-e^{-\varepsilon^2/4}\right) \quad \text{whenever } |y| \geq \varepsilon$$

for all $\varepsilon > 0$. Hence, for this choice of $f$ we have

$$R \geq \frac{1}{2\sqrt{2}}\left(1-e^{-\varepsilon^2/4}\right)\sum_{i=1}^{n} E\xi_i^2 \mathbf{1}_{\{|\xi_i|>\varepsilon\}}. \tag{3.49}$$

It thus remains to show that $R$ and $R_1$ are close enough, after which (3.1), (3.47) and (3.49) complete the proof.

For this step, note that for $f(w) = we^{-w^2/2}$ and $h(w) = f'(w) - wf(w)$ we have

$$c_1 := \int_{-\infty}^{\infty} \left|h'(w)\right| dw \leq 7,$$

$$c_2 := \int_{-\infty}^{\infty} \left|f''(w)\right| dw \leq 4; \quad \text{and} \quad c_3 := \sup_w \left|f'''(w)\right| = 3.$$

Now define an intermediate quantity $R_2$ between $R_1$ and $R$, by

$$R_2 := \sum_{i=1}^{n} E\xi_i^2 g(W', \xi_i),$$

where $W'$ has the same distribution as $W$, but is independent of the $\xi_i$'s. Then

$$\begin{aligned}
R_1 &= -\sum_{i=1}^{n} E\left\{\xi_i^2 \int_0^1 \left[f'\left(W^{(i)}+t\xi_i\right) - f'\left(W^{(i)}\right)\right] dt\right\} \\
&= R_2 + \sum_{i=1}^{n} E\left\{\xi_i^2 \int_0^1 \left[f'\left(W'+t\xi_i\right) - f'\left(W^{(i)}+t\xi_i\right)\right] dt\right\} \\
&\quad - \sum_{i=1}^{n} E\left\{\xi_i^2 \int_0^1 \left[f'\left(W'\right) - f'\left(W^{(i)}\right)\right] dt\right\}.
\end{aligned} \tag{3.50}$$

Now, for any $\theta$, using that $W$ and $W'$ have the same distribution, that $\xi_i$ and $W^{(i)}$ are independent, and that $E\xi_i = 0$,

$$\begin{aligned}
&\left|E\left(f'(W'+\theta) - f'\left(W^{(i)}+\theta\right)\right)\right| \\
&\quad = \left|E\left(f'\left(W^{(i)}+\xi_i+\theta\right) - f'\left(W^{(i)}+\theta\right)\right)\right| \\
&\quad = \left|E\left(f'\left(W^{(i)}+\xi_i+\theta\right) - f'\left(W^{(i)}+\theta\right) - \xi_i f''\left(W^{(i)}+\theta\right)\right)\right| \\
&\quad \leq \frac{1}{2} c_3 \sigma_i^2,
\end{aligned}$$

by Taylor's theorem. Hence, from (3.50),

$$R_1 \geq R_2 - c_3 \sum_{i=1}^{n} \sigma_i^4. \tag{3.51}$$

Similarly,

$$R_2 = R + \sum_{i=1}^{n} E\left\{\xi_i^2 \int_0^1 [f'(Z+t\xi_i) - f'(W'+t\xi_i)]\,dt\right\}$$
$$- \sum_{i=1}^{n} E\left\{\xi_i^2 \int_0^1 [f'(Z) - f'(W')]\,dt\right\},$$

and, for any $\theta$, as $\int_{-\infty}^{\infty} |f''(w)|dw = c_2 < \infty$,

$$\big|Ef'(W'+\theta) - Ef'(Z+\theta)\big|$$
$$= \left|\int_{-\infty}^{\infty} f''(w)\big(P(W' \le w-\theta) - \Phi(w-\theta)\big)\,dw\right| \le c_2\delta,$$

so that

$$R_2 \ge R - 2c_2\delta. \tag{3.52}$$

Combining (3.46) and (3.47) with (3.51) and (3.52), it follows that

$$c_1\delta \ge R_1 - \frac{1}{2}c_3 \sum_{i=1}^{n} \sigma_i^4 \ge R - \frac{3}{2}c_3 \sum_{i=1}^{n} \sigma_i^4 - 2c_2\delta.$$

In view of (3.49), collecting terms, it follows that

$$\delta(c_1 + 2c_2) + \frac{3}{2}c_3 \sum_{i=1}^{n} \sigma_i^4 \ge \frac{1}{2\sqrt{2}}\big(1 - e^{-\varepsilon^2/4}\big) \sum_{i=1}^{n} E\xi_i^2 \mathbf{1}_{\{|\xi_i|>\varepsilon\}} \tag{3.53}$$

for any $\varepsilon > 0$. This proves (3.44), with $C \le 43$. □

# Chapter 4
# $L^1$ Bounds

In this chapter we focus on normal approximation using smooth functions, and the $L^1$ norm in particular. We begin with a discussion of distances induced by function classes. Any class of functions $\mathcal{H}$ mapping $\mathbb{R}$ to $\mathbb{R}$ induces a measure of the separation between the distributions $\mathcal{L}(X)$ and $\mathcal{L}(Y)$ of the random variables $X$ and $Y$ by

$$\left\|\mathcal{L}(X)-\mathcal{L}(Y)\right\|_{\mathcal{H}}=\sup_{h\in\mathcal{H}}\left|Eh(X)-Eh(Y)\right|. \tag{4.1}$$

Certain choices of $\mathcal{H}$ lead to classical distances, for instance, taking

$$\mathcal{H}=\left\{\mathbf{1}(x\le z),\ z\in\mathbb{R}\right\} \tag{4.2}$$

leads to the Kolmogorov, $L^\infty$, or supremum norm distance, while the class of measurable functions

$$\mathcal{H}=\left\{h\colon 0\le h(x)\le 1,\ \forall x\in\mathbb{R}\right\} \tag{4.3}$$

leads to the total variation distance.

Calculations with smooth functions are typically simpler than those with functions such as the discontinuous indicators in (4.2), or the bounded measurable functions in (4.3). Our main focus in this chapter is the $L^1$ distance, given by (4.1) with $\mathcal{H}=\mathfrak{L}$, the collection of Lipschitz functions in (4.7). In Sect. 4.8 we move to the distance $\|\mathcal{L}(W)-\mathcal{L}(Z)\|_{\mathcal{H}_{m,\infty}}$, produced by taking $\mathcal{H}$ to be the collection of functions $\mathcal{H}_{m,\infty}$ defined in (4.183), a class including functions allowed to posses some small number of additional higher order derivatives.

Our $L^1$ examples include: the sums of independent random variables and an associated contraction principle, hierarchical structures, cone measure on the sphere, combinatorial central limit theorems, simple random sampling, coverage processes, and locally dependent random variables. To illustrate our approach for the smooth functions $\mathcal{H}_{m,\infty}$ we show how fast convergence rates may result under a vanishing third moment assumption. The use of Stein's method for $L^1$ approximation was pioneered by Erickson (1974).

We begin now by recalling that the $L^1$ distance between distribution functions $F$ and $G$ is defined by

L.H.Y. Chen et al., *Normal Approximation by Stein's Method*,
Probability and Its Applications,
DOI 10.1007/978-3-642-15007-4_4, 

$$\|F - G\|_1 = \int_{-\infty}^{\infty} \big|F(t) - G(t)\big| dt. \tag{4.4}$$

This distance has a number of equivalent forms, and, perhaps for that reason, is known by many names, including Gini's measure of discrepancy, the Kantarovich metric (see Rachev 1984), and the Wasserstein, Dudley, and the Fortet–Mourier distance (see e.g., Barbour et al. 1992). In addition to writing the $L^1$ distance as in (4.4), we will also let $\|\mathcal{L}(X) - \mathcal{L}(Y)\|_1$ denote the $L^1$ distance between the distributions of random variables $X$ and $Y$.

That zero biasing seems to be particularly suited to produce $L^1$ bounds is evidenced in the following theorem from Goldstein (2004).

**Theorem 4.1** *Let $W$ be a mean zero, variance 1 random variable with distribution function $F$ and let $W^*$ have the $W$-zero biased distribution and be defined on the same space as $W$. Then, with $\Phi$ the cumulative distribution function of the standard normal,*

$$\|F - \Phi\|_1 \le 2E|W^* - W|. \tag{4.5}$$

As there may exist many couplings of $W$ and $W^*$ on a joint space, the challenge in producing good $L^1$ bounds is to find one in which the variables are close.

Before proving Theorem 4.1, we recall some facts about the $L^1$ norm which can be found in Rachev (1984). First, the 'dual form' of the $L^1$ distance is given by

$$\|F - G\|_1 = \inf E|X - Y|, \tag{4.6}$$

where the infimum is over all couplings of $X$ and $Y$ on a joint space with marginal distributions $F$ and $G$, respectively. As $\mathbb{R}$ is a Polish space, this infimum is achieved. A yet equivalent form of the $L^1$ distance is given by (4.1) with $\mathfrak{L}$ the collection of Lipschitz functions

$$\mathfrak{L} = \big\{h : \mathbb{R} \to \mathbb{R}\colon \big|h(y) - h(x)\big| \le |y - x|\big\}, \tag{4.7}$$

that is,

$$\big\|\mathcal{L}(Y) - \mathcal{L}(X)\big\|_1 = \sup_{h \in \mathfrak{L}} \big|Eh(Y) - Eh(X)\big|. \tag{4.8}$$

We will also make use of the fact that the elements in $\mathfrak{L}$ are exactly those absolutely continuous functions whose derivatives are (a.e.) bounded by 1 in absolute value. Though the $L^1$ distance is, therefore, just one example of a metric induced by a collection of smooth functions such as those we will study in Sect. 4.8, its many equivalent forms lead to a rich theory which accommodates numerous examples.

Part (ii) of Proposition 2.4 leads directly to the following proof of Theorem 4.1.

*Proof* First, let $(W, W^*)$ achieve the infimum $\|W - W^*\|_1$ in (4.6). As (2.77) holds with $\Delta = W^* - W$, (2.79) yields

$$\big|Eh(W) - Nh\big| \le 2\|h'\|E|W - W^*| = 2\|W - W^*\|_1.$$

Taking supremum over $h \in \mathcal{L}$ and using (4.8) shows

$$\|F - \Phi\|_1 \le 2\|W - W^*\|_1. \tag{4.9}$$

Now for $(W, W^*)$ any coupling of $W$ to a variable $W^*$ with the $W$-zero bias distribution, inequality (4.6) shows that the right hand side of (4.9) can be no greater than that of (4.5), and the result follows. □

The majority of this chapter is devoted to the exploration of various consequences of this bound, starting with sums of independent random variables.

## 4.1 Sums of Independent Variables

### *4.1.1 $L^1$ Berry–Esseen Bounds*

Continuing the discussion in Sect. 3.1, and Theorem 3.1 in particular, in this section we elaborate on the theme of $L^1$ bounds for sums of independent random variables. In particular, we demonstrate the application of Theorem 4.1 and the construction (2.61) in Lemma 2.8 to produce $L^1$ bounds with small, explicit, and distributionally specific constants for the distance between the distribution of a sum of independent variables and the normal. The utility of Theorem 4.2 below is reflected by the fact that the $L^1$ distance on the left hand side of (4.13) is that of a convolution to the normal, but is bounded on the right by terms which require only the calculation of integrals of the form (4.4) involving marginal distributions.

The proof of Theorem 4.2 requires the following simple proposition. For $H$ a distribution function on $\mathbb{R}$ let

$$H^{-1}(u) = \sup\{x\colon H(x) < u\} \quad \text{for } u \in (0, 1)$$

and let $\mathcal{U}(a, b)$ denote the uniform distribution on $(a, b)$. It is well known that when $U \sim \mathcal{U}[0, 1]$ then $H^{-1}(U)$ has distribution function $H$.

**Proposition 4.1** *For $F$ and $G$ distribution functions and $U \sim \mathcal{U}(0, 1)$,*

$$\|F - G\|_1 = E\left|F^{-1}(U) - G^{-1}(U)\right|.$$

*Further, for any $a \ge 0$ and $b \in \mathbb{R}$, with $F_{a,b}$ and $G_{a,b}$ the distribution functions of $aX + b$ and $aY + b$, respectively, we have*

$$\|F_{a,b} - G_{a,b}\|_1 = a\|F - G\|_1. \tag{4.10}$$

*Proof* The first claim is stated in (iii), Sect. 2.3 of Rachev (1984); the second follows immediately from the dual form (4.6) of the $L^1$ distance. □

Note that one consequence of the proposition is a representation of a pair of variables which achieve the infimum in (4.6).

For $X$ a random variable with finite third absolute moment let

$$B(X) = \frac{2\,\text{Var}(X)\|\mathcal{L}(X^*) - \mathcal{L}(X)\|_1}{E|X|^3}. \quad (4.11)$$

Applying (4.10) we have

$$B(aX) = B(X) \quad \text{for } a \neq 0. \quad (4.12)$$

**Theorem 4.2** *Let $\xi_i, i = 1, \ldots, n$ be independent mean zero random variables with variances $\sigma_i^2 = \text{Var}(\xi_i)$ satisfying $\sum_{i=1}^n \sigma_i^2 = 1$. Then for $F$ the distribution function of*

$$W = \sum_{i=1}^n \xi_i$$

*and $\Phi$ that of the standard normal,*

$$\|F - \Phi\|_1 \le \sum_{i=1}^n B(\xi_i) E|\xi_i|^3. \quad (4.13)$$

*Additionally, when $W = \sum_{i=1}^n X_i/(\sigma\sqrt{n})$ with $X, X_1, \ldots, X_n$ i.i.d. mean zero, variance $\sigma^2$ random variables, then*

$$\|F - \Phi\|_1 \le \frac{1}{\sigma^3\sqrt{n}} B(X) E|X|^3. \quad (4.14)$$

*Proof* Let $U_1, \ldots, U_n$ be mutually independent $\mathcal{U}(0, 1)$ variables and set

$$(\xi_i, \xi_i^*) = \left(G_i^{-1}(U_i), \left(G_i^*\right)^{-1}(U_i)\right), \quad i = 1, \ldots, n,$$

where $G_1^*, \ldots, G_n^*$ are the distribution functions of $\xi_1^*, \ldots, \xi_n^*$, respectively. Then $\xi_i$ and $\xi_i^*$ have distribution functions $G_i$ and $G_i^*$, respectively, and by Proposition 4.1,

$$E\left|\xi_i^* - \xi_i\right| = \left\|G_i^* - G_i\right\|_1.$$

Constructing $W^*$ as in Lemma 2.8 yields $W^* - W = \xi_I^* - \xi_I$, with $I$ having distribution $P(I = i) = \sigma_i^2$, so applying Theorem 4.1 we have

$$\begin{aligned}
\|F - \Phi\|_1 &\le 2E|W^* - W| \\
&= 2E\left|\xi_I^* - \xi_I\right| \\
&= 2\sum_{i=1}^n \sigma_i^2 E\left|\xi_i^* - \xi_i\right| \\
&= 2\sum_{i=1}^n \sigma_i^2 \left\|G_i^* - G_i\right\|_1 \\
&= \sum_{i=1}^n B(\xi_i) E|\xi_i|^3,
\end{aligned}$$

thus proving (4.13).

If $X, X_1, \ldots, X_n$ are i.i.d. with mean zero and variance $\sigma^2$ then applying (4.13) with $\xi_i = X_i/(\sigma\sqrt{n})$, and (4.12), yields the bound

$$\|F - \Phi\|_1 \le \frac{1}{\sigma^3 n^{3/2}} \sum_{i=1}^{n} B\left(\frac{X_i}{\sigma\sqrt{n}}\right) E|X_i|^3 = \frac{B(X)E|X|^3}{\sigma^3\sqrt{n}},$$

proving (4.14). □

Specializing (4.14) to particular cases leads to the following corollary.

**Corollary 4.1** *When $X = (\xi - p)/\sqrt{pq}$ where $\xi$ has the Bernoulli distribution with success probability $1 - q = p \in (0, 1)$,*

$$B(X) = 1 \quad \text{and} \quad \|F - \Phi\|_1 \le \frac{E|X|^3}{\sqrt{n}} = \frac{p^2 + q^2}{\sqrt{npq}} \quad \text{for all } n = 1, 2, \ldots.$$

*When $X$ has the uniform distribution $\mathcal{U}[-\sqrt{3}, \sqrt{3}]$, then*

$$B(X) = 1/3 \quad \text{and} \quad \|F - \Phi\|_1 \le \frac{E|X|^3}{3\sqrt{n}} = \frac{\sqrt{3}}{4\sqrt{n}} \quad \text{for all } n = 1, 2, \ldots.$$

*Proof* In the Bernoulli case, by (2.55), $X^*$ has the uniform distribution function

$$G^*(x) = \sqrt{pq}\,x + p \quad \text{for } x \in \left[\frac{-p}{\sqrt{pq}}, \frac{q}{\sqrt{pq}}\right],$$

that is, $X^* =_d (U - p)/\sqrt{pq}$, where $U \sim \mathcal{U}[0, 1]$. Hence, by Proposition 4.1,

$$\|G^* - G\|_1 = \left\|\frac{U - p}{\sqrt{pq}} - \frac{\xi - p}{\sqrt{pq}}\right\|_1 = \frac{1}{\sqrt{pq}}\|U - \xi\|_1 = \frac{p^2 + q^2}{2\sqrt{pq}}.$$

Calculating $E|X|^3 = (p^2 + q^2)/\sqrt{pq}$ and using $\mathrm{Var}(X) = 1$ gives $B(X) = 1$, and the claimed bound.

For the uniform distribution $\mathcal{U}[-\sqrt{3}, \sqrt{3}]$, (2.55) yields

$$G^*(x) = -\frac{\sqrt{3}x^3}{36} + \frac{\sqrt{3}x}{4} + \frac{1}{2} \quad \text{for } x \in [-\sqrt{3}, \sqrt{3}]$$

and from (4.4) we obtain

$$\|G^* - G\|_1 = \frac{\sqrt{3}}{8}.$$

Calculating $E|X|^3 = 3\sqrt{3}/4$ now gives $B(X) = 1/3$, and the claimed bound. □

Constants $B(X)$ and bounds for other distributions may be calculated in a similar fashion. A universal $L^1$ constant over a class of distributions $\mathcal{F}$, by Theorem 4.2, is given by

$$B(\mathcal{F}) = \sup_{\mathcal{L}(X) \in \mathcal{F}} B(X).$$

The following result, by Goldstein (2010a) and Tyurin (2010), shows that the Bernoulli distribution achieves the worst case $B(X)$.

**Theorem 4.3** *For $\sigma > 0$ let $\mathcal{F}_\sigma$ be the collection of all mean zero distributions with variance $\sigma^2$ and finite absolute third moment. Then*

$$B(\mathcal{F}) = 1 \quad \text{where } \mathcal{F} = \bigcup_{\sigma>0} \mathcal{F}_\sigma.$$

Theorems 4.3 and 4.2 immediately give

**Corollary 4.2** *If $\xi_i, i = 1, \ldots, n$ are independent mean zero random variables with variances $\sigma_i^2 = \mathrm{Var}(\xi_i)$ satisfying $\sum_{i=1}^n \sigma_i^2 = 1$ and $W = \xi_1 + \cdots + \xi_n$, then*

$$\|F - \Phi\|_1 \le \sum_{i=1}^n E|\xi_i|^3.$$

*In particular, if $W = n^{-1/2} \sum X_i$ with $X, X_1, \ldots, X_n$ i.i.d. variables with mean zero and variance 1, then*

$$\|F - \Phi\|_1 \le \frac{E|X|^3}{\sqrt{n}}.$$

Though it may be difficult to achieve the optimal $L^1$ coupling between $X$ and $X^*$ in particular applications, especially those involving dependence, the following proposition shows how to construct a coupling which results in a constant bounded by 1 when $X$ is symmetric. Proposition 4.2 is applied in Theorem 4.7 to improve the leading constant in Goldstein (2007) for projections of cone measure.

**Proposition 4.2** *Let $\chi$ be a random variable with a symmetric distribution, variance $\sigma^2 \in (0, \infty)$ and finite third absolute moment. Let $\overline{X}$ and $\overline{Y}$ be constructed on a joint space with $0 \le \overline{X} \le \overline{Y}$ a.s. having marginal distributions given by $\overline{X} =_d |\chi|$ and $\overline{Y} =_d |\chi^\square|$, where $\chi^\square$ is as defined in Proposition 2.3. Let $V \sim \mathcal{U}[0, 1]$ and $\epsilon$ take the values 1 and $-1$ with equal probability, and be independent of each other and of $\overline{X}$ and $\overline{Y}$. Then $X = \epsilon\overline{X}$ has distribution $\chi$, the variable*

$$X^* = \epsilon V \overline{Y}$$

*has the $\chi$-zero biased distribution, and*

$$\frac{2\sigma^2 E|X^* - X|}{E|X|^3} \le 1. \tag{4.15}$$

*Proof* That $X =_d \chi$ follows by the symmetry of $\chi$. Again, by the symmetry of $\chi$,

$$\sigma^2 Ef(\chi^\square) = E[\chi^2 f(\chi)] = E[(-\chi)^2 f(-\chi)] = E[\chi^2 f(-\chi)] = \sigma^2 Ef(-\chi^\square).$$

Hence $\chi^\square$ is symmetric, and as $\epsilon V \sim \mathcal{U}[-1, 1]$ and is independent of $\overline{Y}$, by Proposition 2.3,

$$X^* = \epsilon V \overline{Y} =_d \epsilon V \chi^\square =_d \chi^*.$$

Now,

$$E|X^* - X| = E|\epsilon V\overline{Y} - \epsilon\overline{X}| = E|V\overline{Y} - \overline{X}| = \int_{x\geq 0, y>0}\int_0^1 |vy - x|dvdF(x,y)$$

where $dF(x,y)$ is the joint distribution of $(\overline{X}, \overline{Y})$. Since $dF(x,y)$ is zero on sets where $x > y$, we may decompose the integral above as

$$\begin{aligned}
&\int_{x\geq 0, y>0}\int_{x/y<v\leq 1}(vy - x)dvdF(x,y) + \int_{x\geq 0, y>0}\int_{0<v<x/y}(x - vy)dvdF(x,y)\\
&= \int_{x\geq 0, y>0}\left(\frac{1}{2}y\left(1 - \left(\frac{x}{y}\right)^2\right) - x\left(1 - \frac{x}{y}\right)\right)dF(x,y)\\
&\quad + \int_{x\geq 0, y>0}\left(x\left(\frac{x}{y}\right) - \frac{1}{2}y\left(\frac{x}{y}\right)^2\right)dF(x,y)\\
&= E\left(\frac{1}{2}\overline{Y} - \overline{X} + \frac{\overline{X}^2}{\overline{Y}}\right).
\end{aligned}$$

As $\overline{X}/\overline{Y} \leq 1$, we have $\overline{X}^2/\overline{Y} \leq \overline{X}$, and therefore

$$E|X^* - X| \leq \frac{1}{2}E\overline{Y} = \frac{1}{2}E|\chi^{\square}| = \frac{1}{2\sigma^2}E|X|^3.$$

Substituting into (4.15) yields the desired inequality. □

Let $X$ be any random variable with mean zero and variance $\sigma^2$, and let $\phi$ be an increasing function on $[0,\infty)$. Since $x^2$ is an increasing function on $[0,\infty)$, $X^2$ will be positively correlated with $\phi(|X|)$, that is,

$$\sigma^2 E\phi(|X^{\square}|) = EX^2\phi(|X|) \geq EX^2 E\phi(|X|) = \sigma^2 E\phi(|X|),$$

showing $|X^{\square}|$ is stochastically larger than $|X|$. Hence there always exists a coupling where $|X^{\square}| \geq |X|$ a.s., even when $X$ is not symmetric. Though an optimal $L^1$ coupling is similarly assured, in principle, by Proposition 4.1, couplings constructed by following Proposition 4.2 seem to be of more practical use; see in particular where this proposition is applied for cone measure in item 3 of Proposition 4.5.

### *4.1.2 Contraction Principle*

In this section we show that the distribution of a standardized sum of i.i.d. variables is closer in $L^1$ to the normal, in a zero bias sense, than the distribution of the summands themselves. This result leads to a type of $L^1$ contraction principle for the CLT. For some additional generality we will consider weighted averages of i.i.d. random variable.

Let $\|\boldsymbol{\alpha}\|$ denote the Euclidean norm of a vector $\boldsymbol{\alpha} \in \mathbb{R}^k$, and when $\boldsymbol{\alpha}$ is nonzero let

$$\varphi(\boldsymbol{\alpha}) = \frac{\sum_{i=1}^k |\alpha_i|^3}{(\sum_{i=1}^k \alpha_i^2)^{3/2}}. \tag{4.16}$$

Inequality (4.17) of Lemma 4.1 says that taking weighted averages of i.i.d. variables is a contraction in the $L^1$ distance to normal in a zero biased sense.

**Lemma 4.1** *For $\boldsymbol{\alpha} \in \mathbb{R}^k$ with $\lambda = \|\boldsymbol{\alpha}\| \neq 0$, let*

$$Y = \sum_{i=1}^k \frac{\alpha_i}{\lambda} W_i,$$

*where $W_i$ are mean zero, variance one, independent random variables distributed as $W$. Then*

$$\left\|\mathcal{L}(Y^*) - \mathcal{L}(Y)\right\|_1 \le \varphi \left\|\mathcal{L}(W^*) - \mathcal{L}(W)\right\|_1 \tag{4.17}$$

*with $\varphi = \varphi(\boldsymbol{\alpha})$ as in* (4.16), *and $\varphi < 1$ if and only if $\boldsymbol{\alpha}$ is not a multiple of a standard basis vector.*

*If $W_0$ is any mean zero, variance one random variable with finite absolute third moment, $\boldsymbol{\alpha}_n$, $n = 0, 1, \ldots$ a sequence of nonzero vectors in $\mathbb{R}^k$, $\lambda_n = \|\boldsymbol{\alpha}_n\|$, $\varphi_n = \varphi(\boldsymbol{\alpha}_n)$, and*

$$W_{n+1} = \sum_{i=1}^k \frac{\alpha_{n,i}}{\lambda_n} W_{n,i} \quad \text{for } n = 0, 1, \ldots \tag{4.18}$$

*where $W_{n,i}$ are i.i.d. copies of $W_n$, then*

$$\left\|\mathcal{L}(W_n^*) - \mathcal{L}(W_n)\right\|_1 \le \left(\prod_{j=0}^{n-1} \varphi_j\right). \tag{4.19}$$

*If $\limsup_n \varphi_n = \varphi < 1$, then for any $\gamma \in (\varphi, 1)$ there exists $C$ such that*

$$\left\|\mathcal{L}(W_n) - \mathcal{L}(Z)\right\|_1 \le C\gamma^n \quad \text{for all } n, \tag{4.20}$$

*while if $\boldsymbol{\alpha}_n = \boldsymbol{\alpha}$ for some $\boldsymbol{\alpha}$ and all $n$, then*

$$\left\|\mathcal{L}(W_n) - \mathcal{L}(Z)\right\|_1 \le 2\varphi^n \quad \text{for all } n, \tag{4.21}$$

*with $\varphi = \varphi(\boldsymbol{\alpha})$.*

We begin the proof of the lemma by studying how $\varphi$ behaves in terms of $\boldsymbol{\alpha}$, and prove a bit more than we need now, saving the additional results for use in Sect. 4.2.

**Lemma 4.2** *For $\boldsymbol{\alpha} \in \mathbb{R}^k$ with $\lambda = \|\boldsymbol{\alpha}\| \neq 0$,*

$$\sum_{i=1}^k \frac{|\alpha_i|^p}{\lambda^p} \le 1 \quad \text{for all } p > 2, \tag{4.22}$$

*with equality if and only if $\boldsymbol{\alpha}$ is a multiple of a standard basis vector. With $\varphi$ as in* (4.16),

$$\frac{1}{\sqrt{k}} \le \varphi \le 1, \tag{4.23}$$

*with equality to the upper bound if and only if $\boldsymbol{\alpha}$ is a multiple of a standard basis vector, and equality to the lower bound if and only if $|\alpha_i| = |\alpha_j|$ for all $i, j$.*

*In addition, when $\alpha_i \ge 0$ and $\sum_{i=1}^n \alpha_i = 1$ then*

$$\lambda \le \varphi, \tag{4.24}$$

*with equality if and only if $\boldsymbol{\alpha}$ is equal to a standard basis vector.*

*Proof* Since $|\alpha_i|/\lambda \le 1$ we have $|\alpha_i|^{p-2}/\lambda^{p-2} \le 1$, yielding

$$\sum_{i=1}^k \frac{|\alpha_i|^p}{\lambda^p} = \sum_{i=1}^k \left(\frac{|\alpha_i|^{p-2}}{\lambda^{p-2}}\right)\frac{\alpha_i^2}{\lambda^2} \le \sum_{i=1}^k \frac{\alpha_i^2}{\lambda^2} = 1,$$

with equality if and only if $|\alpha_i| = \lambda$ for some $i$ and $\alpha_j = 0$ for all $j \ne i$. Specializing to the case $p = 3$ yields the claims about the upper bound in (4.23).

By Hölder's inequality with $p = 3, q = 3/2$, we have

$$\left(\sum_{i=1}^k \alpha_i^2\right)^{3/2} = \left(\sum_{i=1}^k 1 \cdot \alpha_i^2\right)^{3/2} \le \sqrt{k}\sum_{i=1}^k |\alpha_i|^3,$$

giving the lower bound (4.23), with equality if and only if $\alpha_i^2$ is proportional to 1 for all $i$.

The claim (4.24) follows from the inequality

$$(EY)^2 \le EY^2 \quad \text{when } P(Y = \alpha_i) = \alpha_i,$$

which is an equality if and only if the variable $Y$ is constant. □

We may now proceed to the proof of the lemma.

*Proof of Lemma 4.1* Let $F_{W^*}$ and $F_W$ be the distribution functions of $W^*$ and $W$, respectively, and with $U_1, \ldots, U_n$ independent $\mathcal{U}[0, 1]$ variables let

$$\left(W_i^*, W_i\right) = \left(F_{W^*}^{-1}(U_i), F_W^{-1}(U_i)\right) \quad i = 1, \ldots, n.$$

By Proposition 4.1, $E|W_i^* - W_i| = \|\mathcal{L}(W^*) - \mathcal{L}(W)\|_1$ for all $i = 1, \ldots, n$.

By Lemma 2.8 and (2.59), with $I$ a random index independent of all other variables with distribution

$$P(I = i) = \frac{\alpha_i^2}{\lambda^2},$$

the variable

$$Y^* = Y - \frac{\alpha_I}{\lambda}\left(W_I - W_I^*\right) \tag{4.25}$$

has the $Y$-zero biased distribution. Using (4.6) for the first inequality, we now obtain (4.17) by

$$
\begin{aligned}
\|\mathcal{L}(Y^*) - \mathcal{L}(Y)\|_1 &\le E|Y^* - Y| \\
&= E\sum_{i=1}^k \frac{|\alpha_i|}{\lambda}|W_i^* - W_i|\mathbf{1}(I=i) \\
&= \sum_{i=1}^k \frac{|\alpha_i|^3}{\lambda^3} E|W_i^* - W_i| \\
&= \varphi\|\mathcal{L}(W) - \mathcal{L}(W^*)\|_1.
\end{aligned}
$$

That $\varphi < 1$ if and only if $\boldsymbol{\alpha}$ is not a multiple of a standard basis vector was shown in Lemma 4.2.

To obtain (4.19), note that induction and (4.17) yield

$$
\|\mathcal{L}(W_n^*) - \mathcal{L}(W_n)\|_1 \le \left(\prod_{j=0}^{n-1} \varphi_j\right)\|\mathcal{L}(W_0^*) - \mathcal{L}(W_0)\|_1,
$$

and $\|\mathcal{L}(W_0^*) - \mathcal{L}(W_0)\|_1 \le 1$ by Theorem 4.3.

When $\limsup_n \varphi_n = \varphi < \gamma < 1$ there exists $n_0$ such that

$$
\varphi_j \le \gamma \quad \text{for all } j \ge n_0.
$$

Hence, for all $n \ge n_0$

$$
\prod_{j=0}^{n-1} \varphi_j = \left(\prod_{j=0}^{n_0-1} \frac{\varphi_j}{\gamma}\right)\gamma^{n_0} \prod_{j=n_0}^{n-1} \varphi_j \le \left(\prod_{j=0}^{n_0-1} \frac{\varphi_j}{\gamma}\right)\gamma^n.
$$

The bound (4.20) now follows from this inequality and Theorem 4.1.

The last claim (4.21) is immediate from (4.19) and Theorem 4.1. □

We note that the standardized, classical case (4.14) is recovered from (4.17) and Theorem 4.1 when $\alpha_i = 1/\sqrt{n}$. In Sect. 4.2 we study nonlinear versions of recursion (4.18) with applications to physical models.

## 4.2 Hierarchical Structures

For $k \ge 2$ an integer, $\mathcal{D} \subset \mathbb{R}$, and $F : \mathcal{D}^k \to \mathcal{D}$ a given function, every distribution for a random variable $X_0$ with $P(X_0 \in \mathcal{D}) = 1$ generates the sequence of 'hierarchical' distributions through the recursion

$$
X_{n+1} = F(\mathbf{X}_n), \quad n \ge 0, \tag{4.26}
$$

where $\mathbf{X}_n = (X_{n,1}, \ldots, X_{n,k})^\mathsf{T}$ with $X_{n,i}$ independent, each with distribution $X_n$. Such hierarchical variables have been considered extensively in the physics literature (see Li and Rogers 1999 and the references therein), in particular to model conductivity of random media.

The special case where the function $F$ is determined by the conductivity properties of the diamond lattice has been considered in Griffiths and Kaufman (1982) and Schlösser and Spohn (1992). Figure 4.1 shows the progression of the diamond lattice from large to small scale. At the large scale (a), the conductivity of the system can be measured along the bond connecting its top and bottom nodes. Inspection of the lattice on a finer scale reveals that this bond is actually comprised of four smaller bonds, each similar to (a), connected as shown in (b). Inspection on an even finer scaler reveals that each of the four bonds in (b) are constructed in a self-similar way from bonds at a smaller level, giving the successive diagram (c), and so on.

To determine the conductivity function $F$ associated with a given lattice, first recall that conductances add in parallel, that is, if two components with conductances $x_1$ and $x_2$ are placed in parallel, then the net conductance of the system is

$$L_1(x_1, x_2) = x_1 + x_2. \tag{4.27}$$

Similarly, resistances add for components placed in series. Hence, for these same two components in series, as resistance and conductance are inverses, the resulting conductance of the system is

$$L_{-1}(x_1, x_2) = \left(x_1^{-1} + x_2^{-1}\right)^{-1}. \tag{4.28}$$

For the diamond lattice in particular, assume that each bond has a fixed 'baseline' conductivity characteristic $w \geq 0$ such that when a component with conductivity $x \geq 0$ is present along the bond its net conductivity is $wx$. For bonds in the diamond lattice as in (b), we associate conductivities characteristics $\mathbf{w} = (w_1, w_2, w_3, w_4)^{\mathsf{T}}$, numbering bonds from the top and proceeding counter-clockwise. Hence, if $\mathbf{x} = (x_1, x_2, x_3, x_4)^{\mathsf{T}}$ are the conductances of four elements each as in (a) which are present along the bonds in (b), then the two components in series on the left side have conductance $L_{-1}(w_1x_1, w_2x_2)$, and similarly, the conductance for the two components in series on the right is $L_{-1}(w_3x_3, w_4x_4)$. Combining these two subsystems in parallel gives

$$F(\mathbf{x}) = L_1\big(L_{-1}(w_1x_1, w_2x_2), L_{-1}(w_3x_3, w_4x_4)\big), \tag{4.29}$$

that is,

$$F(\mathbf{x}) = \left(\frac{1}{w_1x_1} + \frac{1}{w_2x_2}\right)^{-1} + \left(\frac{1}{w_3x_3} + \frac{1}{w_4x_4}\right)^{-1}. \tag{4.30}$$

Returning to the sequence of distributions generated by the recursion (4.26), conditions on $F$ which imply the weak law

$$X_n \to_p c \tag{4.31}$$

for some constant $c$ have been considered by various authors. Recall that we say $F$ is homogeneous, or positively homogeneous, if

$$F(ax_1, \ldots, ax_k) = a^k F(x_1, \ldots, x_k)$$

hold for all $a \in \mathbb{R}$, or all $a > 0$, respectively. Shneiberg (1986) proves that (4.31) holds if $\mathcal{D} = [a, b]$ and $F$ is continuous, monotonically increasing, positively homogeneous, convex and satisfies the normalization condition $F(\mathbf{1}_k) = 1$ where $\mathbf{1}_k$

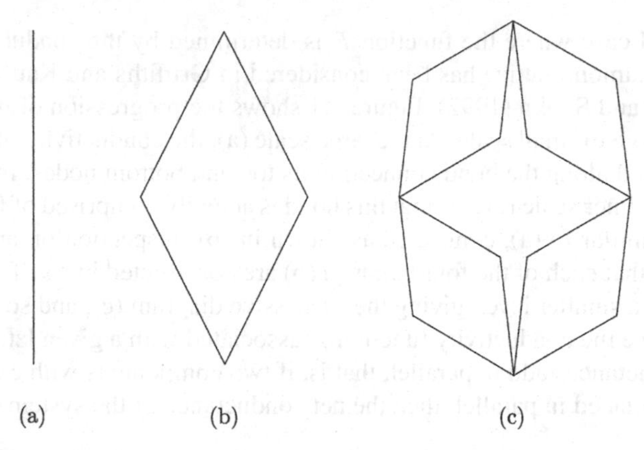


**Fig. 4.1** The diamond lattice

is the vector of all ones in $\mathbb{R}^k$. Li and Rogers (1999) provide rather weak conditions under which (4.31) holds for closed $\mathcal{D} \subset (-\infty, \infty)$. See also Wehr (1997) and Wehr (2001), and Jordan (2002) for an extension of the model to random $F$ and applications of hierarchical structures to computer science.

Letting $X_0$ have mean $c$ and variance $\sigma^2$, the classical central limit theorem can be set in the framework of hierarchical sequences by letting

$$F(x_1, x_2) = \frac{1}{2}(x_1 + x_2), \tag{4.32}$$

which gives

$$X_n =_d \frac{X_{0,1} + \cdots + X_{0,2^n}}{2^n} \tag{4.33}$$

where $X_{0,m}, m = 1, \ldots, 2^n$ are independent and identically distributed as $X_0$. Hence, $X_n \to_p c$ by the weak law of large numbers, and since $X_n$ is the average of $N = 2^n$ i.i.d. variables with finite variance, we have additionally that

$$W_n = \sqrt{N}\left(\frac{X_n - c}{\sigma}\right) \to_d \mathcal{N}(0, 1).$$

Moreover, when $X_0$ has a bounded absolute third moment (4.21) yields

$$\left\|\mathcal{L}(W_n) - \mathcal{L}(Z)\right\|_1 \le C\gamma^n \tag{4.34}$$

with $C = 2$ and $\gamma = 1/\sqrt{2}$.

The function $F$ in (4.32) is a simple average, and one would, therefore, expect normal limiting behavior more generally when the function $F$ averages its inputs in some sense.

**Definition 4.1** We say that $F : \mathcal{D}^k \to \mathcal{D}$ is an averaging function when it satisfies the following three properties on its domain:

1. $\min_i x_i \le F(\mathbf{x}) \le \max_i x_i$.
2. $F(\mathbf{x}) \le F(\mathbf{y})$ whenever $x_i \le y_i$.

3. For all $x < y$ and for any two distinct indices $i_1 \neq i_2$, there exists $x_i \in \{x, y\}$, $i = 1, \ldots, k$ such that $x_{i_1} = x, x_{i_2} = y$ and $x < F(\mathbf{x}) < y$.

We say $F$ is strictly averaging if $F$ satisfies Properties 1 and 2 with strict inequality when $\min_i x_i < \max_i x_i$, and when $x_i < y_i$ for some $i$, respectively.

Properties 1 and 2 say that the 'average' returned by $F$ should lie inbetween the values being 'averaged' and that that 'average' increases with those values. Note that Property 1 says that for $F$ to be an averaging function it is necessary that $F(\mathbf{1}_k) = 1$. Property 3 says that $F$ is sensitive, that is, depends on, all of its coordinates. We note that if $F$ is strictly averaging then $F$ satisfies Property 3 thusly: if $x < y$ and $x_{i_i} = x, x_{i_2} = y$, then any assignment of the values $x, y$ to the remaining coordinates gives $x < F(\mathbf{x}) < y$ by the strict form of Property 1. Hence all strictly averaging functions are averaging. We note that the function $F(\mathbf{x}) = \min_i x_i$ satisfies the first two properties but not the third, and it gives rise to extreme value, rather than normal, limiting behavior.

Normal limits are proved by Wehr and Woo (2001) for sequences $X_n$, $n = 0, 1, \ldots$ determined by the recursion (4.26) when the function $F(\mathbf{x})$ is averaging by showing that such recursions can be treated as the approximate linear recursion around the mean $c_n = EX_n$ with small perturbation $Z_n$,

$$X_{n+1} = \boldsymbol{\alpha}_n \cdot \mathbf{X}_n + Z_n, \quad n \geq 0, \tag{4.35}$$

where $\boldsymbol{\alpha}_n = F'(\mathbf{c}_n)$, the gradient of $F$ at $\mathbf{c}_n$ where $\mathbf{c}_n = (c_n, \ldots, c_n)^\mathsf{T} \in \mathbb{R}^k$. In Sect. 4.2.1 we prove Theorem 4.6, which gives the bound (4.34) for the $L^1$ distance to the normal for sequences generated by the approximate linear recursion (4.35) under Conditions 4.1 and 4.2, which guarantee that $Z_n$ is small relative to $X_n$.

In Sect. 4.2.2 we prove Theorem 4.4 which shows that the normal convergence of the hierarchical sequence $X_n, n = 0, 1, \ldots$ holds with bound (4.34) under mild conditions, and specifies the exponential rate $\gamma$ in an explicit range. Theorem 4.4 is proved by invoking Theorem 4.6 after showing that the required moment conditions are satisfied for a linearization of $X_{n+1} = F(\mathbf{X}_n)$.

**Theorem 4.4** *For some $a < b$ let $X_0$ be a non constant random variable with $P(X_0 \in [a, b]) = 1$ and let*

$$X_{n+1} = F(\mathbf{X}_n), \quad n \geq 0,$$

*where $\mathbf{X}_n = (X_{n,1}, \ldots, X_{n,k})^\mathsf{T}$ with $X_{n,i}$ independent, each with distribution $X_n$ and $F : [a, b]^k \to [a, b]$, twice continuously differentiable. Suppose $F$ is averaging and that $X_n \to_p c$, with $\boldsymbol{\alpha} = F'(\mathbf{1}_k c)$ not a scalar multiple of a standard basis vector. Then with $W_n = (X_n - c_n)/\sqrt{\mathrm{Var}(X_n)}$ and $Z$ a standard normal variable, for all $\gamma \in (\varphi, 1)$ there exists $C$ such that*

$$\big\|\mathcal{L}(W_n) - \mathcal{L}(Z)\big\|_1 \leq C\gamma^n \quad \textit{for all } n \geq 0,$$

*where $\varphi$, given by $\boldsymbol{\alpha}$ through* (4.16), *is a positive number strictly less than* 1. *The value $\varphi$ achieves a minimum of $1/\sqrt{k}$ if and only if the components of $\boldsymbol{\alpha}$ are equal.*

As in (4.33), the variable $X_n$ is a function of $N = k^n$ variables, so achieving the rate $\varphi^n$ exactly corresponds to a 'classical rate' of $N^{-\theta}$ where

$$\varphi^n = N^{-\theta} = k^{-n\theta} \quad \text{or} \quad \theta = -\log_k \varphi. \tag{4.36}$$

Hence when $\varphi$ achieves its minimum value $1/\sqrt{k}$ we have $\theta = -1/2$ and the rate $N^{-1/2}$, and achieving this rate for all $\gamma > \varphi$ therefore corresponds to the rate $N^{-1/2+\epsilon}$ for every $\epsilon > 0$. Further, when $\boldsymbol{\alpha}$ is close to a standard basis vector, $\varphi$ is close to 1, so the bound can have rate $N^{-\theta}$ for $\theta$ arbitrarily close to zero. This behavior is anticipated: for the simple hierarchical sequence generated by the function $F(x_1, x_2) = (1-\epsilon)x_1 + \epsilon x_2$, convergence to the normal will be slow indeed for small $\epsilon > 0$. The condition in Theorem 4.4 that the gradient $\boldsymbol{\alpha} = F'(\mathbf{c})$ of $F$ at the limiting value $\mathbf{c}$ not be a scalar multiple of a standard basis vector rules out cases which behave in the limit degenerately as $F(x_1, x_2) = x_1$.

The function (4.32), and (4.30) when $F(\mathbf{1}_4) = 1$, are examples of averaging functions. To handle multiples, we say that $G(\mathbf{y})$ with $G(\mathbf{1}_k) \neq 0$ is a scaled averaging function if $G(\mathbf{y})/G(\mathbf{1}_k)$ is averaging. Now suppose that $G(\mathbf{y})$ is scaled averaging and homogeneous, and that

$$Y_{n+1} = G(\mathbf{Y}_n) \quad \text{for } n \geq 0,$$

where $Y_0$ is a given random variable and $\mathbf{Y}_n \in \mathbb{R}^k$ is a vector of independent copies of $Y_n$. Then letting

$$a_{n+1} = ka_n + 1 \quad \text{for all } n \geq 0, \text{ and } a_0 = 0,$$

and setting $F(\mathbf{y}) = G(\mathbf{y})/G(\mathbf{1}_k)$, which is an averaging function, and $\mathbf{X}_n = \mathbf{Y}_n/G(\mathbf{1}_k)^{a_n}$ and likewise for $X_n$, we have

$$\begin{aligned} X_{n+1} &= Y_{n+1}/G(\mathbf{1}_k)^{a_{n+1}} \\ &= G(\mathbf{Y}_n)/G(\mathbf{1}_k)^{a_{n+1}} = F(\mathbf{Y}_n)/G(\mathbf{1}_k)^{ka_n} = F(\mathbf{X}_n). \end{aligned}$$

As the scaled and centered and variables $(X_n - EX_n)/\sqrt{\mathrm{Var}(X_n)}$ and $(Y_n - EY_n)/\sqrt{\mathrm{Var}(Y_n)}$ are equal, the conclusion of Theorem 4.4 holds for $Y_n$ when it holds for $X_n$.

Theorem 4.4 is applied in Sect. 4.2.3 to the specific hierarchical variables generated by the diamond lattice conductivity function (4.30), and, in (4.67), the value $\varphi$ determining the range of $\gamma$ is given as an explicit function of the weights $\mathbf{w}$; for the diamond lattice all rates $N^{-\theta}$ for $\theta \in (0, 1/2)$ are exhibited. Interestingly, there appears to be no such formula, simple or otherwise, for the limiting mean or variance of the sequence $X_n$.

To proceed we introduce another equivalent formulation of the $L^1$ distance. With $\mathcal{L}$ as in (4.7), let

$$\mathcal{F} = \{f\colon f \text{ absolutely continuous } f(0) = f'(0) = 0,\ f' \in \mathcal{L}\}. \tag{4.37}$$

Clearly, if $f \in \mathcal{F}$ then $h \in \mathcal{L}$ for $h = f'$. On the other hand, if $h \in \mathcal{L}$ then

$$f \in \mathcal{F} \quad \text{and} \quad f'(y) - f'(x) = h(y) - h(x) \quad \text{for } f(x) = \int_0^x [h(u) - h(0)]du.$$

Then, from (4.8),

$$\left\|\mathcal{L}(Y)-\mathcal{L}(X)\right\|_1=\sup_{f\in\mathcal{F}}\left|E\big(f'(Y)-f'(X)\big)\right|. \tag{4.38}$$

For the application of Theorem 4.4, it is necessary to verify that the function $F(\mathbf{x})$ in (4.26) is averaging. Proposition 3 of Wehr and Woo (2001) shows that the effective conductance of a resistor network is an averaging function of the conductances of its individual components. Theorem 4.5, which shows that strict averaging is preserved under certain compositions, yields an independent proof that, for instance, (4.30) is strictly averaging under natural scaling and positivity conditions on the weights. In addition, Theorem 4.5 provides an additional source of averaging functions to which Theorem 4.4 may be applied.

**Theorem 4.5** *Let $k\ge 1$ and set $I_0=\{1,\dots,k\}$. Suppose subsets $I_i\subset I_0$, $i\in I_0$ satisfy $\bigcup_{i\in I_0}I_i=I_0$. For $\mathbf{x}\in\mathbb{R}^k$ and $i\in I_0$ let $\mathbf{x}_i=(x_{j_1},\dots,x_{j_{|I_i|}})$ where $\{j_1,\dots,j_{|I_i|}\}=I_i$ with $j_1<\dots<j_{|I_i|}$. Let $F_i:\mathbb{R}^{|I_i|}\to\mathbb{R}$ (or $F_i:[0,\infty)^{|I_i|}\to[0,\infty)$), $i=0,\dots,k$. If $F_0,F_1,\dots,F_k$ are strictly averaging and $F_0$ is (positively) homogeneous, then the composition*

$$F_{\mathbf{s}}(\mathbf{x})=F_0\big(s_1F_1(\mathbf{x}_1),\dots,s_kF_k(\mathbf{x}_k)\big)$$

*is strictly averaging for any $\mathbf{s}$ which satisfies $F_0(\mathbf{s})=1$ and $s_i>0$ for all $i$. If $F_0,F_1,\dots,F_k$ are scaled, strictly averaging and $F_0$ is (positively) homogeneous, then*

$$F_{\mathbf{1}}(\mathbf{x})=F_0\big(F_1(\mathbf{x}_1),\dots,F_k(\mathbf{x}_k)\big)$$

*is a scaled strictly averaging function.*

Note that the parallel and series combination rules (4.27) and (4.28) are the $p=1$ and $p=-1$ special cases, respectively, with $w_i=1$, of the weighted $L^p$ norm functions

$$L_p^{\mathbf{w}}(\mathbf{x})=\left(\sum_{i=1}^{k}(w_ix_i)^p\right)^{1/p},\quad \mathbf{w}=(w_1,\dots,w_k)^{\mathsf{T}},\quad w_i\in(0,\infty),$$

which are scaled, strictly averaging, and positively homogeneous on $[0,\infty)^k$ for $p>0$ and on $(0,\infty)$ for $p<0$. Since $F(\mathbf{x})$ in (4.30) is represented by the composition (4.29), Theorem 4.5 obtains to show that $F$ is a scaled, strictly averaging function on $(0,\infty)^4$ for any choice of positive weights. In particular, for positive weights such that $F(\mathbf{1})=1$, the function $F$ is strictly averaging on $(0,\infty)^4$. Theorem 4.4 requires $F$ to have domain $[a,b]^k$. However, if $F$ is an averaging function on, say, $(0,\infty)^4$, then Property 1 implies that $F:[a,b]^k\to[a,b]$ for all $[a,b]\subset(0,\infty)$, and hence $F$ will be averaging on this smaller domain. Note lastly that Theorem 4.5 shows the same conclusion holds when the resistor parallel $L_1$ and series $L_{-1}$ combination rules in this network are replaced by, say, $L_2$ and $L_{-2}$ respectively.

### 4.2.1 Bounds to the Normal for Approximately Linear Recursions

In this section we study sequences $\{X_n\}_{n\geq 0}$ generated by the approximate linear recursion

$$X_{n+1} = \boldsymbol{\alpha}_n \cdot \mathbf{X}_n + Z_n, \quad n \geq 0, \tag{4.39}$$

where $X_0$ is a given nontrivial random variable and the components $X_{n,1}, \ldots, X_{n,k}$ of $\mathbf{X}_n$ are independent copies of $X_n$. We present Theorem 4.6 which shows the exponential bound (4.34) holds when the perturbation term $Z_n$, which measures the departure from linearity, is small. The effective size of $Z_n$ is measured by the quantity $\beta_n$ of (4.42), which will be small when the moment bounds in Conditions 4.1 and 4.2 are satisfied. When the recursion is nearly linear, $X_{n+1}$ will be approximately equal to $\boldsymbol{\alpha}_n \cdot \mathbf{X}_n$, and therefore its variance $\sigma_{n+1}^2$ will be close to $\sigma_n^2 \lambda_n^2$ where $\lambda_n = \|\boldsymbol{\alpha}_n\|$. Iterating, the variance of $X_n$ will grow like a some constant $C^2$ times $\lambda_{n-1}^2 \cdots \lambda_0^2$, so when $\boldsymbol{\alpha}_n \to \boldsymbol{\alpha}$, like $C^2\lambda^{2n}$. Condition 4.1 assures that $Z_n$ is small relative to $X_n$ in that its variance grows at a slower rate. This condition was assumed in Wehr and Woo (2001) for deriving a normal limiting law for the standardized sequence generated by (4.39).

**Condition 4.1** *The nonzero sequence of vectors $\boldsymbol{\alpha}_n \in \mathbb{R}^k$, $k \geq 2$, converges to $\boldsymbol{\alpha}$, not equal to any multiple of a standard basis vector. With $\lambda = \|\boldsymbol{\alpha}\|$, there exist $0 < \delta_1 < \delta_2 < 1$ and positive constants $C_{X,2}, C_{Z,2}$ such that for all $n$,*

$$\begin{aligned}
\operatorname{Var}(X_n) &\geq C_{X,2}^2 \lambda^{2n}(1-\delta_1)^{2n}, \\
\operatorname{Var}(Z_n) &\leq C_{Z,2}^2 \lambda^{2n}(1-\delta_2)^{2n}.
\end{aligned}$$

Bounds on the distance between $X_n$ and the normal can be provided under the following additional conditions on the fourth order moments of $X_n$ and $Z_n$. Condition 4.2 on the higher order moments is satisfied under the same averaging assumption on $F$ used in Wehr and Woo (2001) to guarantee Condition 4.1 for weak convergence to the normal.

**Condition 4.2** *With $\delta_1$ and $\delta_2$ as in Condition 4.1, there exists $\delta_3 \geq 0$ and $\delta_4 \geq 0$ such that*

$$\phi_1 = \frac{(1-\delta_2)(1+\delta_3)^3}{(1-\delta_1)^4} < 1 \quad \textit{and} \quad \phi_2 = \left(\frac{1-\delta_4}{1-\delta_1}\right)^2 < 1,$$

*and constants $C_{X,4}, C_{Z,4}$ such that*

$$\begin{aligned}
E(X_n - EX_n)^4 &\leq C_{X,4}^4 \lambda^{4n}(1+\delta_3)^{4n}, \\
E(Z_n - EZ_n)^4 &\leq C_{Z,4}^4 \lambda^{4n}(1-\delta_4)^{4n}.
\end{aligned}$$

The following is our main result on $L^1$ bounds for approximately linear recursions.

**Theorem 4.6** *Let $X_0$ be a random variable with variance $\sigma_0^2 \in (0,\infty)$ and*

$$X_{n+1} = \boldsymbol{\alpha}_n \cdot \mathbf{X}_n + Z_n \quad \text{for } n \geq 0 \tag{4.40}$$

*with $\boldsymbol{\alpha}_n \in \mathbb{R}^k$, $\lambda_n = \|\boldsymbol{\alpha}_n\| \neq 0$ and $\mathbf{X}_n$ a vector in $\mathbb{R}^k$ with independent components distributed as $X_n$ with mean $c_n$ and finite, non-zero variance $\sigma_n^2$. Set $Y_0 = 0$ and $\mathbf{W}_n = (\mathbf{X}_n - E\mathbf{X}_n)/\sigma_n$, and for $n \geq 0$ let*

$$W_n = \frac{X_n - c_n}{\sigma_n}, \qquad Y_{n+1} = \frac{\boldsymbol{\alpha}_n}{\lambda_n} \cdot \mathbf{W}_n, \tag{4.41}$$

*and*

$$\beta_n = E|W_n - Y_n| + \frac{1}{2}E\left|W_n^3 - Y_n^3\right|. \tag{4.42}$$

*If there exist $(\beta,\varphi) \in (0,1)^2$ such that*

$$\limsup_{n\to\infty} \frac{\beta_n}{\beta^n} < \infty \tag{4.43}$$

*and $\varphi_n = \varphi(\boldsymbol{\alpha}_n)$ in* (4.16) *satisfies*

$$\limsup_{n\to\infty} \varphi_n = \varphi, \tag{4.44}$$

*then with $\gamma = \beta$ when $\beta > \varphi$, and for any $\gamma \in (\varphi, 1)$ when $\beta \leq \varphi$, there exists $C$ such that*

$$\left\|\mathcal{L}(W_n) - \mathcal{L}(Z)\right\|_1 \leq C\gamma^n \quad \text{for all } n \geq 0. \tag{4.45}$$

*Under Conditions* 4.1 *and* 4.2, *the bound* (4.45) *holds for all $\gamma \in (\max(\beta,\varphi),1)$ with $\beta = \max\{\phi_1,\phi_2\} < 1$ and $\varphi = \sum_{i=1}^k |\alpha_i|^3/\lambda^3 < 1$ where $\boldsymbol{\alpha}$ and $\lambda$ are the limiting values of $\boldsymbol{\alpha}_n$ and $\lambda_n$, respectively.*

*Proof* Let $f \in \mathcal{F}$ with $\mathcal{F}$ given by (4.37). Then $f'$ is absolutely continuous with

$$\left|f''(w)\right| \leq 1, \quad \text{and in addition} \quad \left|f'(w)\right| \leq |w| \quad \text{and} \quad \left|f(w)\right| \leq w^2/2.$$

Letting $h$ be given by

$$h(w) = f'(w) - wf(w) \tag{4.46}$$

we have $Nh = 0$ by Lemma 2.1. Differentiation yields

$$h'(w) = f''(w) - wf'(w) - f(w),$$

and therefore

$$\left|h'(w)\right| \leq \left(1 + \frac{3}{2}w^2\right). \tag{4.47}$$

Letting

$$r_n = \frac{\lambda_n \sigma_n}{\sigma_{n+1}} \quad \text{and} \quad T_n = \frac{\sigma_n}{\sigma_{n+1}}\left(\frac{Z_n - EZ_n}{\sigma_n}\right) \tag{4.48}$$

and using (4.41), write the recursion (4.40) as

$$\begin{aligned} W_{n+1} &= \frac{X_{n+1} - EX_{n+1}}{\sigma_{n+1}} \\ &= \frac{\sigma_n}{\sigma_{n+1}}\left(\boldsymbol{\alpha}_n \cdot \frac{\mathbf{X}_n - E\mathbf{X}_n}{\sigma_n} + \frac{Z_n - EZ_n}{\sigma_n}\right) \\ &= \frac{\sigma_n}{\sigma_{n+1}}\left(\boldsymbol{\alpha}_n \cdot \mathbf{W}_n + \frac{Z_n - EZ_n}{\sigma_n}\right) \\ &= r_n Y_{n+1} + T_n. \end{aligned} \tag{4.49}$$

Now by (4.47) and the definition of $\beta_n$ in (4.42),

$$E\big|h(W_n) - h(Y_n)\big| = E\left|\int_{Y_n}^{W_n} h'(u)du\right| \le \beta_n.$$

Now by (2.51), that $\mathrm{Var}(W_{n+1}) = 1$, (4.46) and $Nh = 0$, we have

$$\begin{aligned} \big|Ef'(W_{n+1}) - Ef'\big(W_{n+1}^*\big)\big| &= \big|Ef'(W_{n+1}) - W_{n+1}f(W_{n+1})\big| \\ &= \big|Eh(W_{n+1}) - Nh\big| \\ &= \big|E\big(h(W_{n+1}) - h(Y_{n+1}) + h(Y_{n+1}) - Nh\big)\big| \\ &\le \beta_{n+1} + \big|Eh(Y_{n+1}) - Nh\big| \\ &= \beta_{n+1} + \big|E\big(f'\big(Y_{n+1}^*\big) - f'(Y_{n+1})\big)\big| \\ &\le \beta_{n+1} + \big\|Y_{n+1}^* - Y_{n+1}\big\|_1 \quad \text{by (4.38)} \\ &\le \beta_{n+1} + \varphi_n\big\|W_n^* - W_n\big\|_1 \quad \text{by Lemma 4.1.} \end{aligned}$$

Taking supremum over $f \in \mathcal{F}$ on the left hand side, using (4.38) again and letting $d_n = \|W_n^* - W_n\|_1$ we obtain, for all $n \ge 0$,

$$d_{n+1} \le \beta_{n+1} + \varphi_n d_n.$$

Iteration yields that for all $n, n_0 \ge 0$,

$$d_{n_0+n} \le \sum_{j=n_0+1}^{n_0+n}\left(\prod_{i=j}^{n_0+n-1}\varphi_i\right)\beta_j + \left(\prod_{i=n_0}^{n_0+n-1}\varphi_i\right)d_{n_0}. \tag{4.50}$$

Now suppose the bounds (4.43) and (4.44) hold on $\beta_n$ and $\varphi_n$, respectively, and recall the choice of $\gamma$. When $\beta > \varphi$ take $\overline{\varphi} \in (\varphi, \beta)$ so that $\varphi < \overline{\varphi} < \beta = \gamma$; when $\beta \le \varphi$ take $\overline{\varphi} \in (\varphi, \gamma)$ so that $\beta \le \varphi < \overline{\varphi} < \gamma$. Then for any $\overline{B} > \limsup_n \beta_n/\beta^n$ there exists $n_0$ such that for all $n \ge n_0$

$$\beta_n \le \overline{B}\beta^n \quad \text{and} \quad \varphi_n \le \overline{\varphi}.$$

Applying these inequalities in (4.50) and summing yields, for all $n \ge 0$,

$$d_{n+n_0} \le \overline{B}\beta^{n_0+1}\left(\frac{\beta^n - \overline{\varphi}^n}{\beta - \overline{\varphi}}\right) + \overline{\varphi}^n d_{n_0}.$$

Since $\max(\beta, \overline{\varphi}) \le \gamma$, for some $C$ we have that $d_n \le C\gamma^n$ for all $n \ge n_0$, and by enlarging $C$ if necessary, for all $n \ge 0$. Now (4.45) follows from Theorem 4.1.

To prove the final claim under Conditions 4.1 and 4.2 it suffices to show that (4.43) and (4.44) hold with $\beta = \max\{\phi_1, \phi_2\}$ and $\varphi = \sum_{i=1}^{k} |\alpha_i|^3/\lambda^3 < 1$ where $\boldsymbol{\alpha}$ is the limiting value of $\boldsymbol{\alpha}_n$. Lemma 6 of Wehr and Woo (2001) gives that the limit as $n \to \infty$ of $\sigma_n/(\lambda_0 \cdots \lambda_{n-1})$ exists in $(0, \infty)$, and therefore that

$$\lim_{n\to\infty} r_n = 1 \quad \text{and} \quad \lim_{n\to\infty} \frac{\sigma_{n+1}}{\sigma_n} = \lambda. \tag{4.51}$$

Referring to the definition of $T_n$ in (4.48) and using (4.51) and Conditions 4.1 and 4.2, there exist positive constants $C_{T,2}$, $C_{T,4}$ such that

$$\left(E|T_n|\right)^2 \le ET_n^2 = \mathrm{Var}(T_n) = \left(\frac{\sigma_n}{\sigma_{n+1}}\right)^2 \frac{\mathrm{Var}(Z_n)}{\mathrm{Var}(X_n)} \le C_{T,2}^2 \left(\frac{1-\delta_2}{1-\delta_1}\right)^{2n},$$
$$\text{and} \quad ET_n^4 = \left(\frac{\sigma_n}{\sigma_{n+1}}\right)^4 E\left(\frac{Z_n - EZ_n}{\sigma_n}\right)^4 \le C_{T,4}^4 \left(\frac{1-\delta_4}{1-\delta_1}\right)^{4n}.$$

By independence, a simple bound and Condition 4.2 for the second inequality we have

$$\left(E|Y_n|\right)^2 \le EY_n^2 = \mathrm{Var}(Y_n) = 1, \quad \text{and}$$
$$EY_{n+1}^4 \le 6E\left(\frac{X_n - c_n}{\sigma_n}\right)^4 \le 6C_{X,4}^4 \left(\frac{1+\delta_3}{1-\delta_1}\right)^{4n}.$$

Using the recursion (4.39) and writing $\sigma_{Z_n}^2 = \mathrm{Var}(Z_n)$, we have $\sigma_{n+1} \le \lambda_n \sigma_n + \sigma_{Z_n}$ and $\lambda_n \sigma_n \le \sigma_{n+1} + \sigma_{Z_n}$, hence with $C_{r,1} = C_{T,2}$ we have

$$|\lambda_n \sigma_n - \sigma_{n+1}| \le \sigma_{Z_n} \quad \text{so} \quad |r_n - 1| \le C_{r,1} \left(\frac{1-\delta_2}{1-\delta_1}\right)^n.$$

Now, since $|r_n^p - 1| = |(r_n - 1 + 1)^p - 1| \le \sum_{j=1}^{p} \binom{p}{j} |r_n - 1|^j$ and $0 < \delta_1 < \delta_2 < 1$, there are constants $C_{r,p}$ such that

$$\left|r_n^p - 1\right| \le C_{r,p} \left(\frac{1-\delta_2}{1-\delta_1}\right)^n, \quad p = 1, 2, \ldots.$$

Now considering the first term of $\beta_n$ in (4.42), recalling (4.49),

$$\begin{aligned} E|W_{n+1} - Y_{n+1}| &= E|(r_n - 1)Y_{n+1} + T_n| \\ &\le |r_n - 1| E|Y_{n+1}| + E|T_n| \le (C_{r,1} + C_{T,2}) \left(\frac{1-\delta_2}{1-\delta_1}\right)^n, \end{aligned}$$

which is upper bounded by a constant times $\phi_1^{n+1}$.

For the second term of (4.42) we have

$$E\left|W_{n+1}^3 - Y_{n+1}^3\right| = E\left|(r_n^3 - 1)Y_{n+1}^3 + 3r_n^2 Y_{n+1}^2 T_n + 3r_n Y_{n+1} T_n^2 + T_n^3\right|.$$

Applying the triangle inequality, the first term which results may be bounded as

$$\begin{aligned} \left|r_n^3 - 1\right| E\left|Y_{n+1}^3\right| &\le \left|r_n^3 - 1\right| \left(EY_{n+1}^4\right)^{3/4} \\ &\le 6^{3/4} C_{r,3} C_{X,4}^3 \left(\frac{(1-\delta_2)(1+\delta_3)^3}{(1-\delta_1)^4}\right)^n, \end{aligned}$$

which is smaller than some constant times $\phi_1^{n+1}$.

Since $r_n \to 1$ by (4.51), it suffices to bound the next two terms without the factor of $r_n$. Thus,

$$E|Y_{n+1}^2 T_n| \le \sqrt{EY_{n+1}^4 ET_n^2} \le 6^{1/2} C_{X,4}^2 C_{T,2} \left(\frac{(1-\delta_2)(1+\delta_3)^2}{(1-\delta_1)^3}\right)^n,$$

which is less than a constant times $\phi_1^{n+1}$. Lastly,

$$E|Y_{n+1} T_n^2| \le \sqrt{EY_{n+1}^2 ET_n^4} \le C_{T,4}^2 \left(\frac{1-\delta_4}{1-\delta_1}\right)^{2n} = C_{T,4}^2 \phi_2^n \quad \text{and}$$

$$E|T_n^3| \le (ET_n^4)^{3/4} \le C_{T,4}^3 \left(\frac{1-\delta_4}{1-\delta_1}\right)^{3n} \le C_{T,4}^3 \phi_2^{3n/2}.$$

Hence (4.43) holds with the given $\beta$.

Since $\boldsymbol{\alpha}_n \to \boldsymbol{\alpha}$, we have $\varphi_n \to \varphi$, verifying (4.44). Under Condition 4.1, $\boldsymbol{\alpha}$ is not a scalar multiple of a standard basis vector and hence $\varphi < 1$ by Lemma 4.1. As the first part of the theorem shows that (4.43) and (4.44) imply that (4.45) holds for all $\gamma \in (\max(\beta, \varphi), 1)$, the last claim is shown. □

We note that this proof reverses the way in which the Stein equation is typically applied, where $h$ is given and the properties of $f$ are dependent on those assumed for $h$. In particular, in the proof of Theorem 4.6 the function $f \in \mathcal{F}$ is taken as given, and the function $h$, whose properties are determined by $f$ through (2.4), plays only an auxiliary role.

### 4.2.2 Normal Bounds for Hierarchical Sequences

The following result, extending Proposition 9 of Wehr and Woo (2001) to higher orders, is used to show that the moment bounds of Conditions 4.1 and 4.2 are satisfied under the hypotheses of Theorem 4.4, allowing Theorem 4.6 to be invoked. The dependence of the constants in (4.53) and (4.54) on $\epsilon$ is suppressed for notational simplicity.

**Lemma 4.3** *Let the hypotheses of Theorem 4.4 be satisfied for the recursion*

$$X_{n+1} = F(\mathbf{X}_n) \quad \text{for } n \ge 0.$$

*With $c_n = EX_n$ and $\boldsymbol{\alpha}_n = F'(c_n)$, define*

$$Z_n = F(\mathbf{X}_n) - \boldsymbol{\alpha}_n \cdot \mathbf{X}_n. \tag{4.52}$$

*Then with $\boldsymbol{\alpha}$ the limit of $\boldsymbol{\alpha}_n$ and $\lambda = \|\boldsymbol{\alpha}\|$, for any integer $p \ge 1$ and $\epsilon > 0$, there exists constants $C_{X,p}, C_{Z,p}$ such that*

$$E|Z_n - EZ_n|^p \le C_{Z,p}^p (\lambda + \epsilon)^{2pn} \quad \text{for all } n \ge 0, \tag{4.53}$$

*and*

$$E|X_n - c_n|^p \le C_{X,p}^p (\lambda + \epsilon)^{pn} \quad \text{for all } n \ge 0. \tag{4.54}$$

*Proof* Expanding $F(\mathbf{X}_n)$ around the mean $\mathbf{c}_n = \mathbf{1}_k c_n$ of $\mathbf{X}_n$ yields

$$F(\mathbf{X}_n) = F(\mathbf{c}_n) + \sum_{i=1}^{k} \alpha_{n,i}(X_{n,i} - c_n) + R_2(\mathbf{c}_n, \mathbf{X}_n), \tag{4.55}$$

where

$$R_2(\mathbf{c}_n, \mathbf{X}_n) = \sum_{i,j=1}^{k} \int_0^1 (1-t) \frac{\partial^2 F}{\partial x_i \partial x_j} \big(\mathbf{c}_n + t(\mathbf{X}_n - \mathbf{c}_n)\big)(X_{n,i} - c_n)(X_{n,j} - c_n) dt.$$

Since the second partials of $F$ are continuous on the compact set $\mathcal{D} = [a, b]^k$, with $\|\cdot\|$ the supremum norm on $\mathcal{D}$ we have

$$B = \frac{1}{2} \max_{i,j} \left\| \frac{\partial^2 F}{\partial x_i \partial x_j} \right\| < \infty,$$

and therefore

$$\big|R_2(\mathbf{c}_n, \mathbf{X}_n)\big| \le B \sum_{i,j=1}^{k} \big|(X_{n,i} - c_n)(X_{n,j} - c_n)\big|. \tag{4.56}$$

Using (4.52), (4.55) and (4.56), we have for all $p \ge 1$

$$\begin{aligned}
&E|Z_n - EZ_n|^p \\
&\quad = E\left| F(\mathbf{X}_n) - EF(\mathbf{X}_n) - \sum_{i=1}^{k} \alpha_{n,i}(X_{n,i} - c_n) \right|^p \\
&\quad = E\big|F(\mathbf{c}_n) - EF(\mathbf{X}_n) + R_2(\mathbf{c}_n, \mathbf{X}_n)\big|^p \\
&\quad \le 2^{p-1}\left( \big|F(\mathbf{c_n}) - EF(\mathbf{X}_n)\big|^p + B^p E\left( \sum_{i,j} \big|(X_{n,i} - c_n)(X_{n,j} - c_n)\big| \right)^p \right).
\end{aligned} \tag{4.57}$$

For the first term of (4.57), again using (4.56),

$$\begin{aligned}
\big|F(\mathbf{c_n}) - EF(\mathbf{X_n})\big|^p &= \big|ER_2(\mathbf{c}_n, \mathbf{X}_n)\big|^p \\
&\le B^p \left( E \sum_{i,j} \big|(X_{n,i} - c_n)(X_{n,j} - c_n)\big| \right)^p \\
&\le B^p \left( \sum_{i,j} E(X_n - c_n)^2 \right)^p \\
&= B^p k^{2p} \big[E(X_n - c_n)^2\big]^p \\
&\le B^p k^{2p} E(X_n - c_n)^{2p},
\end{aligned} \tag{4.58}$$

using Jensen's inequality for the final step.

Similarly, for the second term in (4.57),

$$
\begin{aligned}
E\left(\sum_{i,j}\left|(X_{n,i}-c_n)(X_{n,j}-c_n)\right|\right)^p &\le k^p E\left(\sum_{i=1}^k (X_{n,i}-c_n)^2\right)^p \\
&\le k^{2p-1} E\left(\sum_{i=1}^k (X_{n,i}-c_n)^{2p}\right) \\
&= k^{2p} E(X_n-c_n)^{2p}. \qquad (4.59)
\end{aligned}
$$

Applying the bounds (4.58) and (4.59) in (4.57) we obtain for all $p \ge 1$, with $C_p = 2^p B^p k^{2p}$,

$$
E|Z_n - EZ_n|^p \le C_p E(X_n - c_n)^{2p}. \qquad (4.60)
$$

To demonstrate the proposition it therefore suffices to prove (4.54).

Note that since $X_n \to c$ for $X_n \in [a,b]$ the bounded convergence theorem implies that $c_n = EX_n \to c$. Lemma 8 of Wehr and Woo (2001) shows that if $F : [a,b]^k \to [a,b]$ is an averaging function and there exists $c \in [a,b]$ such that $X_n \to_p c$, then

$$
\forall \epsilon \in (0,1) \exists M \quad \text{such that} \quad \text{for all } n \ge 0, \quad P\big(|X_n - c| > \epsilon\big) \le M\epsilon^n. \qquad (4.61)
$$

In particular the large deviation estimate (4.61) holds under the given assumptions, and therefore also with $c$ replaced by $c_n$.

We now show that if $a_n, n = 0, 1, \ldots$ is a sequence such that for every $\epsilon > 0$ there exists $M$ and $n_0 \ge 0$ such that

$$
a_{n+1} \le (\lambda + \epsilon)^p a_n + M(\lambda + \epsilon)^{p(n+1)} \quad \text{for all } n \ge n_0, \qquad (4.62)
$$

then for all $\epsilon > 0$ there exists $C$ such that

$$
a_n \le C(\lambda + \epsilon)^{pn} \quad \text{for all } n \ge 0. \qquad (4.63)
$$

Let $\epsilon > 0$ be given, and let $M$ and $n_0$ be such that (4.62) holds with $\epsilon$ replaced by $\epsilon/2$. Setting

$$
C = \max\left\{\frac{a_{n_0}}{(\lambda+\epsilon)^{n_0}}, \frac{M}{1 - (\frac{\lambda+\epsilon/2}{\lambda+\epsilon})^p}\right\},
$$

it is trivial that (4.63) holds for $n = n_0$, and a direct induction shows (4.63) holds for all $n \ge n_0$. By increasing $C$ if necessary, we have that (4.63) holds for all $n \ge 0$.

Unqualified statements in the remainder of the proof below involving $\epsilon$ and $M$ are to be read to mean that for every $\epsilon > 0$ there exists $M$ such that the statement holds for all $n$; the values of $\epsilon$ and $M$ are not necessarily the same at each occurrence, even from line to line. By (4.61) and that $X_n \in [a,b]$ we have

$$
\begin{aligned}
E(X_n - c_n)^{2p} &= E\big[(X_n - c_n)^{2p}; |X_n - c_n| \le \epsilon\big] + E\big[(X_n - c_n)^{2p}; |X_n - c_n| > \epsilon\big] \\
&\le \epsilon^p E|X_n - c_n|^p + M\epsilon^n.
\end{aligned}
$$

From (4.60), this inequality gives that

$$E|Z_n - EZ_n|^p \le \epsilon^p E|X_n - c_n|^p + M\epsilon^n. \tag{4.64}$$

Since for all $\epsilon > 0$ we have

$$\lim_{x\to\infty} (x+1)^p - (1+\epsilon)x^p = -\infty \quad \text{and therefore}$$
$$\sup_{x\ge 0}(x+1)^p - (1+\epsilon)x^p < \infty,$$

substituting $x = |w|/|z|$ when $z \ne 0$ we see that there exists $M$ such that for all $w, z$ we have

$$|w+z|^p \le (1+\epsilon)|w|^p + M|z|^p,$$

noting that the inequality holds trivially with $M = 1$ for $z = 0$. Now applying definition (4.52),

$$E|X_{n+1} - c_{n+1}|^p \le (1+\epsilon)E\left|\sum_{i=1}^k \alpha_{n,i}(X_{n,i} - c_n)\right|^p + ME|Z_n - EZ_n|^p. \tag{4.65}$$

Specializing (4.65) to the case $p = 2$ gives

$$E(X_{n+1} - c_{n+1})^2 \le (\lambda+\epsilon)^2 E(X_n - c_n)^2 + ME(Z_n - EZ_n)^2.$$

Applying (4.64) with $p = 2$ to this inequality yields

$$\begin{aligned} E(X_{n+1} - c_{n+1})^2 &\le (\lambda+\epsilon)^2 E(X_n - c_n)^2 + M\epsilon^{2n+2} \\ &\le (\lambda+\epsilon)^2 E(X_n - c_n)^2 + M(\lambda+\epsilon)^{2(n+1)}. \end{aligned}$$

Hence inequality (4.62), and therefore (4.63), are true for $a_n = E(X_n - c_n)^2$ and $p = 2$, yielding (4.54) for $p = 2$. Now Hölder's inequality shows that (4.54) is also true for $p = 1$.

Now let $p > 2$ be an integer and suppose that (4.54) is true for all integers $q$, $1 \le q < p$. In expanding the first term in (4.65) we let $\mathbf{p} = (p_1, \dots, p_k)$ denote a multi-index and $|\mathbf{p}| = \sum_i p_i$. Use the induction hypotheses, and (4.22) of Lemma 4.2 in (4.66), to obtain, with $A_{X,p} = \max_{q<p} C_{X,q}$ and $B_{X,p}^p = k^{p-1}A_{X,p}^p$, that

$$\begin{aligned} &E\left|\sum_{i=1}^k \alpha_{n,i}(X_{n,i} - c_n)\right|^p \\ &\le \sum_{i=1}^k |\alpha_{n,i}|^p E|X_{n,i} - c_n|^p + \sum_{|\mathbf{p}|=p, 0\le p_i<p} \binom{p}{\mathbf{p}} E\prod_{i=1}^k |\alpha_{n,i}|^{p_i}|X_{n,i} - c_n|^{p_i} \\ &\le E|X_n - c_n|^p \sum_{i=1}^k |\alpha_{n,i}|^p + \sum_{|\mathbf{p}|=p,\, 0\le p_i<p} \binom{p}{\mathbf{p}} \prod_{i=1}^k |\alpha_{n,i}|^{p_i} C_{X,p_i}^{p_i}(\lambda+\epsilon)^{p_i n} \\ &\le E|X_n - c_n|^p \sum_{i=1}^k |\alpha_{n,i}|^p + A_{X,p}^p(\lambda+\epsilon)^{pn} \sum_{|\mathbf{p}|=p} \binom{p}{\mathbf{p}} \prod_{i=1}^k |\alpha_{n,i}|^{p_i} \end{aligned}$$

$$= E|X_n - c_n|^p \sum_{i=1}^{k} |\alpha_{n,i}|^p + A_{X,p}^p (\lambda + \epsilon)^{pn} \left( \sum_{i=1}^{k} |\alpha_{n,i}| \right)^p$$

$$\leq \sum_{i=1}^{k} |\alpha_{n,i}|^p \left( E|X_n - c_n|^p + B_{X,p}^p (\lambda + \epsilon)^{pn} \right)$$

$$\leq (\lambda + \epsilon)^p E|X_n - c_n|^p + B_{X,p}^p (\lambda + \epsilon)^{p(n+1)}. \tag{4.66}$$

Applying (4.64) and (4.66) in (4.65) gives

$$E|X_{n+1} - c_{n+1}|^p \leq (\lambda + \epsilon)^p E|X_n - c_n|^p + M(\lambda + \epsilon)^{p(n+1)},$$

from which we can conclude that (4.63) holds for $a_n = E|X_n - c_n|^p$, completing the induction on $p$. □

*Proof of Theorem 4.4* By Theorem 4.6 it suffices to show that Conditions 4.1 and 4.2 are satisfied for some $\delta_i$, $i = 1, 2, 3, 4$ satisfying $\beta < \varphi$.

By Property 1 of averaging functions, $F(\mathbf{1}_k c) = c$, and differentiation with respect to $c$ yields $\sum_{i=1}^n \alpha_i = 1$. By Property 2, monotonicity, $\alpha_i \geq 0$, and (4.24) of Lemma 4.2 yields $0 < \lambda < \varphi < 1$, using that $\boldsymbol{\alpha}$ is not a multiple of a standard basis vector.

Let $\delta_4 \in (1 - \varphi, 1 - \lambda)$. Since $\delta_4 < 1 - \lambda$ we have $\lambda^2 < \lambda(1 - \delta_4)$, and therefore there exists $\epsilon > 0$ such that $(\lambda + \epsilon)^2 < \lambda(1 - \delta_4)$. By Lemma 4.3, for $p = 2$ and $p = 4$, for this $\epsilon$ there exists $C_{Z,p}^p$ such that

$$E(Z_n - EZ_n)^p \leq C_{Z,p}^p (\lambda + \epsilon)^{2pn} \leq C_{Z,p}^p \lambda^{pn} (1 - \delta_4)^{pn}.$$

Hence the fourth and second moment bounds in Conditions 4.1 and 4.2 on $Z_n$ are satisfied with $\delta_4$ and $\delta_2 = \delta_4$, respectively.

Since $1 - \delta_4 < \varphi$ there $\delta_1 \in (0, \delta_2)$ and $\delta_3 > 0$ such that $\eta < \varphi$ where

$$\eta = \frac{(1 - \delta_4)(1 + \delta_3)^3}{(1 - \delta_1)^4}.$$

Proposition 10 of Wehr and Woo (2001) shows that under the assumptions of Theorem 4.4, for every $\epsilon > 0$ there exists $C_{X,2}^2$ such that

$$\mathrm{Var}(X_n) \geq C_{X,2}^2 (\lambda - \epsilon)^{2n}.$$

Taking $\epsilon = \lambda\delta_1$, we have $\mathrm{Var}(X_n)$ satisfies the lower bound in Condition 4.1. Applying Lemma 4.3 with $p = 4$ and $\epsilon = \lambda\delta_3$ we see the fourth moment bound on $X_n$ in Condition 4.2 is satisfied.

With these choices for $\delta_i$, $i = 1, \ldots, 4$, as $\eta < \varphi < 1$, we have $\phi_2 < \eta < 1$ and $\phi_1 = \eta < 1$, hence Conditions 4.1 and 4.2 are satisfied. Noting that $\beta = \max\{\phi_1, \phi_2\} = \eta < \varphi$ now completes the proof. □

### 4.2.3 Convergence Rates for the Diamond Lattice

We now apply Theorem 4.4 to hierarchical sequences generated by the diamond lattice conductivity function $F(\mathbf{x})$ in (4.30). We have already argued that Theorem 4.5 implies that $F(\mathbf{x})$ is strictly averaging on, say $[a,b]^4$, for any $0<a<b$ and choice of positive weights satisfying $F(\mathbf{1}_4)=1$, and on this domain such an $F(\mathbf{x})$ is easily seen to be twice continuously differentiable. For all such $F(\mathbf{x})$ the result of Shneiberg (1986) quoted in Sect. 4.2 shows that $X_n$ satisfies a weak law.

We now study the quantity $\varphi$ which determines the exponential decay rate of the upper bound of Theorem 4.4 to zero. The first partial derivative $\partial F(\mathbf{x})/\partial x_1$ has the form

$$\frac{\partial F(\mathbf{x})}{\partial x_1}=\frac{(w_1x_1^2)^{-1}}{((w_1x_1)^{-1}+(w_2x_2)^{-1})^2},$$

and similarly for the other partials. Hence $F'(t\mathbf{x})=F'(\mathbf{x})$ for all $t\neq 0$. As $X_n$ is a random variable on $[a,b]$ we have $c_n=EX_n\neq 0$, and therefore

$$\boldsymbol{\alpha}_n=F'(c_n\mathbf{1}_4)=F'(\mathbf{1}_4)\quad\text{for all } n\geq 0.$$

In particular, $\boldsymbol{\alpha}=\lim_{n\to\infty}\boldsymbol{\alpha}_n$ is given by

$$\boldsymbol{\alpha}=\left[\frac{w_1^{-1}}{(w_1^{-1}+w_2^{-1})^2},\frac{w_2^{-1}}{(w_1^{-1}+w_2^{-1})^2},\frac{w_3^{-1}}{(w_3^{-1}+w_4^{-1})^2},\frac{w_4^{-1}}{(w_3^{-1}+w_4^{-1})^2}\right]^{\mathsf{T}}.$$

Since we are considering the case where all the weights are positive, the vector $\boldsymbol{\alpha}$ is not a scalar multiple of a standard basis vector. Now from (4.16) we compute

$$\varphi=\lambda^{-3}\left(\frac{w_1^{-3}+w_2^{-3}}{(w_1^{-1}+w_2^{-1})^6}+\frac{w_3^{-3}+w_4^{-3}}{(w_3^{-1}+w_4^{-1})^6}\right), \tag{4.67}$$

where

$$\lambda=\left(\frac{w_1^{-2}+w_2^{-2}}{(w_1^{-1}+w_2^{-1})^4}+\frac{w_3^{-2}+w_4^{-2}}{(w_3^{-1}+w_4^{-1})^4}\right)^{1/2}.$$

As an illustration of the bounds provided by Theorem 4.4, first consider the 'side equally weighted network', the one with $\mathbf{w}=(w,w,2-w,2-w)^{\mathsf{T}}$ for $w\in[1,2)$; we recall the weights $\mathbf{w}$ refer to the bonds in the lattice traversed counterclockwise from the top in Fig. 4.1(c). The vector of weights for $w$ in this range are positive and satisfy $F(\mathbf{1}_4)=1$. For $w=1$ all weights are equal and $\boldsymbol{\alpha}=4^{-1}\mathbf{1}_4$, so $\varphi$ achieves its minimum value $1/2=1/\sqrt{k}$ with $k=4$. By Theorem 4.4, for all $\gamma\in(1/2,1)$ there exists a constant $C$ such that $\|W_n-Z\|_1\leq C\gamma^n$. The values of $\gamma$ just above $1/2$ correspond, in view of (4.36), to the rate $N^{-\theta}$ for $\theta$ just below $-\log_4 1/2=1/2$, that is, $N^{-1/2+\epsilon}$ for small $\epsilon>0$, where $N=4^n$, the number of variables at stage $n$. As $w$ increases from 1 to 2, $\varphi$ increases continuously from $1/2$ to $1/\sqrt{2}$, with $w$ approaching 2 from below corresponding to the least favorable rate for the side equally weighted network of $\theta$ just under $-\log_4 1/\sqrt{2}=1/4$, that is, of $N^{-1/4+\epsilon}$ for any $\epsilon>0$.

With only the restriction that the weights are positive and satisfy $F(\mathbf{1}_4) = 1$ consider

$$\mathbf{w} = (1 + 1/t, s, t, 1/t)^{\mathsf{T}}$$
$$\text{where } s = \left[\left(1 - (1/t + t)^{-1}\right)^{-1} - (1 + 1/t)^{-1}\right]^{-1}, \quad t > 0.$$

When $t = 1$ we have $s = 2/3$ and $\varphi = 11\sqrt{2}/27$. As $t \to \infty$, $s/t \to 1/2$ and $\boldsymbol{\alpha}$ tends to the standard basis vector $(1, 0, 0, 0)$, so $\varphi \to 1$. Since $11\sqrt{2}/27 \in (1/2, 1/\sqrt{2})$, the above two examples show that the value of $\gamma$ given by Theorem 4.4 for the diamond lattice can take any value in the range $(1/2, 1)$, corresponding to $N^{-\theta}$ for any $\theta \in (0, 1/2)$.

## 4.3 Cone Measure Projections

In this section we use Stein's method to obtain $L^1$ bounds for the normal approximation of one dimensional projections of the form

$$Y = \boldsymbol{\theta} \cdot \mathbf{X}, \tag{4.68}$$

where for some $p > 0$, the vector $\mathbf{X} \in \mathbb{R}^n$ has the cone measure distribution $C_p^n$ given in (4.71) below, and $\boldsymbol{\theta} \in \mathbb{R}^n$ is of unit length. The normal approximation of projections of random vectors in lesser and greater generality has been studied by many authors, and under a variety of metrics. In the case $p = 2$, when cone measure is uniform on the surface of the unit Euclidean sphere in $\mathbb{R}^n$, Diaconis and Freedman (1987) show that the low dimensional projections of $\mathbf{X}$ are close to normal in total variation. It is particularly easy to see in this case, and true in general, that cone measure $C_p^n$ is *coordinate symmetric*, that is,

$$(X_1, \ldots, X_n) =_d (e_1 X_1, \ldots, e_n X_n) \quad \text{for all } (e_1, \ldots, e_n) \in \{-1, 1\}^n. \tag{4.69}$$

Meckes and Meckes (2007) derive bounds using Stein's method for the normal approximation of random vectors with symmetries in general, including coordinate-symmetry, considering the supremum and total variation norm. Goldstein and Shao (2009) give $L^\infty$ bounds on the projections of coordinate symmetric random vectors of order $1/\sqrt{n}$ without applying Stein's method. Klartag (2009) proves bounds of order $1/n$ on the $L^\infty$ distance under additional conditions on the distribution of $\mathbf{X}$, including that its density be log concave. One special case of note where $\mathbf{X}$ is coordinate symmetric is when its distribution is uniform over a convex set which has symmetry with respect to all coordinate planes. For general results on the projections of vectors sampled uniformly from convex sets, see Klartag (2007) and references therein. Studying here the specific instance of the projections of cone measure allows, naturally, for the sharpening of general results about projections of coordinate symmetric vectors to this particular case.

To define cone measure let

$$S\left(\ell_p^n\right) = \left\{\mathbf{x} \in \mathbb{R}^n \colon \sum_{i=1}^n |x_i|^p = 1\right\} \quad \text{and}$$
$$B\left(\ell_p^n\right) = \left\{\mathbf{x} \in \mathbb{R}^n \colon \sum_{i=1}^n |x_i|^p \le 1\right\}. \tag{4.70}$$

Then with $\mu^n$ Lebesgue measure in $\mathbb{R}^n$, the cone measure of $A \subset S(\ell_p^n)$ is given by

$$C_p^n(A) = \frac{\mu^n([0,1]A)}{\mu^n(B(\ell_p^n))} \quad \text{where } [0,1]A = \{ta\colon\, a \in A,\ 0 \le t \le 1\}. \tag{4.71}$$

The main result in the this section on the projections of $C_p^n$ is the following.

**Theorem 4.7** *Let* $\mathbf{X}$ *have cone measure* $C_p^n$ *on the sphere* $S(\ell_p^n)$ *for some* $p > 0$ *and let*

$$Y = \sum_{i=1}^n \theta_i X_i$$

*be the one-dimensional projection of* $\mathbf{X}$ *along the direction* $\boldsymbol{\theta} \in \mathbb{R}^n$ *with* $\|\boldsymbol{\theta}\| = 1$. *Then with* $\sigma_{n,p}^2 = \mathrm{Var}(X_1)$ *and* $m_{n,p} = E|X_1|^3/\sigma_{n,p}^2$, *given in* (4.84) *and* (4.87), *respectively, and* $F$ *the distribution function of the normalized sum* $W = Y/\sigma_{n,p}$, *we have*

$$\|F - \Phi\|_1 \le \left(\frac{m_{n,p}}{\sigma_{n,p}}\right) \sum_{i=1}^n |\theta_i|^3 + \left(\frac{1}{p} \vee 1\right) \frac{4}{n+2}, \tag{4.72}$$

*where* $\Phi$ *is the cumulative distribution function of the standard normal.*

We note that by the limits in (4.84) and (4.88), the constant $m_{n,p}/\sigma_{n,p}$ that multiplies the sum in the bound (4.72) is of the order of a constant with asymptotic value

$$\lim_{n\to\infty} \frac{m_{n,p}}{\sigma_{n,p}} = \frac{\Gamma(4/p)\sqrt{\Gamma(1/p)}}{\Gamma(3/p)^{3/2}}.$$

Since, for $\boldsymbol{\theta} \in \mathbb{R}^n$ with $\|\boldsymbol{\theta}\| = 1$, we have

$$\sum |\theta_i|^3 \ge \frac{1}{\sqrt{n}},$$

the second term in (4.72) is always of smaller order than the first, so the decay rate of the bound to zero is determined by $\sum_i |\theta_i|^3$. The minimal rate $1/\sqrt{n}$ is achieved when $\theta_i = 1/\sqrt{n}$.

In the special cases $p = 1$ and $p = 2$, $C_p^n$ is uniform on the simplex $\sum_{i=1}^n |x_i| = 1$ and the unit Euclidean sphere $\sum_{i=1}^n x_i^2 = 1$, respectively. By (4.84) and (4.87) for $p = 1$,

$$\sigma_{n,1}^2 = \frac{2}{n(n+1)} \quad \text{and} \quad m_{n,1} = \frac{3}{n+2},$$

and, using also (4.88) for $p = 2$,

$$\sigma_{n,2}^2 = \frac{1}{n} \quad \text{and} \quad m_{n,2} \le \sqrt{\frac{3}{n+2}};$$

these relations yield

$$\frac{m_{n,1}}{\sigma_{n,1}} = 3\sqrt{\frac{n(n+1)}{2(n+2)^2}} \le \frac{3}{\sqrt{2}} \quad \text{and} \quad \frac{m_{n,2}}{\sigma_{n,2}} \le \sqrt{\frac{3n}{n+2}} \le \sqrt{3}.$$

Substituting into (4.72) now gives

$$\|F - \Phi\|_1 \le \frac{3}{\sqrt{p+1}} \sum_{i=1}^{n} |\theta_i|^3 + \frac{4}{n+2} \quad \text{for } p \in \{1, 2\}. \tag{4.73}$$

### 4.3.1 Coupling Constructions for Coordinate Symmetric Variables and Their Projections

We generalize the construction in Proposition 2.3 to coordinate symmetric vectors, beginning by generalizing the notion of square biasing, given there, to square biasing in coordinates.

To begin, note that if $\mathbf{Y}$ is a coordinate symmetric random vector in $\mathbb{R}^n$ and $EY_i^2 < \infty$ for $i = 1, \ldots, n$, then the symmetry condition (4.69) implies

$$EY_i = -EY_i \quad \text{and} \quad EY_iY_j = -EY_iY_j \quad \text{for all } i \ne j,$$

and hence

$$EY_i = 0 \quad \text{and} \quad EY_iY_j = \sigma_i^2\delta_{ij} \quad \text{for all } i, j, \tag{4.74}$$

where $\sigma_i^2 = \mathrm{Var}(Y_i) = EY_i^2$. By removing any component which has zero variance, and lowering the dimension accordingly, we may assume without loss of generality that $\sigma_i^2 > 0$ for all $i = 1, \ldots, n$. For such $\mathbf{Y}$, for all $i = 1, \ldots, n$, we claim there exists a distribution $\mathbf{Y}^i$ such that for all functions $f : \mathbb{R}^n \to \mathbb{R}$ for which the expectation of the left hand side below exists,

$$EY_i^2 f(\mathbf{Y}) = \sigma_i^2 Ef(\mathbf{Y}^i), \tag{4.75}$$

and say that $\mathbf{Y}^i$ has the $\mathbf{Y}$-square bias distribution in direction $i$. In particular, the distribution of $\mathbf{Y}^i$ is absolutely continuous with respect to $\mathbf{Y}$ with

$$dF^i(\mathbf{y}) = \frac{y_i^2}{\sigma_i^2} dF(\mathbf{y}). \tag{4.76}$$

By specializing (4.75) to the case where $f$ depends only on $Y_i$, we see, in the language of Proposition 2.3, that $Y_i^i =_d Y_i^\square$, that is, that $Y_i^i$ has the $Y_i$-square bias distribution.

Proposition 4.3 shows how to construct the zero bias distribution $Y^*$ for the sum $Y$ of the components of a coordinate-symmetric vector in terms of $\mathbf{Y}^i$ and a random index in a way that parallels the construction for size biasing given in Proposition 2.2. Again we let $\mathcal{U}[a, b]$ denote the uniform distribution on $[a, b]$.

**Proposition 4.3** *Let* $\mathbf{Y} \in \mathbb{R}^n$ *be a coordinate-symmetric random vector with* $\mathrm{Var}(Y_i) = \sigma_i^2 \in (0, \infty)$ *for all* $i = 1, 2, \ldots, n$, *and*

$$Y = \sum_{i=1}^{n} Y_i.$$

*Let* $\mathbf{Y}^i, i = 1, \ldots, n$, *have the square bias distribution given in* (4.75), $I$ *a random index with distribution*

$$P(I = i) = \frac{\sigma_i^2}{\sum_{j=1}^{n} \sigma_j^2} \tag{4.77}$$

*and* $U_i \sim \mathcal{U}[-1, 1]$, *with* $\mathbf{Y}^i, I$ *and* $U_i$ *mutually independent for all* $i = 1, \ldots, n$. *Then*

$$Y^* = U_I Y_I^I + \sum_{j \neq I} Y_j^I \tag{4.78}$$

*has the* $Y$*-zero bias distribution.*

*Proof* Let $f$ be an absolutely continuous function with $E|Yf(Y)| < \infty$. Staring with the given form of $Y^*$ then averaging over the index $I$, integrating out the uniform variable $U_i$ and applying (4.75) and (4.69) we obtain

$$\begin{aligned}
\sigma^2 Ef'(Y^*) &= \sigma^2 Ef'\Big(U_I Y_I^I + \sum_{j \neq I} Y_j^I\Big) \\
&= \sigma^2 \sum_{i=1}^{n} \frac{\sigma_i^2}{\sigma^2} Ef'\Big(U_i Y_i^i + \sum_{j \neq i} Y_j^i\Big) \\
&= \sum_{i=1}^{n} \sigma_i^2 E\left(\frac{f(Y_i^i + \sum_{j \neq i} Y_j^i) - f(-Y_i^i + \sum_{j \neq i} Y_j^i)}{2Y_i^i}\right) \\
&= \sum_{i=1}^{n} EY_i\left(\frac{f(Y_i + \sum_{j \neq i} Y_j) - f(-Y_i + \sum_{j \neq i} Y_j)}{2}\right) \\
&= \sum_{i=1}^{n} EY_i f\Big(Y_i + \sum_{j \neq i} Y_j\Big) \\
&= EYf(Y).
\end{aligned}$$

Thus, $Y^*$ has the $Y$-zero bias distribution. □

Factoring (4.76) as

$$dF^i(\mathbf{y}) = dF_i^i(y_i) dF(y_1, \ldots, y_{i-1}, y_{i+1}, \ldots, y_n | y_i)$$
$$\text{where } dF_i^i(y_i) = \frac{y_i^2 dF_i(y_i)}{\sigma_i^2} \tag{4.79}$$

provides an alternate way of seeing that $Y_i^i =_d Y_i^\square$. Moreover, it suggests a coupling between $Y$ and $Y^*$ where, given $\mathbf{Y}$, an index $I = i$ is chosen with weight proportional to the variance $\sigma_i^2$, the summand $Y_i$ is replaced by $Y_i^i$ having that summand's 'square bias' distribution and then multiplied by $U$, and, finally, the remaining variables of $\mathbf{Y}$ are perturbed, so that they achieve their original distribution conditional on the $i$th variable now taking on the value $Y_i^i$. Typically the remaining variables are changed as little as possible in order to make the coupling between $Y$ and $Y^*$ close.

Now let $\mathbf{X} \in \mathbb{R}^n$ be an exchangeable coordinate-symmetric random vector with components having finite second moments and let $\boldsymbol{\theta} \in \mathbb{R}^n$ have unit length. Then, by (4.74), the projection $Y$ of $\mathbf{X}$ along the direction $\boldsymbol{\theta}$,

$$Y = \sum_{i=1}^{n} \theta_i X_i$$

has mean zero and variance $\sigma^2$ equal to the common variance of the components of $\mathbf{X}$. To form $Y^*$ using the construction just outlined, in view of (4.79) in particular, requires a vector of random variables to be 'adjusted' according to their original distribution, conditional on one coordinate taking on a newly chosen, biased, value. Random vectors which have the 'scaling-conditional' property in Definition 4.2 can easily be so adjusted. Let $\mathcal{L}(V)$ and $\mathcal{L}(V|X=x)$ denote the distribution of $V$, and the conditional distribution of $V$ given $X = x$, respectively.

**Definition 4.2** Let $\mathbf{X} = (X_1, \ldots, X_n)$ be an exchangeable random vector and $\mathcal{D} \subset \mathbb{R}$ the support of the distribution of $X_1$. If there exists a function $g : \mathcal{D} \to \mathbb{R}$ such that $P(g(X_1) = 0) = 0$ and

$$\mathcal{L}(X_2, \ldots, X_n | X_1 = a) = \mathcal{L}\left(\frac{g(a)}{g(X_1)}(X_2, \ldots, X_n)\right) \quad \text{for all } a \in \mathcal{D}, \qquad (4.80)$$

we say that $\mathbf{X}$ is scaling $g$-conditional, or simply scaling-conditional.

Proposition 4.4 is an application of Theorem 4.1 and Proposition 4.3 to projections of coordinate symmetric, scaling-conditional vectors.

**Proposition 4.4** *Let $\mathbf{X} \in \mathbb{R}^n$ be an exchangeable, coordinate symmetric and scaling $g$-conditional random vector with finite second moment. For $\boldsymbol{\theta} \in \mathbb{R}^n$ of unit length set*

$$Y = \sum_{i=1}^{n} \theta_i X_i, \quad \sigma^2 = \mathrm{Var}(Y), \quad \textit{and} \quad F(x) = P(Y/\sigma \le x).$$

*Then any construction of $(\mathbf{X}, X_i^i)$ on a joint space for each $i = 1, \ldots, n$ with $X_i^i$ having the $X_i$-square biased distribution provides the upper bound*

$$\|F - \Phi\|_1 \le \frac{2}{\sigma} E\left|\theta_I\left(U_I X_I^I - X_I\right) + \left(\frac{g(X_I^I)}{g(X_I)} - 1\right)\sum_{j \ne I} \theta_j X_j\right|, \qquad (4.81)$$

*where $P(I = i) = \theta_i^2$ and $U_i \sim \mathcal{U}[-1, 1]$ with $\{X_i^i, X_j, j \ne i\}$, $I$ and $U_i$ mutually independent for $i = 1, 2, \ldots, n$.*

*Proof* For all $i = 1, \ldots, n$, since $\mathbf{X}$ is scaling $g$-conditional, given $\mathbf{X}$ and $X_i^i$ with the $X_i$-square bias distribution, by (4.79) and (4.80) the vector

$$\mathbf{X}^i = \left(\frac{g(X_i^i)}{g(X_i)} X_1, \ldots, \frac{g(X_i^i)}{g(X_i)} X_{i-1}, X_i^i, \frac{g(X_i^i)}{g(X_i)} X_{i+1}, \ldots, \frac{g(X_i^i)}{g(X_i)} X_n\right)$$

has the $\mathbf{X}$-square bias distribution in direction $i$ as given in (4.75), that is, for every $h$ for which the expectation on the left-hand side below exists,

$$EX_i^2 h(\mathbf{X}) = EX_i^2 Eh(\mathbf{X}^i). \tag{4.82}$$

We now apply Proposition 4.3 to $\mathbf{Y} = (\theta_1 X_1, \ldots, \theta_n X_n)$. First, the coordinate symmetry of $\mathbf{Y}$ follows from that of $\mathbf{X}$. Next, we claim

$$\mathbf{Y}^i = (\theta_1 X_1^i, \ldots, \theta_n X_n^i)$$

has the $\mathbf{Y}$-square bias distribution in direction $i$. Given $f$, let

$$h(\mathbf{X}) = f(\theta_1 X_1, \ldots, \theta_n X_n).$$

Applying (4.82) we obtain

$$\begin{aligned}
EY_i^2 f(\mathbf{Y}) &= E\theta_i^2 X_i^2 f(\mathbf{Y}) \\
&= \theta_i^2 EX_i^2 h(\mathbf{X}) \\
&= \theta_i^2 EX_i^2 Eh(\mathbf{X}^i) \\
&= E\theta_i^2 X_i^2 Ef(\mathbf{Y}^i) \\
&= EY_i^2 Ef(\mathbf{Y}^i).
\end{aligned}$$

Since $\mathbf{X}$ is exchangeable, the variance of $Y_i$ is proportional to $\theta_i^2$ and the distribution of $I$ in (4.77) specializes to the one claimed. Lastly, as $\mathbf{Y}^i, I$ and $U_i$ are mutually independent for $i = 1, \ldots, n$, Proposition 4.3 yields that

$$Y^* = U_I Y_I^I + \sum_{j \neq I} Y_j^I$$

has the $Y$-zero bias distribution.

The difference $Y^* - Y$ is given by

$$\begin{aligned}
Y^* - Y &= U_I Y_I^I + \sum_{j \neq I} Y_j^I - \sum_{i=1}^n Y_i \\
&= U_I \theta_I X_I^I + \sum_{j \neq I} \theta_j X_j^I - \sum_{j=1}^n \theta_j X_j \\
&= \theta_I (U_I X_I^I - X_I) + \sum_{j \neq I} \theta_j (X_j^I - X_j) \\
&= \theta_I (U_I X_I^I - X_I) + \sum_{j \neq I} \theta_j \left(\frac{g(X_I^I)}{g(X_I)} - 1\right) X_j \\
&= \theta_I (U_I X_I^I - X_I) + \left(\frac{g(X_I^I)}{g(X_I)} - 1\right) \sum_{j \neq I} \theta_j X_j.
\end{aligned}$$

The proof is completed by dividing both sides by $\sigma$, applying (2.59) to yield $Y^*/\sigma = (Y/\sigma)^*$, and invoking Theorem 4.1. □

### 4.3.2 Construction and Bounds for Cone Measure

Proposition 4.5 below shows that Proposition 4.4 can be applied to cone measure. We denote the Gamma and Beta distributions with parameters $\alpha, \beta$ as $\Gamma(\alpha, \beta)$ and $B(\alpha, \beta)$, respectively. That is, with the Gamma function at $\alpha > 0$ given by

$$\Gamma(\alpha) = \int_0^\infty x^{\alpha-1} e^{-x} dx,$$

with $\beta > 0$, the density of the $\Gamma(\alpha, \beta)$ distribution is

$$\frac{x^{\alpha-1} e^{-x/\beta}}{\beta^\alpha \Gamma(\alpha)} \mathbf{1}_{\{x>0\}};$$

the density of the Beta distribution $B(\alpha, \beta)$ is given in (4.90).

**Proposition 4.5** *Let $C_p^n$ denote cone measure as given in* (4.71) *for some $n \in \mathbb{N}$ and $p > 0$.*

1. *Cone measure $C_p^n$ is exchangeable and coordinate-symmetric. For $\{G_j, \epsilon_j, j = 1, \ldots, n\}$ independent variables with $G_j \sim \Gamma(1/p, 1)$ and $\epsilon_j$ taking values $-1$ and $+1$ with equal probability, setting $G_{a,b} = \sum_{i=a}^b G_i$ we have*

$$\mathbf{X} = \left( \epsilon_1 \left( \frac{G_1}{G_{1,n}} \right)^{1/p}, \ldots, \epsilon_n \left( \frac{G_n}{G_{1,n}} \right)^{1/p} \right) \sim C_p^n. \tag{4.83}$$

2. *The common marginal distribution $X_i$ of cone measure is characterized by*

$$X_i =_d -X_i \quad \text{and} \quad |X_i|^p \sim B\big(1/p, (n-1)/p\big),$$

*and the variance $\sigma_{n,p}^2 = \mathrm{Var}(X_i)$ is given by*

$$\sigma_{n,p}^2 = \frac{\Gamma(3/p)\Gamma(n/p)}{\Gamma(1/p)\Gamma((n+2)/p)} \tag{4.84}$$

*and satisfies*

$$\lim_{n\to\infty} n^{2/p} \sigma_{n,p}^2 = \frac{p^{2/p}\Gamma(3/p)}{\Gamma(1/p)}.$$

3. *The square bias distribution $X_i^\square$ of $X_i$ is characterized by*

$$X_i^\square =_d -X_i^\square \quad \text{and} \quad \big|X_i^\square\big|^p \sim B\big(3/p, (n-1)/p\big). \tag{4.85}$$

*Letting $\{G_j, G_j', \epsilon_j, j = 1, \ldots, n\}$ be independent variables with $G_j \sim \Gamma(1/p, 1)$, $G_j' \sim \Gamma(2/p, 1)$ and $\epsilon_j$ taking values $-1$ and $+1$ with equal probability, for each $i = 1, \ldots, n$, a construction of $(\mathbf{X}, X_i^\square)$ on a joint space is given by the representation of $\mathbf{X}$ in* (4.83) *along with*

$$X_i^i = \epsilon_i \left( \frac{G_i + G_i'}{G_{1,n} + G_i'} \right)^{1/p}. \tag{4.86}$$

*The mean* $m_{n,p} = E|X_i^i| = E|X_i^3|/\sigma_{n,p}^2$ *for all* $i = 1, \dots, n$ *is given by*

$$m_{n,p} = \frac{\Gamma(4/p)\Gamma((n+2)/p)}{\Gamma(3/p)\Gamma((n+3)/p)} \tag{4.87}$$

*and satisfies*

$$\lim_{n\to\infty} n^{1/p} m_{n,p} = \frac{p^{1/p}\Gamma(4/p)}{\Gamma(3/p)} \quad \text{and} \quad m_{n,p} \le \left(\frac{3}{n+2}\right)^{1/(p\vee 1)}. \tag{4.88}$$

4. *Cone measure* $C_p^n$ *is scaling* $(1 - |x|^p)^{1/p}$ *conditional.*

The proof of Proposition 4.5 is deferred to the end of this section. Before proceeding to Theorem 4.7, we remind the reader of the following known facts about the Gamma and Beta distributions; see Bickel and Doksum (1977), Theorem 1.2.3 for the case $n = 2$ of the first claim, the extension to general $n$ and the following claim being straightforward. For $\gamma_i \sim \Gamma(\alpha_i, \beta), i = 1, \dots, n$, independent with $\alpha_i > 0$ and $\beta > 0$,

$$\gamma_1 + \gamma_2 \sim \Gamma(\alpha_1 + \alpha_2, \beta), \quad \frac{\gamma_1}{\gamma_1 + \gamma_2} \sim B(\alpha_1, \alpha_2), \tag{4.89}$$

$$\text{and} \quad \left(\frac{\gamma_1}{\sum_{i=1}^n \gamma_i}, \dots, \frac{\gamma_n}{\sum_{i=1}^n \gamma_i}\right) \quad \text{and} \quad \sum_{i=1}^n \gamma_i \quad \text{are independent;}$$

the Beta distribution $B(\alpha, \beta)$ has density

$$p_{\alpha,\beta}(u) = \frac{\Gamma(\alpha+\beta)}{\Gamma(\alpha)\Gamma(\beta)} u^{\alpha-1}(1-u)^{\beta-1}\mathbf{1}_{u\in[0,1]}$$
$$\text{and } \kappa > 0 \text{ moment} \frac{\Gamma(\alpha+\kappa)\Gamma(\alpha+\beta)}{\Gamma(\alpha+\beta+\kappa)\Gamma(\alpha)}. \tag{4.90}$$

*Proof of Theorem 4.7* Using Proposition 4.5, we apply Proposition 4.4 for $\mathbf{X}$ with $g(x) = (1 - |x|^p)^{1/p}$ and the joint construction of $(\mathbf{X}, X_i^i)$ given in item 3. Note that Proposition 4.2 applies, using the notation there, with $V \sim \mathcal{U}[0, 1]$, independent of all other variables, $U_i = \epsilon_i V$, and

$$\overline{X}_i = \left(\frac{G_i}{G_{1,n}}\right)^{1/p} \quad \text{and} \quad \overline{Y}_i = \left(\frac{G_i + G_i'}{G_{1,n} + G_i'}\right)^{1/p}.$$

Applying the triangle inequality on (4.81) yields the bound on the $L^1$ norm $\|F - \Phi\|_1$ of

$$\frac{2}{\sigma_{n,p}}\left(E\left|\theta_I(U_I X_I^I - X_I)\right| + E\left|\left(\frac{g(X_I^I)}{g(X_I)} - 1\right)\sum_{j\neq I}\theta_j X_j\right|\right). \tag{4.91}$$

We begin by averaging the first term over $I$. Note that

$$|X_1| = \left(\frac{G_1}{G_{1,n}}\right)^{1/p} \le \left(\frac{G_1 + G_1'}{G_{1,n} + G_1'}\right)^{1/p} = |X_1^1|,$$

and therefore, recalling $P(I = i) = \theta_i^2$, we may invoke Proposition 4.2 to conclude

$$\begin{aligned} E\left|\theta_I\left(U_I X_I^I - X_I\right)\right| &= \sum_{i=1}^{n} |\theta_i|^3 E\left|U_i X_i^i - X_i\right| \\ &= E\left|U_1 X_1^1 - X_1\right| \sum_{i=1}^{n} |\theta_i|^3 \\ &\le \frac{E|X_1|^3}{2\sigma_{n,p}^2} \sum_{i=1}^{n} |\theta_i|^3 = \frac{m_{n,p}}{2} \sum_{i=1}^{n} |\theta_i|^3. \end{aligned} \tag{4.92}$$

Now, averaging the second term in (4.91) over the distribution of $I$ yields

$$E\left|\left(\frac{g(X_I^I)}{g(X_I)} - 1\right) \sum_{j \ne I} \theta_j X_j\right| = \sum_{i=1}^{n} E\left|\left(\frac{g(X_i^i)}{g(X_i)} - 1\right) \sum_{j \ne i} \theta_j X_j\right| \theta_i^2. \tag{4.93}$$

Using (4.83), (4.86) and $g(x) = (1 - |x|^p)^{1/p}$, we have

$$\frac{g(X_i^i)}{g(X_i)} - 1 = \left(\frac{G_{1,n}}{G_{1,n} + G_i'}\right)^{1/p} - 1. \tag{4.94}$$

Applying (4.89) we have that $\{G_{1,n}, G_i'\}$ are independent of $X_1, \ldots, X_n$; hence, the term (4.94) is independent of the sum it multiplies in (4.93) and therefore (4.93) equals

$$\sum_{i=1}^{n} E\left|\frac{g(X_i^i)}{g(X_i)} - 1\right| E\left|\sum_{j \ne i} \theta_j X_j\right| \theta_i^2. \tag{4.95}$$

To bound the first expectation in (4.95), since $G_{1,n}/(G_{1,n} + G_i') \sim B(n/p, 2/p)$, we have

$$E\left|\frac{g(X_i^i)}{g(X_i)} - 1\right| = E\left(1 - \left(\frac{G_{1,n}}{G_{1,n} + G_i'}\right)^{1/p}\right) \le \left(\frac{1}{p} \vee 1\right) \frac{2}{n+2} \tag{4.96}$$

since for $p \ge 1$, using (4.90) with $\kappa = 1$,

$$\begin{aligned} & E\left(1 - \left(\frac{G_{1,n}}{G_{1,n} + G_i'}\right)^{1/p}\right) \\ & \quad \le E\left(1 - \left(\frac{G_{1,n}}{G_{1,n} + G_i'}\right)\right) = 1 - \frac{n/p}{(n+2)/p} = \frac{2}{n+2}, \end{aligned}$$

while for $0 < p < 1$, using Jensen's inequality and the fact that

$$(1 - x)^{1/p} \ge 1 - x/p \quad \text{for } x \le 1,$$

we have

$$E\left(1-\left(\frac{G_{1,n}}{G_{1,n}+G_i'}\right)^{1/p}\right)$$
$$\le 1-\left(E\left(\frac{G_{1,n}}{G_{1,n}+G_i'}\right)\right)^{1/p}=1-\left(\frac{n}{n+2}\right)^{1/p}\le\frac{2}{p(n+2)}.$$

We may bound the second expectation in (4.95) by $\sigma_{n,p}$ since

$$\left(E\left|\sum_{j\neq i}\theta_j X_j\right|\right)^2$$
$$\le E\left(\sum_{j\neq i}\theta_j X_j\right)^2=\operatorname{Var}\left(\sum_{j\neq i}\theta_j X_j\right)=\sigma_{n,p}^2\sum_{j\neq i}\theta_j^2\le\sigma_{n,p}^2.$$

Neither this bound nor the bound (4.96) depends on $i$, so substituting them into (4.95) and summing over $i$, again using $\sum_i \theta_i^2=1$, yields

$$\sum_{i=1}^n E\left|\frac{g(X_i^i)}{g(X_i)}-1\right|E\left|\sum_{j\neq i}\theta_j X_j\right|\theta_i^2\le\sigma_{n,p}\left(\frac{1}{p}\vee 1\right)\frac{2}{n+2}. \tag{4.97}$$

Adding (4.92) and (4.97) and multiplying by $2/\sigma_{n,p}$ in accordance with (4.81) yields (4.72). □

*Proof of Proposition 4.5*

1. For $A\subset S(\ell_p^n)$, $\mathbf{e}=(e_1,\ldots,e_n)\in\{-1,1\}^n$ and a permutation $\pi\in\mathcal{S}_n$, let

$$A_{\mathbf{e}}=\{\mathbf{x}\colon (e_1x_1,\ldots,e_nx_n)\in A\}$$
$$\text{and}\quad A_\pi=\{\mathbf{x}\colon (x_{\pi(1)},\ldots,x_{\pi(n)})\in A\}.$$

By the properties of Lebesgue measure, $\mu^n([0,1]A_{\mathbf{e}})=\mu^n([0,1]A_\pi)=\mu^n([0,1]A)$, so by (4.71), cone measure is coordinate symmetric and exchangeable.

The coordinate symmetry of $\mathbf{X}$ implies that

$$P(\mathbf{X}\in A)=P(\mathbf{X}\in A_{\mathbf{e}})\quad\text{for all }\mathbf{e}\in\{-1,1\}^n,$$

so with $\epsilon_i$, $i=1,\ldots,n$, i.i.d. variables taking the values 1 and $-1$ with probability $1/2$ and independent of $\mathbf{X}$,

$$P\big((\epsilon_1X_1,\ldots,\epsilon_nX_n)\in A\big)=P(\mathbf{X}\in A_\epsilon)$$
$$=\frac{1}{2^n}\sum_{\mathbf{e}\in\{-1,1\}^n}P(\mathbf{X}\in A_{\mathbf{e}})$$
$$=P(\mathbf{X}\in A),$$

and hence $(\epsilon_1X_1,\ldots,\epsilon_nX_n)=_d(X_1,\ldots,X_n)$. Note that for any $(s_1,\ldots,s_n)\in\{-1,1\}^n$ that

$$(\epsilon_1s_1,\ldots,\epsilon_ns_n)=_d(\epsilon_1,\ldots,\epsilon_n),\quad\text{and is independent of }\mathbf{X}.$$

Hence, since $P(X_i = 0) = 0$, with $s_i = X_i/|X_i|$, the sign of $X_i$, we have

$$\begin{aligned} P\big((\epsilon_1|X_1|, \ldots, \epsilon_n|X_n|) \in A\big) &= P\big((\epsilon_1 s_1 X_1, \ldots, \epsilon_n s_n X_n) \in A\big) \\ &= P\big((\epsilon_1 X_1, \ldots, \epsilon_n X_n) \in A\big) \\ &= P\big((X_1, \ldots, X_n) \in A\big). \end{aligned}$$

We thus obtain (4.83) applying that $\mathbf{X} \sim \mathcal{C}_p^n$ satisfies

$$\big(|X_1|, \ldots, |X_n|\big) =_d \left(\left(\frac{G_1}{G_{1,n}}\right)^{1/p}, \ldots, \left(\frac{G_n}{G_{1,n}}\right)^{1/p}\right) \tag{4.98}$$

shown, for instance, by Schechtman and Zinn (1990).

2. Applying the coordinate symmetry of $\mathbf{X}$ coordinatewise gives $X_i =_d -X_i$ and (4.98) yields $|X_i|^p = G_i/G_{1,n}$, which has the claimed Beta distribution, by (4.89). As $EX_i = 0$, we have

$$\mathrm{Var}(X_i) = EX_i^2 = E\big(|X_i|^p\big)^{2/p} \tag{4.99}$$

and the variance claim in (4.84) follows from (4.90) for $\alpha = 1/p$, $\beta = (n-1)/p$ and $\kappa = 2/p$.

From Stirlings formula, for all $x > 0$,

$$\lim_{m\to\infty} \frac{m^x \Gamma(m)}{\Gamma(m+x)} = 1,$$

so letting $m = n/p$ and $x = k/p$,

$$\lim_{n\to\infty} \frac{n^{k/p}\Gamma(n/p)}{\Gamma((n+k)/p)} = p^{k/p}. \tag{4.100}$$

The limit (4.84) now follows.

3. If $X$ is symmetric with variance $\sigma^2 > 0$ and $X^\square$ has the $X$-square bias distribution, then for all bounded continuous functions $f$

$$\begin{aligned} &\sigma^2 Ef\big(X^\square\big) \\ &\quad = EX^2 f(X) = E\big[(-X)^2 f(-X)\big] = EX^2 f(-X) = \sigma^2 Ef\big(-X^\square\big), \end{aligned}$$

showing $X^\square$ is symmetric.

From (4.90) and a change of variable, a random variable $X$ satisfies

$$|X|^p \sim B(\alpha/p, \beta/p)$$

if and only if the density $p_{|X|}(u)$ of $|X|$ is

$$p_{|X|}(u) = \frac{p\Gamma((\alpha+\beta)/p)}{\Gamma(\alpha/p)\Gamma(\beta/p)} u^{\alpha-1}\big(1-u^p\big)^{\beta/p-1}\mathbf{1}_{u\in[0,1]}. \tag{4.101}$$

Hence, since $|X_i|^p \sim B(1/p, (n-1)/p)$ by item 2, the density $p_{|X_i|}(u)$ of $|X_i|$ is

$$p_{|X_i|}(u) = \frac{p\Gamma(n/p)}{\Gamma(1/p)\Gamma((n-1)/p)}\big(1-u^p\big)^{(n-1)/p-1}\mathbf{1}_{u\in[0,1]}.$$

Multiplying by $u^2$ and renormalizing produces the $|X_i^i|$ density

$$\begin{aligned} p_{|X_i^i|}(u) &= \frac{u^2 p_{|X_i|}(u)}{EX_i^2} \\ &= \frac{p\Gamma((n+2)/p)}{\Gamma(3/p)\Gamma((n-1)/p)} u^2 (1-u^p)^{(n-1)/p-1} \mathbf{1}_{u\in[0,1]}, \end{aligned} \tag{4.102}$$

and comparing (4.102) to (4.101) shows the second claim in (4.85). The representation (4.86) now follows from (4.89) and the symmetry of $X_i^i$. The moment formula (4.87) for $m_{n,p}$ follows from (4.90) for $\alpha = 3/p$, $\beta = (n-1)/p$ and $\kappa = 1/p$, and the limit in (4.88) follows from (4.100).

Regarding the last claim in (4.88), for $p \ge 1$ Hölder's inequality gives

$$m_{n,p} = E|X^1| \le (E|X^1|^p)^{1/p} = \left(\frac{3}{n+2}\right)^{1/p},$$

while for $0 < p < 1$, we have

$$m_{n,p} = E|X^1| = E\left(\frac{G_i + G_i'}{G_{1,n} + G_i'}\right)^{1/p} \le E\left(\frac{G_i + G_i'}{G_{1,n} + G_i'}\right) = \frac{3}{n+2}.$$

4. We consider the conditional distribution on the left-hand side of (4.80), and use the representation, and notation $G_{a,b}$, given in (4.83). The second equality below follows from the coordinate-symmetry of $\mathbf{X}$, and the fourth follows since we may replace $G_{1,n}$ by $G_{2,n}/(1-|a|^p)$ on the conditioning event. Using the notation $a\mathcal{L}(V)$ for the distribution of $aV$, we have

$$\begin{aligned} &\mathcal{L}(X_2, \ldots, X_n | X_1 = a) \\ &= \mathcal{L}\left(\epsilon_2\left(\frac{G_2}{G_{1,n}}\right)^{1/p}, \ldots, \epsilon_n\left(\frac{G_n}{G_{1,n}}\right)^{1/p} \middle| \epsilon_1\left(\frac{G_1}{G_{1,n}}\right)^{1/p} = a\right) \\ &= \mathcal{L}\left(\epsilon_2\left(\frac{G_2}{G_{1,n}}\right)^{1/p}, \ldots, \epsilon_n\left(\frac{G_n}{G_{1,n}}\right)^{1/p} \middle| \left(\frac{G_1}{G_{1,n}}\right)^{1/p} = |a|\right) \\ &= \mathcal{L}\left(\epsilon_2\left(\frac{G_2}{G_{1,n}}\right)^{1/p}, \ldots, \epsilon_n\left(\frac{G_n}{G_{1,n}}\right)^{1/p} \middle| \frac{G_{2,n}}{G_{1,n}} = 1 - |a|^p\right) \\ &= (1-|a|^p)^{1/p} \mathcal{L}\left(\epsilon_2\left(\frac{G_2}{G_{2,n}}\right)^{1/p}, \ldots, \epsilon_n\left(\frac{G_n}{G_{2,n}}\right)^{1/p} \middle| \frac{G_{2,n}}{G_{1,n}} = 1 - |a|^p\right) \\ &= (1-|a|^p)^{1/p} \mathcal{L}\left(\epsilon_2\left(\frac{G_2}{G_{2,n}}\right)^{1/p}, \ldots, \epsilon_n\left(\frac{G_n}{G_{2,n}}\right)^{1/p} \middle| \frac{G_1}{G_{1,n}} = |a|^p\right) \\ &= (1-|a|^p)^{1/p} \mathcal{L}\left(\epsilon_2\left(\frac{G_2}{G_{2,n}}\right)^{1/p}, \ldots, \epsilon_n\left(\frac{G_n}{G_{2,n}}\right)^{1/p}\right) \\ &= g(a)\mathcal{L}\left(\epsilon_2\left(\frac{G_2}{G_{2,n}}\right)^{1/p}, \ldots, \epsilon_n\left(\frac{G_n}{G_{2,n}}\right)^{1/p}\right). \end{aligned} \tag{4.103}$$

In the penultimate step may we remove the conditioning on $G_1/G_{1,n}$ since (4.89) and the independence of $G_1$ from all other variables gives that

$$\left(\frac{G_2}{G_{2,n}}, \ldots, \frac{G_n}{G_{2,n}}\right) \quad \text{is independent of } (G_1, G_{2,n})$$

and therefore independent of $G_1/(G_1+G_{2,n})=G_1/G_{1,n}$.

Regarding the right-hand side of (4.80), using $1-|X_1|^p=\sum_{i=2}^n|X_i|^p$ and the representation (4.83), we obtain

$$\begin{aligned} g(a)(X_2,\ldots,X_n)/g(X_1) &= g(a)\left(\frac{(X_2,\ldots,X_n)}{(|X_2|^p+\cdots+|X_n|^p)^{1/p}}\right) \\ &= g(a)\left(\frac{(\epsilon_2(\frac{G_2}{G_{1,n}})^{1/p},\ldots,\epsilon_n(\frac{G_n}{G_{1,n}})^{1/p})}{((\frac{G_2}{G_{1,n}})+\cdots+(\frac{G_n}{G_{1,n}}))^{1/p}}\right) \\ &= g(a)\left(\frac{(\epsilon_2 G_2^{1/p},\ldots,\epsilon_n G_n^{1/p})}{(G_2+\cdots+G_n)^{1/p}}\right) \\ &= g(a)\left(\epsilon_2\left(\frac{G_2}{G_{2,n}}\right)^{1/p},\ldots,\epsilon_n\left(\frac{G_n}{G_{2,n}}\right)^{1/p}\right) \end{aligned}$$

matching the distribution (4.103). □

In principle, Proposition 4.3 and Theorem 4.1 may be applied to compute bounds to the normal for projections of other coordinate-symmetric vectors when the required couplings, and conditioning, are as tractable as here.

## 4.4 Combinatorial Central Limit Theorems

In this section we apply Theorem 4.1 to derive $L^1$ bounds in the combinatorial central limit theorem, that is, for random variables $Y$ of the form

$$Y=\sum_{i=1}^n a_{i,\pi(i)}, \tag{4.104}$$

where $\pi$ is a permutation distributed uniformly over the symmetric group $\mathcal{S}_n$, and $\{a_{ij}\}_{1\le i,j\le n}$ are the components of a matrix $A\in\mathbb{R}^{n\times n}$.

Random variables of this form are of interest in permutation tests. In particular, given a function $d(x,y)$ which in some sense measures the closeness of two observations $x$ and $y$, given values $x_1,\ldots,x_n$ and $y_1,\ldots,y_n$ and a putative 'matching' permutation $\tau$ that associates $x_i$ to $y_{\tau(i)}$, one can test whether the level of matching given by $\tau$, as measured by

$$y_\tau=\sum_{i=1}^n a_{i\tau(i)} \quad \text{where } a_{ij}=d(x_i,y_j),$$

is unusually high by seeing how large the matching level $y_\tau$ is relative to that provided by a random matching, that is, by seeing whether $P(Y\ge y_\tau)$ is significantly small.

Motivated by these considerations, Wald and Wolfowitz (1944) proved the central limit theorem as $n\to\infty$ when the factorization $a_{ij}=b_ic_j$ holds; Hoeffding (1951) later generalized this result to arrays $\{a_{ij}\}_{1\le i,j\le n}$. Motoo (1957) gave

Lindeberg-type sufficient conditions for the normal limit to hold. In Sect. 6.1 the $L^\infty$ distance to the normal is considered for the case where $\pi$ is uniformly distributed, and also when its distribution is constant on conjugacy classes of $\mathcal{S}_n$.

Letting

$$a_{\bullet\bullet} = \frac{1}{n^2}\sum_{i,j=1}^n a_{ij}, \quad a_{i\bullet} = \frac{1}{n}\sum_{j=1}^n a_{ij} \quad \text{and} \quad a_{\bullet j} = \frac{1}{n}\sum_{i=1}^n a_{ij},$$

straightforward calculations show that when $\pi$ is uniform over $\mathcal{S}_n$ the mean $\mu_A$ and variance $\sigma_A^2$ of $Y$ are given by

$$\begin{aligned} \mu_A &= n a_{\bullet\bullet} \quad \text{and} \\ \sigma_A^2 &= \frac{1}{n-1}\sum_{i,j}\left(a_{ij}^2 - a_{i\bullet}^2 - a_{\bullet j}^2 + a_{\bullet\bullet}^2\right) \\ &= \frac{1}{n-1}\sum_{i,j}(a_{ij} - a_{i\bullet} - a_{\bullet j} + a_{\bullet\bullet})^2. \end{aligned} \tag{4.105}$$

For simplicity, writing $\mu$ and $\sigma^2$ for $\mu_A$ and $\sigma_A^2$, respectively, we prove in (4.124) the following equivalent representation for $\sigma^2$,

$$\sigma^2 = \frac{1}{4n^2(n-1)}\sum_{i,j,k,l}\left[(a_{ik} + a_{jl}) - (a_{il} + a_{jk})\right]^2, \tag{4.106}$$

and assume in what follows that $\sigma^2 > 0$ to rule out trivial cases. By (4.106), $\sigma^2 = 0$ if and only if $a_{il} - a_{i\bullet}$ does not depend on $i$, that is, if and only if the difference between any two rows $\mathbf{a}_i$ and $\mathbf{a}_j$ of $A$ satisfy $\mathbf{a}_i - \mathbf{a}_j = (a_{i\bullet} - a_{j\bullet})(1, \ldots, 1)$.

For each $n \geq 3$, Theorem 4.8 provides an $L^1$ bound between the standardized version of the variable $Y$ given in (4.104) and the normal, with an explicit constant depending on the third-moment-type quantity

$$\gamma = \gamma_A, \quad \text{where } \gamma_A = \sum_{i,j=1}^n |a_{ij} - a_{i\bullet} - a_{\bullet j} + a_{\bullet\bullet}|^3. \tag{4.107}$$

When the elements of $A$ are all of comparable order, $\sigma^2$ is of order $n$ and $\gamma$ of order $n^2$, resulting in a bound of order $n^{-1/2}$.

**Theorem 4.8** *For $n \geq 3$, let $\{a_{ij}\}_{i,j=1}^n$ be the components of a matrix $A \in \mathbb{R}^{n\times n}$, let $\pi$ be a random permutation uniformly distributed over $\mathcal{S}_n$, and let $Y$ be given by* (4.104). *Then, with $\mu$, $\sigma^2$ given in* (4.105), *and $\gamma$ given in* (4.107), *$F$ the distribution function of $W = (Y - \mu)/\sigma$ and $\Phi$ that of the standard normal,*

$$\|F - \Phi\|_1 \leq \frac{\gamma}{(n-1)\sigma^3}\left(16 + \frac{56}{(n-1)} + \frac{8}{(n-1)^2}\right).$$

The proof of this theorem depends on a construction of the zero bias variable using an exchangeable pair, which we now describe.

### 4.4.1 Use of the Exchangeable Pair

We recall that the exchangeable variables $Y'$, $Y''$ form a $\lambda$-Stein pair if

$$E(Y''|Y') = (1-\lambda)Y' \tag{4.108}$$

for some $0 < \lambda < 1$. When $\mathrm{Var}(Y') = \sigma^2 \in (0, \infty)$, Lemma 2.7 yields

$$EY' = 0 \quad \text{and} \quad E(Y' - Y'')^2 = 2\lambda\sigma^2. \tag{4.109}$$

The following proposition is in some sense a two variable version of Proposition 2.3.

**Proposition 4.6** *Let $Y'$, $Y''$ be a $\lambda$-Stein pair with* $\mathrm{Var}(Y') = \sigma^2 \in (0, \infty)$ *and distribution $F(y', y'')$. Then when $Y^\dagger$, $Y^\ddagger$ have distribution*

$$dF^\dagger(y', y'') = \frac{(y' - y'')^2}{2\lambda\sigma^2} dF(y', y''), \tag{4.110}$$

*and $U \sim \mathcal{U}[0, 1]$ is independent of $Y^\dagger$, $Y^\ddagger$, the variable*

$$Y^* = UY^\dagger + (1-U)Y^\ddagger \quad \textit{has the } Y'\textit{-zero biased distribution.} \tag{4.111}$$

*Proof* For all absolutely continuous functions $f$ for which the expectations below exist,

$$\begin{aligned}
\sigma^2 Ef'(Y^*) &= \sigma^2 Ef'\big(UY^\dagger + (1-U)Y^\ddagger\big) \\
&= \sigma^2 E\left(\frac{f(Y^\dagger) - f(Y^\ddagger)}{Y^\dagger - Y^\ddagger}\right) \\
&= \frac{1}{2\lambda} E\left(\left(\frac{f(Y'') - f(Y')}{Y'' - Y'}\right)(Y'' - Y')^2\right) \\
&= \frac{1}{2\lambda} E\big(f(Y'') - f(Y')(Y'' - Y')\big) \\
&= \frac{1}{\lambda} E\big(Y' f(Y') - Y'' f(Y')\big) \\
&= \frac{1}{\lambda} E\big(Y' f(Y') - (1-\lambda)Y' f(Y')\big) \\
&= EY' f(Y').
\end{aligned}$$

□

The following lemma, leading toward the construction of zero bias variables, is motivated by generalizing the framework of Example 2.3, where the Stein pair is a function of some underlying random variables $\xi_\alpha, \alpha \in \chi$ and a random index $\mathbf{I}$.

**Lemma 4.4** *Let $F(y', y'')$ be the distribution of a Stein pair and suppose there exists a distribution*

$$F(\mathbf{i}, \xi_\alpha, \alpha \in \chi) \tag{4.112}$$

*and an* $\mathbb{R}^2$ *valued function* $(y', y'') = \psi(\mathbf{i}, \xi_\alpha, \alpha \in \chi)$ *such that when* $\mathbf{I}$ *and* $\{\Xi_\alpha, \alpha \in \mathcal{X}\}$ *have distribution* (4.112) *then*

$$(Y', Y'') = \psi(\mathbf{I}, \Xi_\alpha, \alpha \in \mathcal{X})$$

*has distribution* $F(y', y'')$. *If* $\mathbf{I}^\dagger, \{\Xi_\alpha^\dagger, \alpha \in \chi\}$ *have distribution*

$$dF^\dagger(\mathbf{i}, \xi_\alpha, \alpha \in \mathcal{X}) = \frac{(y' - y'')^2}{E(Y' - Y'')^2} dF(\mathbf{i}, \xi_\alpha, \alpha \in \mathcal{X}) \tag{4.113}$$

*then the pair*

$$(Y^\dagger, Y^\ddagger) = \psi(\mathbf{I}^\dagger, \Xi_\alpha^\dagger, \alpha \in \mathcal{X})$$

*has distribution* $F^\dagger(y^\dagger, y^\ddagger)$ *satisfying*

$$dF^\dagger(y', y'') = \frac{(y' - y'')^2}{2\lambda\sigma^2} dF(y', y'').$$

*Proof* For any bounded measurable function $f$

$$\begin{aligned}
Ef(Y^\dagger, Y^\ddagger) &= Ef\big(\psi(\mathbf{I}^\dagger, \Xi_\alpha^\dagger, \alpha \in \mathcal{X})\big) \\
&= \int f(\psi(\mathbf{i}, \xi_\alpha, \alpha \in \chi)) dF^\dagger(\mathbf{i}, \xi_\alpha, \alpha \in \chi) \\
&= \int f(y', y'') \frac{(y' - y'')^2}{2\lambda\sigma^2} dF(\mathbf{i}, \xi_\alpha, \alpha \in \chi) \\
&= E\left(\frac{(Y' - Y'')^2}{2\lambda\sigma^2} f(Y', Y'')\right),
\end{aligned}$$

where $(Y', Y'')$ has distribution $F(y', y'')$. ☐

We continue building a general framework around Example 2.3, where the random index is chosen independently of the permutation, so their joint distribution factors, leading to

$$dF(\mathbf{i}, \xi_\alpha, \alpha \in \chi) = P(\mathbf{I} = \mathbf{i}) dF(\xi_\alpha, \alpha \in \chi). \tag{4.114}$$

Moreover, in view of (2.47), that is, that

$$Y'' - Y' = b\big(i, j, \pi(i), \pi(j)\big) \quad \text{where } b(i, j, k, l) = a_{il} + a_{jk} - (a_{ik} + a_{jl}),$$

we will pay special attention to situations where

$$Y'' - Y' = b(\mathbf{I}, \Xi_\alpha, \alpha \in \chi_{\mathbf{I}}) \tag{4.115}$$

where $\mathbf{I}$ and $\chi_{\mathbf{I}}$ are vectors of small dimensions with components in $\mathcal{I}$ and $\chi$, respectively. In other words, we consider situations where the difference between $Y''$ and $Y'$ depends on only a few variables. In such cases, it will be convenient to further decompose $dF(\mathbf{i}, \xi_\alpha, \alpha \in \chi)$ as

$$dF(\mathbf{i}, \xi_\alpha, \alpha \in \chi) = P(\mathbf{I} = \mathbf{i}) dF_{\mathbf{i}}(\xi_\alpha, \alpha \in \chi_{\mathbf{i}}) dF_{\mathbf{i}^c|\mathbf{i}}(\xi_\alpha, \alpha \notin \chi_{\mathbf{i}} | \xi_\alpha, \alpha \in \chi_{\mathbf{i}}), \tag{4.116}$$

where $dF_{\mathbf{i}}(\xi_\alpha, \alpha \in \chi_{\mathbf{i}})$ is the marginal distribution of $\xi_\alpha$ for $\alpha \in \chi_{\mathbf{i}}$, and $dF_{\mathbf{i}^c|\mathbf{i}}(\xi_\alpha, \alpha \notin \chi_{\mathbf{i}}|\xi_\alpha, \alpha \in \chi_{\mathbf{i}})$ the conditional distribution of $\xi_\alpha$ for $\alpha \notin \chi_{\mathbf{i}}$ given $\xi_\alpha$ for $\alpha \in \chi_{\mathbf{i}}$. One notes, however, that the factorization (4.114) guarantees that the marginal distributions of any $\xi_\alpha$ does not depend on $\mathbf{i}$. In terms of generating variables having the specified distributions for the purposes of coupling, the decomposition (4.116) corresponds to first generating $\mathbf{I}$, then $\{\xi_\alpha, \alpha \in \chi_{\mathbf{I}}\}$, and lastly $\{\xi_\alpha, \alpha \notin \chi_{\mathbf{I}}\}$ conditional on $\{\xi_\alpha, \alpha \in \chi_{\mathbf{I}}\}$. In what follows we will continue the slight abuse notation of letting $\{\alpha\colon \alpha \in \chi_{\mathbf{i}}\}$ denote the set of components of the vector $\chi_{\mathbf{i}}$.

We now consider the square bias distribution $F^\dagger$ in (4.113) when the factorization (4.116) of $F$ holds. Letting $\mathbf{I}$ and $\{\Xi_\alpha\colon \alpha \in \chi\}$ have distribution (4.114), by (4.109), (4.115) and independence we obtain

$$2\lambda\sigma^2 = E(Y' - Y'')^2 = Eb^2(\mathbf{I}, \Xi_\alpha, \alpha \in \chi_{\mathbf{I}}) = \sum_{\mathbf{i} \subset \mathcal{I}} P(\mathbf{I} = \mathbf{i}) Eb^2(\mathbf{i}, \Xi_\alpha, \alpha \in \chi_{\mathbf{i}}).$$

In particular, we may define a distribution for a vector of indices $\mathbf{I}^\dagger$ with components in $\mathcal{I}$ by

$$P(\mathbf{I}^\dagger = \mathbf{i}) = \frac{r_{\mathbf{i}}}{2\lambda\sigma^2} \quad \text{with } r_{\mathbf{i}} = P(\mathbf{I} = \mathbf{i}) Eb^2(\mathbf{i}, \Xi_\alpha, \alpha \in \chi_{\mathbf{i}}). \tag{4.117}$$

Hence, substituting (4.115) and (4.116) into (4.113),

$$\begin{aligned} &dF^\dagger(\mathbf{i}, \xi_\alpha, \alpha \in \chi) \\ &= \frac{P(\mathbf{I} = \mathbf{i}) b^2(\mathbf{i}, \xi_\alpha, \alpha \in \chi_{\mathbf{i}})}{2\lambda\sigma^2} dF_{\mathbf{i}}(\xi_\alpha, \alpha \in \chi_{\mathbf{i}}) dF_{\mathbf{i}^c|\mathbf{i}}(\xi_\alpha, \alpha \notin \chi_{\mathbf{i}}|\xi_\alpha, \alpha \in \chi_{\mathbf{i}}) \\ &= \frac{r_{\mathbf{i}}}{2\lambda\sigma^2} \frac{b^2(\mathbf{i}, \xi_\alpha, \alpha \in \chi_{\mathbf{i}})}{Eb^2(\mathbf{i}, \Xi_\alpha, \alpha \in \chi_{\mathbf{i}})} dF_{\mathbf{i}}(\xi_\alpha, \alpha \in \chi_{\mathbf{i}}) dF_{\mathbf{i}^c|\mathbf{i}}(\xi_\alpha, \alpha \notin \chi_{\mathbf{i}}|\xi_\alpha, \alpha \in \chi_{\mathbf{i}}) \\ &= P(\mathbf{I}^\dagger = \mathbf{i}) dF_{\mathbf{i}}^\dagger(\xi_\alpha, \alpha \in \chi_{\mathbf{i}}) dF_{\mathbf{i}^c|\mathbf{i}}(\xi_\alpha, \alpha \notin \chi_{\mathbf{i}}|\xi_\alpha, \alpha \in \chi_{\mathbf{i}}), \end{aligned} \tag{4.118}$$

where

$$dF_{\mathbf{i}}^\dagger(\xi_\alpha, \alpha \in \chi_{\mathbf{i}}) = \frac{b^2(\mathbf{i}, \xi_\alpha, \alpha \in \chi_{\mathbf{i}})}{Eb^2(\mathbf{i}, \Xi_\alpha, \alpha \in \chi_{\mathbf{i}})} dF_{\mathbf{i}}(\xi_\alpha, \alpha \in \chi_{\mathbf{i}}). \tag{4.119}$$

Definition (4.119) represents $dF^\dagger(\mathbf{i}, \xi_\alpha, \alpha \in \chi)$ in a manner parallel to (4.116) for $dF(\mathbf{i}, \xi_\alpha, \alpha \in \chi)$. This representation gives the parallel construction of variables $\mathbf{I}^\dagger$, $\{\Xi_\alpha^\dagger, \alpha \in \chi\}$ with distribution $dF^\dagger(\mathbf{i}, \xi_\alpha, \alpha \in \chi)$ as follows. First generate $\mathbf{I}^\dagger$ according to the distribution $P(\mathbf{I}^\dagger = \mathbf{i})$. Then, when $\mathbf{I}^\dagger = \mathbf{i}$, generate $\{\Xi_\alpha^\dagger, \alpha \in \chi_{\mathbf{i}}\}$ according to $dF_{\mathbf{i}}^\dagger(\xi_\alpha, \alpha \in \chi_{\mathbf{i}})$ and then $\{\Xi_\alpha^\dagger, \alpha \notin \chi_{\mathbf{i}}\}$ according to $dF_{\mathbf{i}^c|\mathbf{i}}(\xi_\alpha, \alpha \notin \chi_{\mathbf{i}}|\xi_\alpha, \alpha \in \chi_{\mathbf{i}})$. As this last factor is the same as the last factor in (4.116) an opportunity for coupling is presented. In particular, it may be possible to set $\Xi_\alpha^\dagger$ equal to $\Xi_\alpha$ for many $\alpha \notin \chi_{\mathbf{i}}$, thus making the pair $Y^\dagger, Y^\ddagger$ close to $Y', Y''$.

### 4.4.2 Construction and Bounds for the Combinatorial Central Limit Theorem

In this section we prove Theorem 4.8 by specializing the construction given in Sect. 4.4.1 to handle the combinatorial central limit theorem, and then applying Theorem 4.1. Recall that by (2.45) we may, without loss of generality, replace $a_{ij}$ by $a_{ij} - a_{i\bullet} - a_{\bullet j} + a_{\bullet\bullet}$, and assume

$$a_{i\bullet} = a_{\bullet j} = a_{\bullet\bullet} = 0, \tag{4.120}$$

noting that by doing so we may now write

$$W = Y/\sigma, \tag{4.121}$$

and that (4.107) becomes $\gamma = \sum_{ij} |a_{ij}|^3$.

Now, denoting $Y$ and $\pi$ by $Y'$ and $\pi'$, respectively, when convenient, the construction given in Example 2.3 applies. That is, given $\pi$, uniform over $\mathcal{S}_n$, take $(I, J)$ independent of $\pi$ with a uniform distribution over all distinct pairs in $\{1, \ldots, n\}$, in other words, with distribution

$$p_1(i, j) = \frac{1}{(n)_2}\mathbf{1}(i \neq j). \tag{4.122}$$

Letting $\tau_{ij}$ be the permutation which transposes $i$ and $j$, set $\pi'' = \pi\tau_{I,J}$ and let $Y''$ be given by (4.104) with $\pi''$ replacing $\pi$. Example 2.3 shows that $(Y, Y'')$ is a $2/(n-1)$-Stein pair, and (2.48) gives

$$Y - Y'' = (a_{I,\pi(I)} + a_{J,\pi(J)}) - (a_{I,\pi(J)} + a_{J,\pi(I)}). \tag{4.123}$$

In particular, averaging over $I, J, \pi(I)$ and $\pi(J)$ we now obtain (4.106) as follows, using (4.109) for the second equality,

$$\begin{aligned}\frac{1}{n^2(n-1)^2}\sum_{i,j,k,l}\big[(a_{ik} + a_{jl}) - (a_{il} + a_{jk})\big]^2 &= E(Y' - Y'')^2 \\ &= 2\lambda\sigma^2 \\ &= \frac{4\sigma^2}{n-1}.\end{aligned} \tag{4.124}$$

We first demonstrate an intermediate result before presenting a coupling construction of $Y', Y''$ to $Y^\dagger, Y^\ddagger$, leading to a coupling of $Y'$ and $Y^*$.

**Lemma 4.5** *Let $\pi$ be chosen uniformly from $\mathcal{S}_n$ and suppose $i \neq j$ and $k \neq l$ are elements of $\{1, \ldots, n\}$. Then*

$$\pi^\dagger = \begin{cases} \pi\tau_{\pi^{-1}(k),j} & \text{if } l = \pi(i),\ k \neq \pi(j), \\ \pi\tau_{\pi^{-1}(l),i} & \text{if } l \neq \pi(i),\ k = \pi(j), \\ \pi\tau_{\pi^{-1}(k),i}\tau_{\pi^{-1}(l),j} & \text{otherwise}, \end{cases} \tag{4.125}$$

*is a permutation that satisfies*

$$\pi^\dagger(m) = \pi(m) \quad \textit{for all } m \notin \{i, j, \pi^{-1}(k), \pi^{-1}(l)\}, \tag{4.126}$$
$$\{\pi^\dagger(i), \pi^\dagger(j)\} = \{k, l\}, \tag{4.127}$$

*and*

$$P\big(\pi^\dagger(m) = \xi_m^\dagger,\ m \notin \{i, j\}\big) = \frac{1}{(n-2)!} \tag{4.128}$$

*for all distinct* $\xi_m^\dagger, m \notin \{i, j\}$ *with* $\xi_m^\dagger \notin \{k, l\}$.

*Proof* That $\pi^\dagger$ satisfies (4.126) is clear from its definition. To show (4.127) and that $\pi^\dagger$ is a permutation, let $A_1$, $A_2$ and $A_3$ denote the three cases of (4.125) in their respective order. Clearly under $A_1$ we have

$$\pi^\dagger(t) = \pi(t) \quad \text{for all } t \notin \{j, \pi^{-1}(k)\}.$$

Hence, as $i \neq j$ and $i = \pi^{-1}(l) \neq \pi^{-1}(k)$, we have $\pi^\dagger(i) = \pi(i) = l$. Also,

$$\pi^\dagger(j) = \pi\tau_{\pi^{-1}(k),j}(j) = \pi\big(\pi^{-1}(k)\big) = k,$$

showing (4.127) holds on $A_1$. As $\pi^\dagger(\pi^{-1}(k)) = \pi(j)$, both $\pi$ and $\pi^\dagger$ map the set $\{j, \pi^{-1}(k)\}$ to $\{\pi(j), k\}$, and, as their images agree on $\{j, \pi^{-1}(k)\}^c$, we conclude that $\pi^\dagger$ is a permutation on $A_1$. As $A_2$ becomes $A_1$ upon interchanging $i$ with $j$ and $k$ with $l$, these conclusions hold also on $A_2$.

Under $A_3$, either $l = \pi(i)$, $k = \pi(j)$ or $l \neq \pi(i)$, $k \neq \pi(j)$. In the first instance $\pi^\dagger = \pi$, so $\pi^\dagger$ is a permutation, and (4.127) is immediate. Otherwise, as $i \neq j$ and $i \neq \pi^{-1}(l)$, we have

$$\pi^\dagger(i) = \pi\tau_{\pi^{-1}(k),i}\tau_{\pi^{-1}(l),j}(i) = \pi\tau_{\pi^{-1}(k),i}(i) = \pi\big(\pi^{-1}(k)\big) = k$$

and similarly, as $j \neq i$ and $j \neq \pi^{-1}(k)$,

$$\pi^\dagger(j) = \pi\tau_{\pi^{-1}(k),i}\tau_{\pi^{-1}(l),j}(j) = \pi\tau_{\pi^{-1}(k),i}\big(\pi^{-1}(l)\big), \tag{4.129}$$

and now, as $l \neq k$ and $l \neq \pi(i)$,

$$\pi\tau_{\pi^{-1}(k),i}\big(\pi^{-1}(l)\big) = \pi\big(\pi^{-1}(l)\big) = l,$$

so (4.127) holds under $A_3$. As both $\pi$ and $\pi^\dagger$ map $\{i, j, \pi^{-1}(k), \pi^{-1}(l)\}$ to $\{\pi(i), \pi(j), k, l\}$, and agree on $\{i, j, \pi^{-1}(k), \pi^{-1}(l)\}^c$, we conclude that $\pi^\dagger$ is a permutation on $A_3$.

We now turn our attention to (4.128). Let $\xi_m^\dagger, m \notin \{i, j\}$ be distinct and satisfy $\xi_m^\dagger \notin \{k, l\}$. Under $A_1$ we have $k \neq \pi(j)$, and have shown that $i \neq \pi^{-1}(k)$. Hence $\pi^{-1}(k) \notin \{i, j\}$ and therefore $\xi_{\pi^{-1}(k)}^\dagger \notin \{k, l\}$. Setting $\xi_i^\dagger = l$, we have

$$\begin{aligned}
&P\big(\pi^\dagger(m) = \xi_m^\dagger,\ m \notin \{i, j\}, A_1\big)\\
&= P\big(\pi^\dagger(m) = \xi_m^\dagger,\ m \notin \{i, j\},\ \pi(i) = l,\ \pi(j) \neq k\big)\\
&= P\big(\pi^\dagger(m) = \xi_m^\dagger,\ m \notin \{j\},\ \pi(j) \neq k\big)\\
&= P\big(\pi^\dagger(m) = \xi_m^\dagger,\ m \notin \{j, \pi^{-1}(k)\},\ \pi(j) \neq k,\ \pi^\dagger\big(\pi^{-1}(k)\big) = \xi_{\pi^{-1}(k)}^\dagger\big)
\end{aligned}$$

$$
\begin{aligned}
&= P\big(\pi(m) = \xi_m^\dagger,\ m \notin \{j, \pi^{-1}(k)\},\ \pi(j) \neq k,\ \pi(j) = \xi_{\pi^{-1}(k)}^\dagger\big) \\
&= P\big(\pi(m) = \xi_m^\dagger,\ m \notin \{j, \pi^{-1}(k)\},\ \pi(j) = \xi_{\pi^{-1}(k)}^\dagger\big) \\
&= \sum_{q \notin \{i,j\}} P\big(\pi(m) = \xi_m^\dagger,\ m \notin \{j, q\},\ \pi(j) = \xi_q^\dagger,\ \pi(q) = k\big) \\
&= \frac{(n-2)}{n!}.
\end{aligned}
$$

Case $A_2$ being the same upon interchanging $i$ with $j$ and $k$ with $l$, we obtain

$$
P\big(\pi^\dagger(m) = \xi_m^\dagger,\ m \notin \{i, j\},\ A_1 \cup A_2\big) = \frac{2(n-2)}{n!}. \tag{4.130}
$$

Under $A_3$ there are subcases depending on

$$
R = \big|\{\pi(i), \pi(j)\} \cap \{k, l\}\big|,
$$

and we let $A_{3,r} = A_3 \cap \{R = r\}$ for $r = 0, 1, 2$. When $R = 0$ the elements $\pi(i), \pi(j), k, l$ are distinct, and so $A_{3,0} = \{R = 0\}$. Additionally $R = 0$ if and only if the inverse images $i, j, \pi^{-1}(k), \pi^{-1}(l)$ under $\pi$ are also distinct, and so

$$
\begin{aligned}
&P\big(\pi^\dagger(m) = \xi_m^\dagger,\ m \notin \{i, j\}, A_{3,0}\big) \\
&\quad = P\big(\pi^\dagger(m) = \xi_m^\dagger,\ m \notin \{i, j, \pi^{-1}(k),\ \pi^{-1}(l)\}, \\
&\qquad\quad \pi^\dagger\big(\pi^{-1}(k)\big) = \xi_{\pi^{-1}(k)}^\dagger,\ \pi^\dagger\big(\pi^{-1}(l)\big) = \xi_{\pi^{-1}(l)}^\dagger, A_{3,0}\big) \\
&\quad = P\big(\pi(m) = \xi_m^\dagger,\ m \notin \{i, j, \pi^{-1}(k), \pi^{-1}(l)\}, \\
&\qquad\quad \pi(i) = \xi_{\pi^{-1}(k)}^\dagger, \pi(j) = \xi_{\pi^{-1}(l)}^\dagger, A_{3,0}\big) \\
&\quad = \sum_{\{q,r\}:\ |\{q,r,i,j\}|=4} P\big(\pi(m) = \xi_m^\dagger,\ k \notin \{i, j, q, r\}, \\
&\qquad\qquad\qquad \pi(i) = \xi_q^\dagger,\ \pi(j) = \xi_r^\dagger,\ \pi(q) = k,\ \pi(r) = l\big) \\
&\quad = \frac{(n-2)(n-3)}{n!}.
\end{aligned} \tag{4.131}
$$

Considering the case $R = 1$, in view of (4.125) we find

$$
A_{3,1} = A_3 \cap \{R = 1\} = A_{3,1a} \cup A_{3,1b},
$$

where

$$
A_{3,1a} = \big\{\pi(i) = k, \pi(j) \neq l\big\}, \quad \text{and} \quad A_{3,1b} = \big\{\pi(i) \neq k, \pi(j) = l\big\}.
$$

Since by appropriate relabeling each of these cases becomes $A_1$, we have

$$
P\big(\pi^\dagger(m) = \xi_m^\dagger,\ m \notin \{i, j\}, A_{3,1}\big) = \frac{2(n-2)}{n!}. \tag{4.132}
$$

For $R = 2$ we have $A_{3,2} = A_{3,2a} \cup A_{3,2b}$ where

$$
A_{3,2a} = \big\{\pi(i) = l,\ \pi(j) = k\big\} \quad \text{and} \quad A_{3,2b} = \big\{\pi(j) = l,\ \pi(i) = k\big\}.
$$

Under $A_{3,2a}$,

$$\begin{aligned}
&P\big(\pi^\dagger(m) = \xi_m^\dagger, m \notin \{i, j\}, A_{3,2a}\big) \\
&\quad = P\big(\pi^\dagger(m) = \xi_m^\dagger,\ m \notin \{i, j\},\ \pi(i) = l,\ \pi(j) = k\big) = \frac{1}{n!},
\end{aligned}$$

and the same holding for $A_{3,2b}$, by symmetry, yields

$$P\big(\pi^\dagger(m) = \xi_m^\dagger,\ m \notin \{i, j\}, A_{3,2}\big) = \frac{2}{n!}. \tag{4.133}$$

Summing the contributions from (4.130), (4.131), (4.132) and (4.133) we obtain

$$P\big(\pi^\dagger(m) = \xi_m^\dagger,\ k \notin \{i, j\}\big) = \frac{4(n-2)}{n!} + \frac{(n-2)(n-3)}{n!} + \frac{2}{n!} = \frac{1}{(n-2)!}$$

as claimed. □

The following lemma shows how to choose the 'special' indices in Lemma 4.5 to form the square bias, and hence, zero bias, distributions. In addition, as values of the $\pi^\dagger$ permutation can be made to coincide with those of a given $\pi$ using (4.125), a coupling of these variables on the same space is achieved. Before stating the lemma we note that (4.134) is a distribution by virtue of (4.106).

**Lemma 4.6** *Let*

$$Y = \sum_{i=1}^{n} a_{i,\pi(i)}$$

*with $\pi$ chosen uniformly from $\mathcal{S}_n$, and let $(I^\dagger, J^\dagger, K^\dagger, L^\dagger)$ be independent of $\pi$ with distribution*

$$p_2(i, j, k, l) = \frac{[(a_{ik} + a_{jl}) - (a_{il} + a_{jk})]^2}{4n^2(n-1)\sigma^2}. \tag{4.134}$$

*Further, let $\pi^\dagger$ be constructed from $\pi$ as in* (4.125) *with $I^\dagger$, $J^\dagger$, $K^\dagger$ and $L^\dagger$ replacing $i, j, k$ and $l$, respectively and $\pi^\ddagger = \pi^\dagger \tau_{I^\dagger, J^\dagger}$. Then*

$$\pi(i) = \pi^\dagger(i) = \pi^\ddagger(i) \quad \textit{for all } i \notin \mathcal{I} \tag{4.135}$$

*where $\mathcal{I} = \{I^\dagger, J^\dagger, \pi^{-1}(K^\dagger), \pi^{-1}(L^\dagger)\}$, the variables*

$$Y^\dagger = \sum_{i=1}^{n} a_{i,\pi^\dagger(i)} \quad \textit{and} \quad Y^\ddagger = \sum_{i=1}^{n} a_{i,\pi^\ddagger(i)} \tag{4.136}$$

*have the square bias distribution* (4.113), *and with $U$ an uniform variable on* $[0, 1]$, *independent of all other variables*

$$Y^* = UY^\dagger + (1-U)Y^\dagger$$

*has the $Y$-zero bias distribution.*

*Proof* The claim (4.135) follows from (4.126) and the definition of $\pi^{\ddagger}$. When $\mathbf{I} = (I, J)$ is independent of $\pi$ with distribution (4.122), $\chi = \{1, \ldots, n\}$ and $\Xi_\alpha = \pi(\alpha)$ for $\alpha \in \chi$, let $\psi$ be the $\mathbb{R}^2$ valued function of $\{\mathbf{I}, \Xi_\alpha, \alpha \in \chi\}$ which yields the exchangeable pair $Y', Y''$ in Example 2.3. In view of Lemma 4.6, to prove the remainder of the claims it suffices to verify the hypotheses of Lemma 4.4, that is, with $\mathbf{I}^\dagger = (I^\dagger, J^\dagger)$ that $\{\mathbf{I}^\dagger, \Xi^\dagger_\alpha, \alpha \in \chi\}$, or equivalently $\{\mathbf{I}^\dagger, \pi^\dagger(\alpha), \alpha \in \chi\}$, has distribution (4.113). Relying on the discussion following Lemma 4.4, we prove this latter claim by considering the factorization (4.116) of $dF(\mathbf{i}, \xi_\alpha, \alpha \in \chi)$ and show that $\{\mathbf{I}^\dagger, \pi^\dagger(\alpha), \alpha \in \chi\}$ follows the corresponding square bias distribution (4.118).

With $\mathbf{i} = (i, j)$ and $P(\mathbf{I} = \mathbf{i})$ already specified by (4.122), we identify the remaining parts of the factorization (4.116) by noting that the distribution $dF_{\mathbf{i}}(\xi_\alpha, \alpha \in \chi_{\mathbf{i}}) = dF_{\mathbf{i}}(\xi_i, \xi_j)$ of the images of $i$ and $j$ under $\pi$ is uniform over all $\xi_i \neq \xi_j$, and, for such $\xi_i, \xi_j$, $dF_{\mathbf{i}^c|\mathbf{i}}(\xi_\alpha, \alpha \notin \{i, j\}|\xi_i, \xi_j)$ is uniform over all distinct elements $\xi_\alpha, \alpha \in \chi$ that do not intersect $\{\xi_i, \xi_j\}$, that is, for such values

$$dF_{\mathbf{i}^c|\mathbf{i}}\big(\xi_\alpha, \alpha \notin \{i, j\}|\xi_i, \xi_j\big) = \frac{1}{(n-2)!}. \tag{4.137}$$

Now consider the corresponding factorization (4.118). First, this expression specifies the joint distribution of the values $\mathbf{I}^\dagger$ and their images $\Xi^\dagger_\alpha$, $\alpha \in \mathbf{I}^\dagger$ under $\pi^\dagger$ by

$$\begin{aligned} &P(\mathbf{I}^\dagger = \mathbf{i})dF^\dagger_{\mathbf{i}}(\xi_\alpha, \alpha \in \chi_{\mathbf{i}}) \\ &\quad = \frac{P(\mathbf{I} = \mathbf{i})}{2\lambda\sigma^2} b^2(\mathbf{i}, \xi_\alpha, \alpha \in \chi_{\mathbf{i}}) dF_{\mathbf{i}}(\xi_\alpha, \alpha \in \chi_{\mathbf{i}}), \end{aligned} \tag{4.138}$$

where from (2.47) for the difference $Y' - Y''$ we have

$$b(i, j, \xi_i, \xi_j) = (a_{i,\xi_i} + a_{j,\xi_j}) - (a_{i,\xi_j} + a_{j,\xi_i}). \tag{4.139}$$

Since the distribution (4.122) of $\mathbf{I}$ is uniform over the range where $i \neq j$, and for such distinct $i$ and $j$, the distribution $dF_{\mathbf{i}}(\xi_\alpha,\ \alpha \in \chi_{\mathbf{i}})$ is uniform over all distinct choices of images $\xi_i$ and $\xi_j$, we conclude that the joint distribution (4.138) of $\mathbf{I}^\dagger$ and their 'biased permutation images' $(\Xi^\dagger_{I^\dagger}, \Xi^\dagger_{J^\dagger})$ is proportional to $\mathbf{1}_{i \neq j, k \neq l} b^2(i, j, k, l)$. This is exactly the distribution $p_2(i, j, k, l)$ from which $I^\dagger, J^\dagger, K^\dagger, L^\dagger$ is chosen. In addition, the values $\{K^\dagger, L^\dagger\}$ are the images of $\{I^\dagger, J^\dagger\}$ under the permutation $\pi^\dagger$ constructed as specified in the statement of the lemma, as follows. By (4.134) $I^\dagger \neq J^\dagger$ and $K^\dagger \neq L^\dagger$ with probability one. As $\{I^\dagger, J^\dagger, K^\dagger, L^\dagger\}$ and $\pi$ are independent, the construction and conclusions of Lemma 4.5 apply, conditional on these indices. Invoking Lemma 4.5, $\pi^\dagger$ is a permutation that maps $\{I^\dagger, J^\dagger\}$ to $\{K^\dagger, L^\dagger\}$.

To show that the remaining values are distributed according to $dF_{\mathbf{i}}(\xi_\alpha, \alpha \in \chi_{\mathbf{i}})$, again by Lemma 4.5, if $\xi^\dagger_m, m \notin \{I^\dagger, J^\dagger\}$ are distinct values not lying in $\{K^\dagger, L^\dagger\}$, then

$$P\big(\pi^\dagger(m) = \xi^\dagger_m, m \notin \{I^\dagger, J^\dagger\}|I^\dagger, J^\dagger, K^\dagger, L^\dagger\big) = \frac{1}{(n-2)!}. \tag{4.140}$$

As (4.140) agrees with (4.137), the proof of the lemma is complete. □

Note that in general even when $\mathbf{I}$ is uniformly distributed, the index $\mathbf{I}^\dagger$ need not be. In fact, from (4.117) it is clear that when $\mathbf{I}$ is uniform the distribution of $\mathbf{I}^\dagger$ is given by $P(\mathbf{I}^\dagger = \mathbf{i}) = 0$ for all $\mathbf{i}$ such that $P(\mathbf{I} = \mathbf{i}) = 0$, and otherwise

$$P(\mathbf{I}^\dagger = \mathbf{i}) = \frac{Eb^2(\mathbf{i}, \Xi_\alpha, \alpha \in \chi_\mathbf{i})}{\sum_\mathbf{i} Eb^2(\mathbf{i}, \Xi_\alpha, \alpha \in \chi_\mathbf{i})}. \tag{4.141}$$

In particular, the distribution (4.134) selects the indices $\mathbf{I}^\dagger = (I^\dagger, J^\dagger)$ jointly with their 'biased permutation' images $(K^\dagger, L^\dagger)$ with probability that preferentially makes the squared difference large. One can see this effect directly by calculating the marginal distribution of $I^\dagger, J^\dagger$, which, by (4.141), is proportional to $[(a_{ik} + a_{jl}) - (a_{il} + a_{jk})]^2$, by expanding and applying (4.120), yielding

$$\begin{aligned}
&\sum_{k,l} \big[(a_{ik} + a_{jl}) - (a_{il} + a_{jk})\big]^2 \\
&\quad = 2\sum_{k,l} \big(a_{ik}^2 + a_{jl}^2 - a_{ik}a_{jk} - a_{jl}a_{il}\big) \\
&\quad = 2n\sum_{k=1}^{n} (a_{ik} - a_{jk})^2,
\end{aligned}$$

and hence the generally nonuniform distribution

$$P(I^\dagger = i, J^\dagger = j) = \frac{\sum_{k=1}^n (a_{ik} - a_{jk})^2}{2n(n-1)\sigma^2}.$$

With the construction of the zero bias variable now in hand, Theorem 4.8 follows from Lemma 4.6, Theorem 4.1, (4.10) of Proposition 4.1, and the following lemma.

**Lemma 4.7** *For $Y$ and $Y^*$ constructed as in Lemma* 4.6

$$\big\|\mathcal{L}(Y^*) - \mathcal{L}(Y)\big\|_1 \le \frac{\gamma}{(n-1)\sigma^2}\left(8 + \frac{28}{(n-1)} + \frac{4}{(n-1)^2}\right).$$

With $\pi$ and the indices $\{I^\dagger, J^\dagger, K^\dagger, L^\dagger\}$ constructed as in Lemma 4.6 the calculation of the bound proceeds by decomposing

$$V = Y^* - Y \quad \text{as } V = V\mathbf{1}_2 + V\mathbf{1}_1 + V\mathbf{1}_0$$

where

$$\mathbf{1}_k = \mathbf{1}(R = k) \quad \text{with } R = \big|\big\{\pi(I^\dagger), \pi(J^\dagger)\big\} \cap \{K^\dagger, L^\dagger\}\big|.$$

The three factors give rise to the three terms of the bound. The proof of the lemma, though not difficult, requires some attention to detail, and can be found in the Appendix to this chapter.

## 4.5 Simple Random Sampling

Theorem 4.9 gives an $L^1$ bound for the exchangeable pair coupling. After proving the theorem, we will record a corollary and use it to prove an $L^1$ bound for simple random sampling. Recall that $(Y, Y')$ is a $\lambda$-Stein pair for $\lambda \in (0, 1)$ if $(Y, Y')$ are exchangeable and satisfy the linear regression condition

$$E(Y'|Y) = (1-\lambda)Y. \tag{4.142}$$

**Theorem 4.9** *Let $W, W'$ be a mean zero, variance* 1, *$\lambda$-Stein pair. Then if $F$ is the distribution function of $W$,*

$$\|F-\Phi\|_1 \le \sqrt{\frac{2}{\pi}} E\left|E\left(1-\frac{(W'-W)^2}{2\lambda}\middle|W\right)\right| + \frac{1}{2\lambda}E|W'-W|^3.$$

*Proof* Letting $\Delta = W - W'$, the result follows directly from Proposition 2.4 and Lemma 2.7, the latter which shows that identity (2.76) is satisfied with $R = 0$, $\hat{K}(t)$ given by (2.38), $\hat{K}_1 = E(\Delta^2|W)/2\lambda$ by (2.39), and

$$\begin{aligned}\hat{K}_2 &= \frac{|\Delta|}{2\lambda}\left(\mathbf{1}_{\{-\Delta\le 0\}}\int_{-\Delta}^{0}(-t)dt + \mathbf{1}_{\{-\Delta>0\}}\int_0^{-\Delta} tdt\right)\\ &= \frac{|\Delta|}{2\lambda}\left(\mathbf{1}_{\{-\Delta\le 0\}}\frac{\Delta^2}{2} + \mathbf{1}_{\{-\Delta>0\}}\frac{\Delta^2}{2}\right) = \frac{|\Delta^3|}{4\lambda}.\end{aligned}$$ □

In many applications calculation of the expectation of the absolute value of the conditional expectation may be difficult. However, by (2.34) we have

$$E\left(\frac{(W'-W)^2}{2\lambda}\right) = 1 \quad \text{so that} \quad E\left(E\left(1-\frac{(W'-W)^2}{2\lambda}\middle|W\right)\right) = 0.$$

Hence, by the Cauchy–Schwarz inequality,

$$\begin{aligned}E\left|E\left(1-\frac{(W'-W)^2}{2\lambda}\middle|W\right)\right| &\le \sqrt{\operatorname{Var}\left(E\left(1-\frac{(W'-W)^2}{2\lambda}\middle|W\right)\right)}\\ &= \frac{1}{2\lambda}\sqrt{\operatorname{Var}\bigl(E\bigl((W'-W)^2|W\bigr)\bigr)}.\end{aligned}$$

Though the variance of the conditional expectation $E((W'-W)^2|W)$ may still be troublesome, the inequality

$$\operatorname{Var}\bigl(E(Y|W)\bigr) \le \operatorname{Var}\bigl(E(Y|\mathcal{F})\bigr) \quad \text{when } \sigma\{W\}\subset\mathcal{F} \tag{4.143}$$

often leads to the computation of a tractable bound, and provides estimates which result in the optimal rate. To show (4.143), first note that the conditional variance formula, for any $X$, yields

$$\operatorname{Var}\bigl[E(X|W)\bigr] \le E\bigl[\operatorname{Var}(X|W)\bigr] + \operatorname{Var}\bigl[E(X|W)\bigr] = \operatorname{Var}(X).$$

However, for $X = E(Y|\mathcal{F})$ we have

$$E(X|W) = E\big(E(Y|\mathcal{F})|W\big) = E(Y|W),$$

and substituting yields (4.143). Hence we arrive at the following corollary to Theorem 4.9.

**Corollary 4.3** *Under the assumptions of Theorem* 4.9, *when $\mathcal{F}$ is any $\sigma$-algebra containing $\sigma\{W\}$,*

$$\|F - \Phi\|_1 \le \frac{1}{\lambda}\left(\frac{1}{\sqrt{2\pi}}\Theta + \frac{1}{2}E|W' - W|^3\right),$$

*where*

$$\Theta = \sqrt{\mathrm{Var}\big(E\big((W' - W)^2|\mathcal{F}\big)\big)}. \tag{4.144}$$

We use Corollary 4.3 to prove an $L^1$ bound for the sum of numerical characteristics of a simple random sample, that is, for a sample of a population $\{1, \ldots, N\}$ drawn so that all subsets of size $n$, with $0 < n < N$, are equally likely. The limiting normal distribution for simple random sampling was obtained by Wald and Wolfowitz (1944) (see also Madow 1948; Erdös and Rényi 1959a; and Hájek 1960).

Let $a_i \in \mathbb{R}$, $i = 1, 2, \ldots, N$ denote the characteristic of interest associated with individual $i$, and let $Y$ be the sum of the characteristics $\{X_1, \ldots, X_n\}$ of the sampled individuals. One can easily verify that the mean $\mu$ and variance $\sigma^2$ of $Y$ are given by

$$\mu = n\bar{a} \quad \text{and} \quad \sigma^2 = \frac{n(N-n)}{N(N-1)}\sum_{i=1}^{N}(a_i - \bar{a})^2 \quad \text{where } \bar{a} = \frac{1}{N}\sum_{i=1}^{N} a_i. \tag{4.145}$$

As we are interested in bounds to the normal for the standardized variable $(Y - \mu)/\sigma$, by replacing $a$ by $(a - \bar{a})/\sqrt{\sum_{b\in\mathcal{A}}(b - \bar{a})^2}$ we may assume in what follows without loss of generality that

$$\bar{a} = 0 \quad \text{and} \quad \sum_{i=1}^{N} a_i^2 = 1. \tag{4.146}$$

For $m = 1, \ldots, n$ let $(n)_m = n(n-1)\cdots(n-m+1)$, the falling factorial of $n$, and

$$f_m = \frac{(n)_m}{(N)_m}. \tag{4.147}$$

**Theorem 4.10** *Let the numerical characteristics $\mathcal{A} = \{a_i,\ i = 1, 2, \ldots, N\}$ of a population of size $N$ satisfy* (4.146), *and let $Y$ be the sum of characteristics in a simple random sample of size $n$ from $\mathcal{A}$ with $1 < n < N$. Let*

$$\sigma^2 = \frac{n(N-n)}{N(N-1)},$$
$$\lambda = \frac{N}{n(N-n)}, \quad A_4 = \sum_{a \in \mathcal{A}} a^4, \quad \textit{and} \quad \gamma = \sum_{a \in \mathcal{A}} |a|^3. \tag{4.148}$$

*Then with $F$ the distribution function of $Y/\sigma$,*

$$\|F - \Phi\|_1 \le \frac{1}{\lambda}\left(\frac{R_1}{\sqrt{2\pi}} + \frac{R_2}{2}\right),$$

*where*

$$R_1 = \frac{1}{n}\sqrt{\frac{2}{\sigma^2}S_1 + \frac{8}{\sigma^4(N-n)^2}S_2}$$

*with*

$$S_1 = A_4 - \frac{1}{N},$$
$$S_2 = A_4(f_1 - 7f_2 + 6f_3 - 6f_4) + 3(f_2 - f_3 + f_4) - \sigma^4 \quad \textit{and}$$
$$R_2 = 8f_1\gamma/\sigma^3.$$

In the usual asymptotic $n$ and $N$ tend to infinity together with the sampling fraction $f_1 = n/N$ bounded away from zero and one; in such cases $\lambda = O(1/n)$ and $f_m = O(1)$. Additionally, if $a \in \mathcal{A}$ satisfy $\sum_{a\in\mathcal{A}} a^2 = 1$ and are of comparable size then $a = O(1/\sqrt{N})$ which implies $A_4 = O(1/n)$ and $\gamma = O(1/\sqrt{n})$. Overall then the bound provided by the theorem in such an asymptotic, which has main contribution from $R_2$, is $O(1/\sqrt{n})$.

Since distinct labels may be appended to $a_i$, $i = 1, \ldots, N$, say as a second coordinate which is neglected when taking sums, we may assume in what follows that elements of $\mathcal{A} = \{a_i,\ i = 1, \ldots, N\}$ are distinct. The first main point of attention is the construction of a Stein pair, which can be achieved as follows. Let $X_1, X_2, \ldots, X_{n+1}$ be a simple random sample of size $n+1$ from the population and let $I$ and $I'$ be two distinct indices drawn uniformly from $\{1, \ldots, n+1\}$. Now set

$$Y = X_I + T \quad \text{and} \quad Y' = X_{I'} + T \quad \text{where } T = \sum_{i \in \{1,\ldots,n+1\}\setminus\{I,I'\}} X_i.$$

As $(X_I, X_{I'}, T) =_d (X_{I'}, X_I, T)$ the variables $Y$ and $Y'$ are exchangeable. By exchangeability and the first condition in (4.146) we have

$$E(X_I|Y) = \frac{1}{n}Y \quad \text{and} \quad E(X_{I'}|Y) = -\frac{1}{N-n}Y,$$

and therefore

$$E(Y'|Y) = E(Y - X_I + X_{I'}|Y) = (1-\lambda)Y$$

where $\lambda \in (0, 1)$ is given by (4.148); the linearity condition (4.142) is satisfied.

Before starting the proof we pause to simplify the required moment calculations for $\mathcal{X} = \{X_1, \ldots, X_n\}$, a simple random sample of $\mathcal{A}$. For $m \in \mathbb{N}$, $\{k_1, \ldots, k_m\} \subset \mathbb{N}$ and $\mathbf{k} = (k_1, \ldots, k_m)$ let

$$[\mathbf{k}] = E\left( \sum_{\{a,b,\ldots,c\}\subset\mathcal{X},\, |\{a,b,\ldots,c\}|=m} a^{k_1} b^{k_2} \cdots c^{k_m} \right)$$

and

$$\langle \mathbf{k} \rangle = \sum_{\{y_1,\ldots,y_m\}\subset\mathcal{A},\, |\{y_1,\ldots,y_m\}|=m} y_1^{k_1} y_2^{k_2} \ldots y_m^{k_m}.$$

Now observe that, with $f_m$ given in (4.147),

$$[\mathbf{k}] = f_m \langle \mathbf{k} \rangle. \tag{4.149}$$

As $[\mathbf{k}]$ and $\langle \mathbf{k} \rangle$ are invariant under any permutation of its components we may always use the canonical representation where $k_1 \ge \cdots \ge k_m$.

Let $e_j^m$ be the $j$th unit vector in $\mathbb{R}^m$. When the population characteristics satisfy (4.146) we have

$$\langle k_1, \ldots, k_{m-1}, 1 \rangle = -\sum_{j=1}^{m-1} \langle (k_1, \ldots, k_{m-1}) + e_j^{m-1} \rangle \quad \text{and}$$

$$\langle k_1, \ldots, k_{m-1}, 2 \rangle = \langle k_1, \ldots, k_{m-1} \rangle - \sum_{j=1}^{m-1} \langle (k_1, \ldots, k_{m-1}) + 2e_j^{m-1} \rangle.$$

Note then that

$$\begin{aligned}
\langle 2 \rangle &= 1 \\
\langle 3, 1 \rangle &= -\langle 4 \rangle \\
\langle 2, 2 \rangle &= \langle 2 \rangle - \langle 4 \rangle \\
\langle 2, 1, 1 \rangle &= -\langle 3, 1 \rangle - \langle 2, 2 \rangle = \langle 4 \rangle - \langle 2 \rangle + \langle 4 \rangle = 2\langle 4 \rangle - \langle 2 \rangle \\
\langle 1, 1, 1, 1 \rangle &= -3\langle 2, 1, 1 \rangle = -6\langle 4 \rangle + 3\langle 2 \rangle.
\end{aligned} \tag{4.150}$$

*Proof of Theorem 4.10* We may assume $n \le N/2$, as otherwise we may replace $Y$, a sample of size $n$ from $\mathcal{A}$, by $-Y$, a sample of size $N - n$; this assumption is used in (4.151).

We apply Corollary 4.3, beginning with the first term in the bound. Letting $\mathcal{X} = \{X_j,\ j \ne I'\}$ and $\mathcal{F} = \sigma(\mathcal{X})$, applying inequality (4.143) yields

$$\begin{aligned}
\mathrm{Var}\big(E\big((Y' - Y)^2 | Y\big)\big) &\le \mathrm{Var}\big(E\big((Y' - Y)^2 | \mathcal{F}\big)\big) \\
&= \mathrm{Var}\big(E\big((X_{I'} - X_I)^2 | \mathcal{F}\big)\big) \\
&= \mathrm{Var}\big(E\big(X_{I'}^2 - 2X_{I'}X_I + X_I^2 | \mathcal{F}\big)\big).
\end{aligned}$$

For these three conditional expectations,

$$E(X_{I'}^2|\mathcal{F}) = \frac{1}{N-n}\sum_{b\notin\mathcal{X}} b^2,$$

$$E(X_{I'}X_I|\mathcal{F}) = \frac{1}{n(N-n)}\sum_{a\in\mathcal{X},b\notin\mathcal{X}} ab \quad \text{and} \quad E(X_I^2|\mathcal{F}) = \frac{1}{n}\sum_{a\in\mathcal{X}} a^2.$$

By the standardization (4.146) we have,

$$\frac{1}{N-n}\sum_{b\notin\mathcal{X}} b^2 = \frac{1}{N-n}\left(1-\sum_{a\in\mathcal{X}} a^2\right)$$

$$\text{and} \quad \frac{1}{n(N-n)}\sum_{a\in\mathcal{X}}\sum_{b\notin\mathcal{X}} ab = -\frac{1}{n(N-n)}\left(\sum_{a\in\mathcal{X}} a\right)^2.$$

Hence, using $\mathrm{Var}(U+V) \le 2(\mathrm{Var}(U)+\mathrm{Var}(V))$,

$$\begin{aligned}
&\mathrm{Var}\big(E\big((Y'-Y)^2|Y\big)\big) \\
&\le \mathrm{Var}\left(\frac{N-2n}{n(N-n)}\sum_{a\in\mathcal{X}} a^2 + \frac{2}{n(N-n)}\left(\sum_{a\in\mathcal{X}} a\right)^2\right) \\
&\le 2\left(\frac{1}{n^2}\mathrm{Var}\left(\sum_{a\in\mathcal{X}} a^2\right) + \left(\frac{2}{n(N-n)}\right)^2 \mathrm{Var}\left(\sum_{a\in\mathcal{X}} a\right)^2\right). \qquad (4.151)
\end{aligned}$$

Calculating the first variance in (4.151), using (4.149), we begin with

$$\left(E\sum_{a\in\mathcal{X}} a^2\right)^2 = [2]^2 = \big(f_1\langle 2\rangle\big)^2 = f_1^2.$$

Next, note

$$\begin{aligned}
E\left(\sum_{a\in\mathcal{X}} a^2\right)^2 &= [4]+[2,2] = f_1\langle 4\rangle + f_2\langle 2,2\rangle \\
&= f_1\langle 4\rangle + f_2\big(\langle 2\rangle - \langle 4\rangle\big) = \frac{n(N-n)}{N(N-1)}\langle 4\rangle + f_2,
\end{aligned}$$

and therefore

$$\mathrm{Var}\left(\sum_{a\in\mathcal{X}} a^2\right) = \frac{n(N-n)}{N(N-1)}\left(\langle 4\rangle - \frac{1}{N}\right) = \sigma^2 S_1.$$

For the second variance in (4.151), using (4.149) and (4.150) we first obtain the expectation

$$E\left(\sum_{a\in\mathcal{X}} a\right)^2 = [2]+[1,1] = f_1 - f_2 = \sigma^2. \qquad (4.152)$$

Similarly, for the second moment we compute

$$E\Big(\sum_{a\in\mathcal{X}} a\Big)^4 = [4]+4[3,1]+3[2,2]+3[2,1,1]+[1,1,1,1]$$
$$= f_1\langle 4\rangle + f_2\big(4\langle 3,1\rangle + 3\langle 2,2\rangle\big) + f_3 3\langle 2,1,1\rangle + f_4\langle 1,1,1,1\rangle$$
$$= \langle 4\rangle(f_1 - 7f_2 + 6f_3 - 6f_4) + 3(f_2 - f_3 + f_4).$$

The variance of this term is now obtained by subtracting the square of the expectation (4.152), resulting in the quantity $S_2$.

Hence, from (4.151),

$$\mathrm{Var}\big(E\big((Y'-Y)^2|Y\big)\big) \le \frac{1}{n^2}\left(2\sigma^2 S_1 + \frac{8}{(N-n)^2}S_2\right),$$

and therefore, with $W = Y/\sigma$ and $W' = Y'/\sigma$, we have

$$\sqrt{\mathrm{Var}\big(E\big((W'-W)^2|W\big)\big)} = \sqrt{\mathrm{Var}\big(E\big((Y'-Y)^2|Y\big)\big)/\sigma^4} = R_1.$$

Regarding the second term in Corollary 4.3, as

$$E|Y'-Y|^3 = E|X_{I'} - X_I|^3 \le 8E|X_I|^3 = 8\frac{n}{N}\sum_{a\in\mathcal{A}}|a|^3 = 8f_1\gamma,$$

we obtain

$$E|W'-W|^3 = 8f_1\gamma/\sigma^3 = R_2.$$

□

## 4.6 Chatterjee's $L^1$ Theorem

The basis of all normal Stein identities is that $Z \sim \mathcal{N}(0,1)$ if and only if

$$E\big[Zf(Z)\big] = E\big[f'(Z)\big] \tag{4.153}$$

for all absolutely continuous functions $f$ for which these expectations exist. For a mean zero, variance one random variable $W$ which may be close to normal, (4.153) may hold approximately, and there may therefore be a related identity which holds exactly for $W$. One way the identity (4.153) may be altered to hold exactly for some given $W$ is to no longer insist that the same variable, $W$, appear on the right hand side as on the left, thus leading to the zero bias identity (2.51)

$$E\big[Wf(W)\big] = E\big[f'(W^*)\big], \tag{4.154}$$

as discussed in Sect. 2.3.3. Insisting that $W$ appear on both sides, one may be lead instead to consider identities of the form

$$E\big[Wf(W)\big] = E\big[f'(W)T\big], \tag{4.155}$$

for some random variable $T$, defined on the same space as $W$. When such a $T$ exists, by conditioning we obtain

$$E\big[f'(W^*)\big] = E\big[Wf(W)\big] = E\big[f'(W)T\big] = E\big[f'(W)E(T|W)\big],$$

which reveals that

$$E(T|W=w) = \frac{dF^*(w)}{dF(w)}$$

is the Radon–Nikodym derivative of the zero bias distribution of $W$ with respect to the distribution of $W$. In particular, as $W^*$ always has an absolutely continuous distribution, for there to exist a $T$ such that (4.155) holds it is necessary for $W$ to be absolutely continuous; naturally, in other cases, considering approximations allows the equality to become relaxed. Identities of the form (4.155), in some generality, were considered in Cacoullos and Papathanasiou (1992), but $T$ was constrained to be a function of $W$. As we will see, much more flexibility is provided by removing this restriction.

Theorem 4.11, of Chatterjee (2008), gives bounds to the normal, in the $L^1$ norm, for a mean zero function $\psi(\mathbf{X})$ of a vector of independent random variables $\mathbf{X} = (X_1, \ldots, X_n)$ taking values in some space $\mathcal{X}$. For the identity (4.155), or an approximate form thereof, to be useful, a viable $T$ must be produced. Towards this goal, with $\mathbf{X}'$ an independent copy of $\mathbf{X}$, and $A \subset \{1, \ldots, n\}$, let $\mathbf{X}^A$ be the random vector with components

$$X_j^A = \begin{cases} X_j' & j \in A, \\ X_j & j \notin A. \end{cases} \tag{4.156}$$

For $i \in \{1, \ldots, n\}$, writing $i$ for $\{i\}$ when notationally convenient, let

$$\Delta_i \psi(\mathbf{X}) = \psi(\mathbf{X}) - \psi(\mathbf{X}^i), \tag{4.157}$$

which measures the sensitivity of the function $\psi$ to the values in its $i$th coordinate. Now, for any $A \subset \{1, \ldots, n\}$, let

$$T_A = \sum_{i \notin A} \Delta_i \psi(\mathbf{X}) \Delta_i \psi(\mathbf{X}^A) \quad \text{and} \quad T = \frac{1}{2} \sum_{\substack{A \subset \{1,\ldots,n\} \\ |A| \neq n}} \frac{T_A}{\binom{n}{|A|}(n-|A|)}. \tag{4.158}$$

**Theorem 4.11** *Let $W = \psi(\mathbf{X})$ be a function of a vector of independent random variables $\mathbf{X} = (X_1, \ldots, X_n)$, and have mean zero and variance* 1. *Then, with $\Delta_i$ as defined in* (4.157) *and $T$ given in* (4.158) *we have that $ET = 1$ and*

$$\left\| \mathcal{L}(W) - \mathcal{L}(Z) \right\|_1 \le \sqrt{2/\pi} \sqrt{\operatorname{Var}\bigl(E(T|W)\bigr)} + \frac{1}{2} \sum_{i=1}^n E\left|\Delta_i \psi(\mathbf{X})\right|^3.$$

We present the proof, from Chatterjee (2008), at the end of this section.

To explore a simple application, let $\psi(\mathbf{X}) = \sum_{i=1}^n X_i$ where $X_1, \ldots, X_n$ are independent with mean zero, variances $\sigma_1^2, \ldots, \sigma_n^2$ summing to one, and fourth moments $\tau_1, \ldots, \tau_n$. For $A \subset \{1, \ldots, n\}$ and $i \notin A$,

$$\begin{aligned} \Delta_i \psi(\mathbf{X}^A) &= \psi(\mathbf{X}^A) - \psi(\mathbf{X}^{A \cup i}) \\ &= \sum_{j \notin A} X_j + \sum_{j \in A} X_j' - \Bigl(\sum_{j \notin A \cup i} X_j + \sum_{j \in A \cup i} X_j'\Bigr) = X_i - X_i'. \end{aligned} \tag{4.159}$$

Hence,

$$T_A = \sum_{i \notin A} \Delta_i \psi(\mathbf{X}) \Delta_i \psi(\mathbf{X}^A) = \sum_{i \notin A} (X_i - X_i')^2,$$

and

$$\begin{aligned}
T &= \frac{1}{2} \sum_{A \subset \{1,\ldots,n\},\, |A| \neq n} \frac{T_A}{\binom{n}{|A|}(n - |A|)} \\
&= \frac{1}{2} \sum_{a=0}^{n-1} \frac{1}{\binom{n}{a}(n-a)} \sum_{A \subset \{1,\ldots,n\},\, |A|=a} T_A \\
&= \frac{1}{2} \sum_{a=0}^{n-1} \frac{1}{\binom{n}{a}(n-a)} \sum_{A \subset \{1,\ldots,n\}, |A|=a} \sum_{i \notin A} (X_i - X_i')^2 \\
&= \frac{1}{2} \sum_{a=0}^{n-1} \frac{1}{\binom{n}{a}(n-a)} \sum_{i=1}^{n} \sum_{A \subset \{1,\ldots,n\},\, |A|=a, A \not\ni i} (X_i - X_i')^2.
\end{aligned}$$

As for each $i \in \{1, \ldots, n\}$ there are $\binom{n-1}{a}$ subsets of $A$ of size $a$ that do not contain $i$, we obtain

$$\begin{aligned}
T &= \frac{1}{2} \sum_{a=0}^{n-1} \frac{1}{\binom{n}{a}(n-a)} \sum_{i=1}^{n} (X_i - X_i')^2 \sum_{A \subset \{1,\ldots,n\},\, |A|=a, A \not\ni i} 1 \\
&= \left( \frac{1}{2} \sum_{i=1}^{n} (X_i - X_i')^2 \right) \left( \sum_{a=0}^{n-1} \frac{1}{\binom{n}{a}(n-a)} \binom{n-1}{a} \right) \\
&= \frac{1}{2} \sum_{i=1}^{n} (X_i - X_i')^2.
\end{aligned}$$

For the first term in the theorem, applying the bound (4.143) with $\mathcal{F}$ the $\sigma$-algebra generated by $\mathbf{X}$ we obtain

$$\mathrm{Var}\big(E(T|W)\big) \le \mathrm{Var}(T) = \frac{1}{4} \sum_{i=1}^{n} \mathrm{Var}\big((X_i - X_i')^2\big) = \frac{1}{2} \sum_{i=1}^{n} (\tau_i + 3\sigma_i^4).$$

From (4.159),

$$\begin{aligned}
\frac{1}{2} \sum_{i=1}^{n} E|\Delta_i \psi(\mathbf{X})|^3 &= \frac{1}{2} \sum_{i=1}^{n} E|X_i - X_i'|^3 \le \frac{1}{2} \sum_{i=1}^{n} \big(E(X_i - X_i')^4\big)^{3/4} \\
&= \frac{1}{2^{1/4}} \sum_{i=1}^{n} (\tau_i + 3\sigma_i^4)^{3/4}.
\end{aligned}$$

Invoking Theorem 4.11 yields,

$$\|\mathcal{L}(W)-\mathcal{L}(Z)\|_1 \le \sqrt{\frac{1}{\pi}\sum_{i=1}^n (\tau_i+3\sigma_i^4)} + \frac{1}{2^{1/4}}\sum_{i=1}^n (\tau_i+3\sigma_i^4)^{3/4}.$$

When $X_1,\dots,X_n$ are independent, mean zero variables having common second and fourth moments, say, $\sigma^2$ and $\tau$, respectively, then applying this result to $W=(X_1+\cdots+X_n)/\sqrt{n}$ yields

$$\|\mathcal{L}(W)-\mathcal{L}(Z)\|_1 \le n^{-1/2}\left(\sqrt{\frac{1}{\pi}(\tau+3\sigma^4)}+\frac{1}{2^{1/4}}(\tau+3\sigma^4)^{3/4}\right).$$

For a different application of Theorem 4.11 we consider normal approximation of quadratic forms. Let $\operatorname{Tr}(A)$ denote the trace of $A$.

**Proposition 4.7** *Let $\mathbf{X}=(X_1,\dots,X_n)$ be a vector of independent variables taking the values $+1,-1$ with equal probability, A a real symmetric matrix and $Y=\sum_{i\le j} a_{ij}X_iX_j$. Then the mean $\mu$ and variance $\sigma^2$ of Y are given by*

$$\mu=\operatorname{Tr}(A) \quad \text{and} \quad \sigma^2=\frac{1}{2}\operatorname{Tr}(A^2), \tag{4.160}$$

*and $W=(Y-\mu)/\sigma$ satisfies*

$$\|\mathcal{L}(W)-\mathcal{L}(Z)\|_1 \le \left(\frac{1}{\pi\sigma^4}\operatorname{Tr}(A^4)\right)^{1/2}+\frac{7}{2\sigma^3}\sum_{i=1}^n\left(\sum_{j=1}^n a_{ij}^2\right)^{3/2}.$$

*Proof* The mean and variance formulas (4.160) can be obtained by specializing Theorems 1.5 and 1.6 of Seber and Lee (2003) to $\mathbf{X}$ with the given distribution. By subtracting the mean and then replacing $a_{ij}$ by $a_{ij}/\sigma$ it suffices to prove the result when $a_{ii}=0$ and $\sigma^2=1$. Letting

$$\psi(\mathbf{x})=\sum_{i<j} a_{ij}x_ix_j$$

for $\mathbf{x}\in\mathbb{R}^n$, with $\mathbf{x}^i$ the vector $\mathbf{x}$ with $x_i'$ replacing $x_i$ and using the symmetry of $A$ we have

$$\begin{aligned}\Delta_i\psi(\mathbf{x}) &= \psi(\mathbf{x})-\psi(\mathbf{x}^i)\\ &= \sum_{j:\, i<j} a_{ij}x_ix_j+\sum_{j:\, j<i} a_{ji}x_jx_i-\sum_{j:\, i<j} a_{ij}x_i'x_j-\sum_{j:\, j<i} a_{ji}x_jx_i'\\ &= (x_i-x_i')\sum_{j=1}^n a_{ij}x_j.\end{aligned}$$

By replacing $\mathbf{x}$ above by $\mathbf{X}^A$, for $i\notin A$ we have

$$\Delta_i\psi(\mathbf{X}^A)=(X_i-X_i')\left(\sum_{j\notin A} a_{ij}X_j+\sum_{j\in A} a_{ij}X_j'\right).$$

We apply the bound $\operatorname{Var}(E(T|W))\le \operatorname{Var}(E(T|\mathbf{X}))$, from (4.143). For the calculation of $E(T|\mathbf{X})$, with $A\subset\{1,\dots,n\}$ and $i\notin A$, using that $X_i, X_i'$ are in $-1, 1$, we have

$$\begin{aligned}
&E\big(\Delta_i\psi(\mathbf{X})\Delta_i\psi(\mathbf{X}^A)|\mathbf{X}\big)\\
&\quad= E\left((X_i - X_i')^2\left(\sum_{j=1}^n a_{ij}X_j\right)\left(\sum_{j\notin A} a_{ij}X_j + \sum_{j\in A} a_{ij}X_j'\right)\middle|\mathbf{X}\right)\\
&\quad= \left(\sum_{j=1}^n a_{ij}X_j\right)E\left((X_i - X_i')^2\left(\sum_{j\notin A} a_{ij}X_j + \sum_{j\in A} a_{ij}X_j'\right)\middle|\mathbf{X}\right)\\
&\quad= 2\left(\sum_{j=1}^n a_{ij}X_j\right)E\left((1 - X_iX_i')\left(\sum_{j\notin A} a_{ij}X_j + \sum_{j\in A} a_{ij}X_j'\right)\middle|\mathbf{X}\right)\\
&\quad= 2\left(\sum_{j=1}^n a_{ij}X_j\right)\left(\sum_{j\notin A} a_{ij}X_j\right),
\end{aligned}$$

where, since $i \notin A$, all the remaining terms have conditional mean zero. Hence we may write

$$E\big(\Delta_i\psi(\mathbf{X})\Delta_i\psi(\mathbf{X}^A)|\mathbf{X}\big) = 2 \sum_{j\in\{1,\ldots,n\},\, k\notin A} a_{ij}a_{ik}X_jX_k.$$

Summing over all $i \notin A$, (4.158) yields

$$E(T_A|\mathbf{X}) = 2\sum_{i\notin A}\sum_{j\in\{1,\ldots,n\},\, k\notin A} a_{ij}a_{ik}X_jX_k.$$

From the definition of $T$, again from (4.158),

$$\begin{aligned}
E(T|\mathbf{X}) &= \frac{1}{2}\sum_{\substack{A\subset\{1,\ldots,n\}\\ |A|\neq n}} \frac{T_A}{\binom{n}{|A|}(n-|A|)}\\
&= \sum_{\substack{A\subset\{1,\ldots,n\}\\ |A|\neq n}} \frac{1}{\binom{n}{|A|}(n-|A|)}\sum_{i\notin A}\sum_{j\in\{1,\ldots,n\},\, k\notin A} a_{ij}a_{ik}X_jX_k\\
&= \sum_{j=1}^n\sum_{\substack{A\subset\{1,\ldots,n\}\\ |A|\neq n}} \frac{1}{\binom{n}{|A|}(n-|A|)}\sum_{i\notin A}\sum_{k\notin A} a_{ij}a_{ik}X_jX_k\\
&= \sum_{1\le i,j,k\le n} a_{ij}a_{ik}X_jX_k \sum_{A\cap\{i,k\}=\emptyset} \frac{1}{\binom{n}{|A|}(n-|A|)}\\
&= \sum_{1\le i,j,k\le n} a_{ij}a_{ik}X_jX_k \sum_{a=0}^{n-2}\sum_{A\cap\{i,k\}=\emptyset,\, |A|=a} \frac{1}{\binom{n}{a}(n-a)}\\
&= \sum_{1\le i,j,k\le n} a_{ij}a_{ik}X_jX_k \sum_{a=0}^{n-2} \frac{\binom{n-2}{a}}{\binom{n}{a}(n-a)}
\end{aligned}$$

$$
\begin{aligned}
&= \sum_{1\le i,j,k\le n} a_{ij}a_{ik}X_jX_k \sum_{a=0}^{n-2} \frac{n-a-1}{n(n-1)} \\
&= \sum_{1\le i,j,k\le n} a_{ij}a_{ik}X_jX_k \left(\frac{1}{n(n-1)} \sum_{a=1}^{n-1} a\right) \\
&= \frac{1}{2} \sum_{1\le i,j,k\le n} a_{ji}a_{ik}X_jX_k \\
&= \frac{1}{2} \mathbf{X}^\mathsf{T} A^2 \mathbf{X}.
\end{aligned}
$$

Letting $b_{jk} = \sum_{1\le i,j\le n} a_{ji}a_{ik}$, the $jk$th element of $A^2$, again using $X_i^2 = 1$,

$$
\mathrm{Var}\big(E(T|\mathbf{X})\big) = \mathrm{Var}\left(\sum_{j<k} b_{jk}X_jX_k\right) = \sum_{j<k} b_{jk}^2 \le \frac{1}{2}\,\mathrm{Tr}\big(A^4\big).
$$

To bound the final term in Theorem 4.11, we apply Khintchine's inequality, see Haagerup (1982), which yields

$$
E\left|\sum_{j=1}^n a_jX_j\right|^p \le B_p^p \left(\sum_{j=1}^n a_j^2\right)^{p/2}
$$

$$
\text{where } B_p = \begin{cases} 1 & 0 < p \le 2, \\ 2^{1/2}(\Gamma((p+1)/2)/\sqrt{\pi})^{1/p} & 2 < p < \infty. \end{cases}
$$

In particular $B_3^3 \le 1.6$, and using the fact that $X_i$ is independent of the event $\{X_i \ne X_i'\}$, we obtain

$$
E\big|\Delta_i\psi(\mathbf{X})\big|^3 = 4E\left|\sum_{j=1}^n a_{ij}X_j\right|^3 \le 7\left(\sum_{j=1}^n a_{ij}^2\right)^{3/2}. \qquad \square
$$

To consider some further examples, we make the following definition. With $\mathcal{X}$ the space in which our random variables take values, given $n \in \mathbb{N}$ suppose there is a map $\mathcal{G}$, or 'graphical rule', which to every $\mathbf{x} \in \mathcal{X}^n$ assigns an undirected graph, that is, a collection of edges $\mathcal{G}(\mathbf{x})$ on the vertices $\{1,\ldots,n\}$. We will say the map $\mathcal{G}$ is symmetric if it respects the action of permutations, that is, if for every permutation $\pi$ of $\{1,\ldots,n\}$ and any $(x_1,\ldots,x_n) \in \mathcal{X}^n$,

$$
\begin{aligned}
&\big\{\{i,j\}\colon \{i,j\} \in \mathcal{G}(x_{\pi(1)},\ldots,x_{\pi(n)})\big\} \\
&\quad = \big\{\{\pi(i),\pi(j)\}\colon \{i,j\} \in \mathcal{G}(x_1,\ldots,x_n)\big\}.
\end{aligned}
$$

Now fixing $m > n$, we say the vector $\mathbf{x} \in \mathcal{X}^n$ is embedded in the vector $\mathbf{y} \in \mathcal{X}^m$ if there exist distinct indices $i_1,\ldots,i_n$ in $\{1,\ldots,m\}$ with $x_k = y_{i_k}$ for $1 \le k \le n$. A graphical rule $\mathcal{G}'$ on $\mathcal{X}^m$ will be called an extension of the rule $\mathcal{G}$ if whenever the vector $\mathbf{x} \in \mathcal{X}^n$ is embedded in $\mathbf{y} \in \mathcal{X}^m$ the graph $\mathcal{G}(\mathbf{x})$ on $\{1,\ldots,n\}$ is the naturally induced subgraph of $\mathcal{G}(\mathbf{y})$ on $\{1,\ldots,m\}$.

Now let $\mathbf{x}$ and $\mathbf{x}'$ be any two elements of $\mathcal{X}^n$. For every $i \in \{1,\ldots,n\}$, let $\mathbf{x}^i$ be the vector obtained by replacing $x_i$ by $x_i'$ in $\mathbf{x}$, and, for $i$ and $j$ distinct elements of

$\{1, \ldots, n\}$, let $\mathbf{x}^{ij}$ be similarly obtained be replacing $x_i$ and $x_j$ in $\mathbf{x}$ by $x_i'$ and $x_j'$, respectively. With $\psi : \chi^n \to \mathbb{R}$, we say the coordinates $i$ and $j$ are non-interacting with respect to the triple $(\psi, \mathbf{x}, \mathbf{x}')$ if

$$\psi(\mathbf{x}) - \psi(\mathbf{x}^j) = \psi(\mathbf{x}^i) - \psi(\mathbf{x}^{ij}).$$

We will say that $\mathcal{G}$ is an interaction rule for a function $\psi$ if for any choice of $\mathbf{x}, \mathbf{x}'$ and $i, j$, the event that $\{i, j\}$ is not an edge in the graphs $\mathcal{G}(\mathbf{x}), \mathcal{G}(\mathbf{x}^i), \mathcal{G}(\mathbf{x}^j), \mathcal{G}(\mathbf{x}^{ij})$ implies that $i$ and $j$ are non-interacting vertices with respect to $(\psi, \mathbf{x}, \mathbf{x}')$. With these definitions in hand, we can now state the following theorem; we present the proof, from Chatterjee (2008), at the end of this section.

**Theorem 4.12** *Let the symmetric map $\mathcal{G}$ be an interaction rule for $\psi : \mathcal{X}^n \to \mathbb{R}$, and $\mathbf{X} = (X_1, \ldots, X_n)$ a vector of i.i.d. $\mathcal{X}$ valued variates such that $W = \psi(\mathbf{X})$ has mean zero and variance* 1. *For each $i \in \{1, \ldots, n\}$ define*

$$\Delta_i \psi(\mathbf{X}) = \psi(\mathbf{X}) - \psi(\mathbf{X}^i)$$

*where $\mathbf{X}'$ is an independent copy of $\mathbf{X}$, and let*

$$M = \max_{i=1,\ldots,n} \left|\Delta_i \psi(\mathbf{X})\right|. \tag{4.161}$$

*Let $\mathcal{G}'$ be any extension of $\mathcal{G}$ on $\mathcal{X}^{n+4}$, and set*

$$\delta = 1 + \text{degree of vertex } 1 \text{ in } \mathcal{G}'(X_1, \ldots, X_{n+4}). \tag{4.162}$$

*Then for some universal constant $C$,*

$$\left\|\mathcal{L}(W) - \mathcal{L}(Z)\right\|_1 \le Cn^{1/2} E\left(M^8\right)^{1/4} E\left(\delta^4\right)^{1/4} + \frac{1}{2}\sum_{i=1}^n E\left|\Delta_i \psi(\mathbf{X})\right|^3.$$

Following Chatterjee (2008), we apply Theorem 4.12 to prove an $L^1$ bound to the normal for two problems which stem from the theory of coverage processes; the volume of the region covered by the union of $n$ balls with random centers and some radius, and the number of such centers that are isolated at some radius; see Hall (1988) and Penrose (2003) for more background. Generally, we may work in a separable metric space $(\mathcal{X}, \rho)$, and for the first case, we take as given one endowed with measure $\lambda$. Let the components of $\mathbf{X} = (X_1, \ldots, X_n)$ be i.i.d. with values in $\mathcal{X}$. For some fixed radius $r > 0$, let $\mathcal{R}$ be given by

$$\mathcal{R} = \mathcal{R}(\mathbf{X}) \quad \text{where } \mathcal{R}(\mathbf{x}) = \bigcup_{i=1}^n B(x_i, r), \tag{4.163}$$

with $B(x, r)$ the closed ball of radius $r$ centered at $x$. Proposition 4.8 gives an $L^1$ bound to the normal for the 'covered volume' $\lambda(\mathcal{R}(\mathbf{X}))$ in terms of

$$K_V = \sup_{u \in \mathcal{X}} \lambda\left(B(u, r)\right), \tag{4.164}$$

an upper bound to the volume of any ball of radius $r$.

By very similar reasoning, we also derive an $L^1$ bound to the normal for the number $\mathcal{S}$ of isolated points, or singletons, given by

$$\mathcal{S} = \mathcal{S}(\mathbf{X}) \quad \text{where } \mathcal{S}(\mathbf{x}) = \sum_{i=1}^{n} \mathbf{1}\big(\{x_1, \ldots, x_n\} \cap B(x_i, 2r) = \{x_i\}\big), \tag{4.165}$$

that is, the number of points of $\mathbf{X}$ such that the ball $B(X_i, r)$ had empty intersection with $B(X_j, r)$ for all $j \neq i$. Proposition 4.8 gives an $L^1$ bound to the normal for $\mathcal{S}$ in terms of

$$K_S = \sup\big\{k: \exists \mathbf{x} \in \mathcal{X}^{k+1} \text{ such that } B(x_{k+1}, r) \cap B(x_i, r) \neq \emptyset, \text{ and} \\ B(x_i, r) \cap B(x_j, r) = \emptyset \text{ for all distinct } 1 \le i, j \le k\big\}, \tag{4.166}$$

which is an upper bound to the number of points in any collection from $\mathcal{X}$ which may become isolated upon the removal of a single point. In Euclidean space, the number $K_S$ is a lower bound to the so called kissing number, the maximum number of spheres of radius 1 that can simultaneously touch the unit sphere at the origin; see Zong (1999), Conway and Sloane (1999), and Leech and Sloane (1971) for estimates on the kissing number. For example, in two dimensions $K_S = 5$, since at most five unit circles can intersect another unit circle without intersecting each other, while the kissing number in two dimensions is 6.

**Proposition 4.8** *With $p = P(\rho(X_1, X_2) \le 2r)$ we have*

$$\big\|\mathcal{L}(W_V) - \mathcal{L}(Z)\big\|_1 \le \frac{Cn^{1/2}K_V^2(1+np)}{\sigma_V^2} + \frac{nK_V^3}{2\sigma_V^3}$$

*for some universal constant $C$, with $\mu_V = EY_V$, $\sigma_V^2 = \mathrm{Var}(Y_V)$ and $W_V = (Y_V - \mu_V)/\sigma_V$, when $Y_V = \lambda(\mathcal{R})$ with $\mathcal{R}$ as given in* (4.163) *and $K_V$ as in* (4.164). *The same bound holds for $Y_S = \mathcal{S}$ in* (4.165) *and $W_S = (Y_V - \mu_S)/\sigma_S$ where $\mu_S = EY_S$ and $\sigma_S^2 = \mathrm{Var}(Y_S)$, with the same constant $C$, upon replacing $\sigma_V$ and $K_V$ by $\sigma_S$ and $K_S$, respectively.*

*Proof* It suffices to prove the theorem when the variables standardized to have mean zero and variance one; we apply Theorem 4.12. First we consider $\mathcal{R}$, and let $\psi(x) = \lambda(\mathcal{R}(\mathbf{x}))$ for $\mathbf{x} \in \mathcal{X}^n$. Let $\mathcal{G}(\mathbf{x})$ be the graph on $\{1, \ldots, n\}$ with edges between points $i$ and $j$ if and only if $\rho(x_i, x_j) \le 2r$. Clearly the graphical rule $\mathcal{G}$ is symmetric, as distances are unchanged by relabeling.

We verify that $\mathcal{G}$ is an interaction rule as follows. With $\mathbf{x}$ and $\mathbf{x}'$ any points in $\mathcal{X}^n$, let $\mathbf{x}^i$ and $\mathbf{x}^{ij}$ be obtained by replacing the $i$th, or both the $i$th and $j$th, coordinate respectively of $\mathbf{x}$ by those of $\mathbf{x}'$. Writing $B_j$ and $B_j'$ for $B(x_j, r)$ and $B(x_j', r)$ respectively, we let

$$R_i = \bigcup_{j \neq i} B_j \quad \text{so that} \quad \psi(\mathbf{x}) = R_i \cup B_i \quad \text{and} \quad \psi(\mathbf{x}') = R_i \cup B_i'.$$

Hence,

$$\begin{aligned}\psi(\mathbf{x})-\psi(\mathbf{x}^i) &= \lambda(R_i\cup B_i)-\lambda(R_i\cup B_i')\\ &= \lambda\big((R_i\cup B_i)\cap(R_i\cup B_i')^c\big)-\lambda\big((R_i\cup B_i')\cap(R_i\cup B_i)^c\big)\\ &= \lambda\big(B_i\cap(B_i')^c\cap R_i^c\big)-\lambda\big(B_i'\cap B_i^c\cap R_i^c\big)\\ &= \lambda\big(B_i\cap R_i^c\big)-\lambda\big(B_i'\cap R_i^c\big),\end{aligned}$$

where we obtain the last inequality by adding and subtracting $\lambda(B_i\cap B_i'\cap R_i^c)$. Hence, with $N_i(\mathbf{x})$ be the set of indices $j\neq i$ of the neighbors of $x_i$ in the graph $\mathcal{G}(\mathbf{x})$,

$$\psi(\mathbf{x})-\psi(\mathbf{x}^i)=\lambda\bigg(B_i\cap\Big(\bigcup_{j\in N_i(\mathbf{x})}B_j\Big)^c\bigg)-\lambda\bigg(B_i'\cap\Big(\bigcup_{j\in N_i(\mathbf{x}^j)}B_j\Big)^c\bigg). \tag{4.167}$$

The pair $\{i,j\}$ fails to be an edge in the graphs $\mathcal{G}(\mathbf{x}),\mathcal{G}(\mathbf{x}^i),\mathcal{G}(\mathbf{x}^j),\mathcal{G}(\mathbf{x}^{ij})$ if and only no member of $\{x_i,x_i'\}$ is a neighbor of $\{x_j,x_j'\}$, in which case $N_i(\mathbf{x})=N_i(\mathbf{x}^j)$ and $N_i(\mathbf{x}^i)=N_i(\mathbf{x}^{ij})$, and $\psi(\mathbf{x})=\psi(\mathbf{x}^i)$ and $\psi(\mathbf{x}^i)=\psi(\mathbf{x}^{ij})$. Thus $\mathcal{G}$ is an interaction rule. In addition, (4.167) shows that for all $\mathbf{x}\in\mathcal{X}^n$ and all $i=1,\dots,n$

$$\big|\Delta_i\psi(\mathbf{x})\big|=\big|\psi(\mathbf{x})-\psi(\mathbf{x}^i)\big|\le\lambda\big(B(x_i,r)\big)\le K_V. \tag{4.168}$$

Hence we may take $M=K_V$ in the first term in the bound of Theorem 4.12, and also apply this same estimate to the second term.

Defining the graph $\mathcal{G}'$ on $(x_1,\dots,x_{n+4})\in\mathcal{X}^{n+4}$ by placing edges between any two points using the same rule as for $\mathcal{G}$, the rule $\mathcal{G}'$ clearly extends $\mathcal{G}$. As each of the $n+3$ points $x_2,\dots,x_{n+4}$ is independently a neighbor of $x_1$ with probability $p_r$, we have that $\delta-1\sim\mathrm{Bin}(n+3,p_r)$. As $E(\delta-1)^4=n^4p^4+O(n^3)$, we may bound $(E\delta^4)^{1/4}$ by some constant times $1+np$, completing the argument for $\mathcal{R}$.

The calculation for $\mathcal{S}$ is similar. Let $\psi(\mathbf{x})=\mathcal{S}(\mathbf{x})$ and take $\mathcal{G}$ to be the same graphical rule as the one used for $\mathcal{R}$. As the removal of a point from $\mathbf{x}\in\mathcal{X}^n$ can cause at most $K_S$ points to become isolated,

$$\big|\Delta_i\psi(\mathbf{x})\big|=\big|\psi(\mathbf{x})-\psi(\mathbf{x}^i)\big|\le K_S.$$

As the graph for $\mathcal{S}$ is the same as for $\mathcal{R}$, the distribution and bounds for the degree $\delta$ are the same as for $\mathcal{R}$. □

To test the quality of the bounds, we specialize to Euclidean space, and in the case of $V$, let $\lambda$ be the Lebesgue measure. Specializing a bit further, we take the points $X_1,\dots,X_n$ uniformly and independently in the cube $C_n=[0,n^{1/d})^d$ in $\mathbb{R}^d$, with periodic boundary conditions. Then letting $v_\rho=\rho^d\pi^{d/2}/\Gamma(1+d/2)$, the volume of the radius $\rho$ ball in dimension $d$, we have $K_V=v_r$. Now assuming $r\le n^{1/d}/2$ we have $p=v_{2r}/n$. By Goldstein and Penrose (2010),

$$\lim_{n\to\infty}n^{-1}\sigma_V^2=g_V \tag{4.169}$$

with an explicit $g_V>0$, showing the bound of Proposition 4.8 to be of order $n^{-1/2}$.

Similar remarks apply to $S$. In particular, $K_S$, as a lower bound on the kissing number, is bounded in any dimension as $n\to\infty$, and (4.169) holds for some

$g_S > 0$ when $\sigma_V^2$ is replaced by $\sigma_S^2$. At the cost of considerable more effort, Goldstein and Penrose (2010) apply Theorem 5.6 to obtain bounds of order $n^{-1/2}$ for the Kolmogorov distance for both the standardized $V$ and $S$, with explicit constants.

Though Chatterjee's approach might at first glance seem to bear little connection to the methods already presented, and (4.158) indeed appears a bit mysterious, Chen and Röllin (2010) have an interpretation which fits it into a general framework that contains a number of previous techniques mentioned, the exchangeable pair and size bias methods in particular. Chen and Röllin (2010) consider an identity of the form

$$E\big[Gf(W') - Gf(W)\big] = E\big[Wf(W)\big], \tag{4.170}$$

for some triple $(W, W', G)$ of square integrable random variables. If $W', W$ is a $\lambda$-Stein pair then by (2.35) identity (4.170) is satisfied with

$$G = \frac{1}{2\lambda}(W' - W).$$

If $Y^s$ is on the same space as $Y$ and has the $Y$-size biased distribution, and if $EY = \mu$ and $\mathrm{Var}(Y) = \sigma^2$, then by (2.64) the variables $W = (Y - \mu)/\sigma$ and $W' = (Y^s - \mu)/\sigma$ satisfy (4.170) with $G = \mu/\sigma$.

Chatterjee's approach is also included in the framework of Chen and Röllin (2010), by the method of 'interpolation to independence', as follows. Suppose $W$ is a mean zero, variance 1 random variable, and for each $i \in \{1, \ldots, n\}$ we have a random variable $W_i'$ which is close in some sense to $W$. Suppose there exists a sequence of random variables $V_0, \ldots, V_n$ such that $V_0 = W$, that $V_0$ and $V_n$ are independent, and that

$$\big((W, V_{i-1}), (W_i', V_i)\big) =_d \big((W_i', V_i), (W, V_{i-1})\big) \quad \text{for all } i = 1, \ldots, n.$$

Note in particular we must therefore have $W =_d W_i'$ and $V_i =_d V_{i-1}$, so all elements of the sequence $V_0, \ldots, V_n$ are equal in distribution, and have mean $E[V_0] = EW = 0$. Given such variables, letting $I$ be uniform over $\{1, \ldots, n\}$ and independent of the remaining variables and

$$G = \frac{n}{2}(V_I - V_{I-1}),$$

we have, by telescoping the sum, using the independence of $V_n$ and $W$ on the first term and taking conditional expectation with respect to $W$ on the second, that

$$\begin{aligned} E\big[Gf(W)\big] &= \frac{1}{2}\sum_{i=1}^{n}(V_i - V_{i-1})f(W) \\ &= \frac{1}{2}(V_n - V_0)f(W) \\ &= -\frac{1}{2}E\big[Wf(W)\big], \end{aligned}$$

while

$$
\begin{aligned}
E\big[Gf(W')\big] &= \frac{1}{2}\sum_{i=1}^{n}(V_i - V_{i-1})f(W') \\
&= -\frac{1}{2}\sum_{i=1}^{n}(V_i - V_{i-1})f(W) \\
&= -\frac{1}{2}E\big[Gf(W)\big] \\
&= \frac{1}{2}E\big[Wf(W)\big].
\end{aligned}
$$

Hence (4.170) is satisfied with $W' = W'_I$.

Now when $W = \psi(\mathbf{X})$, a mean zero, variance one function of i.i.d. variables $X_1, \ldots, X_n$, one can construct the required sequence $V_0, \ldots, V_n$ by setting $V_i$ to be the function $\psi$ evaluated on $X'_1, \ldots, X'_i, X_{i+1}, \ldots, X_n$ where $X'_i$ is an independent copy of $X_i$. Let also $W'_i = \psi(\mathbf{X}^i)$, where $\mathbf{X}^i$ is the vector $\mathbf{X}$ with $X'_i$ replacing $X_i$. It is clear that $V_0 = W$, and is independent of $V_n$. In the notation of (4.156) we have

$$W'_i = \psi\big(\mathbf{X}^i\big) \quad \text{and} \quad V_i = \psi\big(\mathbf{X}^{\{1,\ldots,i\}}\big).$$

Now consider the variation where $\pi$ is a random permutation independent of the remaining variables, and we interpolate to independence in the order determined by $\pi$, that is,

$$W'_i = \psi\big(\mathbf{X}^{\pi(i)}\big) \quad \text{and} \quad V_i = \psi\big(\mathbf{X}^{\{\pi(1),\ldots,\pi(i)\}}\big).$$

Then (4.170) is satisfied with

$$G = \frac{1}{2n}\big(W'_{\pi(I)} - W'_{\pi(I-1)}\big),$$

where $I$ is an independent index chosen uniformly from $\{1, \ldots, n\}$. Moreover, bounds to the normal in this framework involve conditional expectations such as (4.144), and in particular $E(G(W' - W)|\mathbf{X}, \mathbf{X}')$ is the expression (4.158), see Chen and Röllin (2010) for details.

We now present the proof of Theorems 4.11 and 4.12, starting with some preliminary lemmas.

**Lemma 4.8** *Let $\mathbf{X} = (X_1, \ldots, X_n)$ be a random vector with independent $\chi$ valued components. Then, for any functions $\phi, \psi : \chi^n \to \mathbb{R}$ such that $E\phi(\mathbf{X})^2$ and $E\psi(\mathbf{X})^2$ are both finite,*

$$\operatorname{Cov}\big(\phi(\mathbf{X}), \psi(\mathbf{X})\big) = \frac{1}{2}\sum_{\substack{A\subset\{1,\ldots,n\}\\|A|\neq n}} \frac{1}{\binom{n}{|A|}(n-|A|)}\sum_{j\notin A} E\big[\Delta_j\phi(\mathbf{X})\Delta_j\psi\big(\mathbf{X}^A\big)\big].$$

*Proof* First, we claim that

$$\begin{aligned}
&\sum_{\substack{A\subset\{1,\ldots,n\}\\|A|\neq n}} \frac{1}{\binom{n}{|A|}(n-|A|)}\sum_{j\notin A}\Delta_j\psi(\mathbf{X}^A) \\
&\quad= \sum_{\substack{A\subset\{1,\ldots,n\}\\|A|\neq n}} \frac{1}{\binom{n}{|A|}(n-|A|)}\sum_{j\notin A}\big(\psi(\mathbf{X}^A)-\psi(\mathbf{X}^{A\cup j})\big) \\
&\quad= \psi(\mathbf{X})-\psi(\mathbf{X}'). \qquad (4.171)
\end{aligned}$$

In particular, note that for any set $A\subset\{1,\ldots,n\}$, except $A=\{1,\ldots,n\}$, as there are $n-|A|$ elements $j\notin A$, these set appear in (4.171) with a positive sign a total of

$$\frac{1}{\binom{n}{|A|}(n-|A|)}\times\big(n-|A|\big)=\frac{1}{\binom{n}{|A|}}$$

times. Similarly, any set $B\subset\{1,\ldots,n\}$, except $B=\emptyset$, can be represented as $B=A\cup j$ for $|B|$ different sets $A$, so these sets appear with a negative sign a total of

$$\frac{1}{\binom{n}{|B|-1}(n-|B|+1)}\times|B|=\frac{1}{\binom{n}{|B|}}$$

times. Hence only the terms $A=\emptyset$ and $A\cup j=\{1,\ldots,n\}$ do not cancel out, the first one appearing with a coefficient of $1/\binom{n}{0}=1$, and the latter with coefficient $-1/\binom{n}{n}=-1$.

Now, for a fixed $A$ and $j\notin A$ let $U=\phi(\mathbf{X})\Delta_j\psi(\mathbf{X}^A)$, a function of the random vectors $\mathbf{X}$ and $\mathbf{X}'$. Note that upon interchanging $X_j$ and $X_j'$ the joint distribution of $(\mathbf{X},\mathbf{X}')$ is unchanged, while $U$ becomes $U'=-\phi(\mathbf{X}^j)\Delta_j\psi(\mathbf{X}^A)$. Thus,

$$EU=EU'=\frac{1}{2}E(U+U')=\frac{1}{2}\big[\Delta_j\phi(\mathbf{X})\Delta_j\psi(\mathbf{X}^A)\big].$$

Combining these observations yields

$$\begin{aligned}
\operatorname{Cov}\big(\phi(\mathbf{X}),\psi(\mathbf{X})\big) &= E\big[\phi(\mathbf{X})\psi(\mathbf{X})\big]-E\big[\phi(\mathbf{X})\big]E\big[\psi(\mathbf{X})\big] \\
&= E\big[\phi(\mathbf{X})\big(\psi(\mathbf{X})-\psi(\mathbf{X}')\big)\big] \\
&= \sum_{\substack{A\subset\{1,\ldots,n\}\\|A|\neq n}} \frac{1}{\binom{n}{|A|}(n-|A|)}\sum_{j\notin A}E\big[\phi(\mathbf{X})\Delta_j\psi(\mathbf{X}^A)\big] \\
&= \frac{1}{2}\sum_{\substack{A\subset\{1,\ldots,n\}\\|A|\neq n}} \frac{1}{\binom{n}{|A|}(n-|A|)}\sum_{j\notin A}E\big[\Delta_j\phi(\mathbf{X})\Delta_j\psi(\mathbf{X}^A)\big],
\end{aligned}$$

as desired. □

**Lemma 4.9** *Let $W=\psi(\mathbf{X})$ with $EW=0$ and $\operatorname{Var}(W)=1$ where $\mathbf{X}=(X_1,\ldots,X_n)$ is a vector of $\chi$ valued, independent components, and let $T$ be given by (4.158).*

*Then, for any twice continuously differentiable function $f$ with bounded second derivative, we have*

$$\left|E\big(f(W)W\big) - E\big(f'(W)T\big)\right| \le \frac{\|f''\|}{4} \sum_{j=1}^{n} E\left|\Delta_j \psi(\mathbf{X})\right|^3,$$

*where $T$ is given by* (4.158).

*Proof* For each $A \subset \{1, \dots, n\}$ and $j \notin A$, let

$$R_{A,j} = \Delta_j (f \circ \psi)(\mathbf{X}) \Delta_j \big(\psi(\mathbf{X}^A)\big)$$

and

$$\widetilde{R}_{A,j} = f'\big(\psi(\mathbf{X})\big) \Delta_j \psi(\mathbf{X}) \Delta_j \big(\psi(\mathbf{X}^A)\big).$$

By Lemma 4.8 with $g = f \circ \psi$, we have

$$E\big[f(W)W\big] = \frac{1}{2} \sum_{\substack{A \subset \{1,\dots,n\} \\ |A| \neq n}} \frac{1}{\binom{n}{|A|}(n - |A|)} \sum_{j \notin A} E R_{A,j}. \tag{4.172}$$

By the mean value theorem, and Hölder's inequality, we have

$$\begin{aligned} E|R_{A,j} - \widetilde{R}_{A,j}| &\le \frac{\|f''\|}{2} E\left|\big(\Delta_j \psi(\mathbf{X})\big)^2 \Delta_j \big(\psi(\mathbf{X}^A)\big)\right| \\ &\le \frac{\|f''\|}{2} E\left|\Delta_j \big(\psi(\mathbf{X}^A)\big)\right|^3. \end{aligned} \tag{4.173}$$

From the definition of $T$,

$$f'(W)T = \frac{1}{2} \sum_{\substack{A \subset \{1,\dots,n\} \\ |A| \neq n}} \frac{1}{\binom{n}{|A|}(n - |A|)} \sum_{j \notin A} \widetilde{R}_{A,j}. \tag{4.174}$$

Combining (4.172), (4.174) and (4.173), we obtain

$$\begin{aligned} &E\left|f(W)W - Ef'(W)T\right| \\ &\quad = \left|\frac{1}{2} \sum_{\substack{A \subset \{1,\dots,n\} \\ |A| \neq n}} \frac{1}{\binom{n}{|A|}(n - |A|)} \sum_{j \notin A} E(R_{A,j} - \widetilde{R}_{j,A})\right| \\ &\quad \le \frac{\|f''\|}{4} \sum_{\substack{A \subset \{1,\dots,n\} \\ |A| \neq n}} \frac{1}{\binom{n}{|A|}(n - |A|)} \sum_{j \notin A} E\left|\Delta_j \psi(\mathbf{X})\right|^3 \\ &\quad = \frac{\|f''\|}{4} \sum_{j=1}^{n} E\left|\Delta_j \psi(\mathbf{X})\right|^3, \end{aligned}$$

as claimed. □

*Proof of Theorem 4.11* Let $h$ be any absolutely continuous function with $\|h'\| \le 1$, and let $f$ be the solution to the Stein equation for $h$,

$$Eh(W) - Nh = E\big[f'(W) - Wf(W)\big].$$

By (2.13) of Lemma 2.4, we have that $\|f'\| \le \sqrt{2/\pi}$ and $\|f''\| \le 2$. Setting $\phi = \psi$ in Lemma 4.8, we obtain $ET = EW^2 = 1$. Therefore

$$\begin{aligned}
\big|Eh(W) - Nh\big| &\le E\big|f'(W) - Wf(W)\big| \\
&\le E\big|f'(W) - f'(W)T\big| + E\big|f'(W)T - Wf(W)\big| \\
&\le \sqrt{2/\pi}\,E\big|E(T|W) - 1\big| + E\big|f'(W)T - Wf(W)\big| \\
&\le \sqrt{2/\pi}\big[\mathrm{Var}\big(E(T|W)\big)\big]^{1/2} + \frac{1}{2}\sum_{j=1}^{n} E\big|\Delta_j\psi(\mathbf{X})\big|^3,
\end{aligned}$$

by the Cauchy–Schwarz inequality, and Lemma 4.9. The proof is completed by taking supremum over $h$, noting (4.8). □

We now proceed to the proof of Theorem 4.12. By Theorem 4.11, it suffices to bound $\mathrm{Var}(E(T|\mathbf{X}))$. For this reason, the proof of Theorem 4.12 follows quickly from the following upper bound.

**Lemma 4.10** *Let* $\mathbf{X}$ *be a vector of i.i.d. variates,* $A \subset \{1, \ldots, n\}$ *with* $|A| \ne n$, *and* $T_A$, $M$ *and* $\delta$ *given by* (4.158), (4.161) *and* (4.162), *respectively. Then there exists a constant* $C$ *such that*

$$\mathrm{Var}\big(E(T_A|\mathbf{X})\big) \le C\big(EM^8\big)^{1/2}\big(E\delta^4\big)^{1/2}\sqrt{n(n - |A|)}.$$

For the remainder of this section, we make the convention that constants $C$ need not be the same at each occurrence. Deferring the proof of Lemma 4.10, we present the proof of Theorem 4.12.

*Proof* By the definition of $T$ and Minkowski's inequality, we obtain

$$\big[\mathrm{Var}\big(E(T|\mathbf{X})\big)\big]^{1/2} \le \frac{1}{2}\sum_{\substack{A \subset \{1,\ldots,n\} \\ |A| \ne n}} \frac{[\mathrm{Var}(E(T_A|\mathbf{X}))]^{1/2}}{\binom{n}{|A|}(n - |A|)}.$$

Substituting the bound from Lemma 4.10 yields

$$\begin{aligned}
\big[\mathrm{Var}\big(E(T|\mathbf{X})\big)\big]^{1/2} &\le C\big(EM^8\big)^{1/4}\big(E\delta^4\big)^{1/2} \sum_{\substack{A \subset \{1,\ldots,n\} \\ |A| \ne n}} \frac{n^{1/4}(n - |A|)^{1/4}}{\binom{n}{|A|}(n - |A|)} \\
&= C\big(EM^8\big)^{1/4}\big(E\delta^4\big)^{1/4} \sum_{k=1}^{n} n^{1/4}k^{-3/4} \\
&= C\big(EM^8\big)^{1/4}\big(E\delta^4\big)^{1/4} n^{1/2}.
\end{aligned}$$

Now invoking Theorem 4.11 completes the proof. □

It remains to prove Lemma 4.10. We proceed by way of the following preliminary result.

**Lemma 4.11** *Suppose that $\mathcal{G}$ is a symmetric graphical rule on $\chi^n$ and $\mathbf{X} = (X_1, \dots, X_n)$ is a vector of i.i.d. $\chi$-valued random variables. Let $d_1$ be the degree of vertex 1 in $\mathcal{G}(\mathbf{X})$, and, for any $k \le n-1$, let $i, i_1, \dots, i_k$ be any collection of $k+1$ distinct elements of $\{1, \dots, n\}$. Then*

$$P\big(\{i, i_l\} \in \mathcal{G}(\mathbf{X}) \text{ for all } 1 \le l \le k\big) = \frac{E(d_1)_k}{(n-1)_k}, \tag{4.175}$$

*where $(r)_k$ stands for the falling factorial $r(r-1)\cdots(r-k+1)$.*

*Proof* Since $\mathcal{G}$ is a symmetric rule and $X_1, \dots, X_n$ are i.i.d., the probability

$$P\big(\{i, i_l\} \in \mathcal{G}(\mathbf{X}) \text{ for all } 1 \le l \le k\big)$$

does not depend on $i, i_1, \dots, i_k$. Hence

$$\begin{aligned} &P\big(\{i, i_l\} \in \mathcal{G}(\mathbf{X}) \text{ for all } 1 \le l \le k\big) \\ &\quad = \frac{1}{(n-1)_k} \sum_{\substack{\{j_1,\dots,j_k\} \subset \{1,\dots,n\} \setminus \{i\} \\ |\{j_1,\dots,j_k\}| = k}} P\big(\{i, j_l\} \in \mathcal{G}(\mathbf{X}) \text{ for all } 1 \le l \le k\big). \end{aligned}$$

Lastly, note that

$$\sum_{\substack{\{j_1,\dots,j_k\} \subset \{1,\dots,n\} \setminus \{i\} \\ |\{j_1,\dots,j_k\}| = k}} \mathbf{1}\big(\{i, j_l\} \in \mathcal{G}(\mathbf{X}) \text{ for all } 1 \le l \le k\big) = (d_i)_k,$$

where $d_i$ is the degree of vertex $i$. As $d_i$ and $d_1$ have the same distribution, the argument is complete. □

To prove Lemma 4.10 we require the following result, the Efron–Stein inequality, see Efron and Stein (1981), and Steele (1986).

**Lemma 4.12** *Let $U = g(Y_1, \dots, Y_m)$ be a function of independent random objects $Y_1, \dots, Y_m$, and let $Y_i'$ be an independent copy of $Y_i$ for $i = 1, \dots, m$. Then*

$$\operatorname{Var}(U) \le \frac{1}{2} \sum_{i=1}^{m} E\big(g\big(Y_1, \dots, Y_{i-1}, Y_i', Y_{i+1}, \dots, Y_m\big) - g(Y_1, \dots, Y_m)\big)^2.$$

*Proof of Lemma 4.10* Fix $A \subset \{1, \dots, n\}$ with $|A| \ne n$. For each $j \notin A$, let

$$\begin{aligned} R_j &= \Delta_j \psi(\mathbf{X}) \Delta_j \psi(\mathbf{X}^A) \\ &= \big(\psi(\mathbf{X}) - \psi(\mathbf{X}^j)\big)\big(\psi(\mathbf{X}^A) - \psi(\mathbf{X}^{A \cup j})\big). \end{aligned}$$

Let $Y = (Y_1, \dots, Y_n)$ be a copy of $\mathbf{X}$, which is independent of both $\mathbf{X}$ and $\mathbf{X}'$. For a fixed $i \in \{1, \dots, n\}$ let

$$\widetilde{\mathbf{X}} = (X_1, \dots, X_{i-1}, Y_i, X_{i+1}, \dots, X_n).$$

Similarly, for each $B \subset \{1, \ldots, n\}$, let

$$\widetilde{\mathbf{X}}^B = \begin{cases} (X_1^B, \ldots, X_{i-1}^B, Y_i, X_{i+1}^B, \ldots, X_n^B) & \text{if } i \notin B, \\ \mathbf{X}^B & \text{if } i \in B. \end{cases}$$

Now let

$$R_{ji} = \big(\psi(\widetilde{\mathbf{X}}) - \psi(\widetilde{\mathbf{X}}^j)\big)\big(\psi(\widetilde{\mathbf{X}}^A) - \psi(\widetilde{\mathbf{X}}^{A \cup j})\big),$$

and put

$$h_i = E\bigg(\sum_{j \notin A} (R_j - R_{ji})\bigg)^2.$$

It follows from inequality (4.143) and Lemma 4.12 that

$$\operatorname{Var}\big(E(T_A|\mathbf{X})\big) \le \operatorname{Var}(T_A) \le \frac{1}{2}\sum_{i=1}^n h_i. \tag{4.176}$$

Hence, we turn our attention to bounding $h_i$, and note that we need only consider $j \notin A$. When $j \neq i$ let

$$\begin{aligned} d_{ji}^1 &= \mathbf{1}\big(\{i, j\} \in \mathcal{G}(\mathbf{X})\big), \\ d_{ji}^2 &= \mathbf{1}\big(\{i, j\} \in \mathcal{G}(\mathbf{X}^j)\big), \\ d_{ji}^3 &= \mathbf{1}\big(\{i, j\} \in \mathcal{G}(\widetilde{\mathbf{X}})\big) \quad \text{and} \\ d_{ji}^4 &= \mathbf{1}\big(\{i, j\} \in \mathcal{G}(\widetilde{\mathbf{X}}^j)\big). \end{aligned}$$

Suppose in a particular realization we have $d_{ji}^1 = d_{ji}^2 = d_{ji}^3 = d_{ji}^4 = 0$. Since $\mathcal{G}$ is an interaction rule for $\psi$, on this event we have

$$\psi(\mathbf{X}) - \psi(\mathbf{X}^j) = \psi(\widetilde{\mathbf{X}}) - \psi(\widetilde{\mathbf{X}}^j).$$

If we now take $\mathbf{X}^A$ and $\widetilde{\mathbf{X}}^A$ in place of $\mathbf{X}$ and $\widetilde{\mathbf{X}}$, and define $e_{ji}^1, e_{ji}^2, e_{ji}^3$ and $e_{ji}^4$ analogously, then when $e_{ji}^1 = e_{ji}^2 = e_{ji}^3 = e_{ji}^4 = 0$ we have

$$\psi(\mathbf{X}^A) - \psi(\mathbf{X}^{A \cup j}) = \psi(\widetilde{\mathbf{X}}^A) - \psi(\widetilde{\mathbf{X}}^{A \cup j}),$$

whether $i \in A$ or not. Now, let

$$L_i = \max_{j \notin A} \big|\Delta_j \psi(\mathbf{X}) \Delta_j \psi(\mathbf{X}^A) - \Delta_j \psi(\widetilde{\mathbf{X}}) \Delta_j \psi(\widetilde{\mathbf{X}}^A)\big|.$$

From the preceding considerations, when $j \neq i$

$$|R_j - R_{ji}| \le L_i \sum_{k=1}^4 (d_{ji}^k + e_{ji}^k).$$

When $j = i$ then $i \notin A$ and we have $|R_j - R_{ji}| \le L_i$. The Cauchy–Schwarz inequality now yields

$$h_i \le \left[ E L_i^4 E\bigg(\mathbf{1}(i \notin A) + \sum_{j \notin A \cup i} \sum_{k=1}^4 (d_{ji}^k + e_{ji}^k)\bigg)^4 \right]^{1/2}. \tag{4.177}$$

Applying the inequality $(\sum_{i=1}^r a_i)^4 \le r^3 \sum_{i=1}^r a_i^4$, we obtain

$$E\left(\mathbf{1}(i \notin A) + \sum_{j \notin A \cup i} \sum_{k=1}^{4} (d_{ji}^k + e_{ji}^k)\right)^4$$

$$\le 9^3 \mathbf{1}(i \in A) + 9^3 \sum_{k=1}^{4} E\left(\sum_{j \notin A \cup i} d_{ji}^k\right)^4 + 9^3 \sum_{k=1}^{4} E\left(\sum_{j \notin A \cup i} e_{ji}^k\right)^4.$$

To handle the first term in the first sum, from Lemma 4.11, for any $j, k, l$ and $m$,

$$E(d_{ji}^1 d_{ki}^1 d_{li}^1 d_{mi}^1) \le C \frac{E\delta_1^r}{n^r},$$

where $r$ is the number of distinct indices among $j$, $k$, $l$, $m$, and $\delta_1$ is the degree of vertex 1 in $\mathcal{G}(\mathbf{X})$. Recall the definition of $\delta$ from (4.162), and observe that $\delta \ge \delta_1 + 1$. It follows easily that

$$E\left(\sum_{j \notin A \cup i} d_{ji}^1\right)^4 \le CE(\delta^4)\left(\frac{n - |A|}{n}\right).$$

Now we consider bounding $E(d_{ji}^2 d_{ki}^2 d_{li}^2 d_{mi}^2)$. First suppose that $j, k, l, m$ are distinct. Now let $\widetilde{\mathbf{X}}$ be the random vector in $\chi^{n+4}$ given by

$$\widetilde{\mathbf{X}} = (X_1, \dots, X_n, X_j', X_k', X_l', X_m').$$

Note that if $d_{ji}^2 = d_{ki}^2 = d_{li}^2 = d_{mi}^2 = 1$ then $\{i, n+1\}$, $\{i, n+2\}$, $\{i, n+3\}$ and $\{i, n+4\}$ are all edges in the extended graph $\mathcal{G}'(\widetilde{\mathbf{X}})$. Since $\mathcal{G}'$ is a symmetric rule and the components of $\widetilde{\mathbf{X}}$ are i.i.d., it follows from Lemma 4.10 that

$$E(d_{ji}^2 d_{ki}^2 d_{li}^2 d_{mi}^2) \le C \frac{E\delta^4}{n^4}.$$

Now, suppose $j, k, l$ are distinct, and that $m = l$. Let $s \in \{1, \dots, n\}$ be distinct from $j, k$ and $l$, and define

$$\widetilde{\mathbf{X}} = (X_1, \dots, X_n, X_j', X_k', X_l', X_s')$$

and argue as before to conclude that in this case

$$E(d_{ji}^2 d_{ki}^2 d_{li}^2 d_{mi}^2) = E(d_{ji}^2 d_{ki}^2 d_{li}^2) \le C \frac{E\delta^3}{n^3}.$$

In general, if $r$ is the number of distinct elements among $j$, $k$, $l$, $m$, then

$$E(d_{ji}^2 d_{ki}^2 d_{li}^2 d_{mi}^2) \le C \frac{E\delta^r}{n^r}.$$

From this inequality we obtain as before that

$$E\left(\sum_{j \notin A \cup i} d_{ji}^2\right)^4 \le CE(\delta^4)\left(\frac{n - |A|}{n}\right).$$

The $d^3$, $e^1$ and $e^3$ terms can be bounded as the $d^1$ term, while the $d^4$, $e^2$ and $e^4$ terms like the $d^2$ term. Combining, we conclude

$$E\left(\mathbf{1}(i \notin A) + \sum_{j \notin A \cup i} \sum_{k=1}^{4} (d_{ji}^k + e_{ji}^k)\right)^4 \le CE(\delta^4)\left(\mathbf{1}(i \notin A) + \frac{n - |A|}{n}\right).$$

As $M = \max_j |\Delta_j \psi(\mathbf{X})|$ have $EL_i^4 \le CEM^8$, and applying these bounds in (4.177), along with the inequality $\sqrt{x+y} \le \sqrt{x} + \sqrt{y}$ for nonnegative $x$ and $y$, we obtain

$$h_i \le C\big(EM^8\big)^{1/2}\big(E\delta^4\big)^{1/2}\left(\mathbf{1}(i \notin A) + \sqrt{\frac{n - |A|}{n}}\right).$$

Substituting this bound in (4.176), we obtain

$$\begin{aligned}
\operatorname{Var}\big(E(T_A|\mathbf{X})\big) &\le C\big(EM^8\big)^{1/2}\big(E\delta^4\big)^{1/2}\big(n - |A| + \sqrt{n(n - |A|)}\big) \\
&\le C\big(EM^8\big)^{1/2}\big(E\delta^4\big)^{1/2}\sqrt{n(n - |A|)},
\end{aligned}$$

completing the proof. □

## 4.7 Locally Dependent Random Variables

In this section we consider $L^1$ bounds for sums of locally dependent random variables. We being by recalling that an $m$-dependent sequence of random variables $\xi_i$, $i \in \mathbb{N}$, is one with the property that, for each $i$, the sets of random variables $\{\xi_j,\ j \le i\}$ and $\{\xi_j,\ j > i + m\}$ are independent. Independent random variables are the special case of $m$-dependence when $m = 0$. Local dependence generalizes the notion of $m$-dependence to collections of random variables indexed more generally. The concept of local dependence is applicable, for example, to random variables indexed by the vertices of a graph such that the collections $\{\xi_i,\ i \in I\}$ and $\{\xi_j,\ j \in J\}$ are independent whenever $I \cap J = \emptyset$ and the graph contains no edges $\{i, j\}$ with $i \in I$ and $j \in J$.

Let $\mathcal{J}$ be a finite index set of cardinality $n$, and let $\{\xi_i, i \in \mathcal{J}\}$ be a random field, that is, an indexed collection of random variables, with zero means and finite variances. Define $W = \sum_{i \in \mathcal{J}} \xi_i$, and assume that $\operatorname{Var}(W) = 1$. For any $A \subset \mathcal{J}$ let

$$A^c = \{j \in \mathcal{J} \colon j \notin A\} \quad \text{and} \quad \xi_A = \{\xi_i \colon i \in A\}.$$

We introduce the following two conditions, corresponding to different degrees of local dependence.

(LD1) For each $i \in \mathcal{J}$ there exists $A_i \subset \mathcal{J}$ such that $\xi_i$ and $\xi_{A_i^c}$ are independent.
(LD2) For each $i \in \mathcal{J}$ there exist $A_i \subset B_i \subset \mathcal{J}$ such that $\xi_i$ is independent of $\xi_{A_i^c}$ and $\xi_{A_i}$ is independent of $\xi_{B_i^c}$.

Clearly (LD2) implies (LD1). Whenever (LD1) or (LD2) hold we set

$$\eta_i = \sum_{j \in A_i} \xi_j \quad \text{and} \quad \tau_i = \sum_{j \in B_i} \xi_j \tag{4.178}$$

respectively. Note that when $\{\xi_i,\ i \in \mathcal{J}\}$ are independent (LD2) holds with $A_i = B_i = \{i\}$, in which case $\eta_i = \tau_i = \xi_i$.

**Theorem 4.13** *Let* $\{\xi_i,\ i \in \mathcal{J}\}$ *be a random field with mean zero and* $\mathrm{Var}(W) = 1$ *where* $W = \sum_{i \in \mathcal{J}} \xi_i$. *If* (LD1) *holds then, then with* $\eta_i$ *as in* (4.178),

$$\left\| \mathcal{L}(W) - \mathcal{L}(Z) \right\|_1 \le \sqrt{\frac{2}{\pi}} E \left| \sum_{i \in \mathcal{J}} \{\xi_i \eta_i - E(\xi_i \eta_i)\} \right| + \sum_{i \in \mathcal{J}} E \left| \xi_i \eta_i^2 \right|, \tag{4.179}$$

*and if* (LD2) *holds, then with* $\eta_i$ *and* $\tau_i$ *as in* (4.178),

$$\left\| \mathcal{L}(W) - \mathcal{L}(Z) \right\|_1 \le 2 \sum_{i \in \mathcal{J}} \left( E|\xi_i \eta_i \tau_i| + \left| E(\xi_i \eta_i) \right| E|\tau_i| \right) + \sum_{i \in \mathcal{J}} E \left| \xi_i \eta_i^2 \right|. \tag{4.180}$$

We remark that for independent random variables, applying Hölder's inequality to the bound in (4.180) yields $5 \sum_{i \in \mathcal{J}} E|\xi_i|^3$, somewhat larger than the constant of 1 given by Corollary 4.2.

*Proof* Assume (LD1) holds and let $f = f_h$ be the solution of the Stein equation (2.4) for an absolutely continuous function $h$ satisfying $\|h'\| \le 1$. By the independence of $\xi_i$ and $W - \eta_i$, and that $E\xi_i = 0$, we have

$$E\{Wf(W)\} = \sum_{i \in \mathcal{J}} E\xi_i f(W) = \sum_{i \in \mathcal{J}} E\xi_i \left[ f(W) - f(W - \eta_i) \right].$$

Now adding and subtracting yields

$$\begin{aligned} E\{Wf(W)\} &= \sum_{i \in \mathcal{J}} E\left\{\xi_i \left[ f(W) - f(W - \eta_i) - \eta_i f'(W) \right] \right\} \\ &\quad + E\left\{ \left( \sum_{i \in \mathcal{J}} \xi_i \eta_i \right) f'(W) \right\}. \end{aligned} \tag{4.181}$$

Now, using again that $E\xi_i = 0$ for all $i$, from (LD1) it follows that

$$1 = EW^2 = \sum_{i \in \mathcal{J}} \sum_{j \in \mathcal{J}} E\{\xi_i \xi_j\} = \sum_{i \in \mathcal{J}} E\{\xi_i \eta_i\},$$

and so

$$\begin{aligned} E\{f'(W) - Wf(W)\} &= -E\left( \sum_{i \in \mathcal{J}} \{\xi_i \eta_i - E(\xi_i \eta_i)\} f'(W) \right) \\ &\quad - \sum_{i \in \mathcal{J}} E\left\{ \xi_i \left[ f(W) - f(W - \eta_i) - \eta_i f'(W) \right] \right\}. \end{aligned} \tag{4.182}$$

By (2.13), $\|f'\| \le \sqrt{2/\pi}$ and $\|f''\| \le 2$. Therefore it follows from (4.182) and a Taylor expansion that

$$\left|Eh(W) - Eh(Z)\right| \le \sqrt{\frac{2}{\pi}} E\left|\sum_{i\in\mathcal{J}} \{\xi_i \eta_i - E(\xi_i \eta_i)\}\right| + \sum_{i\in\mathcal{J}} E\left|\xi_i \eta_i^2\right|.$$

Now (4.179) follows from (4.8).

When (LD2) is satisfied, $f'(W - \tau_i)$ and $\xi_i \eta_i$ are independent for each $i \in \mathcal{J}$. Hence, using (4.182), we can write

$$\begin{aligned}
&\left|Eh(W) - Eh(Z)\right| \\
&\quad \le \left|E\sum_{i\in\mathcal{J}} \{\xi_i \eta_i - E(\xi_i \eta_i)\}\big(f'(W) - f'(W - \tau_i)\big)\right| + \sum_{i\in\mathcal{J}} E\left|\xi_i \eta_i^2\right| \\
&\quad \le 2\sum_{i\in\mathcal{J}} \big(E|\xi_i \eta_i \tau_i| + \left|E(\xi_i \eta_i)\right| E|\tau_i|\big) + \sum_{i\in\mathcal{J}} E\left|\xi_i \eta_i^2\right|,
\end{aligned}$$

as desired. □

We provide two examples of locally dependent random variables. We refer to Baldi and Rinott (1989), Rinott (1994), Baldi et al. (1989), Dembo and Rinott (1996), and Chen and Shao (2004) for more details.

*Example 4.1* (Graphical dependence) Consider a set of random variables $\{\xi_i, i \in \mathcal{V}\}$ indexed by the vertices of a graph $\mathcal{G} = (\mathcal{V}, \mathcal{E})$. The graph $\mathcal{G}$ is said to be a dependency graph if, for any pair of disjoint sets $\Gamma_1$ and $\Gamma_2$ in $\mathcal{V}$ such that no edge in $\mathcal{E}$ has one endpoint in $\Gamma_1$ and the other in $\Gamma_2$, the sets of random variables $\{\xi_i, i \in \Gamma_1\}$ and $\{\xi_i, i \in \Gamma_2\}$ are independent. Let

$$A_i = \{i\} \cup \big\{j \in \mathcal{V}\colon \{i, j\} \in \mathcal{E}\big\}$$

and $B_i = \bigcup_{j\in A_i} A_j$. Then $\{\xi_i, i \in \mathcal{V}\}$ satisfies (LD2). Hence (4.180) holds.

*Example 4.2* (The number of local maxima on a graph) Consider a graph $\mathcal{G} = (\mathcal{V}, \mathcal{E})$ (which is not necessary a dependency graph) and independent and identically distributed continuous random variables $\{Y_i, i \in \mathcal{V}\}$. For $i \in \mathcal{V}$ define the indicator variable

$$\xi_i = \begin{cases} 1 & \text{if } Y_i > Y_j \text{ for all } j \in \mathcal{N}_i, \\ 0 & \text{otherwise} \end{cases}$$

where $\mathcal{N}_i = \{j \in \mathcal{V}\colon \{i, j\} \in \mathcal{E}\}$. Hence $\xi_i = 1$ indicates that $Y_i$ is a local maximum and $W = \sum_{i\in\mathcal{V}} \xi_i$ is the total number of local maxima. Letting

$$A_i = \{i\} \cup \mathcal{N}_i \cup \bigcup_{j\in\mathcal{N}_i} \mathcal{N}_j \quad \text{and} \quad B_i = \bigcup_{j\in A_i} A_j$$

we find that $\{\xi_i, i \in \mathcal{V}\}$ satisfies (LD2), and therefore (4.180) holds. Bounds in $L^\infty$ for this problem are considered in Example 6.4.

## 4.8 Smooth Function Bounds

In defining a distance $\|\mathcal{L}(X) - \mathcal{L}(Y)\|_{\mathcal{H}}$ through (4.1) one typically chooses $\mathcal{H}$ to be a convergence determining class of functions, that is, a collection of functions such that if $\{X_n\}_{n\geq 0}$ is any sequence of random variables then

$$Eh(X_n) \to Eh(X_0) \quad \text{for all } h \in \mathcal{H} \text{ implies } X_n \to_d X_0.$$

A convergence determining class can consist of functions all of which are very smooth, such as the collection of all infinity differentiable functions with compact support.

To describe the collection of functions we consider in this section, following E.M. Stein (1970), let $L_m^\infty(\mathbb{R})$ be all functions $h : \mathbb{R} \to \mathbb{R}$ satisfying $\|h\|_{L_m^\infty(\mathbb{R})} < \infty$ where

$$\|h\|_{L_m^\infty(\mathbb{R})} = \max_{0\leq k\leq m} \|h^{(k)}\|.$$

That is $L_m^\infty(\mathbb{R})$ consists of all functions possessing $m$ bounded derivatives. Now let $\|\mathcal{L}(W) - \mathcal{L}(Z)\|_{\mathcal{H}_{m,\infty}}$ be the distance which is obtained through (4.1) by setting

$$\mathcal{H}_{m,\infty} = \{h \in L_m^\infty(\mathbb{R}) : \|h\|_{L_m^\infty(\mathbb{R})} \leq 1\}. \tag{4.183}$$

In the following section we show how fast rates of convergence can be obtained under a vanishing third moment assumption when inducing our distance by $\mathcal{H}_{4,\infty}$. In Chap. 12 we prove a smooth function theorem in $\mathbb{R}^p$ using a multidimensional generalization of the distances defined here, and produce bounds in that distance for the problem of counting the number of vertices in a random graph that have specified degree counts.

### *4.8.1 Fast Rates for Smooth Functions*

In this section we first prove Theorem 4.14, a smooth function theorem parallel to Theorem 4.9, for the zero bias coupling as discussed in Sect. 2.3.3. Comparing Theorems 4.9 and 4.14, we see that the latter requires the computation of a conditional expectation of a difference, rather than of a difference squared, and that the second, or remainder term is of a square, rather than a cube. Lastly, Theorem 4.9 requires the linearity condition (4.108) to be satisfied, whereas Theorem 4.14 does not. After the proof we apply Theorem 4.14 in an independent case to show that fast rates of convergence for smooth functions are obtained when fourth moments exist and third moment vanishes.

**Theorem 4.14** *Let $W$ be a mean zero, variance* 1 *random variable and suppose that the pair $(W, W^*)$ is given on a joint probability space so that $W^*$ has the $W$-zero biased distribution. Then*

$$\|\mathcal{L}(W) - \mathcal{L}(Z)\|_{\mathcal{H}_{4,\infty}} \leq \frac{1}{3} E\left|E(W^* - W|W)\right| + \frac{1}{8} E(W^* - W)^2.$$

*Proof* Let $g$ be the solution to (2.19) for a given $h \in \mathcal{H}_{4,\infty}$. By the bounds in Lemma 2.6

$$\|g^{(3)}\| \le \frac{1}{3} \quad \text{and} \quad \|g^{(4)}\| \le \frac{1}{4}. \tag{4.184}$$

By (2.19), (2.51), and Taylor expansion,

$$\begin{aligned} |Eh(W) - Nh| &= |E(g''(W) - Wg'(W))| \\ &= |E(g''(W^*) - g''(W))| \\ &\le |Eg^{(3)}(W)(W^* - W)| + \left| E\int_W^{W^*} g^{(4)}(t)(W^* - t)dt \right|. \end{aligned}$$

Conditioning on $W$ we may bound the first term as

$$|E[g^{(3)}(W)E(W^* - W|W)]| \le \|g^{(3)}\| E|E(W^* - W|W)|.$$

For the second term

$$\left| E\int_W^{W^*} g^{(4)}(t)(W^* - t)dt \right| \le \frac{1}{2}\|g^{(4)}\| E(W^* - W)^2.$$

Applying (4.184) completes the proof. □

We now apply Theorem 4.14 to the sum of independent identically distributed variables and show how the zero bias transformation leads to an error bound for smooth functions of order $n^{-1}$, under additional moment assumptions which include a vanishing third moment.

**Corollary 4.4** *Let $X_1, X_2, \ldots, X_n$ be independent and identically distributed mean zero, variance one random variables with vanishing third moment and $EX^4 < \infty$. Then, for $W = n^{-1/2}\sum_{i=1}^n X_i$,*

$$\|\mathcal{L}(W) - \mathcal{L}(Z)\|_{\mathcal{H}_{4,\infty}} \le \frac{1}{24n}(11 + EX^4).$$

*Proof* For $i = 1, \ldots, n$ let $X_i^*$ have the $X_i$-zero biased distribution and be independent of $X_j, j = 1, \ldots, n$, and $I$ a random index independent of $X_i, X_i^*, i = 1, \ldots, n$ with distribution

$$P(I = i) = 1/n.$$

Then, by Lemma 2.8 and the scaling property (2.59),

$$W^* = W - X_I/\sqrt{n} + X_I^*/\sqrt{n}$$

has the $W$-zero biased distribution.

From substituting $f(x) = x^2/2$ into (2.51), for every $i = 1, \ldots, n$ we have

$$EX_i^* = (1/2)EX_i^3 = 0 \quad \text{and therefore} \quad EX_I^* = 0. \tag{4.185}$$

Next, using that $X_1, \ldots, X_n$'s are i.i.d., and therefore exchangeable, $E(X_I|W) = W/\sqrt{n}$.

Now, by the independence of $X_I^*$ and $W$, and (4.185), we obtain

$$\begin{aligned} E(W^* - W|W) &= n^{-1/2} E(X_I^* - X_I|W) \\ &= n^{-1/2}\big(E(X_I^*) - E(X_I|W)\big) \\ &= -n^{-1/2} E(X_I|W) \\ &= -n^{-1} W. \end{aligned}$$

Therefore

$$E\big|E(W^* - W|W)\big| = n^{-1} E|W| \leq \frac{1}{n}.$$

For the second term in Theorem 4.14, application of (2.51) with $f(x) = x^3/3$ yields

$$E(X_I^*)^2 = E(X_i^*)^2 = \frac{1}{3} E X_i^4.$$

Since $X_I$ and $X_I^*$ are independent, and the latter variable has mean zero,

$$E(W^* - W)^2 = \frac{1}{n} E(X_I^* - X_I)^2 = \frac{1}{n}\big(E(X_I^*)^2 + E X_I^2\big) = \frac{1}{n}\left(\frac{E X^4}{3} + 1\right).$$

Applying Theorem 4.14 now yields the claim. □

Under more special assumptions a fast rates may be obtained for distances induced by classes of non-smooth functions. In particular, Klartag (2009) demonstrates a bound of order $1/n$ for cases which include the sum of independent symmetric random variables whose density is log concave.

## Appendix

*Proof of Lemma 4.7* Let $\pi$ be uniform on $\mathcal{S}_n$ and $I^\dagger, J^\dagger, K^\dagger, L^\dagger$ be independent of $\pi$ with distribution (4.134). Constructing $Y$ from $\pi$ and $Y^\dagger$ and $Y^\ddagger$ from $\pi^\dagger$ and $\pi^\ddagger$ respectively, as in Lemma 4.6, we have

$$\begin{aligned} Y^* - Y &= U Y^\dagger + (1 - U) Y^\ddagger - Y \\ &= U \sum_{i=1}^n a_{i,\pi^\dagger(i)} + (1 - U) \sum_{i=1}^n a_{i,\pi^\ddagger(i)} - \sum_{i=1}^n a_{i,\pi(i)}. \end{aligned}$$

With

$$\mathcal{I} = \big\{I^\dagger, J^\dagger, \pi^{-1}(K^\dagger), \pi^{-1}(L^\dagger)\big\}, \tag{4.186}$$

we see from (4.126) in Lemma 4.5, and from $\pi^\ddagger = \pi^\dagger \tau_{I^\dagger, J^\dagger}$, that if $m \notin \mathcal{I}$, then $\pi(m) = \pi^\dagger(m) = \pi^\ddagger(m)$. Hence, setting $V = Y^* - Y$, we have

$$V = \sum_{i \in \mathcal{I}} \big(U a_{i,\pi^\dagger(i)} + (1 - U) a_{i,\pi^\ddagger(i)} - a_{i,\pi(i)}\big). \tag{4.187}$$

Further, letting

$$R = \left|\{\pi(I^\dagger), \pi(J^\dagger)\} \cap \{K^\dagger, L^\dagger\}\right|$$

and $\mathbf{1}_k = \mathbf{1}(R = k)$, since $P(R \le 2) = 1$, we have

$$\begin{aligned} V &= V\mathbf{1}_2 + V\mathbf{1}_1 + V\mathbf{1}_0, \\ \text{and therefore} \quad E|V| &\le E|V|\mathbf{1}_2 + E|V|\mathbf{1}_1 + E|V|\mathbf{1}_0. \end{aligned} \tag{4.188}$$

The three terms on the right hand side of (4.188) give rise to the three components of the bound in the theorem.

For notational simplicity, the following summations in this section are performed over all indices which appear, whether in the summands or in a (possibly empty) collection of restrictions. In what follows, we will apply equalities and bounds such as

$$\begin{aligned} \sum |a_{il}|\big[(a_{ik} + a_{jl}) - (a_{il} + a_{jk})\big]^2 &= \sum |a_{il}|\big(a_{ik}^2 + a_{jl}^2 + a_{il}^2 + a_{jk}^2\big) \\ &\le 4n^2\gamma. \end{aligned} \tag{4.189}$$

Due to the form of the terms being squared on the left-hand side, if the factors in a cross term agree in their first index, they will have differing second indices, and likewise if their second indices agree. This gives cross terms which are zero by virtue of (4.120), since there will be at least one unpaired index outside the absolute value over which to sum, for instance, the index $k$ in the term $\sum |a_{il}|a_{ik}a_{il}$. Hence the equality. To obtain the inequality, on each of the four terms are argue as for the first,

$$\sum_{i,j,k,l} |a_{i,l}|a_{i,k}^2 \le \Big(\sum_{j,k} |a_{il}|^3\Big)^{1/3} \Big(\sum_{j,l} |a_{ik}|^3\Big)^{2/3} = n^2\gamma. \tag{4.190}$$

Generally, the power of $n$ in such an inequality, in this case 2, will be 2 less than the number of indices of summation, in this case 4.

**Calculation on $R = 2$** On $\mathbf{1}_2$ we have $\{\pi(I^\dagger), \pi(J^\dagger)\} = \{K^\dagger, L^\dagger\}$ and therefore $\mathcal{I} = \{I^\dagger, J^\dagger, \pi^{-1}(K^\dagger), \pi^{-1}(L^\dagger)\} = \{I^\dagger, J^\dagger\}$. As the intersection which gives $R = 2$ can occur in two different ways, we make the further decomposition

$$V\mathbf{1}_2 = V\mathbf{1}_{2,1} + V\mathbf{1}_{2,2},$$

where

$$\begin{aligned} \mathbf{1}_{2,1} &= \mathbf{1}\big(\pi(I^\dagger) = K^\dagger,\ \pi(J^\dagger) = L^\dagger\big) \\ \text{and} \quad \mathbf{1}_{2,2} &= \mathbf{1}\big(\pi(I^\dagger) = L^\dagger,\ \pi(J^\dagger) = K^\dagger\big). \end{aligned}$$

Since $\pi^\dagger = \pi$ on $\mathbf{1}_{2,1}$ by (4.125), following (4.187) we have

$$\begin{aligned}
V\mathbf{1}_{2,1} &= \sum_{i\in\{I^\dagger,J^\dagger\}} \big(Ua_{i,\pi^\dagger(i)} + (1-U)a_{i,\pi^\ddagger(i)} - a_{i,\pi(i)}\big)\mathbf{1}_{2,1} \\
&= \big[U(a_{I^\dagger,\pi^\dagger(I^\dagger)} + a_{J^\dagger,\pi^\dagger(J^\dagger)}) + (1-U)(a_{I^\dagger,\pi^\ddagger(I^\dagger)} + a_{J^\dagger,\pi^\ddagger(J^\dagger)}) \\
&\quad - (a_{I^\dagger,\pi(I^\dagger)} + a_{J^\dagger,\pi(J^\dagger)})\big]\mathbf{1}_{2,1} \\
&= \big[U(a_{I^\dagger,\pi(I^\dagger)} + a_{J^\dagger,\pi(J^\dagger)}) + (1-U)(a_{I^\dagger,\pi(J^\dagger)} + a_{J^\dagger,\pi(I^\dagger)}) \\
&\quad - (a_{I^\dagger,\pi(I^\dagger)} + a_{J^\dagger,\pi(J^\dagger)})\big]\mathbf{1}_{2,1} \\
&= (1-U)(a_{I^\dagger,\pi(J^\dagger)} + a_{J^\dagger,\pi(I^\dagger)} - a_{I^\dagger,\pi(I^\dagger)} - a_{J^\dagger,\pi(J^\dagger)})\mathbf{1}_{2,1} \\
&= (1-U)(a_{I^\dagger,L^\dagger} + a_{J^\dagger,K^\dagger} - a_{I^\dagger,K^\dagger} - a_{J^\dagger,L^\dagger})\mathbf{1}_{2,1}. \qquad (4.191)
\end{aligned}$$

Due to the presence of the indicator $\mathbf{1}_{2,1}$, taking the expectation of (4.191) requires a joint distribution which includes the values taken on by $\pi$ at $I^\dagger$ and $J^\dagger$, say $s$ and $t$, respectively. Since these images can be any two distinct values, and are independent of $I^\dagger$, $J^\dagger$, $K^\dagger$ and $L^\dagger$, we have, with $p_1$ and $p_2$ given in (4.122) and (4.134), respectively,

$$\begin{aligned}
p_3(i,j,k,l,s,t) &= P\big((I^\dagger,J^\dagger,K^\dagger,L^\dagger,\pi(I^\dagger),\pi(J^\dagger)) = (i,j,k,l,s,t)\big) \\
&= p_2(i,j,k,l)p_1(s,t) \\
&= \frac{[(a_{ik}+a_{jl}) - (a_{il}+a_{jk})]^2}{4n^3(n-1)^2\sigma^2}\mathbf{1}(s\neq t). \qquad (4.192)
\end{aligned}$$

Now bounding the absolute value of the first term in (4.191) using (4.189), we obtain

$$\begin{aligned}
E\big|(1-U)a_{I^\dagger,L^\dagger}\big|\mathbf{1}_{2,1} &= \frac{1}{2}\sum |a_{il}|\mathbf{1}(s=k,t=l)p_3(i,j,k,l,s,t) \\
&= \frac{1}{2}\sum |a_{il}|p_3(i,j,k,l,k,l) \\
&= \frac{1}{8n^3(n-1)^2\sigma^2}\sum |a_{il}|\big[(a_{ik}+a_{jl}) - (a_{il}+a_{jk})\big]^2 \\
&\le \frac{\gamma}{2n(n-1)^2\sigma^2}.
\end{aligned}$$

Using the triangle inequality in (4.191) and applying the same reasoning to the remaining three terms shows that $E|V|\mathbf{1}_{2,1} \le 2\gamma/(n(n-1)^2\sigma^2)$. Since by symmetry the term $V\mathbf{1}_{2,2}$ can be handled the same way, we obtain

$$E|V|\mathbf{1}_2 \le \frac{4\gamma}{n(n-1)^2\sigma^2} \le \frac{4\gamma}{(n-1)^3\sigma^2}. \qquad (4.193)$$

**Calculation on $R=1$** As the event $R=1$ can occur in four different ways, depending on which element of $\{\pi(I^\dagger),\pi(J^\dagger)\}$ equals an element of $\{K^\dagger,L^\dagger\}$, we decompose $\mathbf{1}_1$ to yield

$$V\mathbf{1}_1 = V\mathbf{1}_{1,1} + V\mathbf{1}_{1,2} + V\mathbf{1}_{1,3} + V\mathbf{1}_{1,4}, \qquad (4.194)$$

where $\mathbf{1}_{1,1} = \mathbf{1}(\pi(I^\dagger) = K^\dagger \text{ and } \pi(J^\dagger) \neq L^\dagger)$, specifying the remaining three indicators in (4.194) similarly.

On $\mathbf{1}_{1,1}$ we have, from (4.186), that $\mathcal{I} = \{I^\dagger, J^\dagger, \pi^{-1}(L^\dagger)\}$, and from (4.125) that $\pi^\dagger = \pi\tau_{\pi^{-1}(L^\dagger),J^\dagger}$ and so $\pi^\ddagger = \pi\tau_{\pi^{-1}(L^\dagger),J^\dagger}\tau_{J^\dagger,I^\dagger}$, yielding $\pi^\ddagger(\pi^{-1}(L)) = \pi^\dagger(\pi^{-1}(L)) = \pi(J)$. Now, using (4.187),

$$\begin{aligned}
V\mathbf{1}_{1,1} &= \sum_{i\in\{I^\dagger,J^\dagger,\pi^{-1}(L^\dagger)\}} \bigl(Ua_{i,\pi^\dagger(i)} + (1-U)a_{i,\pi^\ddagger(i)} - a_{i,\pi(i)}\bigr)\mathbf{1}_{1,1} \\
&= \bigl[U(a_{I^\dagger,\pi^\dagger(I^\dagger)} + a_{J^\dagger,\pi^\dagger(J^\dagger)} + a_{\pi^{-1}(L^\dagger),\pi^\dagger(\pi^{-1}(L^\dagger))}) \\
&\quad + (1-U)(a_{I^\dagger,\pi^\ddagger(I^\dagger)} + a_{J^\dagger,\pi^\ddagger(J^\dagger)} + a_{\pi^{-1}(L^\dagger),\pi^\ddagger(\pi^{-1}(L^\dagger))}) \\
&\quad - (a_{I^\dagger,\pi(I^\dagger)} + a_{J^\dagger,\pi(J^\dagger)} + a_{\pi^{-1}(L^\dagger),\pi(\pi^{-1}(L^\dagger))})\bigr]\mathbf{1}_{1,1} \\
&= \bigl[U(a_{I^\dagger,K^\dagger} + a_{J^\dagger,L^\dagger} + a_{\pi^{-1}(L^\dagger),\pi(J^\dagger)}) \\
&\quad + (1-U)(a_{I^\dagger,L^\dagger} + a_{J^\dagger,K^\dagger} + a_{\pi^{-1}(L^\dagger),\pi(J^\dagger)}) \\
&\quad - (a_{I^\dagger,K^\dagger} + a_{J^\dagger,\pi(J^\dagger)} + a_{\pi^{-1}(L^\dagger),L^\dagger})\bigr]\mathbf{1}_{1,1} \\
&= \bigl[Ua_{J^\dagger,L^\dagger} + (1-U)(a_{I^\dagger,L^\dagger} + a_{J^\dagger,K^\dagger} - a_{I^\dagger,K^\dagger}) \\
&\quad - a_{J^\dagger,\pi(J^\dagger)} - a_{\pi^{-1}(L^\dagger),L^\dagger} + a_{\pi^{-1}(L^\dagger),\pi(J^\dagger)}\bigr]\mathbf{1}_{1,1}.
\end{aligned} \tag{4.195}$$

For the first term in (4.195), dropping the restriction $t \neq l$ and summing over $t$ to obtain the first inequality, and then applying (4.189) with $|a_{il}|$ replaced by $|a_{jl}|$, we obtain

$$\begin{aligned}
EU|a_{J^\dagger,L^\dagger}|\mathbf{1}_{1,1} &= \frac{1}{2}\sum |a_{jl}|\mathbf{1}(s=k, t\neq l)p_3(i,j,k,l,s,t) \\
&\le \frac{1}{8n^2(n-1)^2\sigma^2}\sum |a_{jl}|\bigl[(a_{ik}+a_{jl}) - (a_{il}+a_{jk})\bigr]^2 \\
&\le \frac{\gamma}{2(n-1)^2\sigma^2}.
\end{aligned} \tag{4.196}$$

The second, third and fourth terms in (4.195) also may be bounded by (4.196) upon replacing $|a_{jl}|$ by $|a_{il}|$, $|a_{jk}|$ and $|a_{ik}|$, respectively, yielding

$$E\bigl|Ua_{J^\dagger,L^\dagger} + (1-U)(a_{I^\dagger,L^\dagger} + a_{J^\dagger,K^\dagger} - a_{I^\dagger,K^\dagger})\bigr|\mathbf{1}_{1,1} \le \frac{2\gamma}{(n-1)^2\sigma^2}. \tag{4.197}$$

For the fifth term in (4.195), that is, for $-a_{J^\dagger,\pi(J^\dagger)}$, reasoning similarly,

$$\begin{aligned}
E|a_{J^\dagger,\pi(J^\dagger)}|\mathbf{1}_{1,1} &= \sum |a_{jt}|\mathbf{1}(s=k, t\neq l)p_3(i,j,k,l,s,t) \\
&\le \frac{1}{4n^3(n-1)^2\sigma^2}\sum |a_{jt}|\bigl[(a_{ik}+a_{jl}) - (a_{il}+a_{jk})\bigr]^2 \\
&\le \frac{\gamma}{(n-1)^2\sigma^2}.
\end{aligned} \tag{4.198}$$

Note that for the final inequality, though the sum being bounded is not of the form (4.189), having the index $t$, the same reasoning applies and that, moreover, the five indices of summation require that $n^2$ in (4.190) be replaced by $n^3$.

To handle the sixth term in (4.195), $-a_{\pi^{-1}(L^\dagger),L^\dagger}$, we need the joint distribution

$$\begin{aligned}&p_4(i,j,k,l,s,t,u)\\&\quad=P\big((I^\dagger,J^\dagger,K^\dagger,L^\dagger,\pi(I^\dagger),\pi(J^\dagger),\pi^{-1}(L^\dagger))=(i,j,k,l,s,t,u)\big),\end{aligned}$$

accounting for the value $u$ taken on by $\pi^{-1}(L^\dagger)$. If $l$ equals $s$ or $t$, then $u$ is already fixed at $i$ or $j$, respectively; otherwise, $\pi^{-1}(L^\dagger)$ is free to take any of the remaining available $n-2$ values, with equal probability. Hence, with $p_3$ given by (4.192), we deduce that

$$p_4(i,j,k,l,s,t,u)=\begin{cases}p_3(i,j,k,l,s,t), & \text{if } (l,u)\in\{(s,i),(t,j)\},\\ p_3(i,j,k,l,s,t)\frac{1}{n-2}, & \text{if } l\notin\{s,t\} \text{ and } u\notin\{i,j\},\\ 0, & \text{otherwise.}\end{cases}$$

Note, for example, that on $\mathbf{1}_{1,1}$, where $\pi(I^\dagger)=K^\dagger$ and $\pi(J^\dagger)\neq L^\dagger$, the value $u$ of $\pi^{-1}(L^\dagger)$ is neither $I^\dagger$ nor $J^\dagger$, so the second case above is the relevant one and the vanishing of the first sum on the third line of the following display is to be expected.

Now, applying the density $p_4$ we may bound the sixth term in (4.195) as follows,

$$\begin{aligned}&E|a_{\pi^{-1}(L^\dagger),L^\dagger}|\mathbf{1}_{1,1}\\&\quad=\sum|a_{ul}|\mathbf{1}(s=k,t\neq l)p_4(i,j,k,l,s,t,u)\\&\quad=\sum_{t\neq l}|a_{ul}|p_4(i,j,k,l,k,t,u)\\&\quad=\sum|a_{ik}|p_3(i,j,k,k,k,t)+\frac{1}{n-2}\sum_{l\notin\{k,t\},u\notin\{i,j\}}|a_{ul}|p_3(i,j,k,l,k,t)\\&\quad=\frac{1}{n-2}\sum_{l\neq t,u\notin\{i,j\}}|a_{ul}|p_2(i,j,k,l)p_1(k,t)\\&\quad=\frac{1}{(n)_3}\sum_{t\notin\{l,k\},u\notin\{i,j\}}|a_{ul}|p_2(i,j,k,l) \qquad (4.199)\\&\quad=\frac{1}{(n)_2}\sum_{u\notin\{i,j\}}|a_{ul}|p_2(i,j,k,l)\\&\quad\le\frac{1}{4n^3(n-1)^2\sigma^2}\sum|a_{ul}|\big[(a_{ik}+a_{jl})-(a_{il}+a_{jk})\big]^2\\&\quad\le\frac{\gamma}{(n-1)^2\sigma^2}, \qquad (4.200)\end{aligned}$$

where the final inequality is achieved using (4.189) in the same way as for (4.198).

The computation for the seventh term in (4.195) begins as that for the sixth, yielding (4.199) with $a_{ut}$ replacing $a_{ul}$, so that

$$\begin{aligned}E|a_{\pi^{-1}(L^\dagger),\pi(J^\dagger)}|\mathbf{1}_{1,1}&=\frac{1}{(n)_3}\sum_{t\notin\{l,k\},u\notin\{i,j\}}|a_{ut}|p_2(i,j,k,l)\\&\le\frac{1}{4(n)_3n^2(n-1)\sigma^2}\sum|a_{ut}|\big[(a_{ik}+a_{jl})-(a_{il}+a_{jk})\big]^2\end{aligned}$$

$$\begin{aligned} &\le \frac{n^2\gamma}{(n)_3(n-1)\sigma^2} \\ &\le \frac{3\gamma}{(n-1)^2\sigma^2}, \end{aligned} \tag{4.201}$$

where we have applied reasoning as in (4.189), replaced $n^2$ by $n^4$ in (4.190) due to the sum over six indices, and recalled our assumption that $n \ge 3$.

Returning to (4.195) and adding the contribution (4.197) from the first four terms together with (4.198), (4.200) and (4.201) from the fifth, sixth and seventh, respectively, we obtain $E|V|\mathbf{1}_{1,1} \le 7\gamma/((n-1)^2\sigma^2)$. Since, by symmetry, all four terms on the right-hand side of (4.194) can be handled in the same way as the first, we obtain the following bound on the event $R = 1$:

$$E|V|\mathbf{1}_1 \le \frac{28\gamma}{(n-1)^2\sigma^2}. \tag{4.202}$$

**Calculation on $R = 0$** We may write the indicator of the event that $R = 0$ as

$$\mathbf{1}_0 = \mathbf{1}\big(\pi(I^\dagger) \notin \{K^\dagger, L^\dagger\}, \pi(J^\dagger) \notin \{K^\dagger, L^\dagger\}\big),$$

and we see from (4.186) that $\mathcal{I} = \{I^\dagger, J^\dagger, \pi^{-1}(K^\dagger), \pi^{-1}(L^\dagger)\}$, a set of size 4, on $R = 0$. Hence, from (4.187),

$$\begin{aligned} V\mathbf{1}_0 &= \sum_{i \in \{I^\dagger, J^\dagger, \pi^{-1}(K^\dagger), \pi^{-1}(L^\dagger)\}} \big(Ua_{i,\pi^\dagger(i)} + (1-U)a_{i,\pi^\ddagger(i)} - a_{i,\pi(i)}\big)\mathbf{1}_0 \\ &= \big[U(a_{I^\dagger,K^\dagger} + a_{J^\dagger,L^\dagger}) + (1-U)(a_{I^\dagger,L^\dagger} + a_{J^\dagger,K^\dagger}) \\ &\quad + a_{\pi^{-1}(K^\dagger),\pi(I^\dagger)} + a_{\pi^{-1}(L^\dagger),\pi(J^\dagger)} \\ &\quad - (a_{I^\dagger,\pi(I^\dagger)} + a_{J^\dagger,\pi(J^\dagger)} + a_{\pi^{-1}(K^\dagger),K^\dagger} + a_{\pi^{-1}(L^\dagger),L^\dagger})\big]\mathbf{1}_0. \end{aligned} \tag{4.203}$$

Since the first four terms in (4.203) have the same distribution, we bound their contribution to $E|V|\mathbf{1}_0$, using (4.189), by

$$\begin{aligned} 4EU|a_{I^\dagger,K^\dagger}|\mathbf{1}_0 \le 4EU|a_{I^\dagger,K^\dagger}| &= 2\sum |a_{ik}|p_2(i,j,k,l) \\ &= \frac{1}{2n^2(n-1)\sigma^2}\sum |a_{ik}|\big[(a_{ik}+a_{jl}) - (a_{il}+a_{jk})\big]^2 \\ &\le \frac{2\gamma}{(n-1)\sigma^2}. \end{aligned} \tag{4.204}$$

The sum of the contributions from the fifth and sixth terms of (4.203) can be bounded as

$$\begin{aligned} &2E|a_{\pi^{-1}(L^\dagger),\pi(J^\dagger)}|\mathbf{1}_0 \\ &\quad = 2\sum_{s \notin \{k,l\}, t \notin \{k,l\}} |a_{ut}|p_4(i,j,k,l,s,t,u) \end{aligned}$$

$$
\begin{aligned}
&= \frac{2}{n-2} \sum_{s \notin \{k,l\}, t \notin \{k,l\}, u \notin \{i,j\}, s \neq t} |a_{ut}| p_3(i,j,k,l,s,t) \\
&\le \frac{n-3}{2(n-2)n^3(n-1)^2\sigma^2} \sum |a_{ut}| \big[(a_{ik}+a_{jl})-(a_{il}+a_{jk})\big]^2 \qquad (4.205) \\
&\le \frac{2n(n-3)\gamma}{(n-2)(n-1)^2\sigma^2} \\
&\le \frac{2\gamma}{(n-1)\sigma^2}, \qquad (4.206)
\end{aligned}
$$

where the second equality follows from the form of $p_4$ and that $l \notin \{s,t\}$ implies $(l,u) \notin \{(s,i),(t,j)\}$, inequality (4.205) is obtained by summing over the $n-3$ choices of $s$ and dropping the remaining restrictions, and the next inequality by following the reasoning of (4.189).

Similarly, for the sum of the contributions from the seventh and eighth terms of (4.203), summing over the $n-3$ choices of $t$ and then dropping the remaining restrictions to obtain the first inequality, we have

$$
\begin{aligned}
2E|a_{I^\dagger,\pi(I^\dagger)}|\mathbf{1}_0 &= 2 \sum_{s \notin \{k,l\}, t \notin \{k,l\}} |a_{is}| p_3(i,j,k,l,s,t) \\
&= \frac{1}{2n^3(n-1)^2\sigma^2} \sum_{s \notin \{k,l\}, t \notin \{k,l\}, s \neq t} |a_{is}| \big[(a_{ik}+a_{jl})-(a_{il}+a_{jk})\big]^2 \\
&\le \frac{n-3}{2n^3(n-1)^2\sigma^2} \sum |a_{is}| \big[(a_{ik}+a_{jl})-(a_{il}+a_{jk})\big]^2 \\
&\le \frac{2(n-3)\gamma}{(n-1)^2\sigma^2} \\
&\le \frac{2\gamma}{(n-1)\sigma^2}. \qquad (4.207)
\end{aligned}
$$

The total contribution of the ninth and tenth terms together can be bounded like the sum of the fifth and sixth, yielding (4.205) with $|a_{ul}|$ replacing $|a_{ut}|$, and then summing over the $n$ choices of $t$ to give

$$
\begin{aligned}
2E|a_{\pi^{-1}(L^\dagger),L^\dagger}|\mathbf{1}_0 &\le \frac{n-3}{2(n-2)n^2(n-1)^2\sigma^2} \sum |a_{ul}| \big[(a_{ik}+a_{jl})-(a_{il}+a_{jk})\big]^2 \\
&\le \frac{2n(n-3)\gamma}{(n-2)(n-1)^2\sigma^2} \\
&\le \frac{2\gamma}{(n-1)\sigma^2}. \qquad (4.208)
\end{aligned}
$$

Adding up the bounds for the first four terms (4.204), the fifth and sixth terms (4.206), the seventh and eighth terms (4.207) and the ninth through tenth terms (4.208) yields

$$
E|V|\mathbf{1}_0 \le \frac{8\gamma}{(n-1)\sigma^2}. \qquad (4.209)
$$

Now, from (4.188), adding up the contributions from (4.193), (4.202) and (4.209) from $R=2$, $R=1$, and $R=0$, respectively, for this coupling of $Y^*$ and $Y$ we find that

$$E|Y^* - Y| \le \frac{\gamma}{(n-1)\sigma^2}\left(8 + \frac{28}{(n-1)} + \frac{4}{(n-1)^2}\right).$$

The proof of the lemma may now be completed by noting that $E|Y^* - Y|$ is an upper bound on the $L^1$ norm $\|\mathcal{L}(Y^*) - \mathcal{L}(Y)\|_1$, by the dual form of the $L^1$ norm 4.6.

□

# Chapter 5
# $L^\infty$ by Bounded Couplings

In this chapter we prove a number of Berry–Esseen type theorems, for a random variable $W$, which may be applied when certain couplings are bounded. For example, the first result here, Theorem 5.1, requires the construction of a variable $W^*$, having the $W$-zero bias distribution, on the same space as $W$ such that $|W^* - W| \le \delta$. The theorem is shown by the use of concentration inequalities. Similar results are shown for the exchangeable pair and size bias couplings. In addition to the Kolmogorov distance, we use smoothing inequalities to derive bounds which hold more generally, for distances given in terms of the supremum over classes of non-smooth functions.

We illustrate some of our bounded coupling results to sums of independent bounded random variables, and apply them in more general situations starting in Chap. 6. In addition, some results given in this chapter can handle situations where the couplings are not bounded. Theorem 5.3, for the exchangeable pair, can be applied in the unbounded case when the term $E(W' - W)^2 \mathbf{1}_{\{|W'-W|>a\}}$ can be usefully upper bounded, with similar remarks applying to Theorem 5.7 for the size biased coupling.

## 5.1 Bounded Zero Bias Couplings

The calculation here is greatly simplified due to the assumption of boundedness. For $W$ a mean zero random variable with variance one, recall definition (2.51) of $W^*$, the $W$-zero biased variable. Theorem 4.1 shows that when $W$ and $W^*$ are close then $W$ is close to normal in the $L^1$ sense. Theorem 5.1 below, the bounded zero bias coupling theorem, provides a corresponding result in $L^\infty$.

**Theorem 5.1** *Let $W$ be a mean zero and variance 1 random variable, and suppose that there exists $W^*$, having the $W$-zero bias distribution, defined on the same space as $W$, satisfying $|W^* - W| \le \delta$. Then*

$$\sup_{z\in\mathbb{R}} \left| P(W \le z) - P(Z \le z) \right| \le c\delta$$

*where $c = 1 + 1/\sqrt{2\pi} + \sqrt{2\pi}/4 \le 2.03$.*

L.H.Y. Chen et al., *Normal Approximation by Stein's Method*,
Probability and Its Applications,
DOI 10.1007/978-3-642-15007-4_5, © Springer-Verlag Berlin Heidelberg 2011

We note that the application of this theorem does not require the sometimes difficult calculation of a variance of a conditional expectation, such as the term $\Theta$ in (4.144) of Corollary 4.3, for the exchangeable pair.

Theorem 5.1 may be directly applied to the sum of independent, mean zero random variables $X_1, \ldots, X_n$ all bounded by some $C$, yielding a bound with an explicit constant that has the order of the inverse of the standard deviation of the sum. In particular, let $\sigma_i^2 = \mathrm{Var}(X_i)$, $B_n^2 = \sum_{i=1}^n \sigma_i^2$ and

$$W = \sum_{i=1}^n \xi_i \quad \text{where } \xi_i = X_i / B_n.$$

Then applying Lemma 2.8, (2.58) to yield $|X_i^*| \le C$, and the scaling property (2.59), we obtain, for $I$ an index independent of $X_1, \ldots, X_n$ with $P(I=i)=\sigma_i^2/B_n^2$,

$$W^* - W = \xi_I^* - \xi_I, \quad \text{so in particular} \quad |W^* - W| \le 2C/B_n.$$

Hence the conclusion of Theorem 5.1 holds with $\delta = 2C/B_n$.

The proof of Theorem 5.1 is similar to that of Theorem 3.5, even though the relation between $W$ and a coupled $W^*$ having the $W$-zero bias distribution cannot be expressed as in (3.23).

*Proof* Let $z \in \mathbb{R}$ and $f$ be the solution to the Stein equation (2.2) with $z$ replaced by $z - \delta$. Then

$$\begin{aligned} f'(W^*) &= \mathbf{1}_{\{W^* \le z-\delta\}} - \Phi(z-\delta) + W^* f(W^*) \\ &\le \mathbf{1}_{\{W \le z\}} - \Phi(z-\delta) + W^* f(W^*). \end{aligned}$$

By taking expectation in this inequality, applying a simple bound on the normal distribution function and the zero bias definition (2.51), we obtain

$$\begin{aligned} P(W \le z) - \Phi(z) &= \big(\Phi(z-\delta) - \Phi(z)\big) + P(W \le z) - \Phi(z-\delta) \\ &\ge -\frac{\delta}{\sqrt{2\pi}} + P(W \le z) - \Phi(z-\delta) \\ &\ge -\frac{\delta}{\sqrt{2\pi}} + E\big[f'(W^*) - W^* f(W^*)\big] \\ &= -\frac{\delta}{\sqrt{2\pi}} + E\big[W f(W) - W^* f(W^*)\big]. \end{aligned} \tag{5.1}$$

Writing $\Delta = W^* - W$ and applying the bound (2.10) yields

$$\begin{aligned} \big|E\big(Wf(W) - W^* f(W^*)\big)\big| &= \big|E\big(Wf(W) - (W+\Delta) f(W+\Delta)\big)\big| \\ &\le E\big(\big(|W| + \sqrt{2\pi}/4\big)|\Delta|\big) \\ &\le \delta\big(1 + \sqrt{2\pi}/4\big). \end{aligned}$$

Using this inequality in (5.1) yields

$$P(W \le z) - \Phi(z) \ge -\delta\left(\frac{1}{\sqrt{2\pi}} + 1 + \frac{\sqrt{2\pi}}{4}\right) \ge -2.03\delta.$$

A similar argument yields the reverse inequality. □

## 5.2 Exchangeable Pairs, Kolmogorov Distance

In this section we provide results that give a bound on the Kolmogorov distance when we can construct $W, W'$, an exact or approximate Stein pair, whose difference $|W'-W|$ is bounded. Theorem 5.3 can also be applied when $W'-W$ is not bounded if the term $E(W'-W)^2\mathbf{1}_{|W'-W|>a}$ can be handled.

For a pair $(W, W')$ and a given $\delta$, some of the results in this section are expressed in terms of $\Delta = W - W'$,

$$\hat{K}_1 = E\left(\int_{|t|\le\delta} \hat{K}(t)dt \,\middle|\, W\right)$$
$$\text{with } \hat{K}(t) = \frac{\Delta}{2\lambda}(\mathbf{1}_{\{-\Delta\le t\le 0\}} - \mathbf{1}_{\{0<t\le-\Delta\}}), \tag{5.2}$$

and additionally

$$B = E\left|1 - E\left(\frac{(W-W')^2}{2\lambda}\,\middle|\, W\right)\right|$$
$$\text{and} \quad \Theta = \sqrt{\mathrm{Var}\big(E\big((W'-W)^2|W\big)\big)}. \tag{5.3}$$

When $(W, W')$ is a $\lambda$-Stein pair with variance 1, then by (2.34) we have that $E\Delta^2 = 2\lambda$, and the Cauchy–Schwarz inequality yields

$$B \le \frac{\Theta}{2\lambda}, \tag{5.4}$$

so in such cases $B$ may be replaced by $\Theta/2\lambda$ in all the upper bounds of this section.

We first present result for the exchangeable pair technique which can handle situations where the linear regression condition (2.33),

$$E(W' \mid W) = (1-\lambda)W$$

is satisfied only approximately. Indeed, given any $W$ and $W'$ with mean zero and variance one, one can always express the conditional expectation of $W'$ given $W$ as the linear regression of $W'$ on $W$, with coefficient $1-\lambda$ equal to the correlation coefficient, plus the difference. If the difference is small then the methods which apply when the conditional expectation is linear should apply approximately. Rinott and Rotar (1997) proved such a result, and applied it to the case of non-degenerate $U$-statistics, obtaining a bound of rate $n^{-1/2}$. The result below, along similar lines, is a consequence of Theorem 3.5.

**Theorem 5.2** *If $W, W'$ are mean zero, variance 1 exchangeable random variables satisfying*

$$E(W - W'|W) = \lambda(W - R) \tag{5.5}$$

*for some $\lambda \in (0,1)$ and random variable $R$, and if $|W'-W| \le \delta$ for some $\delta$, then*

$$\sup_{z\in\mathbb{R}}\big|P(W\le z) - \Phi(z)\big| \le \delta\big(1.1 + E\big[|W|\hat{K}_1\big]\big) + 2.7B + \frac{\sqrt{2\pi}}{4}E|R|,$$

*where $\hat{K}_1$ is given by* (5.2).

*When $W, W'$ is mean zero, variance one $\lambda$-Stein pair as in Definition* 2.1, *then with $B$ as in* (5.4)

$$\sup_{z\in\mathbb{R}}\bigl|P(W\le z)-\Phi(z)\bigr|\le 1.1\delta+\frac{\delta^3}{2\lambda}+2.7B.$$

*Proof* As (5.5) and (2.41) are equivalent, by (2.42) identity (3.23) holds with $R_1 = Rf_z(W)$. Further, as $|W'-W|\le\delta$ we have

$$\hat{K}_1 = E\left(\frac{(W-W')^2}{2\lambda}\,\middle|\, W\right).$$

Now applying (2.9) we invoke Theorem 3.5 with $\delta_0=\delta$ and $\delta_1=(\sqrt{2\pi}/4)E|R|$ to obtain the first conclusion. When $W, W'$ is a Stein pair then $R=0$, and the bound on $|W'-W|$ and that $E|W|\le 1$ yields the second conclusion. □

One significant difference between Theorem 5.1 and 5.2 is that the latter bound, for the exchangeable pair coupling, requires the calculation of $B$, or $\Theta$, in (5.3), which may be difficult in particular cases, whereas such a computation is not required for zero bias couplings. Similar remarks apply to the computation of terms appearing in the bounds for the size biased coupling constructions given in Sect. 5.3. However, exchangeable pair and size bias couplings can be constructed for a broader range of examples than zero bias couplings, generally speaking.

We also present a bound for the exchangeable pair coupling from Shao and Su (2005), depending on the following concentration inequality.

**Lemma 5.1** *Let $W', W$ be a $\lambda$-Stein pair with variance* 1. *Then for any $z\in\mathbb{R}$ and $a>0$,*

$$E(W'-W)^2\mathbf{1}_{\{-a\le W'-W\le 0\}}\mathbf{1}_{\{z-a\le W\le z\}}\le 3\lambda a.$$

*Proof* Let

$$f(w)=\begin{cases}-3a/2 & w\le z-2a,\\ w-z+a/2 & z-2a\le w\le z+a,\\ 3a/2 & w\ge z+a.\end{cases}$$

Then using (2.35),

$$\begin{aligned}3a\lambda &\ge 2\lambda E\bigl(Wf(W)\bigr)\\ &= E(W-W')\bigl(f(W)-f(W')\bigr)\\ &= E\left((W-W')\int_{W'-W}^{0} f'(W+t)dt\right)\\ &\ge E\left((W-W')\int_{W'-W}^{0}\mathbf{1}_{\{|t|\le a\}}\mathbf{1}_{\{z-a\le W\le z\}}f'(W+t)dt\right).\end{aligned}$$

Noting that $f'(w+t)=\mathbf{1}_{\{z-2a\le w+t\le z+a\}}$, we have

$$\mathbf{1}_{\{|t|\le a\}}\mathbf{1}_{\{z-a\le W\le z\}}f'(W+t)=\mathbf{1}_{\{|t|\le a\}}\mathbf{1}_{\{z-a\le W\le z\}},$$

and hence

$$\begin{aligned}
3a\lambda &\ge E\left((W-W')\int_{W'-W}^{0}\mathbf{1}_{\{|t|\le a\}}dt\mathbf{1}_{\{z-a\le W\le z\}}\right)\\
&= E\big(|W-W'|\min\big(a,|W-W'|\big)\mathbf{1}_{\{z-a\le W\le z\}}\big)\\
&\ge E\big((W-W')^2\mathbf{1}_{\{0\le W-W'\le a\}}\mathbf{1}_{\{z-a\le W\le z\}}\big).
\end{aligned}$$

□

**Theorem 5.3** *If $W, W'$ are mean zero, variance* 1 *exchangeable random variables satisfying*

$$E(W-W'|W)=\lambda(W-R)$$

*for some $\lambda\in(0,1)$ and some random variable $R$, then for any $a\ge 0$,*

$$\begin{aligned}
&\sup_{z\in\mathbb{R}}\big|P(W\le z)-P(Z\le z)\big|\\
&\quad\le B+\frac{0.41a^3}{\lambda}+1.5a+\frac{E(W'-W)^2\mathbf{1}_{\{|W'-W|>a\}}}{2\lambda}+\frac{\sqrt{2\pi}}{4}E|R|,
\end{aligned}$$

*where $B$ is as in* (5.3).

*If $W, W'$ is a variance one $\lambda$-Stein pair satisfying $|W'-W|\le\delta$, then*

$$\sup_{z\in\mathbb{R}}\big|P(W\le z)-P(Z\le z)\big|\le B+\frac{0.41\delta^3}{\lambda}+1.5\delta.$$

*Proof* Let $f$ be the solution to the Stein equation (2.2) for some arbitrary $z\in\mathbb{R}$. Following the reasoning in the derivation of (2.35), we find that

$$E\big[Wf(W)\big]=\frac{1}{2\lambda}E\big\{(W-W')\big(f(W)-f(W')\big)\big\}+E\big[f(W)R\big].$$

Hence,

$$\begin{aligned}
&\big|P(W\le z)-\Phi(z)\big|\\
&\quad=\big|E\big(f'(W)-Wf(W)\big)\big|\\
&\quad=\left|E\left(f'(W)-\frac{(W'-W)(f(W')-f(W))}{2\lambda}+f(W)R\right)\right|\\
&\quad=\left|E\left(f'(W)\left(1-\frac{(W'-W)^2}{2\lambda}\right)\right.\right.\\
&\quad\qquad\left.\left.+\frac{f'(W)(W'-W)^2-(f(W')-f(W))(W'-W)}{2\lambda}+f(W)R\right)\right|\\
&\quad:=\big|E(J_1+J_2+J_3)\big|,\quad\text{say,}\\
&\quad\le|EJ_1|+|EJ_2|+|EJ_3|.
\end{aligned}\tag{5.6}$$

For the first term, by conditioning and then taking expectation, using (2.8) we obtain

$$|EJ_1| = \left|E\left(f'(W)E\left(\left(1 - \frac{(W'-W)^2}{2\lambda}\right)\middle|W\right)\right)\right| \le B. \tag{5.7}$$

For the third term, applying (2.9) we have

$$|EJ_3| \le \frac{\sqrt{2\pi}}{4} E|R|.$$

To bound $|EJ_2|$, with $a \ge 0$ write

$$\begin{aligned}
&f'(W)(W'-W)^2 - \big(f(W') - f(W)\big)(W'-W) \\
&= (W'-W)\int_0^{W'-W} \big(f'(W) - f'(W+t)\big)dt \\
&= (W'-W)\mathbf{1}_{|W'-W|>a}\int_0^{W'-W} \big(f'(W) - f'(W+t)\big)dt \\
&\quad + (W'-W)\mathbf{1}_{|W'-W|\le a}\int_0^{W'-W} \big(f'(W) - f'(W+t)\big)dt \\
&:= J_{21} + J_{22}, \quad \text{say.}
\end{aligned} \tag{5.8}$$

By (2.8),

$$|EJ_{21}| \le E(W'-W)^2\mathbf{1}_{|W'-W|>a},$$

yielding the second to last term in the bound of the theorem.

Now express $J_{22}$, using (2.2), as the sum

$$\begin{aligned}
&(W'-W)\mathbf{1}_{|W'-W|\le a}\int_0^{W'-W} \big(Wf(W) - (W+t)f(W+t)\big)dt \\
&+ (W'-W)\mathbf{1}_{|W'-W|\le a}\int_0^{W'-W} (\mathbf{1}_{\{W\le z\}} - \mathbf{1}_{\{W+t\le z\}})dt.
\end{aligned} \tag{5.9}$$

Applying (2.10) to the first term in (5.9) shows that the absolute value of its expectation is bounded by

$$\begin{aligned}
&\left|E(W'-W)\mathbf{1}_{|W'-W|\le a}\int_0^{W'-W}\left(|W| + \frac{\sqrt{2\pi}}{4}\right)|t|dt\right| \\
&\le E\left(\frac{1}{2}|W'-W|^3\mathbf{1}_{|W'-W|\le a}\left(|W| + \frac{\sqrt{2\pi}}{4}\right)\right) \\
&\le \frac{1}{2}a^3\left(1 + \frac{\sqrt{2\pi}}{4}\right) \\
&\le 0.82a^3.
\end{aligned}$$

We break up the expectation of the second term in (5.9) according to the sign of $W'-W$. When $W'-W \le 0$, we have

$$
\begin{aligned}
&E\left((W'-W)\mathbf{1}_{\{-a\le W'-W\le 0\}}\int_0^{W'-W}(\mathbf{1}_{\{W\le z\}}-\mathbf{1}_{\{W+t\le z\}})dt\right)\\
&=E\left((W-W')\mathbf{1}_{\{-a\le W'-W\le 0\}}\int_{W'-W}^{0}\mathbf{1}_{\{z-t<W\le z\}}dt\right)\\
&\le E\big((W-W')^2\mathbf{1}_{\{-a\le W'-W\le 0\}}\mathbf{1}_{\{z-a<W\le z\}}\big)\\
&\le 3a\lambda.
\end{aligned}
$$

As a bound may be similarly produced for the case $W'-W\ge 0$, the proof of the first claim is complete. The second claim follows by choosing $a=\delta$ in the first bound, and noting that $R=0$ for a $\lambda$-Stein pair. □

Next we present two results that are obtained by using the linearly smoothed indicator function $h_{z,\alpha}(w)$, as given in (2.14) for $z\in\mathbb{R}$ and $\alpha>0$, which equals the indicator $h_z(w)=\mathbf{1}_{(-\infty,z]}(w)$ over $(-\infty,z]$, decays to zero linearly over $[z,z+\alpha]$, and equals zero on $(z+\alpha,\infty)$. Let

$$\kappa=\sup\big\{\big|Eh_z(W)-Nh_z\big|:\ z\in\mathbb{R}\big\},\tag{5.10}$$

the Kolmogorov distance, and for $\alpha>0$ set

$$\kappa_\alpha=\sup\big\{\big|Eh_{z,\alpha}(W)-Nh_{z,\alpha}\big|:\ z\in\mathbb{R}\big\}.\tag{5.11}$$

**Theorem 5.4** *If $W',W$ is a variance one $\lambda$-Stein pair that satisfies $|W'-W|\le\delta$ for some $\delta$ then*

$$\sup_{z\in\mathbb{R}}\big|P(W\le z)-P(Z\le z)\big|\le\frac{3\delta^3}{\lambda}+2B\tag{5.12}$$

*where $B$ is given by* (5.3).

If $\delta$ is of order $1/\sigma$, $B$ of order $1/\sigma$, and $\lambda$ of order $1/\sigma^2$, then the bound has order $1/\sigma$. A more careful optimization in the proof leads to the improved bound

$$\sup_{z\in\mathbb{R}}\big|P(W\le z)-P(Z\le z)\big|\le\frac{(\sqrt{11\delta^3+10\lambda B}+2\delta^{3/2})^2}{10\lambda}.\tag{5.13}$$

The bound (5.12) follows from (5.13) and the fact that $(\sqrt{a}+\sqrt{b})^2\le 2(a+b)$.

*Proof* For $z\in\mathbb{R}$ arbitrary and $\alpha>0$ let $f$ be the solution (2.4) to the Stein equation for the function $h_{z,\alpha}$ given in (2.14). Decompose $Eh_{z,\alpha}(W)-Nh_{z,\alpha}$ into $E(J_1+J_2)$ as in the proof of Theorem 5.3, noting that here the term $R$ is zero. By the second inequality in (2.15) of Lemma 2.5 we may again bound $|EJ_1|$ by $B$ as in (5.7). From (5.8) with $a=\delta$ we obtain

$$
\begin{aligned}
|EJ_2|&\le\frac{1}{2\lambda}\left|E(W'-W)\int_0^{W'-W}[f'(W)-f'(W+v)]dv\right|\\
&\le\frac{1}{2\lambda}E\left(|W'-W|\int_{0\wedge(W'-W)}^{0\vee(W'-W)}\big|f'(W)-f'(W+v)\big|dv\right).
\end{aligned}
$$

By applying $|W' - W| \le \delta$ and a simple change of variable in (2.16) of Lemma (2.5), we may bound $|EJ_2|$ by

$$\frac{\delta}{2\lambda} E\left( (1+|W|) \int_{0\wedge(W'-W)}^{0\vee(W'-W)} |v| dv + \frac{1}{\alpha} \int_{-\delta}^{\delta} \int_{v\wedge 0}^{v\vee 0} \mathbf{1}_{\{z \le W+u \le z+\alpha\}} du dv \right).$$

As

$$\int_{0\wedge(W'-W)}^{0\vee(W'-W)} |v| dv = \frac{1}{2}(W'-W)^2 \le \frac{\delta^2}{2} \quad \text{and} \quad E|W| \le 1$$

we obtain

$$|EJ_2| \le \frac{\delta}{2\lambda}\left( \delta^2 + \frac{1}{\alpha} \int_{-\delta}^{\delta} \int_{v\wedge 0}^{v\vee 0} P(z \le W+u \le z+\alpha) du dv \right). \tag{5.14}$$

Now, recalling the definitions of $\kappa$ and $\kappa_\alpha$ in (5.10) and (5.11) respectively, as

$$\begin{aligned} P(a \le W \le b) &= \big(P(W \le b) - \Phi(b)\big) - \big(P(W < a) - \Phi(a)\big) + \big(\Phi(b) - \Phi(a)\big) \\ &\le 2\kappa + (b-a)/\sqrt{2\pi}, \end{aligned}$$

we bound (5.14) by

$$\frac{1}{2\lambda}\left( \delta^3 + \delta\alpha^{-1} \int_{-\delta}^{\delta} \int_{v\wedge 0}^{v\vee 0} (2\kappa + 0.4\alpha) du dv \right) \le \frac{1}{2\lambda}\big(1.4\delta^3 + 2\delta^3\alpha^{-1}\kappa\big).$$

Combining the bounds for $|EJ_1|$ and $|EJ_2|$ and taking supremum over $z \in \mathbb{R}$ we obtain

$$\kappa_\alpha \le B + \frac{1}{2\lambda}\big(1.4\delta^3 + 2\delta^3\alpha^{-1}\kappa\big). \tag{5.15}$$

As

$$\begin{aligned} P(W \le z) - \Phi(z) &\le Eh_{z,\alpha}(Z) - \Phi(z) \\ &= Eh_{z,\alpha}(Z) - Nh_{z,\alpha} - \big(\Phi(z) - Nh_{z,\alpha}\big) \\ &\le \kappa_\alpha + \alpha/\sqrt{2\pi}, \end{aligned}$$

with similar reasoning providing a corresponding lower bound, taking supremum over $z \in \mathbb{R}$ we obtain $\kappa \le \kappa_\alpha + 0.4\alpha$. Now applying the bound (5.15) yields

$$\kappa \le \frac{a\alpha + b}{1 - c/\alpha}, \quad \text{where } a = 0.4,\ b = B + \frac{0.7\delta^3}{\lambda} \text{ and } c = \frac{\delta^3}{\lambda}.$$

Now setting $\alpha = 2c$ yields $4ac + 2b$, the right hand side of (5.12). □

Lastly we present a result of Stein (1986), with an improved constant and slightly extended to allow a nonlinear remainder term; this result has the advantage of not requiring the coupling to be bounded. However, the bound supplied by the theorem is typically not of the best order due to its final term. In particular, if $W$ is the sum of i.i.d. variables taking the values $1/\sqrt{n}$ and $-1/\sqrt{n}$ with equal probability and $W'$ is formed from $W$ by replacing a uniformly chosen variable by an independent copy, then $\lambda = 1/n$ and $E|W' - W|^3 = 4/n^{3/2}$, so that the final term in the bound of the theorem below becomes of order $n^{-1/4}$. Nevertheless, in Sect. 14.1 we present a number of important examples where this final term makes no contribution.

**Theorem 5.5** *If $W$, $W'$ are mean zero, variance 1 exchangeable random variables satisfying*

$$E(W - W'|W) = \lambda(W - R) \tag{5.16}$$

*for some $\lambda \in (0, 1)$ and some random variable $R$, then*

$$\sup_{z\in\mathbb{R}} \big|P(W \le z) - \Phi(z)\big| \le B + (2\pi)^{-1/4}\sqrt{\frac{E|W' - W|^3}{\lambda}} + E|R|,$$

*where $B$ is given by* (5.3).

*Proof* For $z \in \mathbb{R}$ and $\alpha > 0$ let $f$ be the solution to the Stein equation for $h_{z,\alpha}$, the smoothed indicator given by (2.14). Decompose $f'(W) - Wf(W)$ into $J_1 + J_2 + J_3$ as in the proof of Theorem 5.3. Applying the first inequality in (2.15) of Lemma 2.5, we may bound the contribution from $|EJ_3|$ by $E|R|$, and from $|EJ_1|$ by $B$ as in (5.7).

Next we claim that for $J_2$, the second term of (5.6), we have

$$\begin{aligned} J_2 &= \frac{1}{2\lambda}(W' - W)\int_W^{W'} \big(f'(W) - f'(t)\big)dt \\ &= \frac{1}{2\lambda}(W' - W)\int_W^{W'}\int_t^W f''(u)dudt \qquad (5.17) \\ &= \frac{1}{2\lambda}(W' - W)\int_{W'}^W (W' - u)f''(u)du. \qquad (5.18) \end{aligned}$$

We obtain (5.18) by first considering $W \le W'$ and rewriting (5.17) as

$$\begin{aligned} -\frac{1}{2\lambda}(W' - W)\int_W^{W'}\int_W^t f''(u)dudt &= -\frac{1}{2\lambda}(W' - W)\int_W^{W'}\int_u^{W'} f''(u)dtdu \\ &= -\frac{1}{2\lambda}(W' - W)\int_W^{W'} (W' - u)f''(u)dtdu, \end{aligned}$$

which equals (5.18).

When $W' \le W$, similarly we have

$$\begin{aligned} \frac{1}{2\lambda}(W' - W)\int_W^{W'}\int_t^W f''(u)dudt &= -\frac{1}{2\lambda}(W' - W)\int_{W'}^W\int_t^W f''(u)dudt \\ &= -\frac{1}{2\lambda}(W' - W)\int_{W'}^W\int_{W'}^u f''(u)dtdu \\ &= -\frac{1}{2\lambda}(W' - W)\int_{W'}^W \big(u - W'\big)f''(u)du, \end{aligned}$$

which is again (5.18).

Since $W$ and $W'$ are exchangeable, the expectation of (5.18) is the same as that of

$$\frac{1}{2\lambda}(W'-W)\int_{W'}^{W}\left(\frac{W+W'}{2}-u\right)f''(u)du,$$

which we bound by the expectation of

$$\|f''\|\frac{1}{2\lambda}|W'-W|\int_{W\wedge W'}^{W\vee W'}\left|\frac{W+W'}{2}-u\right|du=\|f''\|\frac{1}{2\lambda}\frac{|W'-W|^3}{4}$$
$$\le\frac{|W'-W|^3}{4\alpha\lambda},$$

where for the inequality we used the fact that $|h'_{z,\alpha}(x)|\le 1/\alpha$ for all $x\in\mathbb{R}$, and then applied (2.13).

Collecting the bounds, we obtain

$$\begin{aligned}P(W\le z)&\le Eh_{z,\alpha}(W)\\&\le Nh_{z,\alpha}+B+\frac{E|W'-W|^3}{4\alpha\lambda}+E|R|\\&\le\Phi(z)+\frac{\alpha}{\sqrt{2\pi}}+B+\frac{E|W'-W|^3}{4\alpha\lambda}+E|R|.\end{aligned}$$

Evaluating the expression at the minimizer

$$\alpha=\frac{(2\pi)^{1/4}}{2}\sqrt{\frac{E|W'-W|^3}{\lambda}}$$

yields the inequality

$$P(W\le z)-\Phi(z)\le B+(2\pi)^{-1/4}\sqrt{\frac{E|W'-W|^3}{\lambda}}+E|R|.$$

Proving the corresponding lower bound in a similar manner completes the proof of the theorem. □

## 5.3 Size Biasing, Kolmogorov Bounds

We now present two results employing size biased couplings, Theorems 5.6 and 5.7, which parallel Theorems 5.4 and 5.3, respectively, for the exchangeable pair. In particular, in Theorem 5.6 we focus on deriving bounds in the Kolmogorov distance in situations where bounded size bias couplings exist, that is, where one can couple the nonnegative variable $Y$ to $Y^s$ having the $Y$-size biased distribution, so that $|Y^s-Y|$ is bounded. In Theorem 5.7 we require the bounded coupling to satisfy an additional monotonicity condition. In principle, Theorem 5.7, like Theorem 5.3, may be applied in situations where $Y^s-Y$ is not bounded.

For $Y$ a nonnegative random variable with positive mean $\mu$, recall that $Y^s$ has the $Y$-size bias distribution if

$$E\big[Yf(Y)\big] = \mu Ef\big(Y^s\big) \tag{5.19}$$

for all functions $f$ for which the expectations above exist. When $Y$ has finite positive variance $\sigma^2$, we consider the normalized variables

$$W = (Y - \mu)/\sigma \quad \text{and, with some abuse of notation,} \quad W^s = \big(Y^s - \mu\big)/\sigma. \tag{5.20}$$

Given a size bias coupling of $Y$ to $Y^s$, the resulting bounds will be expressed in terms of the quantities $D$ and $\Psi$ given by

$$D = E\left|E\left(1 - \frac{\mu}{\sigma}\big(W^s - W\big)|W\right)\right| \quad \text{and} \quad \Psi = \sqrt{\operatorname{Var}\big(E\big(Y^s - Y|Y\big)\big)}$$
$$\text{which obey } D \le \frac{\mu}{\sigma^2}\Psi. \tag{5.21}$$

To demonstrate the inequality, note that $EY^s = EY^2/\mu$ by (5.19), hence

$$\frac{\mu}{\sigma}E\big(W^s - W\big) = \frac{\mu}{\sigma^2}\left(\frac{EY^2}{\mu} - \mu\right) = 1,$$

so the Cauchy–Schwarz inequality yields

$$D \le \frac{\mu}{\sigma}\sqrt{\operatorname{Var}\big(E\big(W^s - W|W\big)\big)} = \frac{\mu}{\sigma^2}\Psi. \tag{5.22}$$

Therefore $D$ may be replaced by $\mu\Psi/\sigma^2$ in all the upper bounds in this section and the one following.

Note that we cannot apply Theorem 3.5 here, as for a size biased coupling in general there is no guarantee that the function $\hat{K}(t)$ will be non-negative.

**Theorem 5.6** *Let $Y$ be a nonnegative random variable with finite mean $\mu$ and positive, finite variance $\sigma^2$, and suppose $Y^s$, having the $Y$-size biased distribution, may be coupled to $Y$ so that $|Y^s - Y| \le A$ for some $A$. Then with $W = (Y - \mu)/\sigma$ and $D$ as in* (5.21),

$$\sup_{z\in\mathbb{R}}\big|P(W \le z) - P(Z \le z)\big| \le \frac{6\mu A^2}{\sigma^3} + 2D. \tag{5.23}$$

Following Goldstein and Penrose (2010), a more careful optimization in the proof yields the improved bound

$$\sup_{z\in\mathbb{R}}\big|P(W \le z) - P(Z \le z)\big| \le \frac{\mu}{5\sigma^2}\left(\sqrt{\frac{11A^2}{\sigma} + \frac{5\sigma^2}{\mu}D} + \frac{2A}{\sqrt{\sigma}}\right)^2.$$

Again, as for the bound in Theorem 5.4, inequality (5.23) follows from the one above and the fact that $(\sqrt{a} + \sqrt{b})^2 \le 2(a + b)$.

Usually the mean $\mu$ and variance $\sigma^2$ of $Y$ will grow at the same rate, typically $n$, so the bound will asymptotically have order $O(\sigma^{-1})$ when $D$ is of this same order. In Chap. 6, Theorem 5.6 is applied to counting the occurrences of fixed relatively ordered sub-sequences in a random permutation, such as rising sequences, and to counting the occurrences of color patterns, local maxima, and sub-graphs in finite random graphs.

Here we consider a simple application of Theorem 5.6 when $Y$ is the sum of the i.i.d. variables $X_1, \ldots, X_n$ with mean $\theta$ and variance $v^2$, satisfying $0 \le X_i \le A$. In this case $\mu = n\theta$ and $\sigma^2 = nv^2$ so $\mu/\sigma^2 = \theta/v^2$ a constant. Next, applying the construction in Corollary 2.1 we have $Y^s - Y = X_I^s - X_I$, and now using (2.67) and the fact that $X_i$ and $X_i^s$ are nonnegative we obtain

$$\left|Y^s - Y\right| = \left|X_I^s - X_I\right| \le A.$$

Lastly, by independence and exchangeability

$$\mathrm{Var}\big(E\big(Y^s - Y|Y\big)\big) = \mathrm{Var}\big(E\big(X_I^s - X_I|Y\big)\big) = \mathrm{Var}\big(EX_I^s - Y/n\big) = v^2/n,$$

so $\Psi$ in (5.21), and therefore the resulting bound, is of order $1/\sqrt{n}$, with an explicit constant.

*Proof of Theorem 5.6* Fix $z \in \mathbb{R}$ and $\alpha > 0$, and let $f$ solve the Stein equation (2.4) for the linearly smoothed indicator $h_{z,\alpha}(w)$ given in (2.14). Then, letting $W^s = (Y^s - \mu)/\sigma$, applying (5.19) we have

$$\begin{aligned} &E\big(h_{z,\alpha}(W) - Nh_{z,\alpha}\big) \\ &\quad = E\big(f'(W) - Wf(W)\big) \\ &\quad = E\left(f'(W) - \frac{\mu}{\sigma}\big(f\big(W^s\big) - f(W)\big)\right) \\ &\quad = E\left(f'(W)\left(1 - \frac{\mu}{\sigma}\big(W^s - W\big)\right) - \frac{\mu}{\sigma}\int_0^{W^s - W}\big(f'(W+t) - f'(W)\big)dt\right). \end{aligned} \tag{5.24}$$

For the first term, taking expectation by conditioning and then applying the second inequality in (2.15) of Lemma 2.5, we have

$$\left|E\left\{f'(W)E\left(1 - \frac{\mu}{\sigma}\big(W^s - W\big)|W\right)\right\}\right| \le D$$

where $D$ is given by (5.21). Hence, letting $\delta = A/\sigma$ so that $|W^s - W| \le \delta$, applying a change of variable on (2.16) of Lemma 2.5 for the second inequality, and then proceeding as in the proof of Theorem 5.4 yields

$$\begin{aligned} &\big|E\big(h_{z,\alpha}(W) - Nh_{z,\alpha}\big)\big| \\ &\quad \le D + \frac{\mu}{\sigma}E\int_{(W^s - W)\wedge 0}^{(W^s - W)\vee 0}\big|f'(W+t) - f'(W)\big|dt \end{aligned}$$

$$
\begin{aligned}
&\leq D + \frac{\mu}{\sigma} E \int_{(W^s-W)\wedge 0}^{(W^s-W)\vee 0} \left[ \big(1+|W|\big)|t| + \alpha^{-1} \int_{t\wedge 0}^{t\vee 0} \mathbf{1}_{\{z\leq W+u\leq z+\alpha\}} du \right] dt \\
&\leq D + \frac{\mu}{2\sigma}\big(1+E|W|\big)\delta^2 + \frac{\mu}{\sigma}\alpha^{-1} \int_{-\delta}^{\delta} \int_{t\wedge 0}^{t\vee 0} (2\kappa + 0.4\alpha) du dt \\
&\leq D + 1.4\frac{\mu}{\sigma}\delta^2 + 2\frac{\mu}{\sigma}\delta^2\alpha^{-1}\kappa. \qquad (5.25)
\end{aligned}
$$

Now, with $\kappa$ and $\kappa_\alpha$ given in (5.10) and (5.11), respectively, continuing to parallel the proof of Theorem 5.4, taking supremum we see that $\kappa_\alpha$ is bounded by (5.25), and since $\kappa \leq 0.4\alpha + \kappa_\alpha$, substitution yields

$$
\kappa \leq \frac{a\alpha + b}{1 - c/\alpha}, \quad \text{where } a = 0.4, \ b = D + 1.4\frac{\mu}{\sigma}\delta^2, \text{ and } c = \frac{2\mu\delta^2}{\sigma}.
$$

Now setting $\alpha = 2c$ yields $4ac + 2b$, the right hand side of (5.23). □

We also present Theorem 5.7 which may be applied when the size bias coupling is monotone, that is, when $Y^s \geq Y$ almost surely. The proof depends on the following concentration inequality, which is in some sense the 'size bias' version of Lemma 5.1.

**Lemma 5.2** *Let $Y$ be a nonnegative random variable with mean $\mu$ and finite positive variance $\sigma^2$, and let $Y^s$ be given on the same space as $Y$, with the $Y$-size biased distribution, satisfying $Y^s \geq Y$. Then with*

$$
W = (Y - \mu)/\sigma \quad \text{and} \quad W^s = \big(Y^s - \mu\big)/\sigma,
$$

*for any $z \in \mathbb{R}$ and $a \geq 0$,*

$$
\frac{\mu}{\sigma} E\big(W^s - W\big)\mathbf{1}_{\{W^s - W \leq a\}}\mathbf{1}_{\{z\leq W\leq z+a\}} \leq a.
$$

*Proof* Let

$$
f(w) = \begin{cases} -a & w \leq z, \\ w - z - a & z < w \leq z + 2a, \\ a & w > z + 2a. \end{cases}
$$

Then

$$
\begin{aligned}
a &\geq E\big(Wf(W)\big) \\
&= \frac{1}{\sigma} E(Y-\mu) f\left(\frac{Y-\mu}{\sigma}\right) \\
&= \frac{\mu}{\sigma} E\big(f\big(W^s\big) - f(W)\big) \\
&= \frac{\mu}{\sigma} E \int_0^{W^s - W} f'(W+t) dt \\
&\geq \frac{\mu}{\sigma} E\left(\int_0^{W^s - W} \mathbf{1}_{\{0\leq t\leq a\}}\mathbf{1}_{\{z\leq W\leq z+a\}} f'(W+t) dt\right).
\end{aligned}
$$

Noting that $f'(w+t) = \mathbf{1}_{\{z\le w+t\le z+2a\}}$, we have

$$\mathbf{1}_{\{0\le t\le a\}}\mathbf{1}_{\{z\le W\le z+a\}}f'(W+t) = \mathbf{1}_{\{0\le t\le a\}}\mathbf{1}_{\{z\le W\le z+a\}},$$

and therefore

$$\begin{aligned}
a &\ge \frac{\mu}{\sigma}E\left(\int_0^{W^s-W}\mathbf{1}_{\{0\le t\le a\}}\mathbf{1}_{\{z\le W\le z+a\}}dt\right)\\
&= \frac{\mu}{\sigma}E\big(\min(a, W^s - W)\mathbf{1}_{\{z\le W\le z+a\}}\big)\\
&\ge \frac{\mu}{\sigma}E\big((W^s - W)\mathbf{1}_{\{W^s-W\le a\}}\mathbf{1}_{\{z\le W\le z+a\}}\big). \qquad \square
\end{aligned}$$

With the use of Lemma 5.2 we present the following result for monotone size bias couplings, from Goldstein and Zhang (2010).

**Theorem 5.7** *Let $Y$ be a nonnegative random variable with mean $\mu$ and finite positive variance $\sigma^2$, and let $Y^s$ be given on the same space as $Y$, with the $Y$-size biased distribution, satisfying $Y^s \ge Y$. Then with*

$$W = (Y-\mu)/\sigma \quad \text{and} \quad W^s = \big(Y^s - \mu\big)/\sigma,$$

*for any $a \ge 0$,*

$$\begin{aligned}
&\sup_{z\in\mathbb{R}}\big|P(W\le z) - P(Z\le z)\big|\\
&\quad \le D + 0.82\frac{a^2\mu}{\sigma} + a + \frac{\mu}{\sigma}E\big(W^s - W\big)\mathbf{1}_{\{W^s-W>a\}},
\end{aligned}$$

*where $D$ is as in* (5.21).

*If $W^s - W \le \delta$ with probability* 1,

$$\sup_{z\in\mathbb{R}}\big|P(W\le z) - P(Z\le z)\big| \le D + 0.82\frac{\delta^2\mu}{\sigma} + \delta.$$

*Proof* Let $z\in\mathbb{R}$ and let $f$ be the solution to the Stein equation (2.4) for $h(w) = \mathbf{1}_{\{w\le z\}}$. Decompose $Eh(W) - Nh$ as in (5.24) in proof of Theorem 5.6, and bound, as there, the first term by $D$, noting that (2.8) applies in the present case.

For the remaining term of (5.24) we write

$$\begin{aligned}
&\frac{\mu}{\sigma}\int_0^{W^s-W}\big(f'(W+t) - f'(W)\big)dt\\
&\quad = \frac{\mu}{\sigma}\mathbf{1}_{\{W^s-W>a\}}\int_0^{W^s-W}\big(f'(W+t) - f'(W)\big)dt\\
&\qquad + \frac{\mu}{\sigma}\mathbf{1}_{\{W^s-W\le a\}}\int_0^{W^s-W}\big(f'(W+t) - f'(W)\big)dt\\
&\quad := J_1 + J_2, \quad \text{say.}
\end{aligned}$$

By (2.8),

$$|EJ_1| \le \frac{\mu}{\sigma} E\big(W^s - W\big)\mathbf{1}_{\{W^s - W > a\}},$$

yielding the last term in the first bound of the theorem.

Now express $J_2$ using (2.4) as the sum

$$\frac{\mu}{\sigma}\mathbf{1}_{\{W^s - W \le a\}} \int_0^{W^s - W} \big[(W+t)f(W+t) - Wf(W)\big]dt$$
$$+ \frac{\mu}{\sigma}\mathbf{1}_{\{W^s - W \le a\}} \int_0^{W^s - W} (\mathbf{1}_{\{W+t \le z\}} - \mathbf{1}_{\{W \le z\}})dt. \tag{5.26}$$

Applying (2.10) to the first term in (5.26) shows that the absolute value of its expectation is bounded by

$$\begin{aligned}
&\frac{\mu}{\sigma} E\left(\mathbf{1}_{\{W^s - W \le a\}} \int_0^{W^s - W} \left(|W| + \frac{\sqrt{2\pi}}{4}\right) t\,dt\right) \\
&\le \frac{\mu}{2\sigma} E\left(\big(W^s - W\big)^2 \mathbf{1}_{\{W^s - W \le a\}}\left(|W| + \frac{\sqrt{2\pi}}{4}\right)\right) \\
&\le \frac{\mu}{2\sigma} a^2 \left(1 + \frac{\sqrt{2\pi}}{4}\right) \\
&\le 0.82 \frac{a^2 \mu}{\sigma}.
\end{aligned}$$

Taking the expectation of the absolute value of the second term in (5.26), we have

$$\begin{aligned}
&\frac{\mu}{\sigma} E\left|\mathbf{1}_{\{W^s - W \le a\}} \int_0^{W^s - W} (\mathbf{1}_{\{W+t \le z\}} - \mathbf{1}_{\{W \le z\}})dt\right| \\
&= \frac{\mu}{\sigma} E\left(\mathbf{1}_{\{W^s - W \le a\}} \int_0^{W^s - W} \mathbf{1}_{\{z - t < W \le z\}} dt\right) \\
&\le \frac{\mu}{\sigma} E\big((W^s - W)\mathbf{1}_{\{W^s - W \le a\}} \mathbf{1}_{\{z - a < W \le z\}}\big)
\end{aligned}$$

which is bounded by $a$, by Lemma 5.2, completing the proof of the first claim. The second claim follows immediately by letting $a = \delta$. □

## 5.4 Size Biasing and Smoothing Inequalities

In this section we present one further result which may be applied in situations where bounded size bias couplings exist. The method here, using smoothing inequalities, yields bounds in terms of supremums over function classes $\mathcal{H}$, and are more general than methods which only produce bounds in the Kolmogorov distance. Naturally, we may pay the price in larger constants. We follow the approach of Rinott and Rotar (1997), itself stemming from Götze (1991) and Bhattacharya and Rao (1986).

In order to state our results we now introduce conditions on the function classes $\mathcal{H}$ we consider. Since in Sect. 12.4 we will consider approximation in $\mathbb{R}^p$ we state Condition 5.1 in this generality, and in particular we take $Z$ in (iii) to be a standard normal variable with mean zero and identity covariance matrix in this space. In the present chapter we consider only the one dimensional case $p=1$.

**Condition 5.1** *$\mathcal{H}$ is a class of real valued measurable functions on $\mathbb{R}^p$ such that*

(i) *The functions $h \in \mathcal{H}$ are uniformly bounded in absolute value by a constant, which we take to be* 1 *without loss of generality.*
(ii) *For any real numbers $c$ and $d$, and for any $h(x) \in \mathcal{H}$, the function $h(cx+d) \in \mathcal{H}$.*
(iii) *For any $\epsilon > 0$ and $h \in \mathcal{H}$, the functions $h_\epsilon^+, h_\epsilon^-$ are also in $\mathcal{H}$, and*

$$E\tilde{h}_\epsilon(Z) \le a\epsilon \tag{5.27}$$

*for some constant $a$ that depends only on the class $\mathcal{H}$, where*

$$h_\epsilon^+(x) = \sup_{|y|\le\epsilon} h(x+y),$$
$$h_\epsilon^-(x) = \inf_{|y|\le\epsilon} h(x+y) \quad \text{and} \quad \tilde{h}_\epsilon(x) = h_\epsilon^+(x) - h_\epsilon^-(x). \tag{5.28}$$

Given a function class $\mathcal{H}$ and random variables $X$ and $Y$, let

$$\left\|\mathcal{L}(X) - \mathcal{L}(Y)\right\|_{\mathcal{H}} = \sup_{h\in\mathcal{H}} \left|Eh(X) - Eh(Y)\right|. \tag{5.29}$$

In one dimension, the collection of indicators of all half lines, and indicators of all intervals, each form classes $\mathcal{H}$ that satisfy Condition 5.1 with $a=\sqrt{2/\pi}$ and $a = 2\sqrt{2/\pi}$ respectively (see e.g. Rinott and Rotar 1997); clearly, in the first case the distance (5.29) specializes to the Kolmogorov metric.

**Theorem 5.8** *Let $Y$ be a nonnegative random variable with finite, nonzero mean $\mu$ and variance $\sigma^2 \in (0,\infty)$, and suppose there exists a variable $Y^s$, having the $Y$-size biased distribution, defined on the same space as $Y$, satisfying $|Y^s - Y| \le A$ for some $A \le \sigma^{3/2}/\sqrt{9\mu}$. Then, when $\mathcal{H}$ satisfies Condition* 5.1 *for some constant $a$,*

$$\left\|\mathcal{L}(W) - \mathcal{L}(Z)\right\|_{\mathcal{H}} \le \frac{0.21aA}{\sigma} + \frac{\mu}{\sigma^2}\left((12.4+58.1a)\frac{A^2}{\sigma} + \frac{2.5A^3}{\sigma^2}\right) + 15D,$$

*where $W=(Y-\mu)/\sigma$, $Z$ is a standard normal, and $D$ is given by* (5.21).

Specializing to the case where $\mathcal{H}$ is the collection of indicators of half lines and $a=\sqrt{2/\pi}$ yields the bound

$$\sup_{z\in\mathbb{R}}\left|P(W\le z) - P(Z\le z)\right| \le \frac{0.17A}{\sigma} + \frac{\mu}{\sigma^2}\left(\frac{58.8A^2}{\sigma} + \frac{2.5A^3}{\sigma^2}\right) + 15D,$$

demonstrating, by comparison with, say, the bound in Theorem 5.6, that the consideration of general function classes $\mathcal{H}$ comes at some expense. One reason for the increase in the magnitude of the constants is that general bounds on the solution to the Stein equation, as given by Lemma 2.4, must be applied here, and not, say, the more specialized bounds of Lemma 2.3 which require that the function $h$ be an indicator.

Let $\phi(y)$ denote the standard normal density and for $h \in \mathcal{H}$ and $t \in (0,1)$, define

$$h_t(w) = \int h(w+ty)\phi(y)dy. \tag{5.30}$$

The function $h_t(w)$ is a smoothed version of $h(w)$, with smoothing parameter $t$, and clearly $\|h_t\| \le \|h\|$. Furthermore, in this section, let

$$\kappa = \sup\{|Eh(W) - Nh|: h \in \mathcal{H}\}$$
$$\text{and for } t \in (0,1) \text{ set} \quad \kappa_t = \sup\{|Eh_t(W) - Nh_t|: h \in \mathcal{H}\}. \tag{5.31}$$

**Lemma 5.3** *Let $\mathcal{H}$ be a class of functions satisfying Condition* 5.1 *with constant $a$. Then, for any random variable $W$,*

$$\kappa \le 2.8\kappa_t + 4.7at \quad \text{for all } t \in (0,1). \tag{5.32}$$

*Furthermore, for all $\delta > 0$, $t \in (0,1)$ and $\tilde{h}_\epsilon$ as in Condition* 5.1,

$$E\left(\int \tilde{h}_{\delta+t|y|}(W)|\phi'(y)|dy\right) \le 1.6\kappa + a(\delta + t). \tag{5.33}$$

*Proof* Inequality (5.32) is Lemma 4.1 of Rinott and Rotar (1997), following Lemma 2.11 of Götze (1991) from Bhattacharya and Rao (1986). As in Rinott and Rotar (1997), adding and subtracting to the left hand side of (5.33) we have

$$E\left(\int (\tilde{h}_{\delta+t|y|}(W) - \tilde{h}_{\delta+t|y|}(Z))|\phi'(y)|dy + \int \tilde{h}_{\delta+t|y|}(Z)|\phi'(y)|dy\right)$$
$$\le \int |E\tilde{h}_{\delta+t|y|}(W) - E\tilde{h}_{\delta+t|y|}(Z)||\phi'(y)|dy + \int E\tilde{h}_{\delta+t|y|}(Z)|\phi'(y)|dy$$
$$\le \left(1.6\kappa + \int a(\delta + t|y|)|\phi'(y)|dy\right) \le 1.6\kappa + a(\delta + t),$$

where we have used the definitions of $\tilde{h}_\epsilon$ and $\kappa$ and that $\int |\phi'(y)|dy = \sqrt{2/\pi}$ for the first term, and then additionally (5.27) and $\int |y||\phi'(y)|dy = 1$ for the second. □

**Lemma 5.4** *Let $Y \ge 0$ be a random variable with mean $\mu$ and variance $\sigma^2 \in (0,\infty)$, and let $Y^s$ be defined on the same space as $Y$, with the $Y$-size biased distribution, satisfying $|Y^s - Y|/\sigma \le \delta$ for some $\delta$. Then for all $t \in (0,1)$,*

$$\kappa_t \le 4D + \frac{\mu}{\sigma}\left(\left(3.3 + \frac{1}{2}a\right)\delta^2 + \frac{2}{3}\delta^3 + \frac{1}{2t}(1.6\kappa\delta^2 + a\delta^3)\right), \tag{5.34}$$

*with $D$ as in* (5.21).

*Proof* With $h \in \mathcal{H}$ and $t \in (0,1)$ let $f$ be the solution to the Stein equation (2.4) for $h_t$. Letting $W = (Y - \mu)/\sigma$ and $W^s = (Y^s - \mu)/\sigma$ we have $|W^s - W| \le \delta$. From (5.19) we obtain,

$$EWf(W) = \frac{\mu}{\sigma}\big(f(W^s) - f(W)\big), \tag{5.35}$$

and, so, letting $V = W^s - W$,

$$\begin{aligned} &Eh_t(W) - Nh_t \\ &= E\big(f'(W) - Wf(W)\big) \\ &= E\left(f'(W) - \frac{\mu}{\sigma}\big(f(W^s) - f(W)\big)\right) \\ &= E\left(f'(W) - \frac{\mu}{\sigma}\int_W^{W^s} f'(w)dw\right) \\ &= E\left(f'(W) - \frac{\mu}{\sigma}V\int_0^1 f'(W+uV)du\right) \\ &= E\left(f'(W)\left(1 - \frac{\mu}{\sigma}V\right)\right) + E\left(\frac{\mu}{\sigma}Vf'(W) - \frac{\mu}{\sigma}V\int_0^1 f'(W+uV)du\right). \end{aligned} \tag{5.36}$$

Bounding the first term in (5.36), by (2.12) and that $\|h_t\| \le 1$, and definition (5.21), we have

$$\left|E\left\{f'(W)E\left(1 - \frac{\mu}{\sigma}V|W\right)\right\}\right| \le 4D. \tag{5.37}$$

By (5.30) and a change of variable we may write

$$h_t(w+s) - h_t(w) = \int h(w+tx)\big(\phi(y - s/t) - \phi(y)\big)dy, \tag{5.38}$$

so, for the second term in (5.36), applying the dominated convergence theorem in (5.38) and differentiating the Stein equation (2.4),

$$\begin{aligned} &f''(w) = f(w) + wf'(w) + h_t'(w) \\ &\text{with } h_t'(w) = -\frac{1}{t}\int h(w+ty)\phi'(y)dy. \end{aligned} \tag{5.39}$$

Hence, we may we write the second term in (5.36) as the expectation of

$$\begin{aligned} &\frac{\mu}{\sigma}V\left\{f'(W) - \int_0^1 f'(W+uV)du\right\} \\ &= \frac{\mu}{\sigma}V\int_0^1 \big(f'(W) - f'(W+uV)\big)du \\ &= -\frac{\mu}{\sigma}V\int_0^1\int_W^{W+uV} f''(v)dvdu \\ &= -\frac{\mu}{\sigma}V\int_0^1\int_W^{W+uV} \big(f(v) + vf'(v) + h_t'(v)\big)dvdu. \end{aligned} \tag{5.40}$$

We apply the triangle inequality and bound the three resulting terms separately. For the expectation arising from the first term on the right-hand side of (5.40), by (2.12) and that $\|h_t\| \le 1$ we have

$$\left|E\left\{\frac{\mu}{\sigma}V\int_0^1\int_W^{W+uV}f(v)dvdu\right\}\right|$$
$$\le \sqrt{2\pi}\frac{\mu}{\sigma}E\left\{|V|\int_0^1 u|V|du\right\} \le 1.3\frac{\mu}{\sigma}\delta^2. \tag{5.41}$$

For the second term in (5.40), again applying (2.12),

$$\begin{aligned}
&\left|E\frac{\mu}{\sigma}V\int_0^1\int_W^{W+uV}vf'(v)dvdu\right| \\
&\le \frac{2\mu}{\sigma}E|V|\int_0^1\left|\int_W^{W+uV}2|v|dv\right|du \\
&\le \frac{2\mu}{\sigma}E|V|\int_0^1(2u|WV|+u^2V^2)du \\
&\le \frac{2\mu}{\sigma}\delta\int_0^1(2\delta uE|W|+u^2\delta^2)du \\
&\le \frac{2\mu}{\sigma}\delta(\delta+\delta^2/3).
\end{aligned} \tag{5.42}$$

For the last term in (5.40), beginning with the inner integral, we have

$$\int_W^{W+uV}h_t'(v)dv = uV\int_0^1 h_t'(W+xuV)dx$$

and using (5.39),

$$\int \phi'(y)dy = 0,$$

and Lemma 5.3 we have

$$\begin{aligned}
&\left|\frac{\mu}{\sigma}EV^2\int_0^1\int_0^1 uh_t'(W+xuV)dxdu\right| \\
&= \frac{\mu}{\sigma t}\left|EV^2\int_0^1\int_0^1\int uh(W+xuV+ty)\phi'(y)dydxdu\right| \\
&= \frac{\mu}{\sigma t}\left|EV^2\int_0^1\int_0^1\int u\big[h(W+xuV+ty)-h(W+xuV)\big]\phi'(y)dydxdu\right| \\
&\le \frac{\mu}{\sigma t}E\left(V^2\int\int_0^1 u\big[h^+_{|V|+t|y|}(W)-h^-_{|V|+t|y|}(W)\big]|\phi'(y)|dudy\right) \\
&= \frac{\mu}{2\sigma t}E\left(V^2\int\big[h^+_{|V|+t|y|}(W)-h^-_{|V|+t|y|}(W)\big]|\phi'(y)|dy\right) \\
&\le \frac{\mu}{2\sigma t}\delta^2E\left(\int\tilde{h}_{\delta+t|y|}(W)|\phi'(y)|dy\right)
\end{aligned}$$

$$\begin{aligned}&\le \frac{\mu}{2\sigma t}\delta^2(1.6\kappa + a(\delta + t))\\ &= \frac{\mu}{2\sigma t}(1.6\kappa\delta^2 + a\delta^3) + \frac{\mu}{2\sigma}a\delta^2.\end{aligned} \tag{5.43}$$

Combining (5.37), (5.41), (5.42), and (5.43) completes the proof. □

*Proof of Theorem 5.8* Substituting (5.34) into (5.32) of Lemma 5.3 we obtain

$$\kappa \le 2.8\left(4D + \frac{\mu}{\sigma}\left(\left(3.3 + \frac{1}{2}a\right)\delta^2 + \frac{2}{3}\delta^3 + \frac{1}{2t}(1.6\kappa\delta^2 + a\delta^3)\right)\right) + 4.7at,$$

or,

$$\kappa \le \frac{2.8(4D + (\mu/\sigma)((3.3 + \frac{1}{2}a)\delta^2 + \frac{2}{3}\delta^3 + a\delta^3/2t)) + 4.7at}{1 - 2.24\mu\delta^2/(\sigma t)}. \tag{5.44}$$

Setting $t = 4 \times 2.24\mu\delta^2/\sigma$, which is a number in (0, 1) since $\delta \le (\sigma/(9\mu))^{1/2}$, we obtain

$$\begin{aligned}\kappa &\le \frac{4}{3} \times 2.8\left(4D + \frac{\mu}{\sigma}\left(\left(3.3 + \frac{1}{2}a\right)\delta^2 + \frac{2}{3}\delta^3 + \frac{\sigma}{2(8.96\mu)}a\delta\right)\right)\\ &\quad + \frac{4}{3} \times 4.7a\left(8.96\frac{\mu\delta^2}{\sigma}\right)\\ &\le 0.21a\delta + \frac{\mu}{\sigma}((12.4 + 58.1a)\delta^2 + 2.5\delta^3) + 15D.\end{aligned}$$

Substituting $\delta = A/\sigma$ now completes the proof. □

# Chapter 6
# $L^\infty$: Applications

In this chapter we consider the application of the results of Chap. 5 to obtain $L^\infty$ bounds for the combinatorial central limit theorem, counting the occurrences of patterns, the anti-voter model, and for the binary expansion of a random integer.

## 6.1 Combinatorial Central Limit Theorem

Recall that in the combinatorial central limit theorem we study the distribution of

$$Y = \sum_{i=1}^{n} a_{i,\pi(i)} \tag{6.1}$$

where $A = \{a_{ij}\}_{i,j=1}^{n}$ is a given array of real numbers and $\pi$ a random permutation. This setting was introduced in Example 2.3, and $L^1$ bounds to the normal were derived in Sect. 4.4 for the case where $\pi$ is chosen uniformly from the symmetric group $\mathcal{S}_n$; some further background, motivation, applications, references and history on the combinatorial CLT were also presented in that section.

For $\pi$ chosen uniformly, von Bahr (1976) and Ho and Chen (1978) obtained $L^\infty$ bounds to the normal when the matrix $A$ is random, which yield the correct rate $O(n^{-1/2})$ only under some boundedness conditions. Here we focus on the case where $A$ is non-random. In Sect. 6.1.1 we present the result of Bolthausen (1984), which gives a bound of the correct order in terms of a third-moment quantity of the type (4.107), but with an unspecified constant. In this same section, based on Goldstein (2005), we give bounds of the correct order and with an explicit constant, but in terms of the maximum absolute array value. In Sect. 6.1.2 we also give $L^\infty$ bounds when the distribution of the permutation $\pi$ is constant on cycle type and has no fixed points, expressing the bounds again in terms of the maximum array value.

For the last two results mentioned we make use of Lemma 4.6, which, given $\pi$, constructs permutations $\pi^\dagger$ and $\pi^\ddagger$ on the same space as $\pi$ such that

$$Y^\dagger = \sum_{i=1}^{n} a_{i\pi^\dagger(i)} \quad \text{and} \quad Y^\ddagger = \sum_{i=1}^{n} a_{i\pi^\ddagger(i)}$$

L.H.Y. Chen et al., *Normal Approximation by Stein's Method*,
Probability and Its Applications,
DOI 10.1007/978-3-642-15007-4_6, © Springer-Verlag Berlin Heidelberg 2011

have the square bias distribution as in Proposition 4.6. As noted in Sect. 4.4 for $L^1$ bounds in the uniform case, the permutations $\pi, \pi^\dagger$ and $\pi^\ddagger$ agree on the complement of some small index set $\mathcal{I}$, and hence we may write

$$Y = S + T, \quad Y^\dagger = S + T^\dagger \quad \text{and} \quad Y^\ddagger = S + T^\ddagger, \tag{6.2}$$

where

$$S = \sum_{i \notin \mathcal{I}} a_{i,\pi(i)}, \quad T = \sum_{i \in \mathcal{I}} a_{i,\pi(i)}, \quad T^\dagger = \sum_{i \in \mathcal{I}} a_{i,\pi^\dagger(i)} \quad \text{and} \quad T^\ddagger = \sum_{i \in \mathcal{I}} a_{i,\pi^\ddagger(i)}. \tag{6.3}$$

Now, as $Y^* = UY^\dagger + (1-U)Y^\ddagger$ has the $Y$-zero bias distribution by Proposition 4.6, we have

$$|Y^* - Y| = \left|UT^\dagger + (1-U)T^\ddagger - T\right| \le U\left|T^\dagger\right| + (1-U)\left|T^\ddagger\right| + |T|. \tag{6.4}$$

Hence when $\mathcal{I}$ is almost surely bounded (6.4) gives an upper bound on $|Y^* - Y|$ equal to the largest size of $\mathcal{I}$ times twice the largest absolute array value. Now Theorem 5.1 for bounded zero bias couplings yields an $L^\infty$ norm bound in any instance where such constructions can be achieved.

In the remainder of this section, to avoid trivial cases we assume that $\mathrm{Var}(Y) = \sigma^2 > 0$, and for ease of notation we will write $Y'$ and $\pi'$ interchangeably for $Y$ and $\pi$, respectively.

### *6.1.1 Uniform Distribution on the Symmetric Group*

We approach the uniform permutation case in two different ways, first using zero biasing, then by an inductive method. Using zero biasing, combining the coupling given in Sect. 4.4 with Theorem 5.1 quickly leads to the following result.

**Theorem 6.1** *Let $\{a_{ij}\}_{i,j=1}^n$ be an array of real numbers and let $\pi$ be a random permutation with uniform distribution over $\mathcal{S}_n$. Then, with $Y$ as in (6.1) and $W = (Y-\mu)/\sigma$,*

$$\sup_{z \in \mathbb{R}} \left|P(W \le z) - P(Z \le z)\right| \le 16.3C/\sigma \quad \textit{for } n \ge 3,$$

*where $\mu$ and $\sigma^2 = \mathrm{Var}(Y)$ are given by (4.105), and*

$$C = \max_{1 \le i,j \le n} |a_{ij} - a_{i\bullet} + a_{\bullet j} - a_{\bullet\bullet}|,$$

*with the row, column and overall array averages $a_{i\bullet}, a_{\bullet j}$ and $a_{\bullet\bullet}$ as in (2.44).*

*Proof* By (2.45) we may first replace $a_{ij}$ by $a_{ij} - a_{i\bullet} - a_{\bullet j} + a_{\bullet\bullet}$, and in particular assume $EY = 0$. Following the construction in Lemma 4.6, we obtain the variable $Y^* = UY^\dagger + (1-U)Y^\ddagger$ with the $Y$-zero biased distribution, where $Y, Y^\dagger$ and $Y^\ddagger$ may be written as in (6.2) and (6.3) with $|\mathcal{I}| = |\{I^\dagger, J^\dagger, \pi^{-1}(K^\dagger), \pi^{-1}(L^\dagger)\}| \le 4$

by (4.135). As $W^* = Y^*/\sigma$ has the $W$-zero bias distribution by (2.59), applying (6.4) we obtain

$$E|W^* - W| = E|Y^* - Y|/\sigma \le 8C/\sigma.$$

Our claim now follows from Theorem 5.1 by taking $\delta = 8C/\sigma$. □

With a bit more work, we can use the zero bias variation of Ghosh (2009) on the inductive method in Bolthausen (1984) to prove an $L^\infty$ bound depending on a third moment type quantity of the array, like the $L^1$ bound in Theorem 4.8. On the other hand, the bound in Theorem 6.2 depends on an unspecified constant, whereas the constant in Theorem 6.1 is explicit. Though induction was used in Sect. 3.4.2 for the independent case, the inductive approach taken here has a somewhat different flavor. Bolthausen's inductive method has also been put to use by Fulman (2006) for character ratios, and by Goldstein (2010b) for addressing questions about random graphs.

**Theorem 6.2** *Let $\{a_{ij}\}_{i,j=1}^n$ be an array of real numbers and let $\pi$ be a random permutation with the uniform distribution over $\mathcal{S}_n$. Let $Y$ be as in* (6.1) *and $\mu_A$ and $\sigma_A^2 = \mathrm{Var}(Y)$ be given by* (4.105). *Then, with $W = (Y - \mu_A)/\sigma_A$, there exists a constant $c$ such that*

$$\sup_{z\in\mathbb{R}} \left| P(W \le z) - P(Z \le z)\right| \le c\gamma_A/\left(\sigma_A^3 n\right) \quad \text{for all } n \ge 2,$$

*where $\gamma_A$ is given in* (4.107).

To prepare for the proof we need some additional notation. For $n \in \mathbb{N}$ and an array $E \in \mathbb{R}^{n\times n}$ let

$$W_E = \sum_{i=1}^n e_{i,\pi(i)},$$

and let $E^0$ be the centered array with components

$$e_{ij}^0 = e_{ij} - e_{i\bullet} - e_{\bullet j} + e_{\bullet\bullet} \tag{6.5}$$

where the array averages are given by (2.44). In addition, when $\sigma_E^2 > 0$, let $\widehat{E}$ be the array given by

$$\widehat{e}_{ij} = e_{ij}^0/\sigma_E, \quad \text{and set} \quad \beta_E = \gamma_E/\sigma_E^3. \tag{6.6}$$

Clearly, if $E$ is an array with $\sigma_E^2 > 0$ then

$$\beta_E = \beta_{E^0} = \beta_{\widehat{E}}. \tag{6.7}$$

For any $E \in \mathbb{R}^{n\times n}$ let $E'$ be the truncated array whose components are given by

$$e'_{ij} = e_{ij}\mathbf{1}\left(|e_{ij}| \le 1/2\right). \tag{6.8}$$

For $\beta > 0$ let

$$\widehat{M_n}(\beta) = \{E \in \mathbb{R}^{n\times n}: e_{\cdot j} = e_{i\cdot} = 0 \text{ for all } i, j = 1, \ldots, n, \sigma_E^2 = 1, \beta_E \le \beta\},$$

$$M_n^1(\beta) = \{E \in \widehat{M_n}(\beta) : |e_{ij}| \le 1 \text{ for all } i, j = 1, \ldots, n\},$$

and

$$\widehat{M_n} = \bigcup_{\beta>0} \widehat{M_n}(\beta) \quad \text{and} \quad M_n^1 = \bigcup_{\beta>0} M_n^1(\beta).$$

We note that if $E$ is any $n \times n$ array with $\sigma_E^2 > 0$ then $\widehat{E} \in \widehat{M_n}(\beta)$ for all $\beta \ge \beta_E$, and if $E \in \widehat{M_n}$ then $\widehat{E} = E$. Let

$$\delta^1(\beta, n) = \sup\{|P(W_E \le z) - \Phi(z)|: z \in \mathbb{R},\ E \in M_n^1(\beta)\}. \tag{6.9}$$

The proof of the theorem depends on the following four lemmas, whose proofs are deferred to the end of this section. The first two lemmas are used to control the effects of truncation and scaling.

**Lemma 6.1** *For $n \ge 2$ and $E \in \widehat{M_n}$ let $E'$ be the truncated $E$ array given by* (6.8)*. Then there exists $c_1 \ge 1$ such that*

$$P(W_E \ne W_{E'}) \le c_1\beta_E/n \quad \text{and} \quad |\mu_{E'}| \le c_1\beta_E/n. \tag{6.10}$$

*In addition, there exist constants $\epsilon_1$ and $c_2$ such that when $\beta_E/n \le \epsilon_1$*

$$|\sigma_{E'}^2 - 1| \le c_2\beta_E/n, \quad \widehat{E'} \in M_n^1 \quad \text{and} \quad \beta_{E'} \le c_1\beta_E. \tag{6.11}$$

**Lemma 6.2** *There exist constants $\epsilon_2$ and $c_3$ such that if $E \in \widehat{M_n}$ for some $n \ge 2$ and $\widehat{E'}$ is as in* (6.8), (6.6) *and* (6.5), *then whenever $\beta_E/n \le \epsilon_2$*

$$\sup_{z\in\mathbb{R}}|P(W_E \le z) - \Phi(z)| \le \sup_{z\in\mathbb{R}}|P(W_{\widehat{E'}} \le z) - \Phi(z)| + c_3\beta_E/n.$$

The following lemma handles the effects of deleting rows and columns from an array in $M_n^1$.

**Lemma 6.3** *There exist $n_0 \ge 16$, $\epsilon_3 > 0$ and $c_4 \ge 1$ such that if $n \ge n_0$, $l \le 4$ and $C \in M_n^1$, when $D$ is the $(n-l) \times (n-l)$ array formed by removing the $l$ rows $\mathcal{R} \subset \{1, 2, \ldots, n\}$ and $l$ columns $\mathcal{C} \subset \{1, 2, \ldots, n\}$ from $C$, we have $|\mu_D| \le 8$, and if $\beta_C/n \le \epsilon_3$ then*

$$|\sigma_D^2 - 1| \le 3/4 \quad \text{and} \quad \beta_D \le c_4\beta_C.$$

The proof, being inductive in nature, expresses the distance to normality for a problem of a given size in terms of the distances to normality for the same problem, but of smaller sizes. This last lemma is used to handle the resulting recursion for the relation between these distances.

**Lemma 6.4** *Let $\{s_n\}_{n\geq 1}$ be a sequence of nonnegative numbers and $m \geq 5$ a positive integer such that*

$$s_n \leq d + \alpha \max_{l\in\{2,3,4\}} s_{n-l} \quad \textit{for all } n \geq m, \tag{6.12}$$

*with $d \geq 0$ and $\alpha \in (0,1)$. Then*

$$\sup_{n\geq 1} s_n < \infty.$$

*Proof of Theorem 6.2* In view of (2.45) and (6.7) it suffices to prove the theorem for $W_B$ with $B \in \widehat{M_n}$. Let $\epsilon_1, c_1$ and $c_2$ be as in Lemma 6.1, $\epsilon_2$ and $c_3$ as in Lemma 6.2, and $n_0, \epsilon_3$ and $c_4$ as in Lemma 6.3. Noting that from the lemmas we have $n_0 \geq 16$ and

$$c_1 \geq 1 \quad \text{and} \quad c_4 \geq 1, \tag{6.13}$$

set

$$\epsilon_0 = \min\{1/(2n_0), \epsilon_1/c_1, \epsilon_3/c_1, 3\epsilon_1/(4c_4c_1), 3\epsilon_2/(4c_4c_1)\}.$$

We first demonstrate that it suffices to prove the theorem for $\beta_B/n < \epsilon_0$ and $n > n_0$. By Hölder's inequality and (4.105), for all $n \in \mathbb{N}$,

$$\frac{(n-1)^{1/2}}{n^{1/3}} = \frac{1}{n^{1/3}}\left(\sum_{i,j=1}^{n} b_{ij}^2\right)^{1/2} \leq \left(\sum_{i,j=1}^{n} |b_{ij}|^3\right)^{1/3} = \beta_B^{1/3}. \tag{6.14}$$

As inequality (6.14) implies that $\beta_B \geq 1/2$ for all $n \geq 2$, we have $\beta_B/n \geq \epsilon_0$ for all $2 \leq n \leq n_0$. Hence, taking $c \geq 1/\epsilon_0$ the theorem holds if either $2 \leq n \leq n_0$ or $B$ satisfies $\beta_B/n \geq \epsilon_0$. We may therefore assume $n \geq n_0$ and $\beta_B/n \leq \epsilon_0$.

As $\beta_B/n \leq \epsilon_0$, setting $C = \widehat{B'}$ as in (6.8), (6.6) and (6.5), Lemma 6.2 yields

$$\sup_{z\in\mathbb{R}} \bigl|P(W_B \leq z) - \Phi(z)\bigr| \leq \sup_{z\in\mathbb{R}} \bigl|P(W_C \leq z) - \Phi(z)\bigr| + c_3\beta_B/n. \tag{6.15}$$

By (6.7) and (6.11) of Lemma 6.1 we have that

$$\beta_C/n = \beta_{B'}/n \leq c_1\beta_B/n \leq c_1\epsilon_0, \tag{6.16}$$

and also that $C \in M_n^1$. Hence, by (6.15) and (6.16) it suffices to prove that there exists a constant $c_5$ such that

$$\delta^1(\beta, n) \leq c_5\beta/n \quad \text{for all } n \geq n_0 \text{ and } \beta/n \leq c_1\epsilon_0. \tag{6.17}$$

For $z \in \mathbb{R}$ and $\alpha > 0$ let $h_{z,\alpha}(w)$ be the smoothed indicator function of $(-\infty, z]$, which decays linearly to zero over the interval $[z, z+\alpha]$, as given by (2.14), and set

$$\delta^1(\alpha, \beta, n) = \sup\{\bigl|Eh_{z,\alpha}(W_C) - Nh_{z,\alpha}\bigr|;\ z \in \mathbb{R},\ C \in M_n^1(\beta)\}. \tag{6.18}$$

Also, define

$$h_{z,0}(x) = \mathbf{1}(x \leq z).$$

As the collection of arrays $M_n^1(\beta)$ increases in $\beta$, so therefore does $\delta^1(\alpha, \beta, n)$.

Now, since for any $z, w \in \mathbb{R}$ and $\alpha > 0$,

$$h_{z,0}(w) \le h_{z,\alpha}(w) \le h_{z+\alpha,0}(w),$$

for all $C \in M_n^1(\beta)$ and all $\alpha > 0$ we have

$$\sup_{z\in\mathbb{R}} \left| P(W_C \le z) - \Phi(z) \right| \le \sup_{z\in\mathbb{R}} \left| Eh_{z,\alpha}(W_C) - Eh_{z,\alpha}(Z) \right| + \frac{\alpha}{\sqrt{2\pi}},$$

and taking supremum yields

$$\delta^1(\beta, n) \le \delta^1(\alpha, \beta, n) + \frac{\alpha}{\sqrt{2\pi}}. \tag{6.19}$$

To prove (6.17), for $n \ge n_0$ let $C \in M_n^1$ satisfy $\beta_C/n \le c_1\epsilon_0$, and let $f$ be the solution to the Stein equation (2.4) with $h = h_{z,\alpha}$ as in (2.14), for some fixed $z \in \mathbb{R}$. Following the construction in Lemma 4.6, we obtain the variable $W_C^* = UW_C^\dagger + (1-U)W_C^\ddagger$ with the $W_C$-zero biased distribution. Now, using the bound (2.16) from Lemma 2.5 on the differences of the derivative of $f$, write

$$\begin{aligned} \left| Eh(W_C) - Nh \right| &= \left| E\left(f'(W_C) - W_C f(W_C)\right) \right| \\ &= \left| E\left(f'(W_C) - f'(W_C^*)\right) \right| \\ &\le E\left| f'(W_C^*) - f'(W_C) \right| \le A_1 + A_2 + A_2, \end{aligned} \tag{6.20}$$

where

$$\begin{aligned} A_1 &= E\left| W_C^* - W_C \right|, \qquad A_2 = E\left| W_C\left(W_C^* - W_C\right) \right| \quad \text{and} \\ A_3 &= \frac{1}{\alpha} E\left( \left| W_C^* - W_C \right| \int_0^1 \mathbf{1}_{[z,z+\alpha]}\left(W_C + r\left(W_C^* - W_C\right)\right) dr \right). \end{aligned}$$

First, from the $L^1$ bound in Lemma 4.7, noting that $\gamma$ in the lemma equals $\beta_C$ as $\sigma_C^2 = 1$, we obtain

$$A_1 = E\left| W_C^* - W_C \right| \le c_6 \beta_C / n. \tag{6.21}$$

Next, to estimate $A_2$, note that by (4.135) of Lemma 4.6 we may write $W_C^\dagger$ and $W_C^\ddagger$ as in (6.2) and (6.3) with $\mathcal{I} = \{I^\dagger, J^\dagger, \pi^{-1}(K^\dagger), \pi^{-1}(L^\dagger)\}$ and

$$\begin{aligned} W_C^* - W_C &= \left(UW_C^\dagger + (1-U)W_C^\ddagger\right) - W_C \\ &= \left(U\left(S + T^\dagger\right) + (1-U)\left(S + T^\ddagger\right)\right) - (S + T) \\ &= UT^\dagger + (1-U)T^\ddagger - T. \end{aligned} \tag{6.22}$$

Now let $\mathbf{I} = (I^\dagger, J^\dagger, \pi^{-1}(K^\dagger), \pi^{-1}(L^\dagger), \pi(I^\dagger), \pi(J^\dagger), K^\dagger, L^\dagger)$. By the construction in Lemma 4.5 the right hand side of (6.22), and hence $W_C^* - W_C$, is measurable with respect to $\mathscr{I} = \{\mathbf{I}, U\}$. Furthermore, since $C \in M_n^1$ and $|\mathcal{I}| \le 4$, we have

$$|W_C| = |S + T| \le |S| + |T| \le |S| + \sum_{i\in\mathcal{I}} |c_{i\pi(i)}| \le |S| + 4.$$

Now, using the definition of $A_2$, and that $U$ is independent of $\{S, \mathbf{I}\}$, we obtain

$$\begin{aligned}
A_2 &= E\big(|W_C(W_C^* - W_C)|\big) \\
&= E\big(|W_C^* - W_C|E(|W_C||\mathscr{I})\big) \\
&\le E\big(|W_C^* - W_C|E(|S|+4|\mathscr{I})\big) \\
&\le E\big(|W_C^* - W_C|\sqrt{E(S^2|\mathbf{I})}\big) + 4E|W_C^* - W_C|.
\end{aligned} \tag{6.23}$$

In the following, for $\iota$ a realization of $\mathcal{I}$, let $l$ denote the number of distinct elements of $\iota$. Since $S = \sum_{i \notin \mathcal{I}} c_{i\pi(i)}$ and $\pi$ is chosen uniformly from $\mathcal{S}_n$, we have that

$$\mathcal{L}(S|\mathbf{I}=\mathbf{i}) = \mathcal{L}(W_D), \tag{6.24}$$

where $W_D = \sum_{1 \le i \le n-l} d_{i\theta(i)}$ with $D$ the $(n-l)\times(n-l)$ array formed by removing from $C$ the rows $\{I^\dagger, J^\dagger, \pi^{-1}(K^\dagger), \pi^{-1}(L^\dagger)\}$ and columns $\{\pi(I^\dagger), \pi(J^\dagger), K^\dagger, L^\dagger\}$, and $\theta$ chosen uniformly from $\mathcal{S}_{n-l}$.

Using $l \in \{2,3,4\}$, that $n \ge n_0$ and $\beta_C/n \le c_1\epsilon_0 \le \epsilon_3$, Lemma 6.3 yields $|\mu_D| \le 8$ and that

$$|\sigma_D^2 - 1| \le 3/4, \quad \text{so that} \quad EW_D^2 \le c_7. \tag{6.25}$$

In particular

$$E\big(S^2|\mathbf{I}=\mathbf{i}\big) = EW_D^2 \le c_7 \quad \text{for all } \mathbf{i}, \text{ and hence} \quad E\big(S^2|\mathbf{I}\big) \le c_7.$$

Now using (6.23) and (6.21), we obtain

$$A_2 \le c_8\beta_C/n. \tag{6.26}$$

Finally, we are left with bounding $A_3$. First we note that for any $r \in \mathbb{R}$,

$$\begin{aligned}
&W_C + r(W_C^* - W_C) \\
&\quad = rW_C^* + (1-r)W_C \\
&\quad = r\big(S + UT^\dagger + (1-U)T^\ddagger\big) + (1-r)(S+T) \\
&\quad = S + rUT^\dagger + r(1-U)T^\ddagger + (1-r)T \\
&\quad = S + g_r \quad \text{where } g_r = rUT^\dagger + r(1-U)T^\ddagger + (1-r)T.
\end{aligned}$$

Now, from the definition of $A_3$, again using that $W_C - W_C^*$ is $\mathscr{I}$ measurable,

$$\begin{aligned}
A_3 &= \frac{1}{\alpha}E\left(|W_C - W_C^*|\int_0^1 \mathbf{1}_{[z,z+\alpha]}\big(W_C + r(W_C^* - W_C)\big)dr\right) \\
&= \frac{1}{\alpha}E\left(|W_C - W_C^*|E\left(\int_0^1 \mathbf{1}_{[z,z+\alpha]}\big(W_C + r(W_C^* - W_C)\big)dr|\mathscr{I}\right)\right) \\
&= \frac{1}{\alpha}E\left(|W_C - W_C^*|\int_0^1 P\big(W_C + r(W_C^* - W_C) \in [z,z+\alpha]|\mathscr{I}\big)dr\right) \\
&= \frac{1}{\alpha}E\left(|W_C - W_C^*|\int_0^1 P\big(S + g_r \in [z,z+\alpha]|\mathscr{I}\big)dr\right)
\end{aligned}$$

$$
\begin{aligned}
&= \frac{1}{\alpha} E\left( |W_C - W_C^*| \int_0^1 P\big(S \in [z - g_r, z + \alpha - g_r] | \mathscr{I}\big) dr \right) \\
&\le \frac{1}{\alpha} E\left( |W_C - W_C^*| \int_0^1 \sup_{z \in \mathbb{R}} P\big(S \in [z - g_r, z + \alpha - g_r] | \mathscr{I}\big) dr \right) \\
&= \frac{1}{\alpha} E\left( |W_C - W_C^*| \int_0^1 \sup_{z \in \mathbb{R}} P\big(S \in [z, z + \alpha] | \mathscr{I}\big) dr \right) \qquad (6.27) \\
&= \frac{1}{\alpha} E\left( |W_C - W_C^*| \sup_{z \in \mathbb{R}} P\big(S \in [z, z + \alpha] | \mathscr{I}\big) \right) \\
&= \frac{1}{\alpha} E\left( |W_C - W_C^*| \sup_{z \in \mathbb{R}} P\big(S \in [z, z + \alpha] | \mathbf{I}\big) \right), \qquad (6.28)
\end{aligned}
$$

where to obtain equality in (6.27) we have used the fact that $g_r$ is measurable with respect to $\mathscr{I}$ for all $r$, and the equality in (6.28) follows from the independence of $U$ from $\{S, \mathbf{I}\}$.

Regarding $P(S \in [z, z+\alpha] | \mathbf{I})$, we claim that

$$
\begin{aligned}
\sup_{z \in \mathbb{R}} P\big(S \in [z, z+\alpha] | \mathbf{I} = \mathbf{i}\big) &= \sup_{z \in \mathbb{R}} P\big(W_D \in [z, z+\alpha]\big) \\
&= \sup_{z \in \mathbb{R}} P\big(W_{D^0} \in [z, z+\alpha]\big) \\
&= \sup_{z \in \mathbb{R}} P\left(W_{\widehat{D}} \in \left[\frac{z}{\sigma_D}, \frac{z+\alpha}{\sigma_D}\right]\right) \\
&\le \sup_{z \in \mathbb{R}} P\big(W_{\widehat{D}} \in [z, z+2\alpha]\big). \qquad (6.29)
\end{aligned}
$$

The first equality is (6.24), the second follows from (6.5) and that $\sum_{i=1}^{n-l} d_{i\bullet}$, $\sum_{i=1}^{n-l} d_{\bullet\theta(i)}$ and $\sum_{i=1}^{n-l} d_{\bullet\bullet}$ do not depend on $\theta$, and the next is by definition (6.6) of $\widehat{D}$. The inequality follows from (6.25), which implies $\sigma_D \ge 1/2$.

Let $E = \widehat{D}$. Using that $\beta_C/n \le c_1\epsilon_0 \le \epsilon_3$ and Lemma 6.3, we have

$$\beta_D \le c_4 \beta_C \qquad (6.30)$$

so that by (6.7)

$$\frac{\beta_E}{n-l} = \frac{\beta_D}{n-l} \le \frac{c_4\beta_C}{n-l} \le \frac{nc_4c_1\epsilon_0}{n-l} \le \frac{3}{4}\frac{n}{n-4} \min\{\epsilon_1, \epsilon_2\} \le \min\{\epsilon_1, \epsilon_2\},$$

since $n_0 \ge 16$. Now Lemma 6.1 and (6.7) yield

$$\widehat{E'} \in M_{n-l}^1 \quad \text{and} \quad \beta_{\widehat{E'}} = \beta_{E'} \le c_1\beta_E = c_1\beta_D.$$

Furthermore, Lemma 6.2 and (6.7) may be invoked to yield

$$
\begin{aligned}
&P\big(W_{\widehat{D}} \in [z, z+2\alpha]\big) \\
&\quad = P\big(W_E \in [z, z+2\alpha]\big) \\
&\quad \le \big|P(W_E \le z+2\alpha) - \Phi(z+2\alpha)\big| + \big|\Phi(z+2\alpha) - \Phi(z)\big| \\
&\qquad + \big|\Phi(z) - P(W_E < z)\big|
\end{aligned}
$$

$$
\begin{aligned}
&\le 2\delta^1(c_1\beta_D, n-l) + \frac{2c_3\beta_D}{n-l} + \frac{2\alpha}{\sqrt{2\pi}} \\
&\le 2 \max_{l\in\{2,3,4\}} \delta^1(c_1\beta_D, n-l) + \frac{2c_3\beta_D}{n-l} + \frac{2\alpha}{\sqrt{2\pi}} \\
&\le 2 \max_{l\in\{2,3,4\}} \delta^1(c_9\beta_C, n-l) + \frac{c_{10}\beta_C}{n} + \frac{2\alpha}{\sqrt{2\pi}},
\end{aligned} \tag{6.31}
$$

where in the final inequality we have again invoked (6.30) and set $c_9 = c_1c_4 \ge 1$, by (6.13).

As (6.31) does not depend on $z$ or $\mathbf{i}$, by (6.29), it bounds $\sup_{z\in\mathbb{R}} P(S \in [z, z+\alpha]|\mathbf{I})$. Now using (6.28), (6.31) and (6.21), we obtain

$$
\begin{aligned}
A_3 &\le \frac{1}{\alpha}\left(2 \max_{l\in\{2,3,4\}} \delta^1(c_9\beta_C, n-l) + \frac{c_{10}\beta_C}{n} + \frac{2\alpha}{\sqrt{2\pi}}\right) E\left|W_C - W_C^*\right| \\
&\le \frac{c_6\beta_C}{n\alpha}\left(2 \max_{l\in\{2,3,4\}} \delta^1(c_9\beta_C, n-l) + \frac{c_{10}\beta_C}{n} + \frac{2\alpha}{\sqrt{2\pi}}\right).
\end{aligned} \tag{6.32}
$$

Recalling $h = h_{z,\alpha}$, as the bound on $A_3$ does not depend on $z \in \mathbb{R}$, combining (6.21), (6.26) and (6.32), then taking supremum over $z \in \mathbb{R}$ on the left hand side of (6.20), we obtain,

$$
\begin{aligned}
&\sup_{z\in\mathbb{R}}\left|Eh_{z,\alpha}(W_C) - Nh_{z,\alpha}\right| \\
&\quad\le \frac{c_{11}\beta_C}{n} + \frac{c_6\beta_C}{n\alpha}\left(2 \max_{l\in\{2,3,4\}} \delta^1(c_9\beta_C, n-l) + \frac{c_{10}\beta_C}{n} + \frac{2\alpha}{\sqrt{2\pi}}\right),
\end{aligned}
$$

and now taking supremum over $C \in M_n^1(\beta)$ with $\beta/n \le c_1\epsilon_0$ we have

$$
\delta^1(\alpha, \beta, n) \le \frac{c_{11}\beta}{n} + \frac{c_6\beta}{n\alpha}\left(2 \max_{l\in\{2,3,4\}} \delta^1(c_9\beta, n-l) + \frac{c_{10}\beta}{n} + \frac{2\alpha}{\sqrt{2\pi}}\right).
$$

Recalling (6.19), we obtain

$$
\delta^1(\beta, n) \le \frac{c_{11}\beta}{n} + \frac{c_6\beta}{n\alpha}\left(2 \max_{l\in\{2,3,4\}} \delta^1(c_9\beta, n-l) + \frac{c_{10}\beta}{n} + \frac{2\alpha}{\sqrt{2\pi}}\right) + \frac{\alpha}{\sqrt{2\pi}}.
$$

Setting $\alpha = 4c_6c_9\beta/n$ yields

$$
\delta^1(\beta, n) \le \frac{c_{12}\beta}{n} + \frac{1}{2} \max_{l\in\{2,3,4\}} \frac{\delta^1(c_9\beta, n-l)}{c_9}.
$$

Multiplying by $n/\beta$ we obtain

$$
\frac{n\delta^1(\beta, n)}{\beta} \le c_{12} + \frac{1}{2} \max_{l\in\{2,3,4\}} \frac{n\delta^1(c_9\beta, n-l)}{c_9\beta}.
$$

Taking supremum over positive $\beta$ satisfying $\beta/n \le c_1\epsilon_0$, and using $n_0 \ge 16$ we obtain

$$
\sup_{0<\frac{\beta}{n}\le c_1\epsilon_0} \frac{n\delta^1(\beta, n)}{\beta} \le \frac{2c_{12}}{3} \max_{l\in\{2,3,4\}} \sup_{0<\frac{\beta}{n}\le c_1c_9\epsilon_0} \frac{(n-l)\delta^1(\beta, n-l)}{\beta}. \tag{6.33}
$$

Clearly

$$\sup_{\beta/n>c_1\epsilon_0} \frac{(n-l)\delta^1(\beta, n-l)}{\beta} \le 1/(c_1\epsilon_0),$$

so letting

$$s_n = \sup_{0<\beta/n\le c_1\epsilon_0} \frac{n\delta^1(\beta, n)}{\beta}$$

and, recalling $c_9 \ge 1$, decomposing the supremum on the right hand side of (6.33) over $\beta/n \le c_1\epsilon_0$ and $c_1\epsilon_0 < \beta/n \le c_1c_9\epsilon_0$ we obtain

$$s_n \le c_{13} + \frac{2}{3} \max_{l\in\{2,3,4\}} s_{n-l} \quad \text{for all } n \ge n_0.$$

Lemma 6.4 now yields $\sup_n s_n < \infty$. Taking the value of this supremum to be $c_5$, we obtain (6.17), as desired. □

We now present the proof of the four technical lemmas that were used in the proof of the theorem.

*Proof of Lemma 6.1* Let $\Lambda = \{(i, j): |e_{ij}| > 1/2\}$ and $\Lambda_i = \{j: (i, j) \in \Lambda\}$ for $i = 1, \ldots, n$. By a Chebyshev type argument we may bound the size of $\Lambda$ by

$$|\Lambda| = \sum_{i,j} \mathbf{1}\big(|e_{ij}| > 1/2\big) \le 8 \sum_{i,j} |e_{ij}|^3 = 8\beta_E. \tag{6.34}$$

Now the inclusion

$$\{W_E \ne W_{E'}\} \subset \bigcup_{i=1}^{n} \big\{\big(i, \pi(i)\big) \in \Lambda\big\}$$

implies

$$P(W_E \ne W_{E'}) \le E \sum_{i=1}^{n} \mathbf{1}\big(\big(i, \pi(i)\big) \in \Lambda\big) = \sum_{i=1}^{n} |\Lambda_i|/n = |\Lambda|/n \le 8\beta_E/n,$$

proving the first claim of (6.10) taking $c_1 = 8$.

Hölder's inequality and (6.34) yield that for all $r \in (0, 3]$

$$\sum_{(i,j)\in\Lambda} |e_{ij}|^r \le |\Lambda|^{1-r/3} \Big(\sum_{i,j} |e_{ij}|^3\Big)^{r/3} \le c_1\beta_E. \tag{6.35}$$

Similarly, as

$$|\Lambda_i| = \sum_{j} \mathbf{1}\big(|e_{ij}| > 1/2\big) \le 8 \sum_{j} |e_{ij}|^3,$$

we have

$$\left|\sum_{j \in \Lambda_i} e_{ij}\right| \le |\Lambda_i|^{2/3}\left(\sum_{j \in \Lambda_i} |e_{ij}|^3\right)^{1/3} \le 4\sum_j |e_{ij}|^3 \le c_1\beta_E, \tag{6.36}$$

with the same bound holding when interchanging the roles of $i$ and $j$.

Regarding the mean $\mu_{E'}$, since $\sum_{ij} e_{ij} = 0$, we have

$$\begin{aligned}|\mu_{E'}| &= \left|\frac{1}{n}\sum_{i,j} e'_{ij}\right| = \left|\frac{1}{n}\sum_{(i,j)\in\Lambda^c} e'_{ij}\right| = \left|\frac{1}{n}\sum_{(i,j)\in\Lambda^c} e_{ij}\right| = \left|\frac{1}{n}\sum_{(i,j)\in\Lambda} e_{ij}\right| \\ &\le \frac{1}{n}\sum_{(i,j)\in\Lambda} |e_{ij}| \le c_1\beta_E/n,\end{aligned}$$

by (6.35) with $r = 1$, proving the second claim in (6.10).

To prove the bound on $\sigma^2_{E'}$, recalling the form of the variance in (4.105) we have

$$\begin{aligned}|\sigma^2_{E'} - 1| &= \frac{1}{n-1}\left|\sum_{i,j} e'^2_{ij} - \sum_{i,j} e'^2_{i\bullet} - \sum_{i,j} e'^2_{\bullet j} + \sum_{i,j} e'^2_{\bullet\bullet} - \sum_{i,j} e^2_{ij}\right| \\ &\le \frac{1}{n-1}\left(\sum_{(i,j)\in\Lambda} e^2_{ij} + \sum_{i,j} e'^2_{i\bullet} + \sum_{i,j} e'^2_{\bullet j} + \sum_{i,j} e'^2_{\bullet\bullet}\right).\end{aligned}$$

Since $n \ge 2$ the first term is bounded by $2c_1\beta_E/n$ using (6.35) with $r = 2$. By (6.36), we have that

$$|e'_{i\bullet}| = \left|\frac{1}{n}\sum_j e'_{ij}\right| = \left|\frac{1}{n}\sum_{j\notin\Lambda_i} e_{ij}\right| = \left|\frac{1}{n}\sum_{j\in\Lambda_i} e_{ij}\right| \le \frac{4}{n}\sum_j |e_{ij}|^3 = \frac{4\beta_E}{n}. \tag{6.37}$$

Hence, for $n \ge 2$,

$$\frac{1}{n-1}\sum_{i,j} e'^2_{i\bullet} \le \frac{4\beta_E}{n-1}\sum_i |e'_{i\bullet}| \le \frac{16\beta_E}{n(n-1)}\sum_{i,j} |e_{ij}|^3 \le 32\beta_E^2/n^2,$$

with the same bound holding when $i$ and $j$ are interchanged. In addition, by the second claim in (6.10),

$$|e'_{\bullet\bullet}| = |\mu_{E'}|/n \le c_1\beta_E/n^2, \tag{6.38}$$

and so

$$\frac{1}{n-1}\sum_{i,j} e'^2_{\bullet\bullet} \le \frac{n^2}{n-1}\frac{c_1^2\beta_E^2}{n^4} \le 2c_1^2\beta_E^2/n^3.$$

Hence

$$|\sigma^2_{E'} - 1| \le \frac{\beta_E}{n}\left(2c_1 + 64\beta_E/n + 2c_1^2\beta_E/n^2\right).$$

Now the first claim of (6.11) holds with $c_2 = 2c_1 + 64 + 2c_1^2$ taking any $\epsilon_1 \in (0,1)$. Requiring further that $\epsilon_1 \in (0, 1/(3c_2))$, when $\beta_E/n \le \epsilon_1$ then

$$|\sigma^2_{E'} - 1| \le 1/3,$$

so that $\sigma_{E'}^2 > 2/3$, implying $\sigma_{E'} > 2/3$. Therefore, when $\beta_E/n \le \epsilon_1$ the elements of $\widehat{E}'$ satisfy

$$|e'_{ij} - e'_{i\bullet} - e'_{\bullet j} + e'_{\bullet\bullet}|/\sigma_{E'} \le \frac{3}{4} + \frac{3}{2}\left(|e'_{i\bullet}| + |e'_{\bullet j}| + |e'_{\bullet\bullet}|\right),$$

and by (6.37) and (6.38) there exists $\epsilon_1$ sufficiently small such that the elements of $\widehat{E}'$ are all bounded by 1, thus showing the second claim of (6.11). Lastly, by the lower bound on $\sigma_{E'}$ we have

$$\beta_{E'} = \frac{\sum_{ij} |e'_{ij}|^3}{\sigma_{E'}^3} \le \frac{\sum_{ij} |e_{ij}|^3}{\sigma_{E'}^3} \le c_1 \beta_E,$$

completing the proof of the lemma. □

*Proof of Lemma 6.2* With $\epsilon_1$, $c_1$ and $c_2$ as in Lemma 6.1, set

$$\epsilon_2 = \min\{\epsilon_1, 1/(9c_2)\}$$

and assume $\beta_E/n \le \epsilon_2$.

The first inequality in (6.10) of Lemma 6.1 yields

$$\begin{aligned}
&\sup_{z\in\mathbb{R}} \left|P(W_E \le z) - \Phi(z)\right| \\
&\quad\le \sup_{z\in\mathbb{R}} \left|P(W_{E'} \le z) - \Phi(z)\right| + c_1\beta_E/n \\
&\quad\le \sup_{z\in\mathbb{R}} \left|P(W_{E'} \le z) - \Phi\left(\frac{z-\mu_{E'}}{\sigma_{E'}}\right)\right| + \sup_{z\in\mathbb{R}} \left|\Phi\left(\frac{z-\mu_{E'}}{\sigma_{E'}}\right) - \Phi(z)\right| + c_1\beta_E/n \\
&\quad\le \sup_{z\in\mathbb{R}} \left|P(W_{\widehat{E}'} \le z) - \Phi(z)\right| + \sup_{z\in\mathbb{R}} \left|\Phi\left(\frac{z-\mu_{E'}}{\sigma_{E'}}\right) - \Phi(z)\right| + c_1\beta_E/n.
\end{aligned}$$

Hence we need only show that there exists some $c_{14}$ such that

$$\sup_{z\in\mathbb{R}} \left|\Phi\left(\frac{z-\mu_{E'}}{\sigma_{E'}}\right) - \Phi(z)\right| \le c_{14}\beta_E/n. \tag{6.39}$$

From the first inequality in (6.11) of Lemma 6.1, since $\beta_E/n \le 1/(9c_2)$ we have $|\sigma_{E'}^2 - 1| \le 1/9$ and so $\sigma_{E'} \in [2/3, 4/3]$.

First consider the case where $|z| \ge c_1\beta_E/n$. It is easy to show that

$$\left|z\exp(-az^2/2)\right| \le 1/\sqrt{a} \quad \text{for all } a > 0, z \in \mathbb{R}. \tag{6.40}$$

Hence

$$\begin{aligned}
\left|z\exp\left(-\frac{9}{32}(z-\mu_{E'})^2\right)\right| &\le \left|(z-\mu_{E'})\exp\left(-\frac{9}{32}(z-\mu_{E'})^2\right)\right| + |\mu_{E'}| \\
&\le \frac{4}{3} + |\mu_{E'}| \\
&\le \frac{4}{3}\left(1 + |\mu_{E'}|\right).
\end{aligned} \tag{6.41}$$

Since $\sigma_{E'} \ge 2/3$ and (6.11) gives that $|\sigma_{E'}^2 - 1| \le c_2\beta_E/n$, we find that

$$|\sigma_{E'} - 1| = \frac{|\sigma_{E'}^2 - 1|}{\sigma_{E'} + 1} \le c_2\beta_E/n. \tag{6.42}$$

Letting $\widehat{z} = (z - \mu_{E'})/\sigma_{E'}$, since $|\mu_{E'}| \le c_1\beta_E/n$ by (6.10) of Lemma 6.1, $z$ and $\widehat{z}$ will be on the same side of the origin. Now, using the mean value theorem, that $\sigma_{E'} \in [2/3, 4/3]$, and Lemma 6.1, we obtain

$$\begin{aligned}
&\bigl|\Phi(\widehat{z}) - \Phi(z)\bigr| \\
&\le \max\left(\phi\left(\frac{z-\mu_{E'}}{\sigma_{E'}}\right), \phi(z)\right)\left|\frac{z-\mu_{E'}}{\sigma_{E'}} - z\right| \quad \text{where } \phi = \Phi' \\
&\le \frac{1}{\sqrt{2\pi}} \max\left\{\exp\left(-\frac{9}{32}(z-\mu_{E'})^2\right), \exp\left(-\frac{z^2}{2}\right)\right\}\left|\frac{z(1-\sigma_{E'})}{\sigma_{E'}}\right| \\
&\quad + \frac{1}{\sqrt{2\pi}}\left|\frac{\mu_{E'}}{\sigma_{E'}}\right| \\
&\le \frac{3}{2\sqrt{2\pi}}|\sigma_{E'} - 1| \max\left\{\left|z\exp\left(-\frac{9}{32}(z-\mu_{E'})^2\right)\right|, \left|z\exp\left(-\frac{z^2}{2}\right)\right|\right\} \\
&\quad + \frac{1}{\sqrt{2\pi}}\left|\frac{\mu_{E'}}{\sigma_{E'}}\right| \\
&\le \frac{2}{\sqrt{2\pi}}|\sigma_{E'} - 1|(1 + |\mu_{E'}|) + \frac{3}{4}|\mu_{E'}|.
\end{aligned}$$

This last inequality using (6.41), and (6.40) with $a = 1$. But now, using (6.10) and (6.42), we have

$$\begin{aligned}
&\frac{2}{\sqrt{2\pi}}|\sigma_{E'} - 1|\bigl(1 + |\mu_{E'}|\bigr) + \frac{3}{4}|\mu_{E'}| \\
&\quad \le \frac{2c_2\beta_E}{n\sqrt{2\pi}}\left(1 + \frac{c_1\beta_E}{n}\right) + \frac{3c_1\beta_E}{4n} \\
&\quad \le \left(\frac{2c_2}{\sqrt{2\pi}}(1 + c_1\epsilon_2) + \frac{3c_1}{4}\right)\frac{\beta_E}{n}, \quad \text{since } \beta_E/n \le \epsilon_2.
\end{aligned} \tag{6.43}$$

When $|z| < c_1\beta_E/n$, the bound is easier. Since $\widehat{z}$ lies in the interval with boundary points $3(z - \mu_{E'})/2$ and $3(z - \mu_{E'})/4$, we have

$$|\widehat{z}| \le \frac{3(|z| + |\mu_{E'}|)}{2}. \tag{6.44}$$

Now using that $|z| < c_1\beta_E/n$, and $|\mu_{E'}| \le c_1\beta_E/n$ by (6.10), from (6.44) we obtain

$$\begin{aligned}
\bigl|\Phi(\widehat{z}) - \Phi(z)\bigr| &\le \frac{1}{\sqrt{2\pi}}|\widehat{z} - z| \\
&\le \frac{1}{\sqrt{2\pi}}\bigl(3|z| + 2|\mu_{E'}|\bigr) \\
&\le \frac{5c_1}{\sqrt{2\pi}}\frac{\beta_E}{n}.
\end{aligned} \tag{6.45}$$

The proof of (6.39), and therefore of the lemma, is now completed by letting $c_{14}$ be the maximum of the constants that multiply $\beta_E/n$ in (6.43) and (6.45). □

*Proof of Lemma 6.3* Let $m = n - l$. Since $c_{i\bullet} = 0$, $|c_{ij}| \le 1$ and $l \le 4$ we have

$$|d_{i\bullet}| = \frac{1}{m}\left|\sum_{j=1}^{m} d_{ij}\right| = \frac{1}{m}\left|\sum_{j\notin\mathcal{C}} c_{ij}\right| = \frac{1}{m}\left|\sum_{j\in\mathcal{C}} c_{ij}\right| \le \frac{|\mathcal{C}|}{m} = \frac{4}{m}, \tag{6.46}$$

with the same bound holding when the roles of $i$ and $j$ are interchanged. Similarly, as $c_{\bullet\bullet} = 0$,

$$\begin{aligned}|d_{\bullet\bullet}| = \frac{1}{m^2}\left|\sum_{i,j=1}^{m} d_{ij}\right| &= \frac{1}{m^2}\left|\sum_{\{i\notin\mathcal{R}\}\cap\{j\notin\mathcal{C}\}} c_{ij}\right| \\ &= \frac{1}{m^2}\left|\sum_{\{i\in\mathcal{R}\}\cup\{j\in\mathcal{C}\}} c_{ij}\right| \\ &\le \frac{|\mathcal{R}| + |\mathcal{C}|}{m} = \frac{8}{m},\end{aligned} \tag{6.47}$$

and the first claim now follows, since $\mu_D = m d_{\bullet\bullet}$.

To handle $\sigma_D^2$, recalling $\sigma_C^2 = 1$, by (4.105) there exists some $n_2 \ge 16$ such that for all $l \le 4$

$$\frac{1}{m-1}\sum_{i,j=1}^{n} c_{ij}^2 = \frac{n-1}{m-1} \in \left[1, 1\frac{3}{8}\right] \quad \text{when } n \ge n_2. \tag{6.48}$$

Again from (4.105),

$$\sigma_D^2 = \frac{1}{m-1}\left(\sum_{i,j=1}^{m} d_{ij}^2 - m\sum_{i=1}^{m} d_{i\bullet}^2 - m\sum_{j=1}^{m} d_{\bullet j}^2 + m^2 d_{\bullet\bullet}^2\right).$$

Applying (6.46) and (6.47), when $\beta_C/n \le \epsilon_3$, a value yet to be specified,

$$\begin{aligned}&\left|\sigma_D^2 - \frac{1}{m-1}\sum_{i,j=1}^{n} c_{ij}^2\right| \\ &\quad\le \frac{1}{m-1}\left(\left|\sum_{\{i\notin\mathcal{R}\}\cap\{j\notin\mathcal{C}\}} c_{ij}^2 - \sum_{i,j=1}^{n} c_{ij}^2\right| + m\sum_{i=1}^{m} d_{i\bullet}^2 + m\sum_{j=1}^{m} d_{\bullet j}^2 + m^2 d_{\bullet\bullet}^2\right) \\ &\quad\le \frac{1}{m-1}\left(\sum_{\{i\in\mathcal{R}\}\cup\{j\in\mathcal{C}\}} c_{ij}^2 + 96\right) \\ &\quad\le \frac{1}{m-1}\left(8\epsilon_3^{2/3} n + 96\right),\end{aligned} \tag{6.49}$$

where for (6.49), we have, by Hölder's inequality

$$\sum_{j=1}^{n} c_{ij}^2 \le n^{\frac{1}{3}} \left( \sum_{j=1}^{n} |c_{ij}|^3 \right)^{\frac{2}{3}} \le n^{\frac{1}{3}} \beta_C^{\frac{2}{3}},$$

with the same inequality holding when the roles of $i$ and $j$ are reversed, and so, when $\beta_C / n \le \epsilon_3$,

$$\sum_{\{i\in\mathcal{R}\}\cup\{j\in\mathcal{C}\}} c_{ij}^2 \le \sum_{i\in\mathcal{R}} \sum_{j=1}^{n} c_{ij}^2 + \sum_{j\in\mathcal{C}} \sum_{i=1}^{n} c_{ij}^2 \le 2ln^{\frac{1}{3}} \beta_C^{\frac{2}{3}} \le 8\epsilon_3^{2/3} n.$$

Now choosing $n_3 \ge n_2$ such that $96/(n_3 - 5) \le 3/16$, and then choosing $\epsilon_3$ such that $8\epsilon_3^{2/3} n_3/(n_3 - 5) \le 3/16$, by (6.48) and (6.49) we obtain $|\sigma_D^2 - 1| \le 3/4$ for all $n \ge n_3$, proving the second claim in the lemma for any $n_0 \ge n_3$.

To prove the final claim, first note

$$\begin{aligned}
\sum_{i=1}^{m} |d_{i\cdot}|^3 &= \frac{1}{m^3} \sum_{i=1}^{m} \left| \sum_{j=1}^{m} d_{ij} \right|^3 = \frac{1}{m^3} \sum_{i=1}^{m} \left| \sum_{j\notin\mathcal{C}} c_{ij} \right|^3 \\
&= \frac{1}{m^3} \sum_{i=1}^{m} \left| \sum_{j\in\mathcal{C}} c_{ij} \right|^3 \le \frac{l^2}{m^3} \sum_{i=1}^{m} \sum_{j\in\mathcal{C}} |c_{ij}|^3 \le \frac{l^2 \beta_C}{m^3},
\end{aligned} \tag{6.50}$$

with the same bound holding when $i$ and $j$ are interchanged. Now, since

$$\sum_{\{i\in\mathcal{R}\}\cup\{j\in\mathcal{C}\}} c_{ij} = \sum_{i=1}^{n} \sum_{j\in\mathcal{C}} c_{ij} + \sum_{j=1}^{n} \sum_{i\in\mathcal{R}} c_{ij} - \sum_{\{i\in\mathcal{R}\}\cap\{j\in\mathcal{C}\}} c_{ij},$$

we obtain

$$\begin{aligned}
|d_{\cdot\cdot}|^3 &= \frac{1}{m^6} \left| \sum_{i,j=1}^{m} d_{ij} \right|^3 = \frac{1}{m^6} \left| \sum_{\{i\notin\mathcal{R}\}\cap\{j\notin\mathcal{C}\}} c_{ij} \right|^3 = \frac{1}{m^6} \left| \sum_{\{i\in\mathcal{R}\}\cup\{j\in\mathcal{C}\}} c_{ij} \right|^3 \\
&\le \frac{9}{m^6} \left( \left| \sum_{i\notin\mathcal{R}} \sum_{j\in\mathcal{C}} c_{ij} \right|^3 + \left| \sum_{i\in\mathcal{R}} \sum_{j\notin\mathcal{C}} c_{ij} \right|^3 + \left| \sum_{\{i\in\mathcal{R}\}\cap\{j\in\mathcal{C}\}} c_{ij} \right|^3 \right).
\end{aligned}$$

Hence, using that $|\sum_{i\notin\mathcal{R}} \sum_{j\in\mathcal{C}} c_{ij}|^3 \le (nl)^2 \sum_{i=1}^{n} \sum_{j\in\mathcal{C}} |c_{ij}|^3$, with the same bound holding for the second term and a similar one for the last, we find that for some $n_0 \ge n_3$, for all $n \ge n_0$ we have

$$|d_{\cdot\cdot}|^3 \le \frac{9}{m^6} (2(nl)^2 + l^4) \beta_C \le \frac{28l^2}{m^4} \beta_C. \tag{6.51}$$

Now, when $\beta_D/n \le \epsilon_3$ and $n \ge n_0$, since $\sigma_D^2 \ge 1/4$, by (6.50) and (6.51),

$$\begin{aligned}
\beta_D &= \sigma_D^{-3} \sum_{i,j=1}^{m} |d_{ij}^0|^3 \le 8 \sum_{i,j=1}^{m} |d_{ij}^0|^3 \\
&= 8 \sum_{i,j=1}^{m} |d_{ij} - d_{i\cdot} - d_{\cdot j} + d_{\cdot\cdot}|^3
\end{aligned}$$

$$\le 8 \times 4^2 \left( \sum_{i,j=1}^{n} |c_{ij}|^3 + \sum_{i,j=1}^{m} \left( |d_{i.}|^3 + |d_{.j}|^3 + |d_{..}|^3 \right) \right)$$
$$\le 128 \left( 1 + \frac{30l^2}{m^2} \right) \beta_C$$
$$\le c_4 \beta_C,$$

thus proving the final claim of the lemma. □

*Proof of Lemma 6.4* Let the sequence $\{t_n\}_{n\ge m}$ be given by

$$t_m = \max_{0\le k\le 3} s_{m-k} \quad \text{and} \quad t_{n+1} = d + \alpha t_n \quad \text{for } n \ge m. \tag{6.52}$$

Explicitly solving the recursion yields

$$t_n = d_1\alpha^n + d_2 \quad \text{where } d_1 = \tfrac{(1-\alpha)t_m - d}{\alpha^m(1-\alpha)} \text{ and } d_2 = \tfrac{d}{1-\alpha}.$$

We note that since $\lim_{n\to\infty} t_n = d_2$ the sequence $\{t_n\}_{n\ge m}$ is bounded, and it suffices to prove $s_n \le t_n$ for all $n \ge m$. We consider the two cases, (a) $d_1 < 0$ and (b) $d_1 \ge 0$.

(a) When $d_1 < 0$ the sequence $\{t_n\}_{n\ge m}$ is increasing. By (6.52) we have $s_m \le t_m$. In addition,

$$\begin{aligned} s_{m+1} &\le d + \alpha \max\{s_{m-1}, s_{m-2}, s_{m-3}\} \le d + \alpha t_m = t_{m+1}, \\ s_{m+2} &\le d + \alpha \max\{s_m, s_{m-1}, s_{m-2}\} \le d + \alpha t_m = t_{m+1} \le t_{m+2}, \end{aligned} \tag{6.53}$$

and hence

$$\begin{aligned} s_{m+3} &\le d + \alpha \max\{s_{m+1}, s_m, s_{m-1}\} \le d + \alpha \max\{t_{m+1}, t_m\} \\ &= d + \alpha t_{m+1} = t_{m+2} \le t_{m+3}. \end{aligned}$$

Hence, for $k = 3$,

$$s_n \le t_n \quad \text{for } m \le n \le m + k. \tag{6.54}$$

Assuming now that (6.54) holds for some $k \ge 3$, for $n = m + k$ we have

$$\begin{aligned} s_{n+1} &\le d + \alpha \max\{s_{n-1}, s_{n-2}, s_{n-3}\} \le d + \alpha \max\{t_{n-1}, t_{n-2}, t_{n-3}\} \\ &= d + \alpha t_{n-1} = t_n \le t_{n+1}, \end{aligned}$$

thus completing the inductive step showing that (6.54) holds for all $k \ge 0$ in case (a).

(b) When $d_1 \ge 0$ the sequence $\{t_n\}_{n\ge m}$ is non-increasing. In a similar way we can show that for $k = 5$,

$$s_n \le t_m \quad \text{for } m \le n \le m + k. \tag{6.55}$$

Assuming now that (6.55) holds for some $k \ge 5$, for $n = m + k$ we have

$$s_{n+1} \le d + \alpha \max\{s_{n-1}, s_{n-2}, s_{n-3}\} \le d + \alpha t_m = t_{m+1} \le t_m,$$

thus completing the inductive step showing that (6.55) holds for all $k \ge 0$ in case (b). □

### 6.1.2 Distribution Constant on Conjugacy Classes

In this section we focus on the normal approximation of $Y$ in (6.1) when the distribution of $\pi$ is a function only of its cycle type. This framework includes two special cases of note, one where $\pi$ is a uniformly chosen fixed point free involution, considered by Goldstein and Rinott (2003) and Ghosh (2009), and the other where $\pi$ has the uniform distribution over permutations with a single cycle, considered by Kolchin and Chistyakov (1973) with the additional restriction that $a_{ij} = b_i c_j$. Both Goldstein and Rinott (2003) and Ghosh (2009) obtained an explicit constant, the latter in terms of third moment quantities on the array $a_{ij}$, rather than on its maximum, as in the former. Kolchin and Chistyakov (1973) considered normal convergence for the long cycle case, but did not provide bounds on the error.

As discussed in Sect. 4.4, being able to approximate the distribution of $Y$ is important for performing permutation tests in statistics. In particular, the case where $\pi$ is a fixed point free involution arises when testing if a given pairing of $n = 2m$ observations shows an unusually high level of similarity, as in Schiffman et al. (1978). In this case, the test statistic $y_\tau$ is of the form (6.1) with $\pi$ replaced by a given pairing $\tau$, and where $a_{ij} = d(x_i, x_j)$ measures the similarity between observations $x_i$ and $x_j$. Under the null hypotheses that no pairing is distinguished, the value of $y_\tau$ will tend to lie near the center of the distribution of $Y$ when $\pi$ is an involution having no fixed points, chosen uniformly. This instance is the particular case where $\pi$ is constant on conjugacy classes, as defined below in (6.57), where the probability of any $\pi$ with $m$ 2-cycles is constant, and has probability zero otherwise.

In the involution case Goldstein and Rinott (2003) used an exchangeable pair construction in which $\pi''$ is obtained from $\pi$ by a transformation which preserves the $m$ 2-cycle structure. The construction in Theorem 6.3 preserves the cycle structure in general, and when there are $m$ 2-cycles, specializes to a construction similar, but not equivalent, to that of Goldstein and Rinott (2003). We note that in the case where $\pi$ is a fixed point free involution the sum $Y$ contains both $a_{i\pi(i)}$ and $a_{\pi(i)i}$, making the symmetry assumption of Theorem 6.3 without loss of generality. This assumption is also satisfied in many statistical applications where one wishes to test the equality of the two distributions generating the samples $X_1, \ldots, X_n$ and $Y_1, \ldots, Y_n$, and $a_{ij} = d(X_i, Y_j)$, a symmetric 'distance' function evaluated at the observed data points $X_i$ and $Y_j$.

Consider a permutation $\pi \in \mathcal{S}_n$ represented in cycle form; in $\mathcal{S}_7$ for example, $\pi = ((1,3,7,5), (2,6,4))$ is the permutation consisting of one 4 cycle in which $1 \to 3 \to 7 \to 5 \to 1$, and one 3 cycle where $2 \to 6 \to 4 \to 2$. For $q = 1, \ldots, n$, let $c_q(\pi)$ be the number of $q$ cycles of $\pi$, and let

$$c(\pi) = \big(c_1(\pi), \ldots, c_n(\pi)\big).$$

We say the permutations $\pi$ and $\sigma$ are of the same cycle type if $c(\pi) = c(\sigma)$, and that a distribution $P$ on $\mathcal{S}_n$ is constant on cycle type if $P(\pi)$ depends only on $c(\pi)$, that is,

$$P(\pi) = P(\sigma) \quad \text{whenever } c(\pi) = c(\sigma). \tag{6.56}$$

Equivalently, see Sagan (1991) for instance, $\pi$ and $\sigma$ are of the same cycle type if and only if $\pi$ and $\sigma$ are conjugate, that is, if and only if there exists a permutation $\rho$ such that $\pi = \rho^{-1}\sigma\rho$. Hence, a probability measure $P$ on $\mathcal{S}_n$ is constant over cycle type if and only if

$$P(\pi) = P\big(\rho^{-1}\pi\rho\big) \quad \text{for all } \pi, \rho \in \mathcal{S}_n. \tag{6.57}$$

A special case of a distribution constant on cycle type is one uniformly distributed over all permutations of some fixed type. Letting

$$\mathcal{N}_n = \left\{(c_1, \ldots, c_n) \in \mathbb{N}_0^n \colon \sum_{i=1}^n c_i = n\right\},$$

the set of possible cycle types for a permutation $\pi \in \mathcal{S}_n$, the number $N(c)$ of permutations in $\mathcal{S}_n$ having cycle type $c$ is given by Cauchy's formula

$$N(c) = n! \prod_{j=1}^n \left(\frac{1}{j}\right)^{c_j} \frac{1}{c_j!} \quad \text{for } c \in \mathcal{N}_n. \tag{6.58}$$

For $c \in \mathcal{N}_n$ let $\mathcal{U}(c)$ denote the distribution over $\mathcal{S}_n$ which is uniform on cycle type $c$, that is, the distribution $P$ given by

$$P(\pi) = \begin{cases} 1/N(c) & \text{if } c(\pi) = c, \\ 0 & \text{otherwise.} \end{cases} \tag{6.59}$$

The situations where $\pi$ is chosen uniformly from the set of all fixed point free involutions, and where $\pi$ is chosen uniformly from all permutations having a single cycle, are both distributions of type $\mathcal{U}(c)$, the first with $c = (0, n/2, 0, \ldots, 0)$ the second with $c = (0, \ldots, 0, 1)$.

The following lemma shows that every distribution $P$ that is constant on cycle type is a mixture of $\mathcal{U}(c)$ distributions.

**Lemma 6.5** *If the distribution $P$ on $\mathcal{S}_n$ is constant on cycle type then*

$$P = \sum_{c \in \mathcal{N}_n} \rho_c \mathcal{U}(c) \quad \text{where } \rho_c = P\big(c(\pi) = c\big). \tag{6.60}$$

*Proof* If $c \in \mathcal{N}_n$ is such that $\rho_c \neq 0$ then by (6.56),

$$1 = \sum_{\gamma : c(\gamma) = c} P\big(\gamma | c(\gamma) = c\big) = N(c) P\big(\pi | c(\pi) = c\big),$$

and therefore $P(\pi | c(\pi) = c) = 1/N(c)$. Hence, for any $\pi \in \mathcal{S}_n$, with $c = c(\pi)$,

$$P(\pi) = P\big(\pi | c(\pi) = c\big) P\big(c(\pi) = c\big) = \rho_c / N(c),$$

that is, $P$ is the mixture (6.60). □

For $\{i, j, k\} \subset \{1, \ldots, n\}$ distinct, let

$$A = \big\{\pi \colon \pi(k) = j\big\} \quad \text{and} \quad B = \big\{\pi \colon \pi(i) = j\big\},$$

and let $\tau_{ik}$ be the transposition of $i$ and $k$. Then

$$\pi \in A \quad \text{if and only if} \quad \tau_{ik}^{-1}\pi\tau_{ik} \in B.$$

Hence, if the distribution of $\pi$ is constant on conjugacy classes,

$$P(A) = P\big(\tau_{ik}^{-1}A\tau_{ik}\big) = P(B),$$

so if in addition $\pi$ has no fixed points,

$$1 = \sum_{k:\, k \neq j} P\big(\pi(k) = j\big) = \sum_{k:\, k \neq j} P\big(\pi(i) = j\big)$$

and hence

$$P\big(\pi(i) = j\big) = \frac{1}{n-1} \quad \text{for } i \neq j. \tag{6.61}$$

If $\pi$ has no fixed points with probability one then no $a_{ii}$ appears in the sum (6.1), and we may take $a_{ii} = 0$ for all $i$ for convenience. In this case, letting

$$a_{io} = \frac{1}{n-2}\sum_{j=1}^{n} a_{ij},$$

$$a_{oj} = \frac{1}{n-2}\sum_{i=1}^{n} a_{ij} \quad \text{and} \quad a_{oo} = \frac{1}{(n-1)(n-2)}\sum_{ij} a_{ij},$$

by (6.61) we have

$$EY = \sum_{i=1}^{n} Ea_{i,\pi(i)} = \frac{1}{n-1}\sum_{i=1}^{n}\sum_{j:\, j \neq i} a_{ij} = \frac{1}{n-1}\sum_{i=1}^{n}\sum_{j=1}^{n} a_{ij} = (n-2)a_{oo}.$$

Now note that

$$\sum_{i=1}^{n} a_{io} = (n-1)a_{oo} \quad \text{and} \quad \sum_{i=1}^{n} a_{o\pi(i)} = \sum_{j=1}^{n} a_{oj} = (n-1)a_{oo},$$

the latter equality holding since $\pi$ is a permutation. Letting

$$\widehat{a_{ij}} = \begin{cases} a_{ij} - a_{io} - a_{oj} + a_{oo} & \text{for } i \neq j, \\ 0 & \text{for } i = j, \end{cases} \tag{6.62}$$

where the choice of $\widehat{a_{ii}}$ is arbitrary, when $\pi$ has no fixed points, using that $\{\pi(j):\ j = 1, \ldots, n\} = \{1, \ldots, n\}$, we have

$$\begin{aligned} \sum_{i=1}^{n} \widehat{a}_{i\pi(i)} &= \sum_{i=1}^{n} a_{i\pi(i)} - 2(n-1)a_{oo} + na_{oo} \\ &= \sum_{i=1}^{n} a_{i\pi(i)} - (n-2)a_{oo} = \sum_{i=1}^{n} a_{i\pi(i)} - EY. \end{aligned} \tag{6.63}$$

Additionally, noting

$$\sum_{i:\, i\neq j} a_{ij} = \sum_{i=1}^{n} a_{ij} = (n-2)a_{oj},$$

and

$$\sum_{i:\, i\neq j} a_{io} = \sum_{i=1}^{n} a_{io} - a_{oj} = (n-1)a_{oo} - a_{oj},$$

we have

$$\begin{aligned}\sum_{i=1}^{n} \widehat{a_{ij}} &= \sum_{i:\, i\neq j} \widehat{a_{ij}} \\ &= (n-2)a_{oj} - \left[(n-1)a_{oo} - a_{oj}\right] - (n-1)a_{oj} + (n-1)a_{oo} = 0. \quad (6.64)\end{aligned}$$

In summary, in view of (6.63), (6.64), and the corresponding identity when the roles of $i$ and $j$ are reversed, when $\pi$ has no fixed points, by replacing $a_{ij}$ by $\widehat{a_{ij}}$, we may without loss of generality assume that

$$EY = 0, \quad a_{io} = 0, \quad \text{and} \quad a_{oj} = 0,$$

and in particular, that

$$\sum_{ij} a_{ij} = 0. \tag{6.65}$$

Lastly note that if $a_{ij}$ is symmetric then so is $\widehat{a_{ij}}$, and in this case

$$\widehat{a_{ij}} = \begin{cases} a_{ij} - 2a_{io} + a_{oo} & \text{for } i \neq j, \\ 0 & \text{for } i = j. \end{cases} \tag{6.66}$$

Regarding the variance of $Y$, Lemma 6.7 below shows that when $\pi$ is chosen uniformly over a fixed cycle type $c$ without fixed points, $n \geq 4$ and $a_{ij} = a_{ji}$, the variance $\sigma_c^2 = \mathrm{Var}(Y)$ is given by

$$\sigma_c^2 = \left(\frac{1}{n-1} + \frac{2c_2}{n(n-3)}\right) \sum_{i\neq j} (a_{ij} - 2a_{io} + a_{oo})^2. \tag{6.67}$$

Remarkably, for a given $n$ the variance $\sigma_c^2$ depends on the vector $c$ of cycle types only though $c_2$, the number of 2-cycles. When $\pi$ is uniform over the set of fixed point free involutions, $n$ is even and $c_2 = n/2$, (6.67) yields

$$\sigma_c^2 = \frac{2(n-2)}{(n-1)(n-3)} \sum_{i\neq j} (a_{ij} - 2a_{io} + a_{oo})^2. \tag{6.68}$$

On the other hand, if $\pi$ has no 2-cycles, $c_2 = 0$ and

$$\sigma_c^2 = \frac{1}{n-1} \sum_{i\neq j} (a_{ij} - 2a_{io} + a_{oo})^2. \tag{6.69}$$

When normal approximations hold for $Y$ when $\pi$ has distribution $\mathcal{U}(c)$ for some $c \in \mathcal{N}_n$ it is clear upon comparing (6.68) with (6.69) that the variance of the approximating normal variable depends on $c$. More generally, when the distribution $\pi$ is constant on cycle type, the mixture property (6.60) allows for an approximation of $Y$ in terms of mixtures of mean zero normal variables as in the following theorem.

**Theorem 6.3** *Let $n \ge 5$ and let $\{a_{ij}\}_{i,j=1}^n$ be an array of real numbers satisfying*

$$a_{ij} = a_{ji}. \tag{6.70}$$

*Let $\pi \in \mathcal{S}_n$ be a random permutation with distribution constant on cycle type, having no fixed points. Then, with $Y$ given by* (6.1) *and $W = (Y - EY)/\sigma_\rho$, we have*

$$\sup_{z\in\mathbb{R}} \big|P(W \le z) - P(Z_\rho \le z)\big| \le 40C\left(1 + \frac{1}{\sqrt{2\pi}} + \frac{\sqrt{2\pi}}{4}\right) \sum_{c\in\mathcal{N}_n} \frac{\rho_c}{\sigma_c}$$

*where*

$$\sigma_\rho^2 = \sum_{c\in\mathcal{N}_n} \rho_c \sigma_c^2 \quad \textit{and} \quad \mathcal{L}(Z_\rho) = \sum_{c\in\mathcal{N}_n} \rho_c \mathcal{L}(Z_c/\sigma_\rho) \tag{6.71}$$

*with $Z_c \sim \mathcal{N}(0, \sigma_c^2)$, $\sigma_c^2$ given by* (6.67), *$\rho_c = P(c(\pi) = c)$, and $C = \max_{i\ne j} |a_{ij} - 2a_{io} + a_{oo}|$.*

*In the special case where $\pi$ is uniformly distributed on fixed point free involutions, with $W = (Y - EY)/\sigma_c$ and $\sigma_c^2$ given by* (6.68),

$$\sup_{z\in\mathbb{R}} \big|P(W \le z) - P(Z \le z)\big| \le 24C\left(1 + \frac{1}{\sqrt{2\pi}} + \frac{\sqrt{2\pi}}{4}\right)\Big/\sigma_c$$

*where $Z \sim \mathcal{N}(0, 1)$.*

We note the numerical value of the coefficient of $C$ in the general, and the involution case, are approximately equal to 125.07 and 75.04, respectively.

The proof of the theorem follows fairly quickly from Lemma 6.10, which considers the special case where $\pi$ has distribution $\mathcal{U}(c)$, and the mixture property in Lemma 6.5. The proof of Lemma 6.10 is preceded by a sequence of lemmas. Lemma 6.6 provides a helpful decomposition. Lemma 6.7 gives the variance of $Y$ in (6.1) when $\pi$ has distribution $\mathcal{U}(c)$ for some $c \in \mathcal{N}_n$. Lemma 6.8 records some properties of the difference of the pair $y', y''$, given functions of two fixed permutations $\pi'$ and $\pi''$, related by transpositions. Lemma 6.9 constructs a Stein pair $(Y', Y'')$. Then, Lemma 6.10 is shown by following the outline in Sect. 4.4.1 to construct the appropriate square bias variables, followed by applying Theorem 5.1 to the resulting zero bias coupling.

To better highlight the reason for the imposition of the symmetry condition (6.70) and the exclusion of fixed points, in Lemmas 6.6, 6.8 and the proof of Lemma 6.9 we consider an array satisfying only (6.65) and allow fixed points. For a given permutation $\pi$ and $i, j \in \{1, \ldots, n\}$, write $i \sim j$ if $i$ and $j$ are in the same cycle of $\pi$, and let $|i|$ denote the length of the cycle of $\pi$ containing $i$.

**Lemma 6.6** *Let $\pi$ be a fixed permutation. For any $i \neq j$, distinct elements of $\{1, \ldots, n\}$, the sets $A_0, \ldots, A_5$ form a partition of the space where,*

$$A_0 = \{|\{i, j, \pi(i), \pi(j)\}| = 2\}$$
$$A_1 = \{|i| = 1, |j| \geq 2\}, \quad A_2 = \{|i| \geq 2, |j| = 1\}$$
$$A_3 = \{|i| \geq 3, \pi(i) = j\}, \quad A_4 = \{|j| \geq 3, \pi(j) = i\} \quad \text{and}$$
$$A_5 = \{|\{i, j, \pi(i), \pi(j)\}| = 4\}.$$

*Additionally, the sets $A_{0,1}$ and $A_{0,1}$ partition $A_0$, where*

$$A_{0,1} = \{\pi(i) = i, \pi(j) = j\}, \quad A_{0,2} = \{\pi(i) = j, \pi(j) = i\},$$

*and we may also write*

$$A_1 = \{\pi(i) = i, \pi(j) \neq j\}, \quad A_2 = \{\pi(i) \neq i, \pi(j) = j\}$$
$$A_3 = \{\pi(i) = j, \pi(j) \neq i\}, \quad A_4 = \{\pi(j) = i, \pi(i) \neq j\},$$

*and membership in $A_m, m = 0, \ldots, 5$ depends only on $i, j, \pi(i), \pi(j)$.*

*Lastly, the sets $A_{5,m}, m = 1, \ldots, 4$ partition $A_5$, where*

$$A_{5,1} = \{|i| = 2, |j| = 2, i \not\sim j\}$$
$$A_{5,2} = \{|i| = 2, |j| \geq 3\}, \quad A_{5,3} = \{|i| \geq 3, |j| = 2\}$$
$$A_{5,4} = \{|i| \geq 3, |j| \geq 3\} \cap A_5,$$

*and membership in $A_{5,m}, m = 1, \ldots, 4$ depends only on $i, j, \pi^{-1}(j), \pi^{-1}(i), \pi(i), \pi(j)$.*

*Proof* The sets $A_m, m = 0, \ldots, 5$ are clearly disjoint, so we need only demonstrate that they are exhaustive. Let $s = |\{i, j, \pi(i), \pi(j)\}|$. Since $i \neq j$, we have $2 \leq s \leq 4$. The case $A_0$ is exactly the case $s = 2$.

There are four cases when $s = 3$. Either exactly one of $i$ or $j$ is a fixed point, that is, either we are in case $A_1$ or $A_2$, or neither $i$ nor $j$ is a fixed point, and so $i \neq \pi(i)$ and $j \neq \pi(j)$. As $i \neq j$, and therefore $\pi(i) \neq \pi(j)$, the only equalities among $i, \pi(i), j, \pi(j)$ which are yet possible are $\pi(i) = j$ or $\pi(j) = i$. Both equalities cannot hold at once, as then $s = 2$. The case where only the first equality is satisfied is $A_3$, and only the second is $A_4$. Clearly what remains now is exactly $A_5$.

The sets $A_{0,1}$ and $A_{0,2}$ are clearly disjoint and union to $A_0$, and the alternative ways to express $A_1, A_2, A_3$ and $A_4$ are clear, as, therefore, are the claims about what values are sufficient to determine membership is these sets. The sets $A_{5,m}, m = 1, \ldots, 5$ are also clearly disjoint. If either $i$ or $j$ is a fixed point then $s \leq 3$, so on $A_5$ we must have $|i| \geq 2$ and $|j| \geq 2$. The set $A_{5,1}$ is where both $i$ and $j$ are in 2-cycles, in which case these cycles must be distinct in order that $s = 4$. The sets $A_{5,2}$ and $A_{5,3}$ are the cases where exactly one of $i$ or $j$ is in a 2-cycle, and these are already subsets of $A_5$. The remaining case in $A_5$ is when $i$ and $j$ are both in cycles of length at least 3, yielding $A_{5,4}$. □

We now calculate the variance of $Y$ when $\pi$ in (6.1) is chosen uniformly over all permutations of some fixed cycle type with no fixed points.

**Lemma 6.7** *For $n \geq 4$ let $c \in \mathcal{N}_n$ with $c_1 = 0$, and let $\pi$ be uniformly chosen from all permutations with cycle type $c$. Assume that*

$$a_{ij} = a_{ji}.$$

*Then the variance of $Y$ in* (6.1) *is given by*

$$\sigma_c^2 = \left(\frac{1}{n-1} + \frac{2c_2}{n(n-3)}\right) \sum_{i \neq j} (a_{ij} - 2a_{io} + a_{oo})^2.$$

*Proof* Without loss of generality we may take $a_{ii} = 0$ and then replace $a_{ij}$ by $a_{ij} - 2a_{io} + a_{oo}$ for $i \neq j$ as in (6.66), and so, in particular, we may assume (6.65) holds. In particular $EY = 0$ and $\text{Var}(Y) = EY^2$. Expanding,

$$EY^2 = E\sum_{ij} a_{i,\pi(i)} a_{j,\pi(j)} = E\sum_i a_{i,\pi(i)}^2 + E\sum_{i \neq j} a_{i,\pi(i)} a_{j,\pi(j)}.$$

For the first term, by (6.61), we have

$$\sum_i E a_{i,\pi(i)}^2 = \frac{1}{n-1} \sum_{ij} a_{ij}^2. \tag{6.72}$$

It is helpful to write the second term as

$$E\sum_{i \neq j} a_{i,\pi(i)} a_{j,\pi(j)} = n(n-1) E a_{I,\pi(J)} a_{J,\pi(J)} \tag{6.73}$$

where $I$ and $J$ are chosen uniformly from all distinct pairs, independently of $\pi$. We evaluate this expectation with the help of the decomposition in Lemma 6.6, starting with $A_0$. Noting $A_{0,1}$ is null as $c_1 = 0$, from $A_{0,2}$ we have

$$\begin{aligned} E a_{I,\pi(I)} a_{J,\pi(J)} \mathbf{1}_{A_{0,2}} &= E a_{I,J}^2 \mathbf{1}_{\{\pi(I)=J, \pi(J)=I\}} \\ &= \frac{2c_2}{n^2(n-1)^2} \sum_{i \neq j} a_{ij}^2, \end{aligned} \tag{6.74}$$

noting that there are $n(n-1)$ possibilities for $I$ and $J$, another factor of $n(n-1)$ for the possible values of $\pi(i)$ and $\pi(j)$, and $c_2$ ways that $(i, j)$ can be placed as a 2-cycle, with the same holding for $(j, i)$. As $A_1$ and $A_2$ are null, moving on to $A_3$ we have, by similar reasoning,

$$\begin{aligned} E a_{I,\pi(J)} a_{J,\pi(J)} \mathbf{1}_{A_3} &= E a_{I,J} a_{J,\pi(J)} \mathbf{1}_{\{|I| \geq 3, \pi(I)=J\}} \\ &= \frac{\sum_{b \geq 3} b c_b}{n^2(n-1)^2(n-2)} \sum_{|\{i,j,k\}|=3} a_{ij} a_{jk}. \end{aligned} \tag{6.75}$$

By symmetry the event $A_4$ contributes the same.

Lastly, consider the contributions from $A_5$. Starting with $A_{5,1}$, we have

$$\begin{aligned} E a_{I,\pi(J)} a_{J,\pi(J)} \mathbf{1}_{A_{5,1}} &= E a_{I,\pi(J)} a_{J,\pi(J)} \mathbf{1}_{\{|I|=2, |J|=2, I \not\sim J\}} \\ &= \frac{4c_2(c_2-1)}{n^2(n-1)^2(n-2)(n-3)} \sum_{|\{i,j,k,l\}|=4} a_{ik} a_{jl}, \end{aligned} \tag{6.76}$$

and

$$\begin{aligned} Ea_{I,\pi(J)}a_{J,\pi(J)}\mathbf{1}_{A_{5,2}} &= Ea_{I,\pi(J)}a_{J,\pi(J)}\mathbf{1}_{\{|I|=2,\,|J|\ge 3\}} \\ &= \frac{2c_2\sum_{b\ge 3}bc_b}{n^2(n-1)^2(n-2)(n-3)}\sum_{|\{i,j,k,l\}|=4} a_{ik}a_{jl}. \qquad (6.77) \end{aligned}$$

The contribution from $A_{5,3}$ is the same as that from $A_{5,2}$.

We break $A_{5,4}$ into two subcases, depending on whether or not $I$ and $J$ are in a common cycle. When they are, we obtain

$$\begin{aligned} Ea_{I,\pi(J)}a_{J,\pi(J)}\mathbf{1}_{\{A_{5,4},I\sim J\}} &= Ea_{I,\pi(J)}a_{J,\pi(J)}\mathbf{1}_{\{|I|\ge 3,\,I\sim J,\,A_5\}} \\ &= \frac{\sum_{b\ge 3}bc_b(b-3)}{n^2(n-1)^2(n-2)(n-3)}\sum_{|\{i,j,k,l\}|=4} a_{ik}a_{jl}, \qquad (6.78) \end{aligned}$$

where the term $b-3$ accounts for the fact that on $A_5$ the value of $\pi(j)$ in the cycle of length $b$ cannot lie in $\{i,j,\pi(i)\}$. When $I$ and $J$ are in disjoint cycles we have

$$\begin{aligned} &Ea_{I,\pi(J)}a_{J,\pi(J)}\mathbf{1}_{\{A_{5,4},I\not\sim J\}} \\ &\quad= Ea_{I,\pi(J)}a_{J,\pi(J)}\mathbf{1}_{\{|I|\ge 3,|J|\ge 3,\,I\not\sim J\}} \\ &\quad= \frac{1}{n^2(n-1)^2(n-2)(n-3)}\sum_{b\ge 3}bc_b\Big(\sum_{d\ge 3}dc_d-b\Big)\sum_{|\{i,j,k,l\}|=4} a_{ik}a_{jl}, \qquad (6.79) \end{aligned}$$

where the term $-b$ accounts for the fact that $j$ must lie in a cycle of length at least three, different from the one of length $b\ge 3$ that contains $i$.

To simplify the sums, using that $a_{io}=0$ we obtain

$$\sum_{|\{i,j,k\}|=3} a_{ij}a_{jk} = -\sum_{i\ne j}a_{ij}^2$$

and therefore

$$\begin{aligned} \sum_{|\{i,j,k,l\}|=4} a_{ik}a_{jl} &= -\sum_{|\{i,j,k\}|=3} a_{ik}a_{ji} - \sum_{|\{i,j,k\}|=3} a_{ik}a_{jk} \\ &= \sum_{i\ne j}a_{ij}^2 + \sum_{i\ne k}a_{ik}^2 = 2\sum_{i\ne j}a_{ij}^2. \end{aligned}$$

Summing the contributions to (6.73) from the events $A_0,\dots,A_4$, that is, (6.74) and twice (6.75), using $\sum_{b\ge 2}bc_b = n$ and letting $(n)_k = n(n-1)\cdots(n-k+1)$ denote the falling factorial, yields $\sum_{i\ne j}a_{ij}^2/(n)_4$ times

$$\begin{aligned} (n)_4\bigg(\frac{2c_2}{n(n-1)} - \frac{2\sum_{b\ge 3}bc_b}{n(n-1)(n-2)}\bigg) &= (n-3)\big((n-2)2c_2 - 2(n-2c_2)\big) \\ &= (n-3)2n(c_2-1). \qquad (6.80) \end{aligned}$$

Adding up the contributions to (6.73) from $A_5$, that is, (6.76), twice (6.77), (6.78) and (6.79), yields $\sum_{i\ne j}a_{ij}^2/(n)_4$ times

$$
\begin{aligned}
&8c_2(c_2-1)+8c_2\sum_{b\ge 3}bc_b+2\sum_{b\ge 3}bc_b(b-3)+2\sum_{b\ge 3}bc_b\left(\sum_{d\ge 3}dc_d-b\right) \qquad (6.81)\\
&\quad =8c_2(c_2-1)+8c_2\sum_{b\ge 3}bc_b-6\sum_{b\ge 3}bc_b+2\sum_{b\ge 3}bc_b\sum_{d\ge 3}dc_d\\
&\quad =8c_2(c_2-1)+8c_2(n-2c_2)-6(n-2c_2)+2(n-2c_2)^2\\
&\quad =2n^2-6n+4c_2. \qquad (6.82)
\end{aligned}
$$

Now totalling all contributions, adding (6.80) to (6.82) we obtain

$$(n-3)2n(c_2-1)+2n^2-6n+4c_2=2c_2(n-1)(n-2).$$

Dividing by $(n)_4$ gives the second term in the expression for $\sigma_c^2$. The first term is (6.72). □

We will use $Y'$, $y'$ and $\pi'$ interchangeably for $Y$, $y$ and $\pi$, respectively. Again, for $i\in\{1,\dots,n\}$ we let $|i|$ denote the number of elements in the cycle of $\pi$ that contains $i$. Due to the way that $\pi''$ is formed from $\pi'$ in Lemma 6.8 using two distinct indices $i$ and $j$, the various cases for expressing the difference $Y''-Y'$ depend only on $i$ and $j$ and their pre and post images under $\pi$.

**Lemma 6.8** *Let $\pi$ be a fixed permutation and $i$ and $j$ distinct elements of $\{1,\dots,n\}$. Letting $\pi(-\alpha)=\pi^{-1}(\alpha)$ for $\alpha\in\{1,\dots,n\}$ set*

$$
\begin{aligned}
&\chi_{i,j}=\{-j,-i,i,j\},\\
&\textit{so that}\quad \{\pi(\alpha),\alpha\in\chi_{i,j}\}=\{\pi^{-1}(j),\pi^{-1}(i),\pi(i),\pi(j)\}.
\end{aligned}
$$

*Then, for $\pi''=\tau_{ij}\pi'\tau_{ij}$ with $\tau_{ij}$ the transposition of $i$ and $j$, and $y'$ and $y''$ given by* (6.1) *with $\pi'$ and $\pi''$ replacing $\pi$, respectively,*

$$y''-y'=b\big(i,j,\pi(\alpha),\alpha\in\chi_{i,j}\big)$$

*where*

$$b\big(i,j,\pi(\alpha),\alpha\in\chi_{i,j}\big)=\sum_{m=0}^{5}b_m\big(i,j,\pi(\alpha),\alpha\in\chi_{i,j}\big)\mathbf{1}_{A_m} \qquad (6.83)$$

*with $A_m, m=0,\dots,5$ as in Lemma 6.6, $b_0(i,j,\pi(\alpha),\alpha\in\chi_{i,j})=0$,*

$$
\begin{aligned}
&b_1\big(i,j,\pi(\alpha),\alpha\in\chi_{i,j}\big)=a_{ii}+a_{\pi^{-1}(j),j}+a_{j,\pi(j)}-(a_{jj}+a_{\pi^{-1}(j),i}+a_{i,\pi(j)}),\\
&b_2\big(i,j,\pi(\alpha),\alpha\in\chi_{i,j}\big)=a_{jj}+a_{\pi^{-1}(i),i}+a_{i,\pi(i)}-(a_{ii}+a_{\pi^{-1}(i),j}+a_{j,\pi(i)}),\\
&b_3\big(i,j,\pi(\alpha),\alpha\in\chi_{i,j}\big)=a_{\pi^{-1}(i),i}+a_{ij}+a_{j,\pi(j)}-(a_{\pi^{-1}(i),j}+a_{ji}+a_{i,\pi(j)}),\\
&b_4\big(i,j,\pi(\alpha),\alpha\in\chi_{i,j}\big)=a_{\pi^{-1}(j),j}+a_{ji}+a_{i,\pi(i)}-(a_{\pi^{-1}(j),i}+a_{ij}+a_{j,\pi(i)}),
\end{aligned}
$$

*and*

$$
\begin{aligned}
b_5\big(i,j,\pi(\alpha),\alpha\in\chi_{i,j}\big)=\;&a_{\pi^{-1}(i),i}+a_{i,\pi(i)}+a_{\pi^{-1}(j),j}+a_{j,\pi(j)}\\
&-(a_{\pi^{-1}(i),j}+a_{j,\pi(i)}+a_{\pi^{-1}(j),i}+a_{i,\pi(j)}).
\end{aligned}
$$

*Proof* First we note that equality (6.83) defines a function, as Lemma 6.6 shows that $\mathbf{1}_{A_m}$, $m = 0, \ldots, 5$ depend only on the given variables.

Now considering the difference, under $A_0$ either $\pi(i) = i$ and $\pi(j) = j$, or $\pi(i) = j$ and $\pi(j) = i$; in the both cases $\pi'' = \pi$ and therefore $y'' = y$, and their difference is zero, corresponding to the claimed form for $b_0$. When $A_1$ is true, since $|i| = 1$, and $\pi(j) \neq j$ we have

$$\pi''(j) = \tau_{ij}\pi\tau_{ij}(j) = \tau_{ij}\pi(i) = \tau_{ij}(i) = j \quad \text{and}$$
$$\pi''\left(\pi^{-1}(j)\right) = \tau_{ij}\pi\tau_{ij}\pi^{-1}(j) = \tau_{ij}\pi\left(\pi^{-1}(j)\right) = \tau_{ij}j = i$$

and

$$\pi''(i) = \tau_{ij}\pi\tau_{ij}(i) = \tau_{ij}\pi(j) = \pi(j).$$

If $k \notin \{i, \pi^{-1}(j), j\}$ then $\pi(k) \notin \{i, j\}$ so $\tau_{ij}(k) = k$, and therefore

$$\pi''(k) = \tau_{ij}\pi\tau_{ij}(k) = \tau_{ij}\pi(k) = \pi(k) \quad \text{for all } k \notin \left\{i, \pi^{-1}(j), j\right\}.$$

That is, on $A_1$ the permutations $\pi$ and $\pi''$ only differ in that where $\pi$ has the action $i \to i, \pi^{-1}(j) \to j \to \pi(j)$, leading to the terms

$$a_{ii} + a_{\pi^{-1}(j),j} + a_{j,\pi(j)},$$

the permutation $\pi''$ has the action $j \to j, \pi^{-1}(j) \to i \to \pi(j)$, leading to the terms

$$a_{jj} + a_{\pi^{-1}(j),i} + a_{i,\pi(j)}.$$

Taking the difference now leads to the form claimed for $b_1$ when $A_1$ is true. By symmetry, on $A_2$ we have the same result as for $A_1$ upon interchanging $i$ and $j$.

Similarly, when $A_3$ is true the only difference between $\pi$ and $\pi''$ is that the former has the action $\pi^{-1}(i) \to i \to j \to \pi(j)$, leading to the terms

$$a_{\pi^{-1}(i),i} + a_{ij} + a_{j,\pi(j)},$$

while that latter has $\pi^{-1}(i) \to j \to i \to \pi(j)$, leading to

$$a_{\pi^{-1}(i),j} + a_{ji} + a_{i,\pi(j)}.$$

Again, $A_4$ is the same as $A_3$ with the roles of $i$ and $j$ interchanged.

Lastly, when $|\{i, j, \pi(i), \pi(j)\}| = 4$ the permutation $\pi$ has the action $\pi^{-1}(i) \to i \to \pi(i)$ and $\pi^{-1}(j) \to j \to \pi(j)$ while $\pi''$ has $\pi^{-1}(i) \to j \to \pi(i)$ and $\pi^{-1}(j) \to i \to \pi(j)$, making the form of $b_5$ clear. □

Our next task is the construction of a Stein pair $Y', Y''$, which we accomplish in the following lemma in a manner similar to that in Sect. 4.4.2. We remind the reader that we consider the symbols $\pi$ and $Y$ interchangeable with $\pi'$ and $Y'$, respectively.

**Lemma 6.9** *For $n \geq 5$ let $\{a_{ij}\}_{i,j=1}^n$ be an array of real numbers satisfying*

$$a_{ij} = a_{ji} \quad \text{and} \quad a_{ii} = 0.$$

*Let $\pi \in \mathcal{S}_n$ be a random permutation with distribution constant on cycle type, having no fixed points, and let $Y$ be given by* (6.1). *Further, let $I, J$ be chosen independently of $\pi$, uniformly from all pairs of distinct elements of $\{1, \ldots, n\}$. Then, letting $\pi'' = \tau_{IJ}\pi\tau_{IJ}$ and $Y''$ be given by* (6.1) *with $\pi''$ replacing $\pi$, $(Y, Y'')$ is a $4/n$-Stein pair.*

*Proof* First we show that the pair of permutations $\pi', \pi''$ is exchangeable. For fixed permutations $\sigma', \sigma''$, if $\sigma'' \neq \tau_{IJ}\sigma'\tau_{IJ}$ then

$$P(\pi''=\sigma'', \pi'=\sigma')=0=P(\pi'=\sigma'', \pi''=\sigma').$$

Otherwise $\sigma''=\tau_{IJ}\sigma'\tau_{IJ}$, and using (6.57) for the second equality followed by $\tau_{ij}^{-1}=\tau_{ij}$, we have

$$\begin{aligned} P(\pi''=\sigma'', \pi'=\sigma') &= P(\pi'=\sigma') \\ &= P(\pi'=\tau_{IJ}\sigma'\tau_{IJ}) \\ &= P(\pi''=\sigma') \\ &= P(\pi'=\sigma'', \pi''=\sigma'). \end{aligned}$$

Consequently, $\pi$ and $\pi''$, and therefore $Y$ and $Y''$, given by (6.1) with permutations $\pi$ and $\pi''$, respectively, are exchangeable.

It remains to demonstrate that $Y', Y''$ satisfies the linearity condition (4.108) with $\lambda=4/n$, for which it suffices to show

$$E(Y'-Y''|\pi)=\frac{4}{n}Y'. \tag{6.84}$$

We prove (6.84) by computing the conditional expectation given $\pi$ of the sum $\sum_{m=0}^{5} b_m(i,j,\pi(\alpha),\alpha\in\chi_{i,j})\mathbf{1}_{A_m}$ in (6.83) of Lemma 6.6, with $A_0,\ldots,A_5$ given in Lemma 6.8, with $i, j$ replaced by $I, J$.

First we have that $b_0=0$. Next, we claim that the contribution to $n(n-1)E(Y'-Y''|\pi)$ from $b_1$ and $b_2$ totals to

$$\begin{aligned} &2\big(n-c_1(\pi)\big)\sum_{|i|=1} a_{ii}+4c_1(\pi)\sum_{|i|\geq 2} a_{i,\pi(i)} \\ &\quad -2c_1(\pi)\sum_{|i|\geq 2} a_{ii}-2\sum_{|i|=1,\,|j|\geq 2} a_{ij}-2\sum_{|i|\geq 2,\,|j|=1} a_{ij}. \end{aligned} \tag{6.85}$$

In particular, for the first term $a_{II}$ in the function $b_1$, by summing below over $j$ we obtain

$$E(a_{II}\mathbf{1}_{A_1}|\pi)=\frac{1}{n(n-1)}\sum_{i,j} a_{ii}\mathbf{1}_{\{|i|=1,\,|j|\geq 2\}}=\frac{n-c_1(\pi)}{n(n-1)}\sum_{|i|=1} a_{ii}. \tag{6.86}$$

For the next two terms of $b_1$, noting that the sum of $a_{j,\pi(j)}$ over a given cycle of $\pi$ equals the sum of $a_{\pi^{-1}(j),j}$ over that same cycle, we obtain

$$\begin{aligned} E(a_{\pi^{-1}(J),J}\mathbf{1}_{A_1}|\pi)+E(a_{J,\pi(J)}\mathbf{1}_{A_1}|\pi) &= 2E(a_{J,\pi(J)}\mathbf{1}_{A_1}|\pi) \\ &= \frac{2}{n(n-1)}\sum_{j=1}^{n} a_{j,\pi(j)}\mathbf{1}_{\{|i|=1,|j|\geq 2\}} \\ &= \frac{2c_1(\pi)}{n(n-1)}\sum_{|j|\geq 2} a_{j,\pi(j)}. \end{aligned} \tag{6.87}$$

Moving to the final three terms of $b_1$, we have similarly that

$$E(a_{J,J}\mathbf{1}_{A_1}|\pi) = \frac{c_1(\pi)}{n(n-1)} \sum_{|j|\ge 2} a_{jj},$$

$$E(a_{\pi^{-1}(J),I}\mathbf{1}_{A_1}|\pi) = \frac{1}{n(n-1)} \sum_{|i|=1,|j|\ge 2} a_{\pi^{-1}(j),i} = \frac{1}{n(n-1)} \sum_{|i|=1,|j|\ge 2} a_{ji}$$

and

$$E(a_{I,\pi(J)}\mathbf{1}_{A_1}|\pi) = \frac{1}{n(n-1)} \sum_{|i|=1,\,|j|\ge 2} a_{i,\pi(j)} = \frac{1}{n(n-1)} \sum_{|i|=1,\,|j|\ge 2} a_{ij}.$$

Summing (6.86) and (6.87) and subtracting these last three contributions, and then using the fact that the contribution from $b_2$ is the same as that from $b_1$ by symmetry, we obtain (6.85).

Next, it is easy to see that the first three contributions to $n(n-1)E(Y'-Y''|\pi)$ from $b_3$, on the event $A_3 = \mathbf{1}(\pi(I) = J, |I| \ge 3)$, all equal $\sum_{|i|\ge 3} a_{i,\pi(i)}$, that the fourth and sixth both equal $-\sum_{|i|\ge 3} a_{\pi^{-1}(i),\pi(i)}$, and that the fifth equals $-\sum_{|i|\ge 3} a_{\pi(i),i}$. Combining this quantity with the equal amount from $b_4$ yields

$$6\sum_{|i|\ge 3} a_{i,\pi(i)} - 4\sum_{|i|\ge 3} a_{\pi^{-1}(i),\pi(i)} - 2\sum_{|i|\ge 3} a_{\pi(i),i}. \tag{6.88}$$

Next, write $A_5 = \mathbf{1}\{|I| \ge 2, |J| \ge 2, I \ne J, \pi(I) \ne J, \pi(J) \ne I\}$. The first term in $b_5$, $a_{\pi^{-1}(I),I}$, has conditional expectation given $\pi$ of $(n(n-1))^{-1}$ times

$$\sum a_{\pi^{-1}(i),i}\mathbf{1}\big(|i| \ge 2, |j| \ge 2, i \ne j, \pi(i) \ne j, \pi(j) \ne i\big). \tag{6.89}$$

Write $i \sim j$ when $i$ and $j$ are elements of the same cycle. When $i \sim j$ and $\{i, j, \pi(i), \pi(j)\}$ are distinct, then $|i| \ge 4$ and there are $|i| - 3$ possible choices for $j \sim i$ that satisfy the conditions in the indicator in (6.89). Hence, the case $i \sim j$ contributes

$$\begin{aligned}\sum_{|i|\ge 4} a_{\pi^{-1}(i),i} \sum_{j\sim i} \mathbf{1}\big(i \ne j, \pi(i) \ne j, \pi(j) \ne i\big) &= \sum_{|i|\ge 4} a_{\pi^{-1}(i),i}\big(|i| - 3\big) \\ &= \sum_{|i|\ge 3} \big(|i| - 3\big)a_{i,\pi(i)}.\end{aligned}$$

When $i \not\sim j$ the conditions in the indicator function in (6.89) are satisfied if and only if $|i| \ge 2$, $|j| \ge 2$. For $|i| \ge 2$ there are $n - |i| - c_1(\pi)$ choices for $j$, so the case $i \not\sim j$ contributes

$$\begin{aligned}&\sum_{|i|\ge 2} a_{\pi^{-1}(i),i} \sum_{j\not\sim i,\,|j|\ge 2} 1 \\ &\quad= \sum_{|i|\ge 2} \big(n - |i| - c_1(\pi)\big)a_{i,\pi(i)} \\ &\quad= \big(n - 2 - c_1(\pi)\big)\sum_{|i|=2} a_{i,\pi(i)} + \sum_{|i|\ge 3} \big(n - |i| - c_1(\pi)\big)a_{i,\pi(i)}.\end{aligned}$$

As the first four terms of $b_5$ all yield the same contribution, they account for a total of

$$4\big(n-2-c_1(\pi)\big)\sum_{|i|=2}a_{i,\pi(i)}+4\big(n-3-c_1(\pi)\big)\sum_{|i|\geq 3}a_{i,\pi(i)}. \tag{6.90}$$

Decomposing the contribution from the fifth term $-a_{\pi^{-1}(I),J}$ of $b_5$, according to whether $i\sim j$ or $i\not\sim j$, gives

$$\begin{aligned}
&-\sum_{|i|\geq 2,|j|\geq 2}a_{\pi^{-1}(i),j}\mathbf{1}(i\neq j,\pi(i)\neq j,\pi(j)\neq i)\\
&=-\sum_{|i|\geq 4}\sum_{j\sim i}a_{\pi^{-1}(i),j}\mathbf{1}(i\neq j,\pi(i)\neq j,\pi(j)\neq i)-\sum_{|i|\geq 2,|j|\geq 2}\sum_{j\not\sim i}a_{\pi^{-1}(i),j}\\
&=-\sum_{|i|\geq 4}\sum_{j\sim i}a_{\pi^{-1}(i),j}+\sum_{|i|\geq 4}\big(a_{\pi^{-1}(i),i}+a_{\pi^{-1}(i),\pi(i)}+a_{\pi^{-1}(i),\pi^{-1}(i)}\big)\\
&\quad-\sum_{|i|\geq 2,|j|\geq 2}\sum_{j\not\sim i}a_{ij}\\
&=-\sum_{|i|\geq 4}\sum_{j\sim i}a_{ij}+\sum_{|i|\geq 4}\big(a_{i,\pi(i)}+a_{\pi^{-1}(i),\pi(i)}+a_{ii}\big)-\sum_{|i|\geq 2,|j|\geq 2}\sum_{j\not\sim i}a_{ij}.
\end{aligned}\tag{6.91}$$

To simplify (6.91), let $a\wedge b=\min(a,b)$ and consider a decomposition of the sum $\sum_{ij}a_{ij}$ first by whether $i\sim j$ or not, and then according to cycle sizes, and in the first case further as to whether the length of the common cycle of $i$ and $j$ is greater than 4, and in the second case as to whether the distinct cycles of $i$ and $j$ both have size at least 2. That is, write,

$$\begin{aligned}
\sum_{i,j=1}^{n}a_{ij}&=\sum_{|i|\geq 4}\sum_{j\sim i}a_{ij}+\sum_{|i|\leq 3}\sum_{j\sim i}a_{ij}+\sum_{|i|\geq 2,|j|\geq 2}\sum_{j\not\sim i}a_{ij}\\
&\quad+\sum_{|i|\wedge|j|=1}\sum_{j\not\sim i}a_{ij}.
\end{aligned}\tag{6.92}$$

Since $\sum_{i,j}a_{ij}=0$ by (6.65), we may replace the sum of the first and third terms in (6.91) by the sum of the second and fourth terms on the right hand side of (6.92). Hence, the contribution from $a_{\pi^{-1}(I),J}$ on $A_5$ equals

$$\begin{aligned}
&\sum_{|i|\leq 3}\sum_{j\sim i}a_{ij}+\sum_{|i|\wedge|j|=1}\sum_{j\not\sim i}a_{ij}+\sum_{|i|\geq 4}\big(a_{i,\pi(i)}+a_{\pi^{-1}(i),\pi(i)}+a_{ii}\big)\\
&\quad=\sum_{|i|\leq 2}\sum_{j\sim i}a_{ij}+\sum_{|i|\wedge|j|=1}\sum_{j\not\sim i}a_{ij}+\sum_{|i|\geq 3}\big(a_{i,\pi(i)}+a_{\pi^{-1}(i),\pi(i)}+a_{ii}\big),
\end{aligned}$$

where to obtain the equality we used the fact that $\pi^2(i)=\pi^{-1}(i)$ when $|i|=3$. Dealing similarly with the $|i|=2$, $j\sim i$ term we obtain

$$\begin{aligned}
&\sum_{|i|=1}a_{ii}+\sum_{|i|\wedge|j|=1}\sum_{j\not\sim i}a_{ij}+\sum_{|i|\geq 2}\big(a_{i,\pi(i)}+a_{ii}\big)+\sum_{|i|\geq 3}a_{\pi^{-1}(i),\pi(i)}\\
&\quad=\sum_{|i|\wedge|j|=1}\sum_{j\not\sim i}a_{ij}+\sum_{|i|\geq 2}a_{i,\pi(i)}+\sum_{|i|\geq 1}a_{ii}+\sum_{|i|\geq 3}a_{\pi^{-1}(i),\pi(i)}.
\end{aligned}$$

Combining this contribution with the next three terms of $A_5$, each of which yields the same amount, gives the total

$$4\sum_{|i|\ge 2} a_{i,\pi(i)} + 4\sum_{|i|\ge 3} a_{\pi^{-1}(i),\pi(i)} + 4\sum_{|i|\ge 1} a_{ii} + 4\sum_{|i|\wedge|j|=1}\sum_{j\not\sim i} a_{ij}. \tag{6.93}$$

Combining (6.93) with the contribution (6.90) from the first four terms in $b_5$, the $b_1$ and $b_2$ terms in (6.85) and the $b_3$ and $b_4$ terms (6.88), yields $n(n-1)E(Y'-Y''|\pi')$, which, canceling the terms involving $a_{\pi^{-1}(i),\pi(i)}$ and rearranging to group like terms, can be written

$$4(n-1)\sum_{|i|=2} a_{i,\pi(i)} + (4n-2)\sum_{|i|\ge 3} a_{i,\pi(i)} - 2\sum_{|i|\ge 3} a_{\pi(i),i} \tag{6.94}$$

$$+2\big(n-c_1(\pi)+2\big)\sum_{|i|=1} a_{ii} - 2\big(c_1(\pi)-2\big)\sum_{|i|\ge 2} a_{ii} \tag{6.95}$$

$$+4\sum_{|i|\wedge|j|=1,\, j\not\sim i} a_{ij} - 2\sum_{|i|=1,|j|\ge 2} a_{ij} - 2\sum_{|i|\ge 2,|j|=1} a_{ij}. \tag{6.96}$$

The assumption that $a_{ii}=0$ causes the contribution from (6.95) to vanish, the assumption that there are no 1-cycles causes the contribution from (6.96) to vanish, and the assumption that $a_{ij}=a_{ji}$ allows the combination of the second and third terms in (6.94) to yield

$$\begin{aligned} E(Y'-Y''|\pi') &= \frac{1}{n(n-1)}\Big(4(n-1)\sum_{|i|=2} a_{i,\pi'(i)} + (4n-4)\sum_{|i|\ge 3} a_{i,\pi'(i)}\Big) \\ &= \frac{4}{n}\sum_{i=1}^{n} a_{i,\pi'(i)} = \frac{4}{n}Y'. \end{aligned}$$

Hence, the linearity condition (4.108) is satisfied with $\lambda = 4/n$, completing the argument that $Y', Y''$ is a $4/n$-Stein pair. □

We now prove the special case of Theorem 6.3 when $\pi$ is uniform over cycle type.

**Lemma 6.10** *Let $n\ge 5$ and let $\{a_{ij}\}_{i,j=1}^{n}$ be an array of real numbers satisfying*

$$a_{ij} = a_{ji}.$$

*Let $\pi\in\mathcal{S}_n$ be a random permutation with distribution $\mathcal{U}(c)$, uniform on cycle type $c\in\mathcal{N}_n$, having no fixed points. Then, letting $Y$ be the sum in* (6.1), *$\sigma_c^2$ given by* (6.67) *and $W=(Y-EY)/\sigma_c$,*

$$\sup_{z\in\mathbb{R}}\big|P(W\le z)-P(Z\le z)\big| \le 40C\left(1+\frac{1}{\sqrt{2\pi}}+\frac{\sqrt{2\pi}}{4}\right)\Big/\sigma_c, \tag{6.97}$$

*where $C=\max_{i\ne j}|a_{ij}-2a_{io}+a_{oo}|$ and $Z$ is a standard normal variable.*

*When $\pi$ is uniform over involutions without fixed points, then* 40 *in* (6.97) *may be replaced by* 24, *and $\sigma_c^2$ specializes to the form given in* (6.68).

*Proof* We may set $a_{ii} = 0$, and then by replacing $a_{ij}$ by $a_{ij} - 2a_{io} + a_{oo}$ when $i \neq j$, assume without loss of generality that $a_{io} = a_{oj} = EY = 0$. We write $Y'$ and $\pi'$ interchangeably for $Y$ and $\pi$, respectively. We follow the outline in Sect. 4.4.1 to produce a coupling of $Y$ to a pair $Y^\dagger, Y^\ddagger$ with the square bias distribution as in Proposition 4.6, satisfying (6.2) and (6.3). We then produce a coupling of $Y$ to $Y^*$ having the $Y$-zero bias distribution using the uniform interpolation as in that proposition, and lastly invoke Theorem 5.1 to obtain the bound.

First construct the Stein pair $Y', Y''$ as in Lemma 6.9. Let $\pi'$ be a permutation with distribution $\mathcal{U}(c)$. Then, with $I$ and $J$ having distribution

$$P(I = i, J = j) = \frac{1}{n(n-1)} \quad \text{for } i \neq j,$$

set $\pi'' = \tau_{IJ}\pi'\tau_{IJ}$ where $\tau_{ij}$ is the transposition of $i$ and $j$. Now $Y'$ and $Y''$ are given by (6.1) with $\pi$ replaced by $\pi'$ and $\pi''$, respectively.

To specialize the outline in Sect. 4.4.1 to this case, we let $\mathbf{I} = (I, J)$ and $\Xi_\alpha = \pi(\alpha)$. In keeping with the notation of Lemma 6.8, with $\chi = \{1, \ldots, n\}$ we let $\pi(-j) = \pi^{-1}(j)$ for $j \in \chi$, and with $i$ and $j$ distinct elements of $\chi$ we set $\chi_{i,j} = \{-j, -i, i, j\}$ and

$$p_{i,j}(\xi_\alpha, \alpha \in \chi_{i,j}) = P\big(\pi(\alpha) = \xi_\alpha, \alpha \in \chi_{i,j}\big),$$

the distribution of the pre and post images of $i$ and $j$ under $\pi$. Equality (4.116) gives the factorization of the variables from which $\pi'$ and $\pi''$ are constructed as

$$P(\mathbf{i}, \xi_\alpha, \alpha \in \chi) = P(\mathbf{I} = \mathbf{i}) P_{\mathbf{i}}(\xi_\alpha, \alpha \in \chi_{\mathbf{i}}) P_{\mathbf{i}^c|\mathbf{i}}(\xi_\alpha, \alpha \notin \chi_{\mathbf{i}} | \xi_\alpha, \alpha \in \chi_{\mathbf{i}}).$$

The factorization can be interpreted as saying that first we choose $I, J$, then construct the pre and post images of $I$ and $J$, under $\pi$, then, conditional on what has already been chosen, the values of $\pi$ on the remaining variables.

For the distribution of the pair with the square bias distribution, equality (4.118) gives the parallel factorization,

$$P^\dagger(\mathbf{i}, \xi_\alpha, \alpha \in \chi) = P^\dagger(\mathbf{I} = \mathbf{i}) P_{\mathbf{i}}^\dagger(\xi_\alpha, \alpha \in \chi_{\mathbf{i}}) P_{\mathbf{i}^c|\mathbf{i}}(\xi_\alpha, \alpha \notin \chi_{\mathbf{i}} | \xi_\alpha, \alpha \in \chi_{\mathbf{i}}) \tag{6.98}$$

where $P^\dagger(\mathbf{I} = \mathbf{i})$, the distribution of indices we will label $\mathbf{I}^\dagger$, is given by (4.117) and $P_{\mathbf{i}}^\dagger(\xi_\alpha, \alpha \in \chi_{\mathbf{i}})$ by (4.119). Let $\sigma^\dagger, \sigma^\ddagger$ have distribution given by (6.98), that is, with $I^\dagger, J^\dagger$ and $\Xi_\alpha, \alpha \in \chi$ having distribution (6.98), $\sigma^\dagger(\alpha) = \Xi_\alpha$ and $\sigma^\ddagger = \tau_{I^\dagger, J^\dagger}\sigma^\dagger\tau_{I^\dagger, J^\dagger}$. These permutations do not need to be constructed, we only introduce them so that we can conveniently refer to their distribution, which is the one targeted for $\pi^\dagger, \pi^\ddagger$.

We construct $\pi^\dagger, \pi^\ddagger$, of which $Y^\dagger, Y^\ddagger$ will be a function, in stages, beginning with the indices $I^\dagger, J^\dagger$, and their pre and post images under $\pi^\dagger$. Following (4.117), with $\lambda = 4/n$, let $I^\dagger, J^\dagger$ have distribution

$$P(I^\dagger = i, J^\dagger = j) = \frac{r_{i,j}}{2\lambda\sigma_c^2}$$
$$\text{where } r_{i,j} = P(I = i, J = j) E b^2\big(i, j, \pi(\alpha), \alpha \in \chi_{i,j}\big) \tag{6.99}$$

with $b(i, j, \xi_\alpha, \alpha \in \chi_{i,j})$ as in Lemma 6.8. Next, given $I^\dagger = i$ and $J^\dagger = j$, from (4.119), let the pre and post images $\pi^{-\dagger}(J^\dagger), \pi^{-\dagger}(I^\dagger), \pi^\dagger(I^\dagger), \pi^\dagger(J^\dagger)$ have distribution

$$p^\dagger_{i,j}(\xi_\alpha, \alpha \in \chi_{i,j}) = \frac{b^2(i, j, \xi_\alpha, \alpha \in \chi_{i,j})}{Eb^2(i, j, \pi(\alpha), \alpha \in \chi_{i,j})} p_{i,j}(\xi_\alpha, \alpha \in \chi_{i,j}). \quad (6.100)$$

We will place $I^\dagger$ and $J^\dagger$, along with these generated pre and post images, into cycles of appropriate length. The conditional distribution of the remaining values of $\pi^\dagger$, given $I^\dagger$, $J^\dagger$ and their pre and post images, by (4.118), has the same conditional distribution as that of $\pi'$, which is the uniform distribution over all permutations of cycle type $c$ where $I^\dagger$ and $J^\dagger$ have the specified pre and post images. Hence, to complete the specification of $\pi^\dagger$ we fill in the remaining values of $\pi^\dagger$ uniformly. For this last step we will use the values of $\pi'$ to construct $\pi^\dagger$ in a way that makes $\pi^\dagger$ and $\pi'$ close.

Lemma 6.6 gives that, for $\pi^\dagger$, membership in $A_0, \ldots, A_4$ and $A_{5,1}, \ldots, A_{5,4}$ is determined by

$$I^\dagger, J^\dagger, \pi^{-\dagger}(J), \pi^{-\dagger}(I), \pi^\dagger(I^\dagger), \pi^\dagger(J^\dagger). \quad (6.101)$$

As $b_0 = 0$ from Lemma 6.8 the case $A_0$ has probability zero. Note that the distribution of $\sigma^\dagger, \sigma^\ddagger$ is absolutely continuous with respect to that of $\pi', \pi''$, and therefore the permutations $\sigma^\dagger, \sigma^\ddagger$ have the same cycle structure, namely $c$, as $\pi', \pi''$. In particular, since $\pi'$ has no fixed points, $A_2$ is eliminated and we need only consider the events $A_3$, $A_4$ and $A_5$. For the purpose of conditioning on the values in (6.101), for ease of notation we will write

$$(\alpha, \beta) = (I^\dagger, J^\dagger) \quad \text{and} \quad (\gamma, \delta, \epsilon, \zeta) = (\pi^{-\dagger}(J), \pi^{-\dagger}(I), \pi^\dagger(I^\dagger), \pi^\dagger(J^\dagger)).$$

The specification $\pi^\dagger$ depends on which case, or subcase, of the events $A_3$, $A_4$, $A_5$ is determined by the variables (6.101). In every subcase, however, $\pi^\dagger$ will be specified in terms of $\pi$ by conjugating with transpositions as

$$\pi^\dagger = \tau_{\iota,\iota^\dagger} \pi \tau_{\iota,\iota^\dagger} \quad \text{where } \tau_{\iota,\iota^\dagger} = \prod_{k=1}^{\kappa} \tau_{i_k, i_k^\dagger}, \quad (6.102)$$

for $\iota = (i_1, \ldots, i_\kappa)$ and $\iota^\dagger = (i_1^\dagger, \ldots, i_\kappa^\dagger)$, vectors of disjoint indices of some length $\kappa$. Note that when $\pi^\dagger$ is given by $\pi$ through (6.102) then,

$$\pi^\dagger(k) = \pi(k) \quad \text{for all } k \notin \mathcal{I}_{\iota,\iota^\dagger}, \text{ where}$$
$$\mathcal{I}_{\iota,\iota^\dagger} = \{\pi^{-1}(i_k), i_k, \pi^{-1}(i_k^\dagger), i_k^\dagger \colon k = 1, \ldots, \kappa\}. \quad (6.103)$$

Consider first the case where the generated values determine an outcome in $A_3$, that is, when $J^\dagger = \pi^\dagger(I^\dagger)$ and $\{\pi^{-\dagger}(I^\dagger), I^\dagger, \pi^\dagger(I^\dagger)\}$ are distinct. If $\pi^\dagger(J^\dagger) \in \{\pi^{-\dagger}(I^\dagger), I^\dagger, \pi^\dagger(I^\dagger)\}$ then $\pi^\dagger(J^\dagger) = \pi^{-\dagger}(I^\dagger)$ and the generated values form a 3-cycle. By the symmetry of $a_{ij}$ we have that $b_3 = 0$ if $I^\dagger$, $J^\dagger$ are consecutive element of a 3-cycle, so $A_3$ has probability zero unless $\sum_{b \ge 4} c_b \ge 1$, that is, unless the cycle type $c$ has cycles of length at least 4. Hence, if so, under $A_3$ the elements $\pi^{-\dagger}(I^\dagger), I^\dagger, \pi^\dagger(I^\dagger), \pi^\dagger(J^\dagger)$ must be distinct and form part of a cycle of $\pi^\dagger$ of length at least 4. Conditioning on the values in (6.101), and letting $c(\sigma^\dagger, \alpha)$ be the length of the cycle in $\sigma^\dagger$ containing $\alpha$, select a cycle length $b$ according to the distribution

$$P\big(c(\sigma^\dagger, \alpha) = b | \sigma^{-\dagger}(\alpha) = \delta, \sigma^\dagger(\alpha) = \epsilon, \sigma^{\dagger 2}(\alpha) = \zeta\big)$$

and let $\widehat{I}$ be chosen uniformly, and independently from the $b$-cycles of $\pi'$. Now let $\pi^\dagger$ be given by (6.102) with

$$\iota = \big(\pi^{-1}(\widehat{I}), \widehat{I}, \pi(\widehat{I}), \pi^2(\widehat{I})\big) \quad \text{and} \quad \iota^\dagger = \big(\pi^{-\dagger}(I^\dagger), I^\dagger, \pi^\dagger(I^\dagger), \pi^\dagger(J^\dagger)\big).$$

As the inverse images under $\pi$ of the components in $\iota$ are all again components of this vector, with the possible exception of $\pi^{-1}(I)$, the set (6.103) can have size at most $(4+1) + 2 \times 4 = 13$ in this case. The construction on $A_4$ is analogous, with the roles of $I^\dagger$ and $J^\dagger$ reversed.

Moving on to $A_5$, consider $A_{5,1}$, where if $c_2 \geq 2$, the elements $I^\dagger$ and $J^\dagger$ are to be placed in distinct 2-cycles. Choosing $\widehat{I}$ and $\widehat{J}$ from pairs of indices in distinct 2-cycles, let $\pi^\dagger$ be given by (6.102) with

$$\iota = \big(\widehat{I}, \pi(\widehat{I}), \widehat{J}, \pi(\widehat{J})\big) \quad \text{and} \quad \iota^\dagger = \big(I^\dagger, \pi^\dagger(I^\dagger), J^\dagger, \pi^\dagger(J^\dagger)\big).$$

As $\widehat{I}$ and $\widehat{J}$ are members of 2-cycles of $\pi$, the vector $\iota$ already contains all of its inverse images under $\pi$, and therefore the set (6.103) can have size at most $4 + 2 \times 4 = 12$. When $\pi$ is an involution without fixed points, this is the only case.

Similarly, if $c_2$ and $\sum_{b \geq 3} c_b$ are both nonzero, then the probability of $A_{5,2}$ is positive, and we let $\widehat{I}$ and $\widehat{J}$ be chosen independently, the first uniformly from the 2-cycles of $\pi$, the second uniformly from elements of the $b$-cycles of $\pi$ where $b$ has distribution

$$P\big(c(\sigma^\dagger, \beta) = b | \sigma^{-\dagger}(\beta) = \gamma, \sigma^\dagger(\beta) = \zeta\big). \tag{6.104}$$

Now let $\pi^\dagger$ be given by (6.102) with

$$\iota = \big(\widehat{I}, \pi(\widehat{I}), \pi^{-1}(\widehat{J}), \widehat{J}, \pi(\widehat{J})\big) \quad \text{and} \quad \iota^\dagger = \big(I^\dagger, \pi(I^\dagger), \pi^{-\dagger}(J^\dagger), J^\dagger, \pi(J^\dagger)\big).$$

Arguing as above, as $\widehat{I}$ is in a 2-cycle, the set (6.103) can have size at most $(5 + 1) + 2 \times 5 = 16$. The argument is analogous on $A_{5,3}$.

Before beginning our consideration of the final case, $A_{5,4}$, we note that though the generated values (6.101) are placed in $\pi^\dagger$ according to the correct conditional distributions, such as (6.104), as we are considering a worst case analysis, the actual values of these probabilities never enter our considerations. Hence, on $A_{5,4}$, no matter how $\widehat{I}$ and $\widehat{J}$ are selected to be consistent with $A_{5,4}$, the result will be that $\pi^\dagger$ will be given by (6.102) with

$$\begin{aligned} \iota &= \big(\pi^{-1}(\widehat{I}), \widehat{I}, \pi(\widehat{I}), \pi^{-1}(\widehat{J}), \widehat{J}, \pi(\widehat{J})\big) \\ \text{and} \quad \iota^\dagger &= \big(\pi^{-\dagger}(I^\dagger), I^\dagger, \pi(I^\dagger), \pi^{-\dagger}(J^\dagger), J^\dagger, \pi(J^\dagger)\big). \end{aligned}$$

In this case the set (6.103) can have size at most $(6+2) + 2 \times 6 = 20$.

As $A_0, \ldots, A_5$ is a partition, the construction of $\pi^\dagger$ has been specified in every case. By arguments similar to those in Lemma 4.5, the conditional distribution $P_{\{i,j\}^c|\{i,j\}}(\xi_\alpha, \alpha \notin \chi_{i,j}|\xi_\alpha, \alpha \in \chi_{i,j})$ of the remaining values, given the ones now determined, is uniform, so specifying $\pi^\dagger$ by (6.102) and setting

$$\pi^\ddagger = \tau_{I^\dagger,J^\dagger}\pi^\dagger\tau_{I^\dagger,J^\dagger}$$

results in a collection of variables $I^\dagger, J^\dagger$ and a pair of permutations with the square bias distribution (4.113). Hence, letting $Y, Y^\dagger$ and $Y^\ddagger$ be given by (6.1) with $\pi, \pi^\dagger$ and $\pi^\ddagger$, respectively results in a coupling of $Y$ to the variables $Y^\dagger, Y^\ddagger$ with the square bias distribution. Now with $T, T^\dagger$ and $T^\ddagger$ given by (6.3), we have

$$U|T^\dagger| + (1-U)|T^\ddagger| + |T| \le (2\max|\mathcal{I}_{\iota,\iota^\dagger}|)C$$

where the maximum is over the values of $\iota, \iota^\dagger$ appearing in the possible cases. For fixed point free involutions, $A_{5,1}$ is the only case, giving the coefficient $2 \times 12 = 24$ on $C$. In general, the coefficient is bounded by $2 \times 20 = 40$, determined by the worst case on $A_{5,4}$.

Now (6.4) gives $|Y^* - Y| \le 40C$ in general, and the bound $24C$ for involutions. As

$$|W^* - W| = |Y^* - Y|/\sigma_c$$

by (2.59), invoking Theorem 5.1 with $\delta = 40C/\sigma_c$ and $\delta = 24C/\sigma_c$ now completes the proof. □

Lemma 6.10 and the mixing property of Lemma 6.5 are the key ingredients of the following argument.

*Proof of Theorem 6.3* First, note that the claim in Theorem 6.3 regarding involutions is part of Lemma 6.10. Otherwise, by replacing $a_{ij}$ by $\widehat{a_{ij}}$ given in (6.66) we may without loss of generality assume that $EY = 0$ whenever $Y$ is given by (6.1) with $\pi$ having distribution constant on cycle type. In this case, writing $Y_c$ for the variable given by (6.1) when $\pi \sim \mathcal{U}(c)$, the mixture property of Lemma 6.5 yields

$$P(Y \le z) = \sum_{c \in \mathcal{N}_n} \rho_c P(Y_c \le z),$$

with $\rho_c = P(c(\pi) = c)$, and in addition, from (6.71),

$$P(Z_\rho \le z) = \sum_{c \in \mathcal{N}_n} \rho_c P(Z_c/\sigma_\rho \le z).$$

Hence, with $W = Y/\sigma_\rho$, by changes of variable,

$$\begin{aligned}
\sup_{z\in\mathbb{R}}\bigl|P(W \le z) - P(Z_\rho \le z)\bigr| &= \sup_{z\in\mathbb{R}}\bigl|P(Y \le z) - P(\sigma_\rho Z_\rho \le z)\bigr| \\
&\le \sum_{c\in\mathcal{N}_n} \rho_c\bigl|P(Y_c \le z) - P(Z_c \le z)\bigr| \\
&= \sum_{c\in\mathcal{N}_n} \rho_c\bigl|P(W_c \le z) - P(Z \le z)\bigr|
\end{aligned}$$

where $W_c = Y_c/\sigma_c$. Now applying the uniform bound in Lemma 6.10 completes the proof. □

### *6.1.3 Doubly Indexed Permutation Statistics*

In Sect. 4.4 we observed how the distribution of the permutation statistic $Y$ in (4.104), that is,

$$Y = \sum_{i=1}^{n} a_{i\pi(i)},$$

can be used to test whether there is an unusually high degree of similarity in a particular matching between the observations $x_1, \ldots, x_n$ and $y_1, \ldots, y_n$. In particular, if, say $d(x, y)$ is a function which reflects the similarity between $x$ and $y$ and $a_{ij} = d(x_i, y_j)$, one compares the 'overall similarity' score

$$y_\tau = \sum_{i=1}^{n} a_{i\tau(i)}$$

of the distinguished matching $\tau$ to the distribution of $Y$, that is, to this same similarity score for random matchings.

In spatial or spatio-temporal association, two dimensional generalizations of the permutation test statistic $Y$ become of interest. In particular, if $a_{ij}$ and $b_{ij}$ are two different measures of closeness of $x_i$ and $y_j$, which may or may not be related, then the relevant null distribution is that of

$$W = \sum_{(i,j):\, i \neq j} a_{ij} b_{\pi(i),\pi(j)} \tag{6.105}$$

where the permutation $\pi$ is chosen uniformly from $\mathcal{S}_n$; see, for instance, Moran (1948) and Geary (1954) for applications in geography, Knox (1964) and Mantel (1967) in epidemiology, as well as the book of Hubert (1987). Following some initial results which yield the asymptotic normality of $W$, see Barbour and Chen (2005a) for history and references, much less restrictive conditions were given in Barbour and Eagleson (1986). Theorem 6.4 of Barbour and Chen (2005a) provides a Berry–Esseen bound for this convergence; to state it we first need to introduce some notation.

As the diagonal elements play no role, we may set $a_{ii} = b_{ii} = 0$. For such an array $\{a_{ij}\}_{i,j=1}^{n}$, let

$$A_0 = \frac{1}{n(n-1)} \sum_{(i,j):\, i \neq j} a_{ij}, \quad A_{12} = n^{-1} \sum_{i=1}^{n} (a_i^*)^2,$$

$$A_{22} = \frac{1}{n(n-1)} \sum_{(i,j):\, i \neq j} \tilde{a}_{ij}^2 \quad \text{and} \quad A_{13} = \frac{1}{n} \sum_{i=1}^{n} |a_i^*|^3$$

where

$$a_i^* = \frac{1}{n-2}\sum_{j:j\neq i}(a_{ij} - A_0) \quad \text{and} \quad \tilde{a}_{ij} = a_{ij} - a_i^* - a_j^* - A_0,$$

and let the analogous definitions hold for $\{b_{ij}\}$. In addition, let

$$\mu = \frac{1}{n(n-1)}\sum_{(i,j),(l,m):\, i\neq j, l\neq m} a_{ij}b_{lm} \quad \text{and} \quad \sigma^2 = \frac{4n^2(n-2)^2}{n-1}A_{12}B_{12}.$$

**Theorem 6.4** *For $W$ as given in* (6.105) *with $A$ and $B$ symmetric arrays, we have*

$$\sup_{z\in\mathbb{R}}\bigl|P(W-\mu\le\sigma z) - \Phi(z)\bigr| \le (2+c)\delta + 12\delta^2 + (1+\sqrt{2})\tilde{\delta}_2,$$

*where* $\delta = 128n^4\sigma^{-3}A_{13}B_{13}$,

$$\tilde{\delta}_2^2 = \frac{(n-1)^3}{2n(n-2)^2(n-3)}\frac{A_{22}B_{22}}{A_{12}B_{12}}$$

*and $c$ is the constant in Theorem* 6.2.

It turns out that statistics such as $W$ can be expressed as a singly indexed permutation statistic upon which known bounds may be applied, plus a remainder term which may be handled using concentration inequalities and exploiting exchangeability, somewhat similar to the way that some non-linear statistics are handled in Chap. 10. The bounds of Theorem 6.4 compare favorably with those of Zhao et al. (1997).

## 6.2 Patterns in Graphs and Permutations

In this section we will prove and apply corollaries of Theorem 5.6 to evaluate the quality of the normal approximation for various counts that arise in graphs and permutations, in particular, coloring patterns, local maxima, and the occurrence of subgraphs of finite random graphs, and for the number of occurrences of fixed, relatively ordered sub-sequences, such as rising sequences, of random permutations.

We explore the consequences of Theorem 5.6 under a local dependence condition on a collection of random variables

$$\mathbf{X} = \{X_\alpha, \alpha\in\mathcal{A}\},$$

over some arbitrary, finite, index set $\mathcal{A}$. In particular, we consider situations where for every $\alpha\in\mathcal{A}$ there exists a dependency neighborhood $\mathcal{B}_\alpha\subset\mathcal{A}$ of $X_\alpha$, containing $\alpha$, such that

$$X_\alpha \quad \text{and} \quad \{X_\beta:\ \beta\notin\mathcal{B}_\alpha\} \quad \text{are independent.} \tag{6.106}$$

First recalling the definition of size biasing in a coordinate direction given in (2.68) in Sect. 2.3.4, we begin with the following corollary of Theorem 5.6.

**Corollary 6.1** *Let* $\mathbf{X} = \{X_\alpha, \alpha \in \mathcal{A}\}$ *be a finite collection of random variables with values in* $[0, M]$ *and let*

$$Y = \sum_{\alpha \in \mathcal{A}} X_\alpha.$$

*Let* $\mu = \sum_{\alpha \in \mathcal{A}} EX_\alpha$ *denote the mean of* $Y$ *and assume that the variance* $\sigma^2 = \mathrm{Var}(Y)$ *is positive and finite. Let*

$$p_\alpha = EX_\alpha / \sum_{\beta \in \mathcal{A}} EX_\beta \quad \text{and} \quad \overline{p} = \max_{\alpha \in \mathcal{A}} p_\alpha. \tag{6.107}$$

*Next, for each* $\alpha \in \mathcal{A}$ *let* $\mathcal{B}_\alpha \subset \mathcal{A}$ *be a dependency neighborhood of* $X_\alpha$ *such that* (6.106) *holds, and let*

$$b = \max_{\alpha \in \mathcal{A}} |\mathcal{B}_\alpha|. \tag{6.108}$$

*For each* $\alpha \in \mathcal{A}$, *let* $(\mathbf{X}, \mathbf{X}^\alpha)$ *be a coupling of* $\mathbf{X}$ *to a collection of random variables* $\mathbf{X}^\alpha$ *having the* $\mathbf{X}$*-size biased distribution in direction* $\alpha$ *such that for some* $\mathcal{F} \supset \sigma\{Y\}$ *and* $\mathcal{D} \subset \mathcal{A} \times \mathcal{A}$,

$$\begin{aligned} &\text{if } (\alpha_1, \alpha_2) \notin \mathcal{D}, \text{ then for all } (\beta_1, \beta_2) \in \mathcal{B}_{\alpha_1} \times \mathcal{B}_{\alpha_2} \\ &\quad \mathrm{Cov}\big(E\big(X_{\beta_1}^{\alpha_1} - X_{\beta_1} | \mathcal{F}\big), E\big(X_{\beta_2}^{\alpha_2} - X_{\beta_2} | \mathcal{F}\big)\big) = 0. \end{aligned} \tag{6.109}$$

*Then with* $W = (Y - \mu)/\sigma$,

$$\sup_{z \in \mathbb{R}} \big| P(W \le z) - P(Z \le z) \big| \le \frac{6\mu b^2 M^2}{\sigma^3} + \frac{2\mu \overline{p} b M \sqrt{|\mathcal{D}|}}{\sigma^2}.$$

*Proof* In view of Theorem 5.6 and (5.21), it suffices to couple $Y^s$, with the $Y$-size biased distribution, to $Y$ such that

$$\big| Y^s - Y \big| \le bM \quad \text{and} \quad \Psi \le \overline{p} b M \sqrt{|\mathcal{D}|}. \tag{6.110}$$

Assume without loss of generality that $EX_\alpha > 0$ for each $\alpha \in \mathcal{A}$. Note that for every $\alpha \in \mathcal{A}$ the distribution $dF(\mathbf{x})$ of $\mathbf{X}$ factors as

$$\begin{aligned} &dF_\alpha(x_\alpha) dF_{\mathcal{B}_\alpha^c | \alpha}(x_\beta, \beta \notin \mathcal{B}_\alpha | x_\alpha) \\ &\quad \times dF_{\mathcal{B}_\alpha \setminus \{\alpha\} | \{\alpha\} \cup \mathcal{B}_\alpha^c}\big(x_\gamma, \gamma \in \mathcal{B}_\alpha \setminus \{\alpha\} | x_\alpha, x_\beta, \beta \notin \mathcal{B}_\alpha\big), \end{aligned}$$

which, by the independence condition (6.106) we may write as

$$\begin{aligned} &dF_\alpha(x_\alpha) dF_{\mathcal{B}_\alpha^c}(x_\beta, \beta \notin \mathcal{B}_\alpha) \\ &\quad \times dF_{\mathcal{B}_\alpha \setminus \{\alpha\} | \{\alpha\} \cup \mathcal{B}_\alpha^c}\big(x_\gamma, \gamma \in \mathcal{B}_\alpha \setminus \{\alpha\} | x_\alpha, x_\beta, \beta \notin \mathcal{B}_\alpha\big). \end{aligned}$$

Hence, as in (2.73), the coordinate size biased distribution $dF^\alpha(\mathbf{x})$ may be factored as

$$\begin{aligned} dF^\alpha(\mathbf{x}) &= dF_\alpha^\alpha(x_\alpha) dF_{\mathcal{B}_\alpha^c}(x_\beta, \beta \notin \mathcal{B}_\alpha) \\ &\quad \times dF_{\mathcal{B}_\alpha \setminus \{\alpha\} | \{\alpha\} \cup \mathcal{B}_\alpha^c}\big(x_\gamma, \gamma \in \mathcal{B}_\alpha \setminus \{\alpha\} | x_\alpha, x_\beta, \beta \notin \mathcal{B}_\alpha\big), \end{aligned}$$

where

$$dF_\alpha^\alpha(x_\alpha) = \frac{x_\alpha dF_\alpha(x_\alpha)}{EX_\alpha}. \tag{6.111}$$

Given a realization of $\mathbf{X}$, this factorization shows that we can construct $\mathbf{X}^\alpha$ by first choosing $X_\alpha^\alpha$ from the $X_\alpha$-size bias distribution (6.111), then the variables $X_\beta$ for $\beta \in \mathcal{B}_\alpha^c$ according to their original distribution, and so in particular set

$$X_\beta^\alpha = X_\beta \quad \text{for all } \beta \in \mathcal{B}_\alpha^c,$$

and finally the variables $X_\beta^\alpha, \beta \in \mathcal{B} \setminus \{\alpha\}$ using their original conditional distribution given the variables $\{X_\alpha^\alpha, X_\beta, \beta \in \mathcal{B}^c\}$.

As the distribution of $\mathbf{X}^\alpha$ is absolutely continuous with respect to that of $\mathbf{X}$, we have $X_\beta^\alpha \in [0, M]$ for all $\alpha, \beta$, and therefore

$$\left|X_\beta^\alpha - X_\beta\right| \le M \quad \text{for all } \alpha, \beta \in \mathcal{A}. \tag{6.112}$$

By Proposition 2.2, $Y^s = \sum_{\beta \in \mathcal{A}} X_\beta^I$ has the $Y$-size biased distribution, where the random index $I$ has distribution $P(I = \alpha) = p_\alpha$ and is chosen independently of $\{(\mathbf{X}, \mathbf{X}^\alpha), \alpha \in \mathcal{A}\}$ and $\mathcal{F}$. In particular

$$Y^s - Y = \sum_{\beta \in \mathcal{B}_I} \left(X_\beta^I - X_\beta\right), \tag{6.113}$$

yielding the first inequality in (6.110).

Recalling the definition of $\Psi$ in (5.21), since $\sigma\{Y\} \subset \mathcal{F}$, by (4.143),

$$\Psi^2 = \operatorname{Var}\bigl(E\bigl(Y^s - Y|Y\bigr)\bigr) \le \operatorname{Var}\bigl(E\bigl(Y^s - Y|\mathcal{F}\bigr)\bigr).$$

Taking conditional expectation with respect to $\mathcal{F}$ in (6.113) yields,

$$E\bigl(Y^s - Y|\mathcal{F}\bigr) = \sum_{\alpha \in \mathcal{A}} p_\alpha \sum_{\beta \in \mathcal{B}_\alpha} E\bigl(X_\beta^\alpha - X_\beta|\mathcal{F}\bigr),$$

and therefore,

$$\begin{aligned}
&\operatorname{Var}\bigl(E\bigl(Y^s - Y|\mathcal{F}\bigr)\bigr) \\
&\quad = E \sum_{\substack{(\alpha_1,\alpha_2) \in \mathcal{A} \times \mathcal{A} \\ (\beta_1,\beta_2) \in \mathcal{B}_{\alpha_1} \times \mathcal{B}_{\alpha_2}}} p_{\alpha_1} p_{\alpha_2} \operatorname{Cov}\bigl(E\bigl(X_{\beta_1}^{\alpha_1} - X_{\beta_1}|\mathcal{F}\bigr), E\bigl(X_{\beta_2}^{\alpha_2} - X_{\beta_2}|\mathcal{F}\bigr)\bigr) \\
&\quad = E \sum_{\substack{(\alpha_1,\alpha_2) \in \mathcal{D} \\ (\beta_1,\beta_2) \in \mathcal{B}_{\alpha_1} \times \mathcal{B}_{\alpha_2}}} p_{\alpha_1} p_{\alpha_2} \operatorname{Cov}\bigl(E\bigl(X_{\beta_1}^{\alpha_1} - X_{\beta_1}|\mathcal{F}\bigr), E\bigl(X_{\beta_2}^{\alpha_2} - X_{\beta_2}|\mathcal{F}\bigr)\bigr),
\end{aligned}$$

where we have applied (6.109) to obtain the last equality. By (6.112), the covariances are bounded by $M^2$, hence

$$\Psi^2 \le \operatorname{Var}\bigl(E\bigl(Y^s - Y|\mathcal{F}\bigr)\bigr) \le M^2 \sum_{\substack{(\alpha_1,\alpha_2) \in \mathcal{D} \\ (\beta_1,\beta_2) \in \mathcal{B}_{\alpha_1} \times \mathcal{B}_{\alpha_2}}} p_{\alpha_1} p_{\alpha_2}$$

$$= M^2 \sum_{(\alpha_1,\alpha_2)\in\mathcal{D}} p_{\alpha_1} p_{\alpha_2} |\mathcal{B}_{\alpha_1}||\mathcal{B}_{\alpha_2}|$$

$$\le M^2 \sum_{(\alpha_1,\alpha_2)\in\mathcal{D}} \overline{p}^2 b^2 = \overline{p}^2 b^2 M^2 |\mathcal{D}|,$$

by (6.107) and (6.108), thus yielding the second inequality in (6.110). □

Though Corollary 6.1 provides bounds for finite problems, asymptotically, when the mean and variance of $Y$ grow such that $\mu/\sigma^2$ is bounded, and when $b$ and $M$ stay bounded, then the first term in the bound of the corollary is of order $1/\sigma$. Additionally, if $X_\alpha$ have comparable expectations, so that $\overline{p}$ is of order $1/|\mathcal{A}|$, and if the 'dependence diagonal' $\mathcal{D} \subset \mathcal{A} \times \mathcal{A}$ has size comparable to that of $\mathcal{A}$, then the second term will also be of order $1/\sigma$.

We next specialize to the case where the summand variables $\{X_\alpha, \alpha \in \mathcal{A}\}$ are functions of independent random variables.

**Corollary 6.2** *With $\mathcal{G}$ and $\mathcal{A}$ index sets, let $\{C_g, g \in \mathcal{G}\}$ be a collection of independent random elements taking values in an arbitrary set $C$, let $\{\mathcal{G}_\alpha, \alpha \in \mathcal{A}\}$ be a finite collection of subsets of $\mathcal{G}$, and, for $\alpha \in \mathcal{A}$, let*

$$X_\alpha = X_\alpha(C_g\colon g \in \mathcal{G}_\alpha)$$

*be a real valued function of the variables $\{C_g, g \in \mathcal{G}_\alpha\}$, taking values in $[0, M]$. Then for $Y = \sum_\alpha X_\alpha$ with mean $\mu$ and finite, positive variance $\sigma^2$, the variable $W = (Y - \mu)/\sigma$ satisfies*

$$\sup_{z\in\mathbb{R}} \bigl|P(W \le z) - P(Z \le z)\bigr| \le \frac{6\mu b^2 M^2}{\sigma^3} + \frac{2\mu \overline{p} b M \sqrt{|\mathcal{D}|}}{\sigma^2},$$

*where $\overline{p}$ and $b$ are given in* (6.107) *and* (6.108), *respectively, for any*

$$\mathcal{B}_\alpha \supset \{\beta \in \mathcal{A}\colon \mathcal{G}_\beta \cap \mathcal{G}_\alpha \neq \emptyset\}, \tag{6.114}$$

*and any $\mathcal{D}$ for which*

$$\mathcal{D} \supset \bigl\{(\alpha_1, \alpha_2)\colon \text{there exists } (\beta_1, \beta_2) \in \mathcal{B}_{\alpha_1} \times \mathcal{B}_{\alpha_2} \text{ with } \mathcal{G}_{\beta_1} \cap \mathcal{G}_{\beta_2} \neq \emptyset\bigr\}. \tag{6.115}$$

*Proof* We apply Corollary 6.1. Since $X_\alpha$ and $X_\beta$ are functions of disjoint sets of independent variables whenever $\mathcal{G}_\alpha \cap \mathcal{G}_\beta = \emptyset$, the independence condition (6.106) holds when the dependency neighborhoods satisfy (6.114).

To verify the remaining conditions of Corollary 6.1, for each $\alpha \in \mathcal{A}$ we consider the following coupling of $\mathbf{X}$ and $\mathbf{X}^\alpha$. We may assume without loss of generality that $EX_\alpha > 0$. Given $\{C_g, g \in \mathcal{G}\}$ upon which $\mathbf{X}$ depends, for every $\alpha \in \mathcal{A}$ let $\{C_g^{(\alpha)}, g \in \mathcal{G}_\alpha\}$ be independent of $\{C_g, g \in \mathcal{G}\}$ and have distribution

$$dF^\alpha(c_g, g \in \mathcal{G}_\alpha) = \frac{X_\alpha(c_g, g \in \mathcal{G}_\alpha)}{EX_\alpha(C_g, g \in \mathcal{G}_\alpha)} dF(c_g, g \in \mathcal{G}_\alpha),$$

so that the random variables $\{C_g^\alpha, g \in \mathcal{G}_\alpha\} \cup \{C_g, g \notin \mathcal{G}\}$ have distribution $dF^\alpha(c_g, g \in \mathcal{G}_\alpha) dF(c_g, g \notin \mathcal{G}_\alpha)$. Thus, letting $\mathbf{X}^\alpha$ have coordinates given by

$$X_\beta^\alpha = X_\beta\big(C_g, g \in \mathcal{G}_\beta \cap \mathcal{G}_\alpha^c, C_g^{(\alpha)}, g \in \mathcal{G}_\beta \cap \mathcal{G}_\alpha\big), \quad \beta \in \mathcal{A}$$

for any bounded continuous function $f$ we find

$$\begin{aligned} EX_\alpha f(\mathbf{X}) &= \int x_\alpha f(\mathbf{x}) dF(c_g, g \in \mathcal{G}) \\ &= EX_\alpha \int f(\mathbf{x}) \frac{x_\alpha dF(c_g, g \in \mathcal{G}_\alpha)}{EX_\alpha(C_g, g \in \mathcal{G}_\alpha)} dF(c_g, g \notin \mathcal{G}_\alpha) \\ &= EX_\alpha \int f(\mathbf{x}) dF^\alpha(c_g, g \in \mathcal{G}_\alpha) dF(c_g, g \notin \mathcal{G}_\alpha) \\ &= EX_\alpha Ef\big(\mathbf{X}^\alpha\big). \end{aligned}$$

That is, $\mathbf{X}^\alpha$ has the $\mathbf{X}$ distribution biased in direction $\alpha$, as defined in (2.68).

Lastly, taking $\mathcal{F} = \{C_g : g \in \mathcal{G}\}$, so that $Y$ is $\mathcal{F}$ measurable, we verify (6.109). Since $X_\beta^\alpha$ and $\{C_g, g \in \mathcal{G}_\beta\}$ are independent of $\{C_g, g \notin \mathcal{G}_\beta\}$,

$$E\big(X_\beta^\alpha|\mathcal{F}\big) = E\big(X_\beta^\alpha|C_g, g \in \mathcal{G}_\beta, C_g, g \notin \mathcal{G}_\beta\big) = E\big(X_\beta^\alpha|C_g, g \in \mathcal{G}_\beta\big),$$

and, since

$$E(X_\beta|\mathcal{F}) = X_\beta = E(X_\beta|C_g, g \in \mathcal{G}_\beta),$$

the difference $E(X_\beta^\alpha - X_\beta|\mathcal{F})$ is a function of $\{C_g, g \in \mathcal{G}_\beta\}$ only. By choice of $\mathcal{D}$, if $(\alpha_1, \alpha_2) \notin \mathcal{D}$ then for all $\beta_1 \in \mathcal{B}_{\alpha_1}$ and $\beta_2 \in \mathcal{B}_{\alpha_2}$ we have $\mathcal{G}_{\beta_1} \cap \mathcal{G}_{\beta_2} = \emptyset$, and so $E(X_{\beta_1}^{\alpha_1} - X_{\beta_1}|\mathcal{F})$ and $E(X_{\beta_2}^{\alpha_2} - X_{\beta_2}|\mathcal{F})$ are independent, yielding (6.109). The verification of the conditions of Corollary 6.1 is now complete. □

With the exception of Example 6.2, in the remainder of this section we consider graphs $\mathcal{G} = (\mathcal{V}, \mathcal{E})$ having random elements $\{C_g\}_{g \in \mathcal{V} \cup \mathcal{E}}$ assigned to their vertices $\mathcal{V}$ and edges $\mathcal{E}$, and applications of Corollary 6.2 to the sum $Y = \sum_{\alpha \in \mathcal{A}} X_\alpha$ of bounded functions $X_\alpha = X_\alpha(C_g, g \in \mathcal{V}_\alpha \cup \mathcal{E}_\alpha)$, where $\mathcal{G}_\alpha = (\mathcal{V}_\alpha, \mathcal{E}_\alpha), \alpha \in \mathcal{A}$ is a given finite family of subgraphs of $\mathcal{G}$. We abuse notation slightly in that a graph $\mathcal{G}$ is replaced by $\mathcal{V} \cup \mathcal{E}$ when used as an index set for the underlying variables $C_g$. When applying Corollary 6.2 in this setting, in (6.114) and (6.115) the intersection of the two graphs $(\mathcal{V}_1, \mathcal{E}_1)$ and $(\mathcal{V}_2, \mathcal{E}_2)$ is the graph $(\mathcal{V}_1 \cap \mathcal{V}_2, \mathcal{E}_1 \cap \mathcal{E}_2)$.

Given a metric $d$ on $\mathcal{V}$, for every $v \in \mathcal{V}$ and $r \geq 0$ we can consider the restriction $\mathcal{G}_{v,r}$ of $\mathcal{G}$ to the vertices at most a distance $r$ from $v$, that is, the graph with vertex and edge sets

$$\mathcal{V}_{v,r} = \big\{w \in \mathcal{V}: d(v, w) \leq r\big\} \quad \text{and} \quad \mathcal{E}_{v,r} = \big\{\{w, u\} \in \mathcal{E}: w, u \in \mathcal{V}_{v,r}\big\} \quad (6.116)$$

respectively. We say that a graph $\mathcal{G}$ is *distance $r$-regular* if $\mathcal{G}_{v,r}$ is isomorphic to some graph $(\mathcal{V}_r, \mathcal{E}_r)$ for all $v$. This notion of distance $r$-regular is related to, but not the same as, the notion of a distance-regular graph as given in Biggs (1993) and Brouwer et al. (1989). A graph of constant degree with no cliques of size 3 is distance 1-regular.

When $\mathcal{V}_\alpha, \alpha \in \mathcal{V}$ is given by (6.116) for some fixed $r$, regarding the choice of the dependency neighborhoods $\mathcal{B}_\alpha, \alpha \in \mathcal{A}$, we note that if $d(\alpha_1, \alpha_2) > 2r$ and $(\beta_1, \beta_2) \in \mathcal{V}_{\alpha_1} \times \mathcal{V}_{\alpha_2}$, then rearranging yields

$$2r < d(\alpha_1, \alpha_2) \le d(\alpha_1, \beta_1) + d(\beta_1, \beta_2) + d(\beta_2, \alpha_2),$$

and using that $d(\alpha_i, \beta_i) \le r$ implies $d(\beta_1, \beta_2) > 0$, hence

$$d(\alpha_1, \alpha_2) > 2r \quad \text{implies} \quad \mathcal{V}_{\alpha_1} \bigcap \mathcal{V}_{\alpha_2} = \emptyset. \tag{6.117}$$

Natural families of graphs in $\mathbb{R}^p$ can be generated using the vertex set $\mathcal{V} = \{1, \dots, n\}^p$ with componentwise addition modulo $n$, and $d(\alpha, \beta)$ given by e.g. some $L^p$ distance between $\alpha$ and $\beta$. We apply the following result when the subgraphs are indexed by some subset of the vertices only, in which case we take $\mathcal{A} \subset \mathcal{V}$.

**Corollary 6.3** *Let $\mathcal{G}$ be a finite graph with a family of isomorphic subgraphs $\{\mathcal{G}_\alpha, \alpha \in \mathcal{A}\}$ for some $\mathcal{A} \subset \mathcal{V}$, let $d$ be a metric on $\mathcal{A}$, and set*

$$\rho = \min\{\varrho\colon d(\alpha, \beta) > \varrho \text{ implies } \mathcal{V}_\alpha \cap \mathcal{V}_\beta = \emptyset\}. \tag{6.118}$$

*For each $\alpha \in \mathcal{A}$, let $X_\alpha$ be given by*

$$X_\alpha = X(C_g, g \in \mathcal{G}_\alpha)$$

*for a fixed function $X$ taking values in $[0, M]$, and let $\{C_g, g \in \mathcal{G}\}$ be a collection of independent variables such that the distribution of $\{C_g : g \in \mathcal{G}_\alpha\}$ is the same for all $\alpha \in \mathcal{A}$.*

*If $\mathcal{G}$ is a distance-$3\rho$-regular graph, then with $Y = \sum_{\alpha \in \mathcal{A}} X_\alpha$ having mean $\mu$ and finite, positive variance $\sigma^2$, the variable $W = (Y - \mu)/\sigma$ satisfies*

$$\sup_{z \in \mathbb{R}} \bigl|P(W \le z) - P(Z \le z)\bigr| \le \frac{6\mu V(\rho)^2 M^2}{\sigma^3} + \frac{2\mu V(\rho) M}{\sigma^2 |\mathcal{A}|^{1/2}} \sqrt{V(3\rho)},$$

*where*

$$V(r) = |\mathcal{V}_r|. \tag{6.119}$$

*Proof* We verify that conditions (6.114) and (6.115) of Corollary 6.2 are satisfied with

$$\mathcal{B}_\alpha = \{\beta\colon d(\alpha, \beta) \le \rho\} \quad \text{and} \quad \mathcal{D} = \{(\alpha_1, \alpha_2)\colon d(\alpha_1, \alpha_2) \le 3\rho\}. \tag{6.120}$$

First note that to show the intersection of two graphs is empty it suffices to show that the vertex sets of the graphs do not intersect. Since for any $\alpha \in \mathcal{A}$, by (6.118),

$$\mathcal{B}_\alpha^c = \{\beta\colon d(\beta, \alpha) > \rho\} \subset \{\beta\colon \mathcal{V}_\beta \cap \mathcal{V}_\alpha = \emptyset\},$$

we see that condition (6.114) is satisfied.

To verify (6.115), note that rearranging

$$d(\alpha_1, \alpha_2) \le d(\alpha_1, \beta_1) + d(\beta_1, \beta_2) + d(\beta_2, \alpha_2)$$

gives, for $(\alpha_1, \alpha_2) \notin \mathcal{D}$ and $(\beta_1, \beta_2) \in \mathcal{B}_{\alpha_1} \times \mathcal{B}_{\alpha_2}$,

$$d(\beta_1,\beta_2)\ge d(\alpha_1,\alpha_2)-\big(d(\alpha_1,\beta_1)+d(\alpha_2,\beta_2)\big)\ge d(\alpha_1,\alpha_2)-2\rho>\rho,$$

and hence

$$\mathcal{V}_{\beta_1}\cap\mathcal{V}_{\beta_2}=\emptyset.$$

As $EX_\alpha$ is constant we have $\overline{p}=\max_\alpha p_\alpha=1/|\mathcal{A}|$, and in addition, that

$$b=\max_{\alpha\in\mathcal{A}}|\mathcal{B}_\alpha|=V(\rho)\quad\text{and}\quad|\mathcal{D}|=|\mathcal{A}|V(3\rho).$$

Substituting these quantities into the bound of Corollary 6.2 now yields the result. □

*Example 6.1* (Sliding $m$-window) For $n\ge m\ge 1$, let $\mathcal{A}=\mathcal{V}=\{1,\ldots,n\}$ with addition modulo $n$, $\{C_g:g\in\mathcal{G}\}$ i.i.d. real valued random variables, and for each $\alpha\in\mathcal{A}$ set $\mathcal{G}_\alpha=(\mathcal{V}_\alpha,\mathcal{E}_\alpha)$ where

$$\mathcal{V}_\alpha=\{v\in\mathcal{V}\colon \alpha\le v\le\alpha+m-1\}\quad\text{and}\quad\mathcal{E}_\alpha=\emptyset. \tag{6.121}$$

Then for $X:\mathbb{R}^m\to[0,1]$, Corollary 6.3 may be applied to the sum $Y=\sum_{\alpha\in\mathcal{A}}X_\alpha$ of the $m$-dependent sequence $X_\alpha=X(C_\alpha,\ldots,C_{\alpha+m-1})$, formed by applying the function $X$ to the variables in the '$m$-window' $\mathcal{V}_\alpha$. In this example, taking $d(\alpha,\beta)=|\alpha-\beta|$ the bound of Corollary 6.3 obtains with $\rho=m-1$ by (6.118) and $V(r)\le 2r+1$ by (6.119).

In Example 6.2 the underlying variables are not independent, so we turn to Corollary 6.1.

*Example 6.2* (Relatively ordered sub-sequences of a random permutation) For $n\ge m\ge 1$, let $\mathcal{V}$ and $(\mathcal{G}_\alpha,\mathcal{V}_\alpha),\alpha\in\mathcal{V}$ be as specified in (6.121). For $\pi$ and $\tau$ permutations of $\{1,\ldots,n\}$ and $\{1,\ldots,m\}$, respectively, we say the pattern $\tau$ appears at location $\alpha$ if the values $\{\pi(v)\}_{v\in\mathcal{V}_\alpha}$ and $\{\tau(v)\}_{v\in\mathcal{V}_1}$ are in the same relative order. Equivalently, the pattern $\tau$ appears at $\alpha$ if and only if $\pi(\tau^{-1}(v)+\alpha-1),v\in\mathcal{V}_1$ is an increasing sequence. Letting $\pi$ be chosen uniformly from all permutations of $\{1,\ldots,n\}$, and setting $X_\alpha$ to be the indicator that $\tau$ appears at $\alpha$, we may write

$$X_\alpha\big(\pi(v),v\in\mathcal{V}_\alpha\big)=\mathbf{1}\big(\pi\big(\tau^{-1}(1)+\alpha-1\big)<\cdots<\pi\big(\tau^{-1}(m)+\alpha-1\big)\big),$$

and the sum $Y=\sum_{\alpha\in\mathcal{V}}X_\alpha$ counts the number of $m$-element-long segments of $\pi$ that have the same relative order as $\tau$.

For $\alpha\in\mathcal{V}$ we may generate $\mathbf{X}^\alpha=\{X_\beta^\alpha,\beta\in\mathcal{V}\}$ with the $\mathbf{X}=\{X_\beta,\beta\in\mathcal{V}\}$ distribution biased in direction $\alpha$ as follows. Let $\sigma_\alpha$ be the permutation of $\{1,\ldots,m\}$ for which

$$\pi\big(\sigma_\alpha(1)+\alpha-1\big)<\cdots<\pi\big(\sigma_\alpha(m)+\alpha-1\big),$$

and set

$$\pi^\alpha(v)=\begin{cases}\pi(\sigma_\alpha(\tau(v-\alpha+1))+\alpha-1) & v\in\mathcal{V}_\alpha,\\ \pi(v) & v\notin\mathcal{V}_\alpha.\end{cases}$$

In other words $\pi^\alpha$ is the permutation $\pi$ with values $\pi(v)$, $v \in \mathcal{V}_\alpha$ reordered so that the values of $\pi^\alpha(\gamma)$ for $\gamma \in \mathcal{V}_\alpha$ are in the same relative order as $\tau$. Now let

$$X_\beta^\alpha = X_\beta\big(\pi^\alpha(v), v \in \mathcal{V}_\beta\big),$$

the indicator that $\tau$ appears at position $\beta$ in the reordered permutation $\pi^\alpha$.

Since the relative order of non-overlapping segments of the values of $\pi$ are independent, (6.106) holds for $X_\alpha, \alpha \in \mathcal{V}$ with

$$\mathcal{B}_\alpha = \big\{\beta\colon |\beta - \alpha| \le m - 1\big\}.$$

Next, note that with $\mathcal{F} = \sigma\{\pi\}$, for $\beta \in \mathcal{B}_\alpha$ the random variables $E(X_\beta^\alpha|\mathcal{F})$ and $X_\beta$ depend only on the relative order of $\pi(v)$ for $v \in \bigcup_{\beta \in \mathcal{B}_\alpha} \mathcal{B}_\beta$. Since

$$\Big(\bigcup_{\beta_1 \in \mathcal{B}_{\alpha_1}} \mathcal{B}_{\beta_1}\Big) \bigcap \Big(\bigcup_{\beta_2 \in \mathcal{B}_{\alpha_2}} \mathcal{B}_{\beta_2}\Big) = \emptyset \quad \text{when } |\alpha_1 - \alpha_2| > 3(m-1),$$

for such $\alpha_1, \alpha_2$, and $(\beta_1, \beta_2) \in \mathcal{B}_{\alpha_1} \times \mathcal{B}_{\alpha_1}$, the variables $E(X_{\beta_1}^{\alpha_1}|\mathcal{F}) - X_{\beta_1}$ and $E(X_{\beta_2}^{\alpha_2}|\mathcal{F}) - X_{\beta_2}$ are independent. Hence (6.109) holds with

$$\mathcal{D} = \big\{(\alpha_1, \alpha_2) : |\alpha_1 - \alpha_2| \le 3(m-1)\big\},$$

and Corollary 6.1 gives bounds of the same form as for Example 6.1.

When $\tau = \iota_m$, the identity permutation of length $m$, we say that $\pi$ has a rising sequence of length $m$ at position $\alpha$ if $X_\alpha = 1$. Rising sequences were studied in Bayer and Diaconis (1992) in connection with card tricks and card shuffling. Due to the regular-self-overlap property of rising sequences, namely that a non-empty intersection of two rising sequences is again a rising sequence, some improvement on the constant in the bound can be obtained by a more careful consideration of the conditional variance.

*Example 6.3* (Coloring patterns and subgraph occurrences on a finite graph $\mathcal{G}$) With $n, p \in \mathbb{N}$, let $\mathcal{V} = \mathcal{A} = \{1, \ldots, n\}^p$, again with addition modulo $n$, and for $\alpha, \beta \in \mathcal{V}$ let $d(\alpha, \beta) = \|\alpha - \beta\|$ where $\|\cdot\|$ denotes the supremum norm. Further, let $\mathcal{E} = \{\{w, v\}\colon d(w, v) = 1\}$, and, for each $\alpha \in \mathcal{A}$, let $\mathcal{G}_\alpha = (\mathcal{V}_\alpha, \mathcal{E}_\alpha)$ where

$$\mathcal{V}_\alpha = \big\{v\colon d(v, \alpha) \le 1\big\} \quad \text{and} \quad \mathcal{E}_\alpha = \big\{\{v, w\}\colon v, w \in \mathcal{V}_\alpha,\ d(w, v) = 1\big\}.$$

Let $\mathcal{C}$ be a set (of e.g. colors) from which is formed a given pattern $\{c_g\colon g \in \mathcal{G}_0\}$, let $\{C_g, g \in \mathcal{G}\}$ be independent variables in $\mathcal{C}$ with $\{C_g\colon g \in \mathcal{G}_\alpha\}_{\alpha \in \mathcal{A}}$ identically distributed, and let

$$X(C_g, g \in \mathcal{G}_0) = \prod_{g \in \mathcal{G}_0} \mathbf{1}(C_g = c_g), \tag{6.122}$$

and $X_\alpha = X(C_g, g \in \mathcal{G}_\alpha)$. Then $Y = \sum_{\alpha \in \mathcal{A}} X_\alpha$ counts the number of times the pattern appears in the subgraphs $\mathcal{G}_\alpha$. Taking $\rho = 2$ by (6.117) the conclusion of Corollary 6.3 holds with $M = 1$, $V(r) = (2r+1)^p$ and $|\mathcal{A}| = n^p$.

Such multi-dimensional pattern occurrences are a generalization of the well-studied case in which one-dimensional sequences are scanned for pattern occurrences; see, for instance, Glaz et al. (2001) and Naus (1982) for scan and window statistics, see Huang (2002) for applications of the normal approximation in this context to molecular sequence data, and see also Darling and Waterman (1985, 1986), where higher-dimensional extensions are considered.

Occurrences of subgraphs can be handled as a special case. For example, with $(\mathcal{V}, \mathcal{E})$ the graph above, let $G$ be the random subgraph with vertex set $\mathcal{V}$ and random edge set $\{e \in \mathcal{E} : C_e = 1\}$ where $\{C_e\}_{e\in\mathcal{E}}$ are independent and identically distributed Bernoulli variables. Then letting the function $X(C_g, g \in \mathcal{G}_0)$ in (6.122) be the indicator of the occurrence of a distinguished subgraph of $\mathcal{G}_0$, sum $Y = \sum_{\alpha\in\mathcal{A}} X_\alpha$ counts the number of times that copies of the subgraph appear in the random graph $G$; the same bounds hold as above.

*Example 6.4* (Local extremes) For a given graph $\mathcal{G}$, let $\mathcal{G}_\alpha, \alpha \in \mathcal{A}$, be a collection of subgraphs of $\mathcal{G}$ isomorphic to some subgraph $\mathcal{G}_0$ of $\mathcal{G}$, and let $v \in \mathcal{V}_0$ be a distinguished vertex in $\mathcal{G}_0$. Let $\{C_g, g \in \mathcal{V}\}$ be a collection of independent and identically distributed random variables, and let $X_\alpha = X(C_\beta, \beta \in \mathcal{V}_\alpha)$ where

$$X(C_\beta, \beta \in \mathcal{V}_0) = \mathbf{1}(C_v \ge C_\beta, \beta \in \mathcal{V}_0).$$

Then the sum $Y = \sum_{\alpha\in\mathcal{A}} X_\alpha$ counts the number of times the vertex in $\mathcal{G}_\alpha$, the one corresponding under the isomorphism to the distinguished vertex $v \in \mathcal{V}_0$, is a local maxima. Corollary 6.3 holds with $M = 1$; the other quantities determining the bound are dependent on the structure of $\mathcal{G}$.

Consider, for example, the hypercube $\mathcal{V} = \{0, 1\}^n$ and $\mathcal{E} = \{\{v, w\}\colon \|v - w\| = 1\}$, where $\|\cdot\|$ is the Hamming distance (see also Baldi et al. 1989 and Baldi and Rinott 1989). Let $v = \mathbf{0}$ be the distinguished vertex, $\mathcal{A} = \mathcal{V}$, and, for each $\alpha \in \mathcal{A}$, let $\mathcal{V}_\alpha = \{\beta\colon \|\beta - \alpha\| \le 1\}$ and $\mathcal{E}_\alpha = \{\{v, w\}\colon v, w \in \mathcal{V}_\alpha, \|v - w\| = 1\}$. Corollary 6.3 applies with $\rho = 2$ by (6.117), $V(r) = \sum_{j=0}^{r} \binom{n}{j}$, and $|\mathcal{A}| = 2^n$.

## 6.3 The Lightbulb Process

The following problem arises from a study in the pharmaceutical industry on the effects of dermal patches designed to activate targeted receptors. An active receptor will become inactive, and an inactive one active, if it receives a dose of medicine released from the dermal patch. Let the number of receptors, all initially inactive, be denoted by $n$. On study day $i$ over a period of $n$ days, exactly $i$ randomly selected receptors each will receive one dose of medicine, thus changing their status between inactive and active.

The problem has the following, somewhat more colorful, though equivalent, formulation. Consider $n$ toggle switches, each being connected to a lightbulb. Pressing the toggle switch connected to a bulb changes its status from off to on and vice versa. At each stage $i = 1, \dots, n$, exactly $i$ of the $n$ switches are randomly pressed.

Interest centers on the random variable $Y$, which records the number of lightbulbs that are on at the terminal time $n$.

The problem of determining the properties of $Y$ was first considered in Rao et al. (2007) where the following expressions for the mean $\mu = EY$ and variance $\sigma^2 = \mathrm{Var}(Y)$ were derived,

$$\mu = \frac{n}{2}\left(1 - \prod_{i=1}^{n}\left(1 - \frac{2i}{n}\right)\right), \tag{6.123}$$

and

$$\sigma^2 = \frac{n}{4}\left[1 - \prod_{i=1}^{n}\left(1 - \frac{4i}{n} + \frac{4i(i-1)}{n(n-1)}\right)\right] + \frac{n^2}{4}\left[\prod_{i=1}^{n}\left(1 - \frac{4i}{n} + \frac{4i(i-1)}{n(n-1)}\right) - \prod_{i=1}^{n}\left(1 - \frac{2i}{n}\right)^2\right]. \tag{6.124}$$

Other results, for instance, recursions for determining the exact finite sample distribution of $Y$, are derived in Rao et al. (2007). In addition, approximations to the distribution of $Y$, including by the normal, are also considered there, though the question of the asymptotic normality of $Y$ was left open. Note that when $n$ is even then $\mu = n/2$ exactly, as the product in (6.123), containing the term $i = n/2$, is zero. By results in Rao et al. (2007), in the odd case $\mu = (n/2)(1 + O(e^{-n}))$, and in both the even and odd cases $\sigma^2 = (n/2)(1 + O(e^{-n}))$.

The following theorem of Goldstein and Zhang (2010) provides a bound to the normal which holds for all finite $n$, and which tends to zero as $n$ tends to infinity at the rate $n^{-1/2}$, thus showing the asymptotic distribution of $Y$ is normal as $n \to \infty$. Though the results of Goldstein and Zhang (2010) provide a bound no matter the parity of $n$, for simplicity we only consider the case where $n$ even.

**Theorem 6.5** *With $Y$ the number of bulbs on at the terminal time $n$ and $W = (Y - \mu)/\sigma$ where $\mu = n/2$ and $\sigma^2$ is given by* (6.124), *for all $n$ even*

$$\sup_{z\in\mathbb{R}}\left|P(W \le z) - P(Z \le z)\right| \le \frac{n}{2\sigma^2}\Psi + 1.64\frac{n}{\sigma^3} + \frac{2}{\sigma}$$

*where*

$$\Psi \le \frac{1}{2\sqrt{n}} + \frac{1}{2n} + e^{-n/2} \quad \text{for } n \ge 6. \tag{6.125}$$

We now more formally describe the random variable $Y$. Let $\mathbf{Y} = \{Y_{ri}\colon r, i = 1, \dots, n\}$ be the Bernoulli 'switch' variables which have the interpretation

$$Y_{ri} = \begin{cases} 1 & \text{if the status of bulb } i \text{ is changed at stage } r, \\ 0 & \text{otherwise.} \end{cases}$$

We continue to suppress the dependence of $Y$, and also of $Y_{ri}$, on $n$. As the set of $r$ bulbs which have their status changed at stage $r$ is chosen uniformly over all sets of size $r$, and as the stages are independent of each other, with $e_1, \ldots, e_n \in \{0, 1\}$ the joint distribution of $Y_{r1}, \ldots, Y_{rn}$ is given by

$$P(Y_{r1} = e_1, \ldots, Y_{rn} = e_n) = \begin{cases} \binom{n}{r}^{-1} & \text{if } e_1 + \cdots + e_n = r, \\ 0 & \text{otherwise,} \end{cases}$$

with the collections $\{Y_{r1}, \ldots, Y_{rn}\}$ independent for $r = 1, \ldots, n$.

Clearly, for each stage $r$, the variables $(Y_{r1}, \ldots, Y_{rn})$ are exchangeable, and the marginal distribution for each $r, i = 1, \ldots, n$ is given by

$$P(Y_{ri} = 1) = \frac{r}{n} \quad \text{and} \quad P(Y_{ri} = 0) = 1 - \frac{r}{n}.$$

For $r, i = 1, 2, \ldots, n$ the quantity $(\sum_{s=1}^r Y_{si}) \bmod 2$ is the indicator that bulb $i$ is on at time $r$, and therefore

$$Y = \sum_{i=1}^n Y_i \quad \text{where } Y_i = \left(\sum_{r=1}^n Y_{ri}\right) \bmod 2 \tag{6.126}$$

is the number of bulbs on at the terminal time.

The lightbulb process, where the $n$ individual states evolve according to the same marginal Markov chain, is a special case of a certain class of multivariate chains studied in Zhou and Lange (2009), termed 'Composition Markov chains of multinomial type.' As shown there, such chains admit explicit full spectral decompositions, and in particular, each transition matrix of the lightbulb process can be simultaneously diagonalized by a Hadamard matrix. These properties were, in fact, put to use in Rao et al. (2007) for the calculation of the moments needed for (6.123) and (6.124).

We now describe the coupling given by Goldstein and Zhang (2010), which shows that when $n$ is even, $Y$ may be coupled monotonically to a variable $Y^s$ having the $Y$-size bias distribution, in particular, such that

$$Y \le Y^s \le Y + 2. \tag{6.127}$$

For every $i \in \{1, \ldots, n\}$ construct the collection of variables $\mathbf{Y}^i$ from $\mathbf{Y}$ as follows. If $Y_i = 1$, that is, if bulb $i$ is on, let $\mathbf{Y}^i = \mathbf{Y}$. Otherwise, with $\mathcal{J}^i$ a uniformly chosen index over the set $\{j : Y_{n/2,j} = 1 - Y_{n/2,i}\}$, let $\mathbf{Y}^i = \{Y^i_{rk} : r, k = 1, \ldots, n\}$ where

$$Y^i_{rk} = \begin{cases} Y_{rk} & r \ne n/2, \\ Y_{n/2,k} & r = n/2, k \notin \{i, J^i\}, \\ Y_{n/2,J^i} & r = n/2, k = i, \\ Y_{n/2,i} & r = n/2, k = J^i, \end{cases}$$

and let $Y^i = \sum_{k=1}^n Y^i_k$ where

$$Y^i_k = \left(\sum_{j=1}^n Y^i_{jk}\right) \bmod 2.$$

In other words, if bulb $i$ is off, then the switch variable $Y_{n/2,i}$ of bulb $i$ at stage $n/2$ is interchanged with that of a variable whose switch variable at this stage has the opposite status.

With $I$ uniformly chosen from $\{1,\dots,n\}$ and independent of all other variables, it is shown in Goldstein and Zhang (2010) that the mixture $Y^s = Y^I$ has the $Y$ size biased distribution, essentially due to the fact that

$$\mathcal{L}(\mathbf{Y}^i) = \mathcal{L}(\mathbf{Y}|Y_i = 1) \quad \text{for all } i = 1,\dots,n.$$

It is not difficult to see that $Y^s$ satisfies (6.127). If $Y_I = 1$ then $\mathbf{X}^I = \mathbf{X}$, and so in this case $Y^s = Y$. Otherwise $Y_I = 0$, and we obtain $\mathbf{Y}^I$ by interchanging, at stage $n/2$, the unequal switch variables $Y_{n/2,I}$ and $Y_{n/2,J^I}$, which changes the status of both bulbs $I$ and $J^I$. If bulb $J^I$ was on, that is, if $Y_{J^I} = 1$, then after the interchange $Y_I^I = 1$ and $Y_{J^I}^I = 0$, in which case $Y^s = Y$. Otherwise bulb $J^I$ was off, that is, $Y_{J^I} = 0$, in which case after the interchange we have $Y_I^I = 1$ and $Y_{J^I}^I = 1$, yielding $Y^s = Y + 2$.

As the coupling is both monotone and bounded, by (6.127) Theorem 5.7 may be invoked with $\delta = 2/\sigma$. In fact, the first two terms of the bound in Theorem 6.5 arise directly from Theorem 5.7 with this $\delta$. The bound (6.125) is calculated in Goldstein and Zhang (2010), making heavy use of the spectral decomposition provided by Zhou and Lange (2009) to determine various joint probabilities of fourth, but no higher, order.

## 6.4 Anti-voter Model

The anti-voter model was introduced by Matloff (1977) on infinite lattices. Donnelly and Welsh (1984), and Aldous and Fill (1994) consider, as we do here, the case of finite graphs; see also Liggett (1985), and references there. The treatment below closely follows Rinott and Rotar (1997), who deal with a discrete time version.

Let $\mathcal{G} = (\mathcal{V}, \mathcal{E})$, a graph with $n$ vertices $\mathcal{V}$ and edges $\mathcal{E}$, which was assume to be $r$-regular, that is, all vertices $v \in \mathcal{V}$ have degree $r$. Consider the following transition rule for a Markov chain $\{\mathbf{X}^{(t)}, t = 0, 1, \dots\}$ with state space $\{-1, 1\}^{\mathcal{V}}$. At each time $t$, a vertex $v$ is chosen uniformly from $\mathcal{V}$, and then a different vertex $w$ is chosen uniformly from the set

$$N_v = \{w \colon \{v, w\} \in \mathcal{E}\}$$

of neighbors of $v$, and then we let

$$X_u^{(t+1)} = \begin{cases} X_u^{(t)} & u \neq v, \\ -X_w^{(t)} & u = v. \end{cases}$$

That is, the configuration at time $t+1$ is the same at time $t$, but that vertex $v$ takes the sign opposite of its randomly chosen neighbor $w$.

Following Donnelly and Welsh (1984), and Aldous and Fill (1994), when $\mathcal{G}$ is neither an $n$ cycle nor bipartite, the chain is irreducible on the state space consisting of the $2^n - 2$ configurations which exclude those where $X_v^{(t)}$ are identical, and has a stationary distribution supported on this set. We suppose the distribution of $\mathbf{X}^{(0)}$, the chain at time zero, is this stationary distribution.

The exchangeable pair coupling yields the following result on the quality of the normal approximation to the distribution of the standardized net sign of the stationary configuration.

**Theorem 6.6** *Let* $\mathbf{X}$ *have the stationary distribution of the anti voter chain on an* $n$ *vertex,* $r$*-regular graph* $\mathcal{G}$*, neither an* $n$ *cycle nor bipartite, and let* $W$ *be the standardized net sign* $U$ *of the configuration* $\mathbf{X}$*, that is, with* $\sigma^2 = \mathrm{Var}(U)$ *let*

$$W = U/\sigma \quad \text{where } U = \sum_{v \in \mathcal{V}} X_v. \tag{6.128}$$

*Then, if* $U'$ *is the net sign obtained by applying the one step transition to the configuration* $\mathbf{X}$*,* $(U, U')$ *is a* $2/n$*-Stein pair that satisfies*

$$|U' - U| \le 2 \quad \text{and} \quad E\big[(U' - U)^2 | \mathbf{X}\big] = 8(a+b)/(rn) \tag{6.129}$$

*where* $a$ *and* $b$ *are the number of edges that are incident on vertices both of which are in state* $+1$*, or* $-1$*, respectively.*

*In addition,*

$$\sup_{z \in \mathbb{R}} \big| P(W \le z) - P(Z \le z) \big| \le \frac{12n}{\sigma^3} + \frac{\sqrt{\mathrm{Var}(Q)}}{r\sigma^2}$$

*where*

$$Q = \sum_{v \in \mathcal{V}} \sum_{w \in N_v} X_v X_w. \tag{6.130}$$

When $\sigma^2$ and $\mathrm{Var}(Q)$ are of order $n$, the bound given in the theorem has order $1/\sqrt{n}$. Use of (5.13) results in a somewhat more complex, but superior bound.

The first order of business in proving Theorem 6.6 is the construction of an exchangeable pair. It is immediate that if $\mathbf{X}^{(t)}$ is a reversible Markov chain in stationarity then for any measurable function $f$ on the chain, $(f(\mathbf{X}^{(s)}), f(\mathbf{X}^{(t)}))$ is exchangeable, for any $s$ and $t$. Even when a chain is not reversible, as is the case for the anti-voter model, the following lemma may be invoked for functions of chains whose increments are the same as that of a birth death process.

**Lemma 6.11** *Let* $\{\mathbf{X}^{(t)}, t = 0, 1, \ldots\}$ *be a stationary process, and suppose that* $T(\mathbf{X}^{(t)})$ *assumes nonnegative integer values such that*

$$T\big(\mathbf{X}^{(t+1)}\big) - T\big(\mathbf{X}^{(t)}\big) \in \{-1, 0, 1\} \quad \text{for all } t = 0, 1, \ldots. \tag{6.131}$$

*Then for any measurable function* $f$*,*

$$\big(W, W'\big) = \big(f\big(T\big(\mathbf{X}^{(t)}\big)\big), f\big(T\big(\mathbf{X}^{(t+1)}\big)\big)\big)$$

*is an exchangeable pair.*

*Proof* The process $T^{(t)} = T(\mathbf{X}^{(t)})$ is stationary and has values in the nonnegative integers. For integers $i$, $j$ in the range of $T(\cdot)$, set $\pi_i = P(T^{(t)} = i)$ and $p_{ij} = P(T^{(t+1)} = j \mid T^{(t)} = i)$. By stationarity, these probabilities do not depend on $t$. Using stationarity to obtain the second equality, and setting $\pi_i$ and $p_{ij} = 0$ for all $i < 0$, we have for all nonnegative integers $j$,

$$\pi_j = P\bigl(T^{(t)} = j\bigr) = P\bigl(T^{(t+1)} = j\bigr) = \sum_{i \in \mathbb{N}_0} P\bigl(T^{(t+1)} = j, T^{(t)} = i\bigr)$$

$$= \sum_{i \in \mathbb{N}_0} P\bigl(T^{(t+1)} = j | T^{(t)} = i\bigr) P\bigl(T^{(t)} = i\bigr) = \sum_{i:\, |i-j| \le 1} \pi_i p_{ij},$$

where we have restricted the sum in the last equality due to the condition imposed by (6.131). This same system of equations arises in birth and death chains and it is well-known that if it has a solution then it is unique, can be written explicitly, and satisfies $\pi_i p_{ij} = \pi_j p_{ji}$ (which implies reversibility for birth and death chains). Here, the latter relation is equivalent to

$$P\bigl(T^{(t)} = i, T^{(t+1)} = j\bigr) = P\bigl(T^{(t)} = j, T^{(t+1)} = i\bigr),$$

implying that $(T^{(t)}, T^{(t+1)})$ is an exchangeable pair. □

With this result in hand, we may now proceed to the

*Proof of Theorem 6.6* We apply Theorem 5.4. By Lemma 6.11,

$$\bigl(W, W'\bigr) = \bigl(W\bigl(\mathbf{X}^{(t)}\bigr), W\bigl(\mathbf{X}^{(t+1)}\bigr)\bigr)$$

is an exchangeable pair when $W(\mathbf{X})$ is the standardized net sign of the configuration $\mathbf{X}$, as in (6.128). With $U$ and $U'$ the net signs of $\mathbf{X}^{(t)}$ and $\mathbf{X}^{(t+1)}$, respectively, since at most a single 1 becomes $-1$, or a $-1$ a 1 in a one step transition, clearly the first claim of (6.129) holds.

We next verify that $(W, W')$ satisfies the linearity condition (2.33) with $\lambda = 2/n$. Let

$$T = \sum_{v \in \mathcal{V}} \mathbf{1}_{\{X_v = 1\}},$$

the number of vertices with sign 1, and let

$$a = \sum_{\{u,v\} \in \mathcal{E}} \mathbf{1}_{\{X_u = X_v = 1\}},$$

$$b = \sum_{\{u,v\} \in \mathcal{E}} \mathbf{1}_{\{X_u = X_v = -1\}} \quad \text{and} \quad c = \sum_{\{u,v\} \in \mathcal{E}} \mathbf{1}_{\{X_u \ne X_v\}},$$

the number of edges both of whose incident vertices take the value 1, the value $-1$, or both these values, respectively. For an $r$-regular graph,

$$r\mathbf{1}_{\{X_v = 1\}} = \sum_{w \in N_v} \mathbf{1}_{\{X_v = 1, X_w = 1\}} + \sum_{w \in N_v} \mathbf{1}_{\{X_v = 1, X_w = -1\}},$$

hence summing over $v \in \mathcal{V}$ yields

$$rT = \sum_{v\in\mathcal{V}, w\in N_v} \mathbf{1}_{\{X_v=1, X_w=1\}} + \sum_{v\in\mathcal{V},\, w\in N_v} \mathbf{1}_{\{X_v=1, X_w=-1\}} = 2a + c,$$

and so

$$T = (2a+c)/r \quad \text{and likewise} \quad n - T = (2b+c)/r. \tag{6.132}$$

Note $U = 2T - n$ and $U' = 2T' - n$ are the net signs of the configurations $\mathbf{X}^{(t)}$ and $\mathbf{X}^{(t+1)}$, respectively. When making a transition one first chooses a vertex uniformly, then one of its neighbors, uniformly, and so since the graph is regular the edge so chosen is uniform. As the net sign $U$ decreases by 2 in a transition if and only if a 1 becomes a $-1$, and this event occurs if and only if one of the $rn/2$ edges counted by $a$ is chosen, we have

$$P(U' - U = -2|\mathbf{X}) = \frac{2a}{rn} \quad \text{and likewise} \quad P(U' - U = 2|\mathbf{X}) = \frac{2b}{rn}, \tag{6.133}$$

and therefore, by (6.132),

$$E\big(U' - U|\mathbf{X}\big) = \frac{4b}{rn} - \frac{4a}{rn} = \frac{2(n-2T)}{n} = -\frac{2}{n}U.$$

Hence, (2.33) is satisfied for $W = U/\sigma$ with $\lambda = 2/n$, and Theorem 5.4 obtains with this value of $\lambda$ and, as $|U' - U| \le 2$, with $\delta = 2/\sigma$.

Next we bound $\Theta$ in (5.3). By (6.133) we have

$$E\big[(U' - U)^2|\mathbf{X}\big] = 4\left(\frac{2a}{rn} + \frac{2b}{rn}\right),$$

proving the second claim in (6.129). Next, recalling the definition of $Q$ in (6.130), note the relations

$$2a + 2b + 2c = rn \quad \text{and} \quad 2a + 2b - 2c = Q,$$

imply

$$4(a+b) = Q + rn.$$

Hence $E[(U' - U)^2|\mathbf{X}] = 2(Q + rn)/(rn)$, and therefore, using that $W$ is a function of $\mathbf{X}$ and (4.143),

$$\Theta \le \sqrt{\mathrm{Var}\big(E\big[(W' - W)^2|\mathbf{X}\big]\big)} = \sqrt{\mathrm{Var}\left(\frac{2Q}{rn\sigma^2}\right)} = \frac{2}{rn\sigma^2}\sqrt{\mathrm{Var}(Q)}.$$

Applying Theorem 5.4 along with (5.4) and the computed upper bound for $\Theta$, and that $\lambda = 2/n$ and $\delta = 2/\sigma$, the proof of the theorem is complete. □

The quantities $\sigma$ and $\mathrm{Var}(Q)$ depend heavily on the particular graph under consideration. For details on how these quantities may be bounded for graphs having certain regularity properties, and examples which include the Hamming graph and the $k$-bipartite graph, see Rinott and Rotar (1997).

## 6.5 Binary Expansion of a Random Integer

Let $n \ge 2$ be a natural number and $x$ an integer in the set $\{0, 1, \dots, n-1\}$. For $m = [\log_2(n-1)] + 1$, consider the binary expansion of $x$

$$x = \sum_{i=1}^{m} x_i 2^{m-i}.$$

Clearly any leading zeros contribute nothing to the sum $x$. With $X$ uniformly chosen from $\{0, 1, \dots, n-1\}$, the sum $S = X_1 + \cdots + X_m$ is the number of ones in the expansion of $X$. When $n = 2^m$ a uniform random integer between 0 and $2^m - 1$ may be constructed by choosing its $m$ binary digits to be zeros and ones with equal probability, and independently, so the distribution of $S$ in this case has the symmetric binomial distribution with $m$ trials, which of course can be well approximated by the normal. Theorem 6.7 shows that the same is true for any large $n$, and provides an explicit bound. We follow Diaconis (1977).

We approach the problem using exchangeable pairs. For $x$ an integer in $\{0, \dots, n-1\}$ let $Q(x, n)$ be the number of zeros in the $m$ long expansion of $x$ which, when changed to 1, result in an integer $n$ or larger, that is,

$$\begin{aligned} &Q(x,n) = |J_x| \\ &\text{where } J_x = \big\{ i \in \{1, \dots, m\} \colon x_i = 0, x + 2^{m-i} \ge n \big\}. \end{aligned} \tag{6.134}$$

For example, $Q(10, 5) = 1$. With $I$ a random index on $\{1, \dots, m\}$ let

$$X' = \begin{cases} X + (1 - 2X_I)2^{m-I} & \text{if } I \notin J_X, \\ X & \text{otherwise.} \end{cases}$$

That is, the $I^{th}$ digit of $X$ is changed from $X_I$ to $1 - X_I$, if doing so produces a number between $\{0, \dots, n-1\}$. Clearly $(X, X')$ are exchangeable, and $S'$, the number of ones in the expansion of $X'$, is given by

$$S' = \begin{cases} S + 1 - 2X_I & \text{if } I \notin J_X, \\ S & \text{otherwise.} \end{cases}$$

As we see from the following lemma $(S, S')$ is not a Stein pair, as it fails to satisfy the linearity condition. Nevertheless, Theorem 3.5 applies. The lemma also provides the mean and variance of $S$.

**Lemma 6.12** *For $n \ge 2$ let $X$ be uniformly chosen from $\{0, 1, \dots, n-1\}$ and $Q = Q(X, n)$. Then*

$$E(S' - S|X) = \left(1 - \frac{2S}{m}\right) - \frac{Q}{m}, \quad E\big((S' - S)^2|X\big) = 1 - \frac{Q}{m},$$

*and*

$$ES = \frac{1}{2}(m - EQ) \quad \text{and} \quad \mathrm{Var}(S) = \frac{m}{4}\left(1 - \frac{EQ + 2\,\mathrm{Cov}(S, Q)}{m}\right).$$

*Proof* To derive the first identity, write

$$\begin{aligned}
E(S'-S|X) &= P(S'-S=1|X) - P(S'-S=-1|X) \\
&= P\big(X_I=0, X'_I=1|X\big) - P(X_I=1|X) \\
&= P(X_I=0|X) - P\big(X_I=0, X'_I=0|X\big) - P(X_I=1|X) \\
&= 1 - P\big(X_I=0, X'_I=0|X\big) - 2P(X_I=1|X) \\
&= 1 - \frac{1}{m}\sum_{i=1}^{m} \mathbf{1}_{\{X_i=0, X+2^{m-i}\ge n\}} - \frac{2}{m}\sum_{i=1}^{m} X_i \\
&= 1 - \frac{2S}{m} - \frac{Q}{m}.
\end{aligned} \tag{6.135}$$

The expectation of $S$ can now be calculated using that $E(S'-S)=0$.

Similarly, since $S'-S \in \{-1,0,1\}$,

$$\begin{aligned}
E\big((S'-S)^2|X\big) &= P(S'-S=1|X) + P(S'-S=-1|X) \\
&= P\big(X_I=0, X'_I=1|X\big) + P(X_I=1|X) \\
&= P(X_I=0|X) - P\big(X_I=0, X'_I=0|X\big) + P(X_I=1|X) \\
&= 1 - P\big(X_I=0, X'_I=0|X\big) \\
&= 1 - \frac{1}{m}\sum_{i=1}^{m} \mathbf{1}_{\{X_i=0, X+2^{m-i}\ge n\}} \\
&= 1 - \frac{Q}{m}.
\end{aligned} \tag{6.136}$$

To calculate the variance, note first

$$0 = E(S'-S)(S'+S) = E\big(2(S'-S)S + (S'-S)^2\big).$$

Now taking expectation in (6.136), using identity (6.135) and that the quantities involved have mean zero, we obtain

$$\begin{aligned}
E\left(1-\frac{Q}{m}\right) &= E(S'-S)^2 = 2E\big(S(S-S')\big) \\
&= 2E\left(S\left(\frac{2S}{m} - 1 + \frac{Q}{m}\right)\right) \\
&= \frac{4\,\mathrm{Var}(S)}{m} + 2\,\mathrm{Cov}\left(S, \frac{Q}{m}\right).
\end{aligned}$$

Solving for the variance now completes the proof. □

**Theorem 6.7** *For $n \ge 2$ let $X$ be a random integer chosen uniformly from $\{0,\dots,n-1\}$. Then with $S$ the number of ones in the binary expansion of $X$ and $m=[\log_2(n-1)]+1$, the random variable*

$$W = \frac{S - m/2}{\sqrt{m/4}} \tag{6.137}$$

*satisfies*

$$\sup_{z\in\mathbb{R}}\big|P(W\le z)-P(Z\le z)\big|\le\frac{6.9}{\sqrt{m}}+\frac{5.4}{m}.$$

*Proof* With $W'$ given by (6.137) with $S$ replaced by $S'$, the pair $(W,W')$ is exchangeable, and, with $\lambda=2/m$, Lemma 6.12 yields

$$E(W-W'|W)=\lambda\left(W+\frac{E(Q|W)}{\sqrt{m}}\right) \tag{6.138}$$

and

$$\frac{1}{2\lambda}E\big((W-W')^2|W\big)=1-\frac{E(Q|W)}{m}. \tag{6.139}$$

Further, we have

$$\begin{aligned} EQ&=\sum_{i=1}^{m}P\big(X_i=0,X+2^{m-i}\ge n\big)\\ &\le\sum_{i=1}^{m}P\big(X\ge n-2^{m-i}\big)\\ &=\sum_{i=1}^{m}\frac{2^{m-i}}{n}=\frac{2^m-1}{n}\\ &\le\frac{2^m-1}{2^{m-1}}\le 2. \end{aligned} \tag{6.140}$$

Since $W$, $W'$ are exchangeable, for any function $f$ for which the expectations below exist, identity (2.32) yields

$$\begin{aligned} 0&=E\big\{(W-W')\big(f(W')+f(W)\big)\big\}\\ &=E\big\{(W-W')\big(f(W')-f(W)\big)\big\}+2E\big\{f(W)(W-W')\big\}\\ &=E\big\{(W-W')\big(f(W')-f(W)\big)\big\}+2E\big\{f(W)E\big(W-W'\mid W\big)\big\}\\ &=E\big\{(W-W')\big(f(W')-f(W)\big)\big\}+2\lambda E\left\{\left(W+\frac{E(Q|W)}{\sqrt{m}}\right)f(W)\right\}. \end{aligned}$$

Solving for $EWf(W)$ and then reasoning as in the proof of Lemma 2.7 we obtain

$$\begin{aligned} EWf(W)&=\frac{1}{2\lambda}E\big((W'-W)\big(f(W')-f(W)\big)\big)-\frac{1}{\sqrt{m}}E\big(E(Q|W)f(W)\big)\\ &=E\int_{-\infty}^{\infty}f'(W+t)\hat{K}(t)dt-\frac{1}{\sqrt{m}}E\big(E(Q|W)f(W)\big) \end{aligned}$$

where

$$\hat{K}(t)=\frac{\Delta}{2\lambda}(\mathbf{1}_{\{-\Delta\le t\le 0\}}-\mathbf{1}_{\{0<t\le-\Delta\}})\quad\text{with }\Delta=W-W'.$$

Now let $z\in\mathbb{R}$ and $f_z$ be the solution to (2.2). Then, as $|S'-S|\le 1$, we may write

$$EWf_z(W)=E\int_{|t|\le 2/\sqrt{m}}f_z'(W+t)\hat{K}(t)dt+E(R_1)$$

where

$$R_1 = -\frac{1}{\sqrt{m}}\big(E(Q|W)f_z(W)\big).$$

Now, using $Q \geq 0$, inequality (2.9) and (6.140) we have

$$|ER_1| \leq \sqrt{\pi/(2m)}.$$

We invoke Theorem 3.5 with

$$\delta_0 = 2/\sqrt{m} \quad \text{and} \quad \delta_1 = \sqrt{\pi/(2m)}. \tag{6.141}$$

First, by (6.139),

$$\hat{K}_1 = E\left(\int_{|t|\leq\delta_0} \hat{K}(t)dt \middle| W\right) = E\left(\frac{(W-W')^2}{2\lambda}\middle| W\right) = 1 - \frac{E(Q|W)}{m}.$$

Rewriting (6.138),

$$E(W'|W) = (1-\lambda)W - \lambda\frac{E(Q|W)}{\sqrt{m}} \quad \text{so}$$
$$E(WW') = (1-\lambda)EW^2 - \lambda\frac{E(QW)}{\sqrt{m}}.$$

Therefore, taking expectation in (6.139) yields

$$1 - \frac{E(Q)}{m} = \frac{1}{2\lambda}E(W'-W)^2 = \frac{1}{\lambda}E\big(W^2 - WW'\big) = EW^2 + \frac{E(QW)}{\sqrt{m}},$$

or

$$EW^2 = 1 - \frac{E(Q)}{m} - \frac{E(QW)}{\sqrt{m}} \leq 1 - \frac{E(QW)}{\sqrt{m}}.$$

Now, since $0 \leq S \leq m$, we have $|W| \leq \sqrt{m}$, and therefore

$$EW^2 \leq 1 + \frac{E|W|Q}{\sqrt{m}} \leq 1 + EQ \leq 3.$$

Applying in addition the fact that

$$E|1 - \hat{K}_1| = EQ/m \leq 2/m,$$

and that $0 \leq \hat{K}_1 \leq 1$ so that $E|W|\hat{K}_1 \leq E|W| \leq \sqrt{EW^2}$, Theorem 3.5 along with (6.141) yields the bound

$$\begin{aligned}
&\delta_0\big(1.1 + E(|W|\hat{K}_1)\big) + 2.7E|1-\hat{K}_1| + \delta_1 \\
&\leq \delta_0(1.1+\sqrt{3}) + \frac{5.4}{m} + \delta_1 \\
&\leq 2.8\delta_0 + \frac{5.4}{m} + \delta_1 \\
&\leq \frac{6.9}{\sqrt{m}} + \frac{5.4}{m}
\end{aligned}$$

as claimed. □

# Chapter 7
# Discretized Normal Approximation

The very first use of normal approximation was for the binomial. However, glancing at the histogram depicting the mass function of, say, the binomial $S \sim B(16, 1/2)$, with the normal density drawn on the same scale, one sees that integer boundaries split the mass of the binomial into halves. Hence, to make the approximation more accurate, it is often recommended that continuity correction be used, that is, that $1/2$ be added at upper limits, and $1/2$ subtracted at lower ones. For instance, for $S \sim B(16, 1/2)$, one has $P(S \le 4) = 0.0384$. The straightforward normal approximation yields $\Phi((4-8)/2) = \Phi(-2) = 0.023$, while continuity correction gives the much more accurate value, $\Phi((4.5-8)/2) = \Phi(-1.75) = 0.040$. For a discussion of continuity correction, see Feller (1968a), page 185, and for some additional justification, also Cox (1970).

In this chapter, we provide a total variation bound between the Poisson binomial distribution and the integer valued distribution obtained by discretizing the normal according to the continuity correction rule, and then, a more general result for the discretized normal approximation of an independent sum of integer valued random variables. The results in this chapter are due to Chen and Leong (2010).

To study approximation using continuity correction, for $\mu \in \mathbb{R}$ and $\sigma^2 > 0$, let $Z_{\mu,\sigma^2}$ be the discretized normal with distribution given by

$$P(Z_{\mu,\sigma^2} = k) = P\left(\frac{k-\mu-1/2}{\sigma} < Z \le \frac{k-\mu+1/2}{\sigma}\right) \quad \text{for } k \in \mathbb{Z}. \tag{7.1}$$

Recall that the total variation distance between the distributions of two random variables $X$ and $Y$ can be defined by either of the two following equivalent forms,

$$\begin{aligned} &\left\|\mathcal{L}(X) - \mathcal{L}(Y)\right\|_{\mathrm{TV}} \\ &\quad = \sup_{A \subset \mathbb{R}} \left|P(X \in A) - P(Y \in A)\right| = \frac{1}{2} \sup_{\|h\| \le 1} \left|Eh(X) - Eh(Y)\right|. \end{aligned} \tag{7.2}$$

One can verify that if both $X$ and $Y$ are integer valued random variables, then

$$\left\|\mathcal{L}(X) - \mathcal{L}(Y)\right\|_{\mathrm{TV}} = \frac{1}{2} \sum_{k=-\infty}^{\infty} \left|P(X = k) - P(Y = k)\right|, \tag{7.3}$$

L.H.Y. Chen et al., *Normal Approximation by Stein's Method*,
Probability and Its Applications,
DOI 10.1007/978-3-642-15007-4_7, © Springer-Verlag Berlin Heidelberg 2011

while, similarly, if $X$ and $Y$ have densities $p_X$ and $p_Y$, respectively, then

$$\|\mathcal{L}(X)-\mathcal{L}(Y)\|_{\mathrm{TV}}=\frac{1}{2}\int_{-\infty}^{\infty}|p_X(u)-p_Y(u)|du. \tag{7.4}$$

## 7.1 Poisson Binomial

The following theorem gives a total variation bound on the approximation of a Poisson binomial random variable $S$ and a continuity corrected discretized normal. Without loss of generality we assume in what follows that the summands $X_1,\ldots,X_n$ of $S$ are non-trivial.

**Theorem 7.1** *Let $X_1,\ldots,X_n$ be independent Bernoulli variables with distribution $P(X_i=1)=p_i$ and $P(X_i=0)=1-p_i$, $i=1,\ldots,n$, and let $S=\sum_{i=1}^n X_i$, $\mu=\sum_{i=1}^n p_i$ and $\sigma^2=\sum_{i=1}^n p_i(1-p_i)$. Then*

$$\|\mathcal{L}(S)-\mathcal{L}(Z_{\mu,\sigma^2})\|_{\mathrm{TV}}\le\frac{7.6}{\sigma}.$$

We begin with a general result which gives the total variation distance between a zero biased variable $W^*$ and the normal, when $W$ and $W^*$ can be constructed on a joint space.

**Theorem 7.2** *Let $W$ be a mean zero, variance one random variable and suppose $W^*$, on the same space as $W$, has the $W$-zero bias distribution. Then for all functions $h$ with $\|h\|\le 1$,*

$$|Eh(W^*)-Nh|\le 4E|W(W-W^*)|+\sqrt{2\pi}E|W-W^*|.$$

*Proof* With $f$ the solution of the Stein equation for the given $h$, we have

$$\begin{aligned}|Eh(W^*)-Nh|&=|E(f'(W^*)-W^*f(W^*))|\\&=|E(Wf(W)-W^*f(W^*))|\\&\le|E(W(f(W)-f(W^*)))|+|E(f(W^*)(W-W^*))|\\&\le\|f'\|E|W(W-W^*)|+\|f\|E|W-W^*|.\end{aligned}$$

Applying the bounds in Lemma 2.4 yields the inequality. □

We will also consider $\overline{Z}_{\mu,\sigma^2}$, the discretized version of the distribution of $Z$ given by

$$P(\overline{Z}_{\mu,\sigma^2}=k)=P\left(\frac{k-\mu}{\sigma}<Z\le\frac{k-\mu+1}{\sigma}\right), \tag{7.5}$$

and $W^*_{\mu,\sigma^2}$, the similarly discretized version of $W^*$ with distribution

$$P(W^*_{\mu,\sigma^2}=k)=P\left(\frac{k-\mu}{\sigma}<W^*\le\frac{k+1-\mu}{\sigma}\right). \tag{7.6}$$

When $W$ is the sum of independent variables, Theorem 7.2 has the following corollary.

**Corollary 7.1** *Let $X_1, \ldots, X_n$ be independent random variables with $EX_i = \mu_i$, $\mathrm{Var}(X_i) = \sigma_i^2$ and finite third absolute central moments*

$$\gamma_i = E|X_i - \mu_i|^3/\sigma^3,$$

*where $\sigma^2 = \sum_{i=1}^n \sigma_i^2$. Suppose that*

$$W = \sum_{i=1}^n \xi_i \quad \text{where } \xi_i = \frac{X_i - \mu_i}{\sigma}$$

*and that $W^*$ has the $W$ zero bias distribution. Then*

$$\left\|\mathcal{L}(W^*) - \mathcal{L}(Z)\right\|_{\mathrm{TV}} \le (5 + 3\sqrt{\pi/8})\gamma \tag{7.7}$$

*where $\gamma = \sum_{i=1}^n \gamma_i$.*

*If $\overline{Z}_{\mu,\sigma^2}$ and $W^*_{\mu,\sigma^2}$ have distributions* (7.5) *and* (7.8), *respectively, then*

$$\left\|\mathcal{L}(W^*_{\mu,\sigma^2}) - \mathcal{L}(\overline{Z}_{\mu,\sigma^2})\right\|_{\mathrm{TV}} \le (5 + 3\sqrt{\pi/8})\gamma. \tag{7.8}$$

*Proof* By Lemma 2.8 we may construct $W$ and $W^*$ on a joint space by letting

$$W^* = W - \xi_I + \xi_I^*$$

where $\xi_i^*$ has the $\xi_i$ zero bias distribution and $I$ is a random index with distribution (2.60), with $\xi_i^*$ independent of $\xi_i$ and $I$ independent of all other variables.

We apply Lemma 7.2. For the first term, by independence and the Cauchy–Schwarz inequality,

$$E\left|W(W^* - W)\right| = E\left|W(\xi_I^* - \xi_I)\right| \le E\left|\xi_I^*\right| + E|W\xi_I|.$$

Noting that $\mathrm{Var}(\xi_i) = \sigma_i^2/\sigma^2$ and using the distribution (2.60) of $I$ and (2.57), we have

$$E\left|\xi_I^*\right| = \sum_{i=1}^n \frac{\sigma_i^2}{\sigma^2} E\left|\xi_i^*\right| = \sum_{i=1}^n \frac{1}{2} E|\xi_i|^3 = \gamma/2 \quad \text{and} \quad E|\xi_I| = \sum_{i=1}^n \frac{\sigma_i^2}{\sigma^2} E|\xi_i| \le \gamma.$$

Next,

$$\begin{aligned} E|W\xi_I| &\le \sum_{i=1}^n \frac{\sigma_i^2}{\sigma^2} E\left(|W - \xi_i||\xi_i| + \xi_i^2\right) \le \sum_{i=1}^n \frac{\sigma_i^2}{\sigma^2}\left(E|\xi_i| + \frac{\sigma_i^2}{\sigma^2}\right) \\ &\le \gamma + \sum_{i=1}^n \frac{\sigma_i^3}{\sigma^3} \le 2\gamma. \end{aligned}$$

For the second term, by the triangle inequality,

$$E|W^* - W| = E\left|\xi_I^* - \xi_I\right| \le \frac{3}{2}\gamma.$$

Collecting terms in the bound, taking supremum over $\|h\| \le 1$, and applying Definition 7.2 now yields the first conclusion.

Applying (7.7), (7.3), and the definition of total variation distance in (7.2) with

$$h(w) = \sum_{k \in \mathbb{Z}} h_k(w) \mathbf{1}\left(\frac{k-\mu}{\sigma} < w \le \frac{k+1-\mu}{\sigma}\right)$$

where

$$h_k(w) = \begin{cases} +1 & P(W^*_{\mu,\sigma^2} = k) \ge P(Z_{\mu,\sigma^2} = k), \\ -1 & P(W^*_{\mu,\sigma^2} = k) < P(Z_{\mu,\sigma^2} = k) \end{cases}$$

yields the second conclusion. □

We will also make use of the following fact.

**Lemma 7.1** *If* $\overline{Z}_{\mu,\sigma^2}$ *and* $Z_{\mu,\sigma^2}$ *have distributions* (7.5) *and* (7.1), *respectively, then*

$$\left\| \mathcal{L}(\overline{Z}_{\mu,\sigma^2} + 1) - \mathcal{L}(Z_{\mu,\sigma^2}) \right\|_{\mathrm{TV}} \le \frac{1}{2\sqrt{2\pi}\sigma}.$$

*Proof* For $Z$ a standard Gaussian random variable, from (7.5),

$$P(\overline{Z}_{\mu,\sigma^2} + 1 = k) = P\left(k - 1/2 < \sigma\left(Z + \frac{1}{2\sigma}\right) + \mu < k + 1/2\right),$$

while from (7.1),

$$P(Z_{\mu,\sigma^2} = k) = P(k - 1/2 < \sigma Z + \mu < k + 1/2).$$

Therefore,

$$\begin{aligned} \left\| \mathcal{L}(\overline{Z}_{\mu,\sigma^2} + 1) - \mathcal{L}(Z_{\mu,\sigma^2}) \right\|_{\mathrm{TV}} &\le \left\| \mathcal{L}\left(Z + \frac{1}{2\sigma}\right) - \mathcal{L}(Z) \right\|_{\mathrm{TV}} \\ &= 1 - 2\left(1 - \Phi\left(\frac{1}{4\sigma}\right)\right) \\ &= \Phi\left(\frac{1}{4\sigma}\right) - \Phi\left(-\frac{1}{4\sigma}\right) \\ &\le \frac{1}{2\sqrt{2\pi}\sigma}, \end{aligned}$$

where we have applied (7.4) to obtain the first equality. □

To prove Theorem 7.1 we require the following result.

**Lemma 7.2** *Let* $S = \sum_{i=1}^n X_i$ *where* $X_1, \ldots, X_n$ *are independent indicator random variables with* $P(X_i = 1) = p_i$ *for* $i = 1, \ldots, n$, *having variance* $\sigma^2 = \mathrm{Var}(S)$. *If* $I$ *is a random index with distribution*

$$P(I=i)=\frac{p_i(1-p_i)}{\sigma^2},$$

*then*

$$Ef(S)=Ef\left(S^{(I)}+1\right)-\frac{1}{\sigma^2}E\left(\sum_{i=1}^{n}(1-p_i)(X_i-p_i)\right)f(S),$$

*where* $S^{(i)}=S-X_i$ *for* $i=1,\dots,n$.

Assuming Lemma 7.2, the proof of Theorem 7.1 is somewhat direct.

*Proof of Theorem 7.1* For the application of Corollary 7.1 when the variables $X_1,\dots,X_n$ are indicators, note that

$$\gamma=\sum_{i=1}^{n}\frac{E|X_i-p_i|^3}{\sigma^3}\le\sum_{i=1}^{n}\frac{E(X_i-p_i)^2}{\sigma^3}=\frac{1}{\sigma}. \tag{7.9}$$

Standardizing $S$, write $W=(S-\mu)/\sigma$, that is,

$$W=\sum_{i=1}^{n}\xi_i \quad \text{where } \xi_i=\frac{X_i-p_i}{\sigma}.$$

With $I$ a random index with distribution $P(I=i)=\sigma_i^2/\sigma^2$, and $U$ a uniform variable on $[0,1]$, each independent of the other and of the remaining variables, by Lemma 2.8 and (2.52), the variable

$$W^*=W^{(I)}+\xi_I^* \quad \text{with } \xi_i^*=\frac{U-p_i}{\sigma}$$

has the $W$ zero bias distribution, where $W^{(i)}=W-\xi_i$. But now, by (7.6),

$$\begin{aligned}
P\left(W^*_{\mu,\sigma^2}=k\right)&=P\left(\frac{k-\mu}{\sigma}<W^*\le\frac{k+1-\mu}{\sigma}\right)\\
&=P\left(k-\mu<S^{(I)}-\mu_I+p_I+U-p_I\le k+1-\mu\right)\\
&=P\left(k<S^{(I)}+U\le k+1\right)\\
&=P\left(S^{(I)}=k\right),
\end{aligned}$$

that is, $W^*_{\mu,\sigma^2}=_d S^{(I)}$. Now, by (7.8) of Corollary 7.1 and (7.9), we obtain

$$\left\|\mathcal{L}\left(S^{(I)}\right)-\mathcal{L}(\overline{Z}_{\mu,\sigma^2})\right\|_{\mathrm{TV}}\le\left(5+3\sqrt{\frac{\pi}{8}}\right)/\sigma. \tag{7.10}$$

Applying Lemma 7.2 with $f(s)=\mathbf{1}(s=k)$ yields

$$P(S=k)=P\left(S^{(I)}=k-1\right)-\frac{1}{\sigma^2}E\left(\sum_{i=1}^{n}(1-p_i)(X_i-p_i)\right)\mathbf{1}(S=k),$$

and in particular

$$\begin{aligned}
&\sum_{k=0}^{\infty}\left|P(S=k)-P\big(S^{(I)}+1=k\big)\right| \\
&\quad\le \sum_{k=0}^{\infty}\frac{1}{\sigma^2}E\left|\sum_{i=1}^{n}(1-p_i)(X_i-p_i)\right|\mathbf{1}(S=k) \\
&\quad\le \frac{1}{\sigma^2}\left(\sum_{i=1}^{n}(1-p_i)^2E(X_i-p_i)^2\right)^{1/2} \\
&\quad\le \frac{1}{\sigma}.
\end{aligned}$$

Hence, summing and applying (7.3) we have

$$\big\|\mathcal{L}(S)-\mathcal{L}\big(S^{(I)}+1\big)\big\|_{\mathrm{TV}}\le\frac{1}{2\sigma}. \tag{7.11}$$

The proof is now completed by using (7.11), (7.10) and Lemma 7.1, along with the triangle inequality, to yield

$$\begin{aligned}
&\big\|\mathcal{L}(S)-\mathcal{L}(Z_{\mu,\sigma^2})\big\|_{\mathrm{TV}} \\
&\quad\le \big\|\mathcal{L}(S)-\mathcal{L}\big(S^{(I)}+1\big)\big\|_{\mathrm{TV}}+\big\|\mathcal{L}\big(S^{(I)}+1\big)-\mathcal{L}(\overline{Z}_{\mu,\sigma^2}+1)\big\|_{\mathrm{TV}} \\
&\qquad+\big\|\mathcal{L}(\overline{Z}_{\mu,\sigma^2}+1)-\mathcal{L}(Z_{\mu,\sigma^2})\big\|_{\mathrm{TV}} \\
&\quad\le \frac{1}{\sigma}\left(\frac{1}{2}+5+3\sqrt{\frac{\pi}{8}}+\frac{1}{2\sqrt{2\pi}}\right)\le 7.6/\sigma. \qquad\square
\end{aligned}$$

It remains to prove Lemma 7.2.

*Proof of Lemma 7.2* Using the fact that $X_1,\ldots,X_n$ are independent indicator variables,

$$\begin{aligned}
ESf(S) &= E\sum_{i=1}^{n}X_if(S)=E\sum_{i=1}^{n}p_if\big(S^{(i)}+1\big) \\
&= E\sum_{i=1}^{n}p_i(1-p_i)f\big(S^{(i)}+1\big)+E\sum_{i=1}^{n}p_i^2f\big(S^{(i)}+1\big) \\
&= E\sum_{i=1}^{n}p_i(1-p_i)f\big(S^{(i)}+1\big)+E\sum_{i=1}^{n}p_iX_if\big(S^{(i)}+1\big) \\
&= \sigma^2Ef\big(S^{(I)}+1\big)+E\left(\sum_{i=1}^{n}p_iX_if(S)\right),
\end{aligned}$$

while also

$$\begin{aligned}
ESf(S) &= E\big((S-\mu)f(S)+\mu f(S)\big) \\
&= E(S-\mu)f(S)+E\sum_{i=1}^{n}p_i(1-p_i)f(S)+E\sum_{i=1}^{n}p_i^2f(S).
\end{aligned}$$

Equating these expressions yields

$$\begin{aligned}
&\sum_{i=1}^{n} p_i(1-p_i)Ef(S)\\
&\quad = \sigma^2 Ef\big(S^{(I)}+1\big) + E\sum_{i=1}^{n} p_i X_i f(S) - E(S-\mu)f(S) - \sum_{i=1}^{n} p_i^2 Ef(S)\\
&\quad = \sigma^2 Ef\big(S^{(I)}+1\big) - E\left(\sum_{i=1}^{n} -p_i X_i + X_i - p_i + p_i^2\right) f(S)\\
&\quad = \sigma^2 Ef\big(S^{(I)}+1\big) - E\left(\sum_{i=1}^{n} (1-p_i)(X_i-p_i)\right) f(S),
\end{aligned}$$

and now dividing by $\sigma^2$ yields the result. □

## 7.2 Sum of Independent Integer Valued Random Variables

We now consider the more general situation where $S$ is the sum of independent integer-valued random variables $X_1, \ldots, X_n$ with distribution

$$P(X_i = k) = p_{ik} \quad \text{for } i = 1, \ldots, n \text{ and } k \in \mathbb{Z}.$$

The following result plays the role of Corollary 7.1 in the more general setting.

**Theorem 7.3** *Let $X_1, \ldots, X_n$ be independent integer valued random variables with $EX_i = \mu_i$, $\mathrm{Var}(X_i) = \sigma_i^2$, and finite third absolute central moments*

$$\gamma_i = E|X_i - \mu_i|^3/\sigma^3.$$

*Further, let $S = \sum_{i=1}^n X_i$, $\mu = \sum_{i=1}^n \mu_i$, $\sigma^2 = \sum_{i=1}^n \sigma_i^2$ and $\gamma = \sum_{i=1}^n \gamma_i$.*

*Then for each $i \in \{1, 2, \ldots, n\}$, the values $\alpha_{il}, l \in \mathbb{Z}$ given by*

$$\alpha_{il} = \begin{cases} \sum_{k=l+1}^{\infty} (k-\mu_i) p_{ik}/\sigma_i^2, & l \geq 0, \\ \sum_{k=-\infty}^{l} (\mu_i - k) p_{ik}/\sigma_i^2, & l \leq -1 \end{cases} \tag{7.12}$$

*are nonnegative and sum to one, and for $J(i)$ a random integer such that $J(1), J(2), \ldots, J(n), X_1, \ldots, X_n$ are independent with distribution*

$$P\big(J(i) = l\big) = \alpha_{il},$$

*when $I$ is a random index independent of $J(1), J(2), \ldots, J(n), X_1, \ldots, X_n$ with distribution*

$$P(I = i) = \frac{\sigma_i^2}{\sigma^2},$$

*we have*

$$\|\mathcal{L}(S^{(I)} + J(I)) - \mathcal{L}(\overline{Z}_{\mu,\sigma^2})\|_{\mathrm{TV}} \le \left(5 + 3\sqrt{\frac{\pi}{8}}\right)\gamma,$$

*where* $S^{(i)} = S - X_i$ *for* $i = 1, 2, \ldots, n$.

*Proof* Standardizing, write $W = (S - \mu)/\sigma$, so

$$W = \sum_{i=1}^{n} \xi_i \quad \text{where } \xi_i = \frac{X_i - \mu_i}{\sigma}.$$

Note that as $E(X_i - \mu_i) = 0$, for any integer $l$ we have

$$\begin{aligned}\sum_{k=-\infty}^{l} (\mu_i - k)p_{ik} &= -E(X_i - \mu_i)\mathbf{1}(X \le l)\\ &= E(X_i - \mu_i)\mathbf{1}(X \ge l + 1)\\ &= \sum_{k=l+1}^{\infty} (\mu_i - k)p_{ik}.\end{aligned} \tag{7.13}$$

For $i = 1, \ldots, n$ let $\xi_i^*$ have the $\xi_i$-zero biased distribution. From the formula (2.54) we see that the density of $\xi_i^*$ is constant over intervals of the form $((l - \mu_i)/\sigma_i, (l - \mu_i + 1)/\sigma_i)$ for any integer $l$, and so, again in view of (2.54), and applying (7.13) also, we have

$$\begin{aligned}P\big(l < \sigma\xi_i^* + \mu_i < l + 1\big) &= P\left(\frac{l - \mu_i}{\sigma} < \xi_i^* < \frac{l - \mu_i + 1}{\sigma}\right)\\ &= \frac{1}{\sigma}\frac{E\xi_i \mathbf{1}(\xi_i > (l - \mu_i)/\sigma)}{\sigma_i^2/\sigma^2}\\ &= \frac{1}{\sigma}\frac{\sum_{k=l+1}^{\infty}((k - \mu_i)/\sigma)P(\xi_i = (k - \mu_i)/\sigma)}{\sigma_i^2/\sigma^2}\\ &= \alpha_{il}.\end{aligned}$$

Hence the sequence $\alpha_{il}$ is nonnegative, sums to one over $l \in \mathbb{Z}$ for every $i = 1, \ldots, n$, and

$$P\big(l < \sigma\xi_i^* + \mu_i < l + 1\big) = P\big(J(i) = l\big). \tag{7.14}$$

By (7.8) of Corollary 7.1,

$$\|\mathcal{L}(W^*_{\mu,\sigma^2}) - \mathcal{L}(\overline{Z}_{\mu,\sigma^2})\|_{\mathrm{TV}} \le \left(5 + 3\sqrt{\frac{\pi}{8}}\right)\gamma,$$

where the discretized distributions $W^*_{\mu,\sigma^2}$ and $\overline{Z}_{\mu,\sigma^2}$ are given in (7.6) and (7.5), respectively. Hence, the theorem is proved upon noting that $W^* = W^{(I)} + \xi_I^*$ by Lemma 2.8, and therefore, by (7.14),

$$\begin{aligned} P\big(W^*_{\mu,\sigma^2}=k\big) &= P\left(\frac{k-\mu}{\sigma} < W^{(I)}+\xi_I^* \le \frac{k+1-\mu}{\sigma}\right) \\ &= P\big(k \le S^{(I)}+\sigma\xi_I^*+\mu_I < k+1\big) \\ &= P\big(S^{(I)}+J(I)=k\big). \end{aligned}$$ □

The next theorem gives a total variation bound between $\mathcal{L}(S)$ and $\mathcal{L}(Z_{\mu,\sigma^2})$.

**Theorem 7.4** *Let $X_1,\ldots,X_n$ be independent integer valued random variables with $EX_i=\mu_i$, $\mathrm{Var}(X_i)=\sigma_i^2$, and finite third absolute central moments*

$$\gamma_i = E|X_i-\mu_i|^3/\sigma^3.$$

*Further, let $S=\sum_{i=1}^n X_i$, $\mu=\sum_{i=1}^n \mu_i$, $\sigma^2=\sum_{i=1}^n \sigma_i^2$ and $\gamma=\sum_{i=1}^n \gamma_i$. Then*

$$\begin{aligned} &\big\|\mathcal{L}(S)-\mathcal{L}(Z_{\mu,\sigma^2})\big\|_{\mathrm{TV}} \\ &\quad\le \frac{3}{2}\sigma\sum_{i=1}^n\left(\gamma_i+\frac{2\sigma_i^2}{3\sigma^3}\right)\big\|\mathcal{L}\big(S^{(i)}\big)-\mathcal{L}\big(S^{(i)}+1\big)\big\|_{\mathrm{TV}}+\big(5+3\sqrt{\pi/8}\big)\gamma \\ &\qquad+\frac{1}{2\sqrt{2\pi}\sigma}. \end{aligned} \tag{7.15}$$

*Proof* First, we compute a total variation bound between the distributions of $S$ and $S^{(I)}+J(I)$ as follows,

$$\begin{aligned} &2\big\|\mathcal{L}(S)-\mathcal{L}\big(S^{(I)}+J(I)\big)\big\|_{\mathrm{TV}} \\ &\quad=\sum_{k\in\mathbb{Z}}\big|P\big(S^{(I)}+X_I=k\big)-P\big(S^{(I)}+J(I)=k\big)\big| \\ &\quad=\sum_{i=1}^n\frac{\sigma_i^2}{\sigma^2}\sum_{k\in\mathbb{Z}}\big|P\big(S^{(i)}+X_i=k\big)-P\big(S^{(i)}+J(i)=k\big)\big| \\ &\quad\le\sum_{i=1}^n\frac{\sigma_i^2}{\sigma^2}\sum_{k,k_1,k_2\in\mathbb{Z}}\big|P\big(S^{(i)}+k_1=k\big)-P\big(S^{(i)}+k_2=k\big)\big|P\big(X_i=k_1,J(i)=k_2\big) \\ &\quad\le\sum_{i=1}^n\frac{\sigma_i^2}{\sigma^2}2\big\|\mathcal{L}\big(S^{(i)}\big)-\mathcal{L}\big(S^{(i)}+1\big)\big\|_{\mathrm{TV}}\sum_{k_1,k_2\in\mathbb{Z}}|k_1-k_2|P\big(X_i=k_1,J(i)=k_2\big) \\ &\quad=\sum_{i=1}^n\frac{\sigma_i^2}{\sigma^2}2\big\|\mathcal{L}\big(S^{(i)}\big)-\mathcal{L}\big(S^{(i)}+1\big)\big\|_{\mathrm{TV}}E\big|X_i-J(i)\big|. \end{aligned} \tag{7.16}$$

Now, by (7.14), with $\xi_i^*$ a variable with the $\xi_i$-zero bias distribution,

$$\begin{aligned} &E\big|X_i-J(i)\big| \\ &\quad=\sigma E\left|\xi_i-\frac{J(i)-\mu_i}{\sigma}\right| \end{aligned}$$

$$\begin{aligned}
&\leq \sigma E|\xi_i| + \sigma E\left|\frac{J(i)-\mu_i}{\sigma}\right| \\
&= \sigma E|\xi_i| + \sigma \sum_{l\in\mathbb{Z}}\left|\frac{l-\mu_i}{\sigma}\right| P\bigl(J(i)=l\bigr) \\
&= \sigma E|\xi_i| + \sigma E \sum_{l\in\mathbb{Z}}\left|\frac{l-\mu_i}{\sigma}\right| \mathbf{1}\left(\frac{l-\mu_i}{\sigma} < \xi_i^* < \frac{l+1-\mu_i}{\sigma}\right) \\
&\leq \sigma E|\xi_i| + \sigma E \sum_{l\in\mathbb{Z}}\left(|\xi_i^*| + \frac{1}{\sigma}\right) \mathbf{1}\left(\frac{l-\mu_i}{\sigma} < \xi_i^* < \frac{l+1-\mu_i}{\sigma}\right) \\
&= \sigma E|\xi_i| + \sigma E|\xi_i^*| + 1,
\end{aligned}$$

where we obtain the last inequality by noting that $(l-\mu_i)/\sigma < \xi_i^* < (l+1-\mu_i)/\sigma$ implies

$$\frac{l-\mu_i}{\sigma} < |\xi_i^*| \quad \text{and} \quad -\left(\frac{l-\mu_i}{\sigma}\right) < |\xi_i^*| + \frac{1}{\sigma}.$$

Hence, from (7.16), we have

$$\begin{aligned}
&2\bigl\|\mathcal{L}(S) - \mathcal{L}\bigl(S^{(I)} + J(I)\bigr)\bigr\|_{\mathrm{TV}} \\
&\quad \leq \sum_{i=1}^n \frac{\sigma_i^2(\sigma E|\xi_i| + \sigma E|\xi_i^*| + 1)}{\sigma^2} 2\bigl\|\mathcal{L}\bigl(S^{(i)}\bigr) - \mathcal{L}\bigl(S^{(i)}+1\bigr)\bigr\|_{\mathrm{TV}} \\
&\quad \leq 3\sigma \sum_{i=1}^n \gamma_i \bigl\|\mathcal{L}\bigl(S^{(i)}\bigr) - \mathcal{L}\bigl(S^{(i)}+1\bigr)\bigr\|_{\mathrm{TV}} + \sum_{i=1}^n \frac{\sigma_i^2}{\sigma^2} 2\bigl\|\mathcal{L}\bigl(S^{(i)}\bigr) - \mathcal{L}\bigl(S^{(i)}+1\bigr)\bigr\|_{\mathrm{TV}},
\end{aligned} \tag{7.17}$$

where, for the final inequality, we note that

$$\frac{\sigma_i^2}{\sigma^2} E|\xi_i| = E\xi_i^2 E|\xi_i| \leq \gamma_i,$$

while from (2.57),

$$\frac{\sigma_i^2}{\sigma^2} E|\xi_i^*| = \mathrm{Var}(\xi_i) E|\xi_i^*| = \frac{1}{2} E|\xi_i^3| = \frac{1}{2}\gamma_i.$$

The proof may now be completed by applying the bound (7.17), along with Theorem 7.3, Lemma 7.1 and the triangle inequality, to yield

$$\begin{aligned}
&\bigl\|\mathcal{L}(S) - \mathcal{L}(Z_{\mu,\sigma^2})\bigr\|_{\mathrm{TV}} \\
&\quad \leq \bigl\|\mathcal{L}(S) - \mathcal{L}\bigl(S^{(I)} + J(I)\bigr)\bigr\|_{\mathrm{TV}} + \bigl\|\mathcal{L}\bigl(S^{(I)} + J(I)\bigr) - \mathcal{L}(\overline{Z}_{\mu,\sigma^2})\bigr\|_{\mathrm{TV}} \\
&\qquad + \bigl\|\mathcal{L}(\overline{Z}_{\mu,\sigma^2}) - \mathcal{L}(Z_{\mu,\sigma^2})\bigr\|_{\mathrm{TV}},
\end{aligned}$$

and hence the terms in (7.15). □

In order to apply Theorem 7.4, a bound on $\|\mathcal{L}(S^{(i)}) - \mathcal{L}(S^{(i)}+1)\|_{\mathrm{TV}}$ is required. If $\mathcal{L}(S^{(i)})$ is unimodel, then

$$
\begin{aligned}
&\big\|\mathcal{L}\big(S^{(i)}\big)-\mathcal{L}\big(S^{(i)}+1\big)\big\|_{\mathrm{TV}} \\
&\quad\le 2\max_{k\in\mathbb{Z}} P\big(S^{(i)}=k\big) \\
&\quad\le 2\max_{k\in\mathbb{Z}}\left(\left|P\big(S^{(i)}\le k\big)-\Phi\left(\frac{k-ES^{(i)}}{\sqrt{\sigma^2-\sigma_i^2}}\right)\right|\right. \\
&\qquad+\left|P\big(S^{(i)}\le k-1\big)-\Phi\left(\frac{k-1-ES^{(i)}}{\sqrt{\sigma^2-\sigma_i^2}}\right)\right| \\
&\qquad+\left.\left|\Phi\left(\frac{k-ES^{(i)}}{\sqrt{\sigma^2-\sigma_i^2}}\right)-\Phi\left(\frac{k-1-ES^{(i)}}{\sqrt{\sigma^2-\sigma_i^2}}\right)\right|\right) \\
&\quad\le 16.4\sum_{j\ne i}\gamma_j\frac{\sigma^3}{(\sigma^2-\sigma_i^2)^{3/2}}+\frac{2}{\sqrt{2\pi}}\frac{1}{\sqrt{\sigma^2-\sigma_i^2}}, \qquad (7.18)
\end{aligned}
$$

applying the Berry–Esseen theorem, twice, with the constant 4.1 from Chen and Shao (2001).

Recall that we say that the distribution $\mathcal{L}(X)$ is strongly unimodal if the convolution of $\mathcal{L}(X)$ and any unimodal distribution is unimodal. Theorem 4.8 in Dharmadhikari and Joag-Dev (1998) states that the distribution $\mathcal{L}(X)$ of an integer valued random variable $X$ is strongly unimodel if and only if $P(X=k)\ge P(X=k-1)P(X=k+1)$ for all integers $k$. Clearly then, the Bernoulli distributions are strongly unimodel, and hence also the Poisson binomial. In particular, for $S$ the sum of indicators $X_1,\ldots,X_n$ as in Theorem 7.1, and $S^i=S-X_i$ , by inequality (7.18) and the fact that $\gamma\le 1/\sigma$, from (7.9) in the proof of Theorem 7.1, we have

$$
\big\|\mathcal{L}\big(S^{(i)}\big)-\mathcal{L}\big(S^{(i)}+1\big)\big\|_{\mathrm{TV}}\le\frac{16.4\sigma^2}{(\sigma^2-1)^{3/2}}+\frac{2}{\sqrt{2\pi}}\frac{1}{\sqrt{\sigma^2-1}},
$$

which is of order $O(1/\sigma)$. Applying this inequality in the bound (7.15) of Theorem 7.4, we obtain a bound for $\|\mathcal{L}(S)-\mathcal{L}(Z_{\mu,\sigma^2})\|_{\mathrm{TV}}$ of the same order as in Theorem 7.1.

By Proposition 4.6 of Barbour and Xia (1999), for $S$ the sum of independent integer valued random variables,

$$
\begin{aligned}
&\max_{1\le i\le n}\big\|\mathcal{L}\big(S^{(i)}\big)-\mathcal{L}\big(S^{(i)}+1\big)\big\|_{\mathrm{TV}} \\
&\quad\le\left(\sum_{j\ne i}\min\big\{1-\|\mathcal{L}(X_j)-\mathcal{L}(X_j+1)\|_{\mathrm{TV}},1/2\big\}\right)^{-1/2},
\end{aligned}
$$

which is of order $1/\sqrt{n}$ if $\|\mathcal{L}(X_i)-\mathcal{L}(X_i+1)\|_{\mathrm{TV}}\le\alpha<1$ for $i=1,\ldots,n$.

A slightly better bound, for this same case, is given by Corollary 1.6 of Mattner and Roos (2007), which states that

$$\begin{aligned}
&\|\mathcal{L}(S^{(i)}) - \mathcal{L}(S^{(i)}+1)\|_{\mathrm{TV}} \\
&\quad \le \sqrt{\frac{2}{\pi}} \left( \frac{1}{4} + \sum_{j \neq i} \left(1 - \|\mathcal{L}(X_j), \mathcal{L}(X_j+1)\|_{\mathrm{TV}}\right) \right)^{-1/2}.
\end{aligned}$$

Under the hypotheses of Theorem 7.1,

$$1 - \|\mathcal{L}(X_j), \mathcal{L}(X_j+1)\|_{\mathrm{TV}} = \frac{1 - |2p_j - 1|}{2} \ge \sigma_j^2,$$

so $\|\mathcal{L}(S^{(i)}) - \mathcal{L}(S^{(i)}+1)\|_{\mathrm{TV}}$ has an upper bound of order $O(1/\sigma)$, which again leads to a bound on $\|\mathcal{L}(S) - \mathcal{L}(Z_{\mu,\sigma^2})\|_{\mathrm{TV}}$ of the same order as in Theorem 7.1.

# Chapter 8
# Non-uniform Bounds for Independent Random Variables

Throughout this chapter $\xi_1, \xi_2, \dots, \xi_n$ will denote independent random variables with zero means satisfying $\sum_{i=1}^n E\xi_i^2 = 1$, and $W$ will be their sum $\sum_{i=1}^n \xi_i$. Our goal is to prove non-uniform Berry–Esseen bounds of the form

$$\left|P(W \le z) - \Phi(z)\right| \le C\left(1 + |z|\right)^{-3}\gamma \quad \text{for all } z \in \mathbb{R}, \tag{8.1}$$

where $C$ is an absolute constant and $\gamma = \sum_{i=1}^n E|\xi_i|^3$. Non-uniform bounds were first obtained by Esseen (1945) for i.i.d. random variables, and the bound was later improved to $CnE|\xi_1|^3$ by Nagaev (1965) for the identically distributed case. That (8.1) holds for independent random variables was proved by Bikjalis (1966) using Fourier methods.

In this chapter, we follow Chen and Shao (2001, 2005) and approach (8.1) using Stein's method. To use ideas similar to those in the proof of Theorem 3.6 we first need to develop a non-uniform concentration inequality.

## 8.1 A Non-uniform Concentration Inequality

To prove our non-uniform concentration inequality we consider the truncated variables and the sums

$$\bar{xi}_i = \xi_i \mathbf{1}_{\{\xi_i \le 1\}}, \quad \bar{W} = \sum_{i=1}^n \bar{xi}_i \quad \text{and} \quad \bar{W}^{(i)} = \bar{W} - \bar{xi}_i \tag{8.2}$$

and recall the quantities given in (3.5),

$$\beta_2 = \sum_{i=1}^n E\xi_i^2 \mathbf{1}_{\{|\xi_i|>1\}} \quad \text{and} \quad \beta_3 = \sum_{i=1}^n E|\xi_i|^3 \mathbf{1}_{\{|\xi_i|\le 1\}}. \tag{8.3}$$

**Proposition 8.1** *For all real $a < b$ and $i = 1, \dots, n$,*

$$P\left(a \le \bar{W}^{(i)} \le b\right) \le 6\left(\min(1, b-a) + \beta_2 + \beta_3\right)e^{-a/2}. \tag{8.4}$$

L.H.Y. Chen et al., *Normal Approximation by Stein's Method*,
Probability and Its Applications,
DOI 10.1007/978-3-642-15007-4_8, © Springer-Verlag Berlin Heidelberg 2011

We prove Proposition 8.1 using the Bennett–Hoeffding inequality.

**Lemma 8.1** *For some $\alpha > 0$ let $\eta_1, \ldots, \eta_n$ be independent random variables satisfying $E\eta_i \le 0$, $\eta_i \le \alpha$ for all $1 \le i \le n$, and $\sum_{i=1}^n E\eta_i^2 \le B^2$. Then, with $S_n = \sum_{i=1}^n \eta_i$,*

$$Ee^{tS_n} \le \exp(\alpha^{-2}(e^{t\alpha} - 1 - t\alpha)B^2) \quad \text{for } t > 0, \tag{8.5}$$

$$P(S_n \ge x) \le \exp\left(-\frac{B^2}{\alpha^2}\left\{\left(1 + \frac{\alpha x}{B^2}\right)\log\left(1 + \frac{\alpha x}{B^2}\right) - \frac{\alpha x}{B^2}\right\}\right), \tag{8.6}$$

*and*

$$P(S_n \ge x) \le \exp\left(-\frac{x^2}{2(B^2 + \alpha x)}\right) \tag{8.7}$$

*for $x > 0$.*

*Proof* First, one may prove that $(e^s - 1 - s)/s^2$ is an increasing function for $s \in \mathbb{R}$, from which it follows that

$$e^{ts} \le 1 + ts + (ts)^2(e^{t\alpha} - 1 - t\alpha)/(t\alpha)^2 \tag{8.8}$$

for $s \le \alpha$, and $t > 0$. Using the properties of the $\eta_i$'s, for $t > 0$ we have

$$\begin{aligned} Ee^{tS_n} &= \prod_{i=1}^n Ee^{t\eta_i} \\ &\le \prod_{i=1}^n \left(1 + tE\eta_i + \alpha^{-2}(e^{t\alpha} - 1 - t\alpha)E\eta_i^2\right) \\ &\le \prod_{i=1}^n \left(1 + \alpha^{-2}(e^{t\alpha} - 1 - t\alpha)E\eta_i^2\right) \\ &\le \exp(\alpha^{-2}(e^{t\alpha} - 1 - t\alpha)B^2), \end{aligned}$$

proving (8.5).

To prove inequality (8.6), with $x > 0$ let

$$t = \frac{1}{\alpha}\log\left(1 + \frac{\alpha x}{B^2}\right).$$

Then by (8.5),

$$\begin{aligned} P(S_n \ge x) &\le e^{-tx}Ee^{tS_n} \\ &\le \exp(-tx + \alpha^{-2}(e^{t\alpha} - 1 - t\alpha)B^2) \\ &= \exp\left(-\frac{B^2}{\alpha^2}\left\{\left(1 + \frac{\alpha x}{B^2}\right)\log\left(1 + \frac{\alpha x}{B^2}\right) - \frac{\alpha x}{B^2}\right\}\right), \end{aligned}$$

demonstrating (8.6). Lastly, in view of the fact that

$$(1+s)\log(1+s)-s\geq\frac{s^2}{2(1+s)}$$

for $s>0$, (8.7) now follows. □

Although the hypotheses of Lemma 8.1 require that $\eta_i$ be bounded by above for all $i=1,\ldots,n$, the following result shows how the lemma may nevertheless be applied to unbounded variables.

**Lemma 8.2** *Let $\eta_1,\eta_2,\ldots,\eta_n$ be independent random variables satisfying $E\eta_i\leq 0$ for $1\leq i\leq n$ and $\sum_{i=1}^n E\eta_i^2\leq B^2$. Then,*

$$P(S_n\geq x)$$
$$\leq P\Bigl(\max_{1\leq i\leq n}\eta_i>\alpha\Bigr)+\exp\biggl(-\frac{B^2}{\alpha^2}\biggl\{\biggl(1+\frac{\alpha x}{B^2}\biggr)\log\biggl(1+\frac{\alpha x}{B^2}\biggr)-\frac{\alpha x}{B^2}\biggr\}\biggr), \qquad (8.9)$$

*for all $\alpha>0$ and $x>0$. In particular,*

$$P(S_n\geq x)\leq P\biggl(\max_{1\leq i\leq n}\eta_i>\frac{x\vee B}{p}\biggr)+e^p\biggl(1+\frac{x^2}{pB^2}\biggr)^{-p} \qquad (8.10)$$

*for $x>0$ and $p\geq 1$.*

*Proof* Letting $\bar{\eta}_i=\eta_i\mathbf{1}_{\{\eta_i\leq\alpha\}}$ we have

$$P(S_n\geq x)\leq P\Bigl(\max_{1\leq i\leq n}\eta_i>\alpha\Bigr)+P\Biggl(\sum_{i=1}^n\bar{\eta}_i\geq x\Biggr).$$

As $\bar{\eta}_i\leq\alpha$, $E\bar{\eta}_i=E\eta_i-E\eta_i\mathbf{1}_{\{\eta_i>\alpha\}}\leq 0$ and $\sum_{i=1}^n E\bar{\eta}_i^2\leq\sum_{i=1}^n E\eta_i^2\leq B^2$ we may now apply Lemma 8.1 to yield (8.9).

Inequality (8.10) is trivial when $0<x<B$ and $p\geq 1$, since then $e^p(1+\frac{x^2}{pB^2})^{-p}\geq 1$. For $x>B$, taking $\alpha=x/p$ in the second term in (8.9) yields

$$\begin{aligned}&\exp\biggl(-\frac{B^2}{\alpha^2}\biggl\{\biggl(1+\frac{\alpha x}{B^2}\biggr)\log\biggl(1+\frac{\alpha x}{B^2}\biggr)-\frac{\alpha x}{B^2}\biggr\}\biggr)\\&\quad\leq\exp\biggl(-\frac{x}{\alpha}\log\biggl(1+\frac{\alpha x}{B^2}\biggr)+p\biggr)=e^p\biggl(1+\frac{x^2}{pB^2}\biggr)^{-p}.\end{aligned}$$

This proves (8.10). □

We now proceed to the proof of Proposition 8.1, the non-uniform concentration inequality.

*Proof* As $\bar{\xi}_i, i=1,\ldots,n$ satisfies the hypotheses of Lemma 8.1 with $\alpha=1$ and $B^2=1$, it follows from (8.5) with $t=1/2$ that

$$\begin{aligned}P\bigl(a\leq\bar{W}^{(i)}\leq b\bigr)&\leq P\bigl(0\leq\bigl(\bar{W}^{(i)}-a\bigr)/2\bigr)\\&\leq e^{-a/2}Ee^{\bar{W}^{(i)}/2}\leq e^{-a/2}\exp\biggl(e^{1/2}-\frac{3}{2}\biggr)\leq 1.2\,e^{-a/2}.\end{aligned}$$

Thus, (8.4) holds if $6(\beta_2+\beta_3)\ge 1.2$ or $b-a\ge 1$. Hence, it suffices to prove the claim assuming that $(\beta_2+\beta_3)\le 1.2/6=0.2$ and $b-a<1$.

Similar to the proof of the concentration inequality in Lemma 3.1, define $\delta=(\beta_2+\beta_3)/2$ and set

$$f(w)=\begin{cases}0 & \text{if } w<a-\delta,\\ e^{w/2}(w-a+\delta) & \text{if } a-\delta\le w\le b+\delta,\\ e^{w/2}(b-a+2\delta) & \text{if } w>b+\delta.\end{cases} \tag{8.11}$$

Let

$$\bar{M}_i(t)=\xi_i(\mathbf{1}_{\{-\bar{x}i_i\le t\le 0\}}-\mathbf{1}_{\{0<t\le -\bar{x}i_i\}}) \quad \text{and} \quad \bar{M}^{(i)}(t)=\sum_{j\ne i}\bar{M}_j(t).$$

Clearly, $\bar{M}^{(i)}(t)\ge 0$, $f'(w)\ge 0$ and $f'(w)\ge e^{w/2}$ for $a-\delta<w<b+\delta$. Using these inequalities and the independence of $\xi_j$ and $\bar{W}^{(i)}-\bar{x}i_j$,

$$\begin{aligned}
E\{W^{(i)}f(\bar{W}^{(i)})\}&=\sum_{j\ne i}E\{\xi_j(f(\bar{W}^{(i)})-f(\bar{W}^{(i)}-\bar{x}i_j))\}\\
&=\sum_{j\ne i}E\int_{-\infty}^{\infty}f'(\bar{W}^{(i)}+t)\bar{M}_j(t)\,dt\\
&=E\int_{-\infty}^{\infty}f'(\bar{W}^{(i)}+t)\bar{M}^{(i)}(t)\,dt\\
&\ge E\mathbf{1}_{\{a\le\bar{W}^{(i)}\le b\}}\int_{|t|\le\delta}f'(\bar{W}^{(i)}+t)\bar{M}^{(i)}(t)\,dt\\
&\ge Ee^{(\bar{W}^{(i)}-\delta)/2}\mathbf{1}_{\{a\le\bar{W}^{(i)}\le b\}}\int_{|t|\le\delta}\bar{M}^{(i)}(t)\,dt\\
&=Ee^{(\bar{W}^{(i)}-\delta)/2}\mathbf{1}_{\{a\le\bar{W}^{(i)}\le b\}}\sum_{j\ne i}|\xi_j|\min(\delta,|\bar{x}i_j|)\\
&\ge e^{-\delta/2}(H_1-H_2),
\end{aligned} \tag{8.12}$$

where

$$H_1=E\{e^{\bar{W}^{(i)}/2}\mathbf{1}_{\{a\le\bar{W}^{(i)}\le b\}}\}\sum_{j\ne i}E\{|\xi_j|\min(\delta,|\bar{x}i_j|)\} \quad \text{and}$$

$$H_2=E\left\{e^{\bar{W}^{(i)}/2}\left|\sum_{j\ne i}|\xi_j|\min(\delta,|\bar{x}i_j|)-E|\xi_j|\min(\delta,|\bar{x}i_j|)\right|\right\}.$$

Applying the inequality $\min(x,y)\ge y-y^2/4x$ for $x>0$, $y>0$, we obtain

$$\begin{aligned}
\sum_{j\ne i}E|\xi_j|\min(\delta,|\bar{x}i_j|)&\ge\sum_{j\ne i}E|\xi_j|\mathbf{1}_{\{|\xi_j|\le 1\}}\min(\delta,|\xi_j|\mathbf{1}_{\{|\xi_j|\le 1\}})\\
&\ge\sum_{j=1}^{n}E|\xi_j|\mathbf{1}_{\{|\xi_j|\le 1\}}\min(\delta,|\xi_j|\mathbf{1}_{\{|\xi_j|\le 1\}})-\delta E|\xi_i|\mathbf{1}_{\{|\xi_i|\le 1\}}
\end{aligned}$$

$$\begin{aligned}
&\geq \sum_{j=1}^{n} \left\{ E\xi_j^2 \mathbf{1}_{\{|\xi_j|\leq 1\}} - \frac{E|\xi_j|^3 \mathbf{1}_{\{|\xi_j|\leq 1\}}}{4\delta} \right\} - \delta\beta_3^{1/3} \\
&= 1 - \frac{4\delta\beta_2 + \beta_3}{4\delta} - \delta\beta_3^{1/3} \\
&\geq 0.5 - 0.1(0.2)^{1/3} \geq 0.44,
\end{aligned} \tag{8.13}$$

where we have used $\delta = (\beta_2 + \beta_3)/2$, $\delta \leq 0.1$ and $\beta_2 + \beta_3 \leq 0.2$ in the final inequality. Hence

$$H_1 \geq 0.44\, e^{a/2} P\big(a \leq \bar{W}^{(i)} \leq b\big). \tag{8.14}$$

Turning now to $H_2$, by the Bennett–Hoeffding inequality (8.5) with $\alpha = t = B = 1$,

$$Ee^{\bar{W}^{(i)}} \leq \exp(e-2). \tag{8.15}$$

Hence, by the Cauchy–Schwarz inequality,

$$\begin{aligned}
H_2 &\leq \big(Ee^{\bar{W}^{(i)}}\big)^{1/2} \bigg( \mathrm{Var}\bigg( \sum_{j\neq i} |\xi_j| \min\big(\delta, |\bar{x}i_j|\big) \bigg) \bigg)^{1/2} \\
&\leq \exp(e/2 - 1) \bigg( \sum_{j\neq i} E\xi_j^2 \min\big(\delta, |\bar{x}i_j|\big)^2 \bigg)^{1/2} \\
&\leq \exp(e/2 - 1)\delta \bigg( \sum_{j\neq i} E\xi_j^2 \bigg)^{1/2} \\
&\leq \exp(e/2 - 1)\delta \leq 1.44\delta.
\end{aligned} \tag{8.16}$$

As to the left hand side of (8.12), we have

$$\begin{aligned}
E\big\{W^{(i)} f\big(\bar{W}^{(i)}\big)\big\} &\leq (b - a + 2\delta) E\big\{\big|W^{(i)}\big| e^{\bar{W}^{(i)}/2}\big\} \\
&\leq (b - a + 2\delta)\big(E\big|W^{(i)}\big|^2\big)^{1/2} \big(Ee^{\bar{W}^{(i)}}\big)^{1/2} \\
&\leq (b - a + 2\delta)\exp(e/2 - 1) \leq 1.44(b - a + 2\delta).
\end{aligned}$$

Combining the above inequalities and applying the bound $\delta \leq 0.1$ yields

$$\begin{aligned}
P\big(a \leq \bar{W}^{(i)} \leq b\big) &\leq e^{-a/2}\big(e^{\delta/2} 1.44(b - a + 2\delta) + 1.44\delta\big)/0.44 \\
&\leq e^{-a/2}\big(4(b - a) + 12\delta\big) \\
&\leq e^{-a/2}\big(4(b - a) + 6(\beta_2 + \beta_3)\big),
\end{aligned}$$

completing the proof of (8.4). □

## 8.2 Non-uniform Berry–Esseen Bounds

We begin our development of non-uniform bounds with the following lemma.

**Lemma 8.3** *Let $\xi_1, \ldots, \xi_n$ be independent random variables with mean zero satisfying $\sum_{i=1}^n \operatorname{Var}(\xi_i) = 1$ and let $W = \sum_{i=1}^n \xi_i$. Then for $z \ge 2$ and $p \ge 2$,*

$$
\begin{aligned}
&P\Big(W \ge z, \max_{1\le i\le n} \xi_i > 1\Big) \\
&\quad \le 2 \sum_{1\le i\le n} P\bigg(|\xi_i| > \frac{z}{2p}\bigg) + e^p\big(1 + z^2/(4p)\big)^{-p}\beta_2
\end{aligned}
$$

*whenever $\beta_2$, given by* (8.3), *is bounded by* 1.

*Proof* Beginning with the left hand side, we have

$$
\begin{aligned}
&P\Big(W \ge z, \max_{1\le i\le n} \xi_i > 1\Big) \\
&\quad \le \sum_{i=1}^n P(W \ge z, \xi_i > 1) \\
&\quad \le P\Big(\max_{1\le i\le n} \xi_i > z/2\Big) + \sum_{i=1}^n P\big(W^{(i)} \ge z/2, \xi_i > 1\big) \\
&\quad \le \sum_{i=1}^n P(|\xi_i| > z/p) + \sum_{i=1}^n P\big(W^{(i)} \ge z/2\big) P(\xi_i > 1) \\
&\quad \le \sum_{i=1}^n P(|\xi_i| > z/p) \\
&\qquad + \sum_{i=1}^n \Big\{P\Big(\max_{1\le i\le n} |\xi_i| > z/(2p)\Big) + e^p\big(1 + z^2/(4p)\big)^{-p}\Big\} P(\xi_i > 1) \\
&\qquad \text{[by (8.10)]} \\
&\quad \le \sum_{i=1}^n P(|\xi_i| > z/p) + \Big\{P\Big(\max_{1\le i\le n} |\xi_i| > z/(2p)\Big) + e^p\big(1 + z^2/(4p)\big)^{-p}\Big\}\beta_2 \\
&\quad \le 2 \sum_{1\le i\le n} P\big(|\xi_i| > z/(2p)\big) + e^p\big(1 + z^2/(4p)\big)^{-p}\beta_2, \qquad (8.17)
\end{aligned}
$$

since $\beta_2 \le 1$. □

We are now ready to prove the following non-uniform Berry–Esseen inequality, generalizing (8.1).

**Theorem 8.1** *For every $p \ge 2$ there exists a finite constant $C_p$ depending only on $p$ such that for all $z \in \mathbb{R}$*

$$
\begin{aligned}
&\big|P(W \le z) - \Phi(z)\big| \\
&\quad \le 2 \sum_{i=1}^n P\bigg(|\xi_i| > \frac{1 \vee |z|}{2p}\bigg) + C_p\big(1 + |z|\big)^{-p}(\beta_2 + \beta_3), \qquad (8.18)
\end{aligned}
$$

*where $\beta_2$ and $\beta_3$ are given in* (8.3).

Inequality (8.1) follows from Theorem 8.1 using $(1+|z|)/2 \le 1 \vee |z|$, and that we may bound the first sum by

$$\sum_{i=1}^{n} P\left(|\xi_i| > \frac{1+|z|}{4p}\right) \le \left(\frac{4p}{1+|z|}\right)^p \sum_{i=1}^{n} E|\xi_i|^p,$$

and the remaining terms, since $p \ge 2$, using

$$\beta_2 = \sum_{i=1}^{n} E\xi_i^2 \mathbf{1}_{\{|\xi_i|>1\}} \le \sum_{i=1}^{n} E|\xi_i|^p \mathbf{1}_{\{|\xi_i|>1\}} \le \sum_{i=1}^{n} E|\xi_i|^p,$$

and

$$\beta_3 = \sum_{i=1}^{n} E|\xi_i|^3 \mathbf{1}_{\{|\xi_i|\le 1\}} \le \sum_{i=1}^{n} E|\xi_i|^{3\wedge p} \mathbf{1}_{\{|\xi_i|\le 1\}} \le \sum_{i=1}^{n} E|\xi_i|^{3\wedge p}.$$

Hence, Theorem 8.1 implies that there exists $C_p$ such that

$$\left|P(W \le z) - \Phi(z)\right| \le C_p\left(1+|z|\right)^{-p} \sum_{i=1}^{n} \left\{E|\xi_i|^p + E|\xi_i|^{3\wedge p}\right\}. \tag{8.19}$$

*Proof* By replacing $W$ by $-W$ it suffices to consider $z \ge 0$. As $W$ is the sum of independent variables with mean zero and $\mathrm{Var}(W) \le 1$, by (8.10) with $B = 1$, for all $p \ge 2$ we obtain

$$P(W > z) \le P\left(\max_{1\le i\le n} |\xi_i| > \frac{z \vee 1}{p}\right) + e^p\left(1 + \frac{z^2}{p}\right)^{-p}.$$

Thus (8.18) holds if $\beta_2 + \beta_3 \ge 1$, and it suffices to prove the claim assuming that $\beta_2 + \beta_3 < 1$. We may also assume the lower bound $z \ge 2$ holds, as the fact that (8.18) holds for $z$ over any bounded range follows from the uniform bound (3.31) by choosing $C_p$ sufficiently large.

Let $\bar{x}i_i$, $\bar{W}$ and $\bar{W}^{(i)}$ be defined as in (8.2). We first prove that $P(W > z)$ and $P(\bar{W} > z)$ are close and then prove a non-uniform bound for $\bar{W}$. Observing that

$$\begin{aligned}
\{W > z\} &= \left\{W > z, \max_{1\le i\le n} \xi_i > 1\right\} \cup \left\{W > z, \max_{1\le i\le n} \xi_i \le 1\right\} \\
&\subset \left\{W > z, \max_{1\le i\le n} \xi_i > 1\right\} \cup \{\bar{W} > z\},
\end{aligned} \tag{8.20}$$

we obtain the upper bound

$$P(W > z) \le P(\bar{W} > z) + P\left(W > z, \max_{1\le i\le n} \xi_i > 1\right). \tag{8.21}$$

Clearly $W \ge \bar{W}$, yielding the lower bound,

$$P(\bar{W} > z) \le P(W > z). \tag{8.22}$$

Hence, in view of Lemma 8.3, to prove (8.18) it suffices to show that

$$\left|P(\bar{W} \le z) - \Phi(z)\right| \le Ce^{-z/2}(\beta_2 + \beta_3) \tag{8.23}$$

for some absolute constant $C$.

For $z \in \mathbb{R}$ let $f_z$ be the solution to the Stein equation (2.2), and define

$$\bar{K}_i(t) = E\left\{\bar{x}i_i(\mathbf{1}_{\{0 \le t \le \bar{x}i_i\}} - \mathbf{1}_{\{\bar{x}i_i \le t < 0\}})\right\}.$$

Reasoning as in the proof of (2.27), noting here that $\bar{x}i_i \le 1$, and $E\bar{x}i_i \le 0$, and does not equal zero in general, using the independence of $\bar{x}i_i$ and $\bar{W}^{(i)}$, we obtain

$$E\left\{\bar{W} f_z(\bar{W})\right\} = \sum_{i=1}^n \int_{-\infty}^1 Ef_z'\left(\bar{W}^{(i)} + t\right)\bar{K}_i(t)\,dt + \sum_{i=1}^n E\bar{x}i_i Ef_z\left(\bar{W}^{(i)}\right).$$

Additionally, from

$$\sum_{i=1}^n \int_{-\infty}^1 \bar{K}_i(t)\,dt = \sum_{i=1}^n E\bar{x}i_i^2 = 1 - \sum_{i=1}^n E\xi_i^2 \mathbf{1}_{\{\xi_i > 1\}}, \tag{8.24}$$

recalling $E\xi_i = 0$ we have

$$\begin{aligned}
&P(\bar{W} \le z) - \Phi(z)\\
&\quad = Ef_z'(\bar{W}) - E\left\{\bar{W} f_z(\bar{W})\right\}\\
&\quad = \sum_{i=1}^n E\left\{\xi_i^2 \mathbf{1}_{\{\xi_i > 1\}}\right\} Ef_z'(\bar{W})\\
&\qquad + \sum_{i=1}^n \int_{-\infty}^1 E\left[f_z'\left(\bar{W}^{(i)} + \bar{x}i_i\right) - f_z'\left(\bar{W}^{(i)} + t\right)\right]\bar{K}_i(t)\,dt\\
&\qquad + \sum_{i=1}^n E\{\xi_i \mathbf{1}_{\{\xi_i > 1\}}\} Ef_z\left(\bar{W}^{(i)}\right)\\
&\quad := R_1 + R_2 + R_3.
\end{aligned} \tag{8.25}$$

By (2.80), (2.8) and (8.5),

$$\begin{aligned}
E\left|f_z'(\bar{W})\right| &= E\left\{\left|f_z'(\bar{W})\right|\mathbf{1}_{\{\bar{W} \le z/2\}}\right\} + E\left\{\left|f_z'(\bar{W})\right|\mathbf{1}_{\{\bar{W} > z/2\}}\right\}\\
&\le \left(\sqrt{2\pi}(z/2)e^{z^2/8} + 1\right)\left(1 - \Phi(z)\right) + P(\bar{W} > z/2)\\
&\le \left(\sqrt{2\pi}(z/2)e^{z^2/8} + 1\right)\left(1 - \Phi(z)\right) + e^{-z/2}Ee^{\bar{W}}\\
&\le Ce^{-z/2}
\end{aligned}$$

by a standard tail bound on $1 - \Phi(z)$, and hence

$$|R_1| \le C\beta_2 e^{-z/2}. \tag{8.26}$$

Similarly, using (2.3) we obtain $Ef_z(\bar{W}^{(i)}) \le Ce^{-z/2}$ and

$$|R_3| \le C\beta_2 e^{-z/2}. \tag{8.27}$$

To estimate $R_2$, use (2.2) to write

$$R_2 = R_{2,1} + R_{2,2},$$

where

$$R_{2,1} = \sum_{i=1}^{n} \int_{-\infty}^{1} E(\mathbf{1}_{\{\bar{W}^{(i)}+\bar{x}i_i \le z\}} - \mathbf{1}_{\{\bar{W}^{(i)}+t \le z\}}) \bar{K}_i(t)\,dt$$

and

$$R_{2,2} = \sum_{i=1}^{n} \int_{-\infty}^{1} E\big[\big(\bar{W}^{(i)} + \bar{x}i_i\big) f_z\big(\bar{W}^{(i)} + \bar{x}i_i\big) - \big(\bar{W}^{(i)} + t\big) f_z\big(\bar{W}^{(i)} + t\big)\big] \bar{K}_i(t)\,dt.$$

By Proposition 8.1, with $C$ not necessarily the same at each occurrence,

$$\begin{aligned}
R_{2,1} &\le \sum_{i=1}^{n} \int_{-\infty}^{1} E\big\{\mathbf{1}_{\{\bar{x}i_i \le t\}} P\big(z - t < \bar{W}^{(i)} \le z - \bar{x}i_i \,|\, \bar{x}i_i\big)\big\} \bar{K}_i(t)\,dt \\
&\le C \sum_{i=1}^{n} \int_{-\infty}^{1} e^{-(z-t)/2} E\big(\min\big(1, |\bar{x}i_i| + |t|\big) + \beta_2 + \beta_3\big) \bar{K}_i(t)\,dt \\
&\le C e^{-z/2} \sum_{i=1}^{n} \int_{-\infty}^{1} \big(E\big(\min\big(1, |\bar{x}i_i| + |t|\big)\big) + \beta_2 + \beta_3\big) \bar{K}_i(t)\,dt.
\end{aligned}$$

From (8.24),

$$C e^{-z/2} \sum_{i=1}^{n} \int_{-\infty}^{1} (\beta_2 + \beta_2) \bar{K}_i(t) dt \le C e^{-z/2} (\beta_2 + \beta_3).$$

Hence, to prove

$$R_{2,1} \le C e^{-z/2} (\beta_2 + \beta_3) \tag{8.28}$$

it suffices to show that

$$\sum_{i=1}^{n} \int_{-\infty}^{1} E\big(\min\big(1, |\bar{x}i_i| + |t|\big)\big) \bar{K}_i(t)\,dt \le C(\beta_2 + \beta_3). \tag{8.29}$$

As

$$\mathbf{1}_{\{0 \le t \le \bar{x}i_i\}} + \mathbf{1}_{\{\bar{x}i_i \le t < 0\}} \le \mathbf{1}_{\{|t| \le |\bar{x}i_i|\}}$$

we have

$$\bar{K}_i(t) \le E\big(|\bar{x}i_i| \mathbf{1}_{\{|t| \le |\bar{x}i_i|\}}\big).$$

Since both

$$\min\big(1, |\bar{x}i_i| + |t|\big) \quad \text{and} \quad |\bar{x}i_i| \mathbf{1}_{\{|t| \le |\bar{x}i_i|\}}$$

are increasing functions of $|\bar{x}i_i|$ they are positively correlated, therefore

$$\begin{aligned}
E\big(\min(1,|\bar{\xi}_i|+|t|)\big)\bar{K}_i(t) &\le E\big(\min(1,|\bar{\xi}_i|+|t|)\big)E\big(|\bar{\xi}_i|\mathbf{1}_{\{|t|\le|\bar{\xi}_i|\}}\big)\\
&\le E\big(\min(1,|\bar{\xi}_i|+|t|)|\bar{\xi}_i|\mathbf{1}_{\{|t|\le|\bar{\xi}_i|\}}\big)\\
&\le 2E\big(\min(1,|\bar{\xi}_i|)|\bar{\xi}_i|\mathbf{1}_{\{|t|\le|\bar{\xi}_i|\}}\big).
\end{aligned}$$

Hence

$$\begin{aligned}
\sum_{i=1}^n\int_{-\infty}^1 E\big(\min(1,|\bar{\xi}_i|+|t|)\big)\bar{K}_i(t)\,dt &\le 4\sum_{i=1}^n E\min\big(1,|\bar{\xi}_i|\big)|\bar{\xi}_i|^2\\
&\le 4\sum_{i=1}^n E\min\big(1,|\xi_i|\big)|\xi_i|^2\\
&= 4\left(\sum_{i=1}^n E\xi_i^2\mathbf{1}_{|\xi_i|>1}+\sum_{i=1}^n E\big|\xi_i^3\big|\mathbf{1}_{|\xi_i|\le 1}\right)\\
&= 4(\beta_2+\beta_3),
\end{aligned}$$

proving (8.29), and therefore (8.28). Similarly we may show

$$R_{2,1}\ge -Ce^{-z/2}(\beta_2+\beta_3).$$

Using Lemma 8.4 below for the second inequality, by the monotonicity of $wf_z(w)$ provided by (2.6), it follows that

$$\begin{aligned}
R_{2,2} &\le \sum_{i=1}^n\int_{-\infty}^1 E\big\{\mathbf{1}_{\{t\le\bar{\xi}_i\}}\big[E\big(\{\bar{W}^{(i)}+\bar{\xi}_i\}f_z\big(\bar{W}^{(i)}+\bar{\xi}_i\big)\,|\,\bar{\xi}_i\big)\\
&\qquad - E\big(\bar{W}^{(i)}+t\big)f_z\big(\bar{W}^{(i)}+t\big)\big]\big\}\bar{K}_i(t)\,dt\\
&\le Ce^{-z/2}\sum_{i=1}^n\int_{-\infty}^1 E\big(\min\big(1,|\bar{\xi}_i|+|t|\big)\big)\bar{K}_i(t)\,dt\\
&\le Ce^{-z/2}(\beta_2+\beta_3),
\end{aligned}\tag{8.30}$$

where we have applied (8.29) for the last inequality. Therefore

$$R_2\le Ce^{-z/2}(\beta_2+\beta_3).\tag{8.31}$$

Similarly, we may demonstrate the lower bound

$$R_2\ge -Ce^{-z/2}(\beta_2+\beta_3),$$

thus proving the theorem. □

It remains to prove the following lemma.

**Lemma 8.4** *With $\bar{W}^{(i)}$ as in* (8.2) *and $f_z$ the solution to the Stein equation* (2.2) *for $z>0$, for all $s\le t\le 1$*

$$\begin{aligned}
&E\big\{\big(\bar{W}^{(i)}+t\big)f_z\big(\bar{W}^{(i)}+t\big)-\big(\bar{W}^{(i)}+s\big)f_z\big(\bar{W}^{(i)}+s\big)\big\}\\
&\quad\le Ce^{-z/2}\min\big(1,|s|+|t|\big).
\end{aligned}\tag{8.32}$$

*Proof* Let $g(w) = (wf_z(w))'$. Then

$$\begin{aligned} &E\{(\bar{W}^{(i)}+t)f_z(\bar{W}^{(i)}+t) - (\bar{W}^{(i)}+s)f_z(\bar{W}^{(i)}+s)\} \\ &\quad = \int_s^t Eg(\bar{W}^{(i)}+u)du. \end{aligned} \tag{8.33}$$

From the formula (2.81) for $g(w)$ and the bound

$$\sqrt{2\pi}(1+w^2)e^{w^2/2}\Phi(w) + w \le \frac{2}{1+|w|^3} \quad \text{for } w \le 0, \tag{8.34}$$

from (5.4) of Chen and Shao (2001), we obtain the $w \le 0$ case of the first inequality in

$$g(w) \le \begin{cases} \frac{4(1+z^2)(1+z^3)}{1+|w|^3} e^{z^2/8}(1-\Phi(z)) & \text{if } w \le z/2, \\ 8(1+z^2)e^{z^2/2}(1-\Phi(z)) & \text{if } z/2 < w < z \text{ or } w > z. \end{cases}$$

For $0 < w \le z/2$ we apply the simpler inequality

$$\sqrt{2\pi}(1+w^2)e^{w^2/2}\Phi(w) + w \le 3(1+z^2)e^{z^2/8} + z \le 4(1+z^2)e^{z^2/8},$$

and the same reasoning yields the case $z/2 \le w < z$. For $w > z$ we apply (8.34) with $-w$ replacing $w$, and the inequality

$$1-\Phi(z) \ge \frac{e^{-z^2/2}}{4(1+z^2)} \quad \text{for } z > 0.$$

Hence, for any $u \in [s,t]$, we have

$$\begin{aligned} Eg(\bar{W}^{(i)}+u) &= E\{g(\bar{W}^{(i)}+u)\mathbf{1}_{\{\bar{W}^{(i)}+u \le z/2\}}\} \\ &\quad + E\{g(\bar{W}^{(i)}+u)\mathbf{1}_{\{\bar{W}^{(i)}+u > z/2\}}\} \\ &\le E\left(\frac{4(1+z^2)(1+z^3)}{1+|\bar{W}^{(i)}+u|^3}\right)e^{z^2/8}(1-\Phi(z)) \\ &\quad + 8(1+z^2)e^{z^2/2}(1-\Phi(z))P(\bar{W}^{(i)}+u > z/2) \\ &\le Ce^{-z/2}E(1+|\bar{W}^{(i)}+u|^3)^{-1} + C(1+z)e^{-z+2u}Ee^{2\bar{W}^{(i)}} \\ &\le Ce^{-z/2}E(1+|\bar{W}^{(i)}+u|^3)^{-1} + C(1+z)e^{-z+2u} \quad \text{[by (8.5)]} \\ &\le Ce^{-z/2} \end{aligned} \tag{8.35}$$

since $u \le t \le 1$.

Hence, for all $s \le t \le 1$ we have

$$\int_s^t Eg(\bar{W}^{(i)}+u)du \le Ce^{-z/2}(t-s) \le Ce^{-z/2}(|t|+|s|), \tag{8.36}$$

while also, now using that $g(w) \ge 0$ by (2.6), from (8.35),

$$
\begin{aligned}
\int_s^t Eg(\bar{W}^{(i)} + u)du &\le \int_{-\infty}^1 Eg(\bar{W}^{(i)} + u)du \\
&\le Ce^{-z/2} E\int_{-\infty}^1 \frac{1}{1+|\bar{W}^{(i)}+u|^3}du + Ce^{-z/2}\int_{-\infty}^1 e^{2u}du \\
&\le Ce^{-z/2}. \qquad (8.37)
\end{aligned}
$$

Using (8.36) and (8.37) in (8.33) now completes the proof. □

# Chapter 9
# Uniform and Non-uniform Bounds Under Local Dependence

In this chapter we continue the study of Stein's method under the types of local dependence that was first considered in Sect. 4.7 for the $L^1$ distance, and also in Sect. 6.2 for the $L^\infty$ distance. We follow the work of Chen and Shao (2004), with the aim of establishing both uniform and non-uniform Berry–Esseen bounds having optimal asymptotic rates under various local dependence conditions.

Throughout this chapter, $\mathcal{J}$ will denote an index set of cardinality $n$ and $\{\xi_i, i \in \mathcal{J}\}$ a random field, that is, an indexed collection of random variables, with zero means and finite variances. Define $W = \sum_{i\in\mathcal{J}} \xi_i$ and assume that $\operatorname{Var}(W) = 1$. For $A \subset \mathcal{J}$, let $\xi_A = \{\xi_i, i \in A\}$, $A^c = \{j \in \mathcal{J}\colon\ j \notin A\}$ and $|A|$ the cardinality of $A$. We introduce the following four local dependence conditions, the first two of which appeared in Sect. 4.7. In each, the set $A_i$ can be thought of as a neighborhood of dependence for $\xi_i$.

(LD1) For each $i \in \mathcal{J}$ there exists $A_i \subset \mathcal{J}$ such that $\xi_i$ and $\xi_{A_i^c}$ are independent.

(LD2) For each $i \in \mathcal{J}$ there exist $A_i \subset B_i \subset \mathcal{J}$ such that $\xi_i$ is independent of $\xi_{A_i^c}$ and $\xi_{A_i}$ is independent of $\xi_{B_i^c}$.

(LD3) For each $i \in \mathcal{J}$ there exist $A_i \subset B_i \subset C_i \subset \mathcal{J}$ such that $\xi_i$ is independent of $\xi_{A_i^c}$, $\xi_{A_i}$ is independent of $\xi_{B_i^c}$, and $\xi_{B_i}$ is independent of $\xi_{C_i^c}$.

(LD4*) For each $i \in \mathcal{J}$ there exist $A_i \subset B_i \subset B_i^* \subset C_i^* \subset D_i^* \subset \mathcal{J}$ such that $\xi_i$ is independent of $\xi_{A_i^c}$, $\xi_{A_i}$ is independent of $\xi_{B_i^c}$, $\xi_{A_i}$ is independent of $\{\xi_{A_j}, j \in B_i^{*c}\}$, $\{\xi_{A_j}, j \in B_i^*\}$ is independent of $\{\xi_{A_j}, j \in C_i^{*c}\}$, and $\{\xi_{A_j}, j \in C_i^*\}$ is independent of $\{\xi_{A_j}, j \in D_i^{*c}\}$.

It is clear that each condition is implied by the one that follows it, that is, that (LD4*) $\Rightarrow$ (LD3) $\Rightarrow$ (LD2) $\Rightarrow$ (LD1). Roughly speaking, (LD4*) is a version of (LD3) obtained by considering $\{\xi_{A_i}, i \in \mathcal{J}\}$ as the basic random variables in the field. Though the conditions listed are increasingly more restrictive, in many cases the weakest one, (LD1), actually implies that (LD2), (LD3) or (LD4*) hold upon taking $B_i = \bigcup_{j\in A_i} A_j$, $C_i = \bigcup_{j\in B_i} A_j$, $B_i^* = \bigcup_{j\in A_i} B_j$, $C_i^* = \bigcup_{j\in B_i^*} B_j$ and $D_i^* = \bigcup_{j\in C_i^*} B_j$. For example, (LD1) implies (LD4*) when $\{\xi_i, i \in \mathcal{J}\}$ is the $m$-dependent random field considered at the end of the next section. We note that

L.H.Y. Chen et al., *Normal Approximation by Stein's Method*,
Probability and Its Applications,
DOI 10.1007/978-3-642-15007-4_9, © Springer-Verlag Berlin Heidelberg 2011

Bulinski and Suquet (2002) obtain results for random fields having both negative and positive dependence by Stein's method.

## 9.1 Uniform and Non-uniform Berry–Esseen Bounds

We first present a general uniform Berry–Esseen bound under assumption (LD2). Recall that $|\mathcal{J}| = n$.

**Theorem 9.1** *Let $p \in (2, 4]$ and assume that there exists some $\kappa$ such that* (LD2) *is satisfied with $|\mathcal{N}(B_i)| \le \kappa$ for all $i \in \mathcal{J}$, where $\mathcal{N}(B_i) = \{j \in \mathcal{J}: B_j \cap B_i \ne \emptyset\}$. Then*

$$\sup_{z\in\mathbb{R}} \left| P(W \le z) - \Phi(z) \right| \\ \le (13 + 11\kappa) \sum_{i\in\mathcal{J}} \left( E|\xi_i|^{3\wedge p} + E|\eta_i|^{3\wedge p} \right) + 2.5\left( \kappa \sum_{i\in\mathcal{J}} \left( E|\xi_i|^p + E|\eta_i|^p \right) \right)^{1/2},$$

*where $\eta_i = \sum_{j\in A_i} \xi_j$. In particular, if there is some $\theta > 0$ such that $E|\xi_i|^p + E|\eta_i|^p \le \theta^p$ for all $i \in \mathcal{J}$, then*

$$\sup_{z\in\mathbb{R}} \left| P(W \le z) - \Phi(z) \right| \le (13 + 11\kappa) n \theta^{3\wedge p} + 2.5 \theta^{p/2} \sqrt{\kappa n}.$$

In typical asymptotic regimes $\kappa$ is bounded and $\theta$ is of order of $n^{-1/2}$, yielding the order $\kappa n \theta^{3\wedge p} + \theta^{p/2}\sqrt{\kappa n} = O(n^{-(p-2)/4})$. When fourth moments exist we may take $p = 4$ and obtain the best possible order of $n^{-1/2}$. Assuming the stronger local dependence condition (LD3) allows us to relax the moment assumptions.

**Theorem 9.2** *Let $p \in (2, 3]$ and assume that there exists some $\kappa$ such that* (LD3) *is satisfied with $|\mathcal{N}(C_i)| \le \kappa$ for all $i \in \mathcal{J}$, where $\mathcal{N}(C_i) = \{j \in \mathcal{J}: C_i \cap B_j \ne \emptyset\}$. Then*

$$\sup_{z\in\mathbb{R}} \left| P(W \le z) - \Phi(z) \right| \le 75 \kappa^{p-1} \sum_{i\in\mathcal{J}} E|\xi_i|^p.$$

Under the strongest condition (LD4*) we have the following general non-uniform bound.

**Theorem 9.3** *Let $p \in (2, 3]$ and assume that* (LD4*) *is satisfied with $\kappa = \max_{i\in\mathcal{J}} \max(|D_i^*|, |\{j: i \in D_j^*\}|)$. Then for all $z \in \mathbb{R}$,*

$$\left| P(W \le z) - \Phi(z) \right| \le C \kappa^p (1 + |z|)^{-p} \sum_{i\in\mathcal{J}} E|\xi_i|^p,$$

*where $C$ is an absolute constant.*

The above results can immediately be applied to $m$-dependent random fields, indexed, for example, by elements of $\mathbb{N}^d$, the $d$-dimensional space of positive integers. Letting $|i-j|$ denote the $L^\infty$ distance

$$|i-j| = \max_{1\le l\le d} |i_l - j_l|$$

between two points $i=(i_1,\ldots,i_d)$ and $j=(j_1,\ldots,j_d)$ in $\mathbb{N}^d$, define the distance $\rho(A,B)$ between two subsets $A$ and $B$ of $\mathbb{N}^d$ by

$$\rho(A,B) = \inf\{|i-j|: i\in A, j\in B\}.$$

For a given subset $\mathcal{J}\subset\mathbb{N}^d$, a set of random variables $\{\xi_i, i\in\mathcal{J}\}$ is said to be an $m$-dependent random field if $\{\xi_i, i\in A\}$ and $\{\xi_j, j\in B\}$ are independent whenever $\rho(A,B)>m$, for any subsets $A$ and $B$ of $\mathcal{J}$.

It is readily verified that if $\{\xi_i, i\in\mathcal{J}\}$ is an $m$-dependent random field then (LD3) and (LD4*) are satisfied by choosing $A_i=\{j\in\mathcal{J}: |j-i|\le m\}$, $B_i=\{j\in\mathcal{J}: |j-i|\le 2m\}$, $C_i=\{j\in\mathcal{J}: |j-i|\le 3m\}$, $B_i^*=\{j\in\mathcal{J}: |j-i|\le 3m\}$, $C_i^*=\{j\in\mathcal{J}: |j-i|\le 5m\}$, and $D_i^*=\{j\in\mathcal{J}: |j-i|\le 7m\}$. Hence, Theorems 9.2 and 9.3 yield the following uniform and non-uniform bounds.

**Theorem 9.4** *If $\{\xi_i, i\in\mathcal{J}\}$ is a zero mean $m$-dependent random field then for all $p\in(2,3]$*

$$\sup_{z\in\mathbb{R}}\bigl|P(W\le z)-\Phi(z)\bigr| \le 75(10m+1)^{(p-1)d}\sum_{i\in\mathcal{J}} E|\xi_i|^p \tag{9.1}$$

*and for all $z\in\mathbb{R}$,*

$$\bigl|P(W\le z)-\Phi(z)\bigr| \le C\bigl(1+|z|\bigr)^{-p}(14m+1)^{pd}\sum_{i\in\mathcal{J}} E|\xi_i|^p \tag{9.2}$$

*where $C$ is an absolute constant.*

## 9.2 Outline of Proofs

The main ideas behind the proofs of the results in Sect. 9.1 are similar to those in Sects. 3.4.1 and 8.2. First a Stein identity is derived, followed by uniform and non-uniform concentration inequalities. We outline the main steps under the local dependence condition (LD1), referring the reader to Chen and Shao (2004) for further details.

Assume that (LD1) is satisfied and let $\eta_i=\sum_{j\in A_i}\xi_j$. Define

$$\begin{aligned} &\hat K_i(t)=\xi_i\bigl\{\mathbf{1}(-\eta_i\le t<0)-\mathbf{1}(0\le t\le -\eta_i)\bigr\}, \qquad K_i(t)=E\hat K_i(t),\\ &\hat K(t)=\sum_{i\in\mathcal{J}}\hat K_i(t), \quad \text{and} \quad K(t)=E\hat K(t). \end{aligned} \tag{9.3}$$

We first derive a Stein identity for $W$. Let $f$ be a bounded absolutely continuous function. Then, by the independence of $\xi_i$ and $W - \eta_i$,

$$\begin{aligned} E\{Wf(W)\} &= \sum_{i\in\mathcal{J}} E\{\xi_i(f(W) - f(W-\eta_i))\} \\ &= \sum_{i\in\mathcal{J}} E\left\{\xi_i \int_{-\eta_i}^{0} f'(W+t)\,dt\right\} \\ &= \sum_{i\in\mathcal{J}} E\left\{\int_{-\infty}^{\infty} f'(W+t)\hat{K}_i(t)\,dt\right\} \\ &= E\left\{\int_{-\infty}^{\infty} f'(W+t)\hat{K}(t)\,dt\right\}. \end{aligned} \tag{9.4}$$

Now, by virtue of the fact that

$$\begin{aligned} \int_{-\infty}^{\infty} K(t)\,dt &= E\sum_{i\in\mathcal{J}} \xi_i\eta_i \\ &= E\sum_{i\in\mathcal{J},\, j\in A_i} \xi_i\xi_j = E\sum_{i\in\mathcal{J},\, j\in\mathcal{J}} \xi_i\xi_j = EW^2 = 1, \end{aligned} \tag{9.5}$$

we have

$$\begin{aligned} &Ef'(W) - EWf(W) \\ &\quad = E\int_{-\infty}^{\infty} f'(W)K(t)\,dt - E\int_{-\infty}^{\infty} f'(W+t)\hat{K}(t)\,dt. \end{aligned}$$

Let

$$\begin{aligned} r_1 &= \sum_{i\in\mathcal{J}} E|\xi_i\eta_i|\mathbf{1}_{\{|\eta_i|>1\}}, \\ r_2 &= \sum_{i\in\mathcal{J}} E|\xi_i|(\eta_i^2 \wedge 1), \quad \text{and} \quad r_3 = \int_{|t|\le 1} \mathrm{Var}(\hat{K}(t))\,dt. \end{aligned} \tag{9.6}$$

We record some useful inequalities involving integrals of the functions $K(t)$ and $\hat{K}(t)$ in the following lemma, the verification of which follows by simple computations, and are therefore omitted.

**Lemma 9.1** *Let $K(t)$ and $\hat{K}(t)$ be given by* (9.3). *Then*

$$\left|\int_{|t|>1} K(t)dt\right| \le \int_{|t|>1} |K(t)|dt \le \int_{|t|>1} E|\hat{K}(t)|dt \le r_1$$

*and*

$$\int_{|t|\le 1} |tK(t)|dt \le E\int_{|t|\le 1} |t\hat{K}(t)|dt \le 0.5r_2.$$

The concentration inequality given by Proposition 9.1 is used in the proof of Theorem 9.1. Similar ideas are applied to prove Theorems 9.2 and 9.3, requiring conditional and non-uniform concentration inequalities, respectively. In the following, sometimes without mention, we will make use of the inequality

$$ab \le \frac{1}{2}\left(ca^2 + b^2/c\right) \quad \text{for all } c > 0. \tag{9.7}$$

Inequality (9.7) is an immediate consequence of the inequality resulting from replacing $a$ and $b$ in the simpler special case when $c = 1$ by $\sqrt{c}a$, and $b/\sqrt{c}$, respectively.

**Proposition 9.1** *Assume that* (LD1) *is satisfied. Then for any real numbers* $a < b$,

$$P(a \le W \le b) \le 0.625(b-a) + 4r_1 + 2.125r_2 + 4r_3, \tag{9.8}$$

*where* $r_1, r_2$ *and* $r_3$ *are given in* (9.6).

*Proof* Since $\hat{K}(t)$ is not necessary non-negative we cannot use the function defined in (3.32) and must consider a modification. For $a < b$ arbitrary and $\alpha = r_2$ define

$$f(w) = \begin{cases} -(b-a+\alpha)/2 & \text{for } w \le a-\alpha, \\ \frac{1}{2\alpha}(w-a+\alpha)^2 - (b-a+\alpha)/2 & \text{for } a-\alpha < w \le a, \\ w - (a+b)/2 & \text{for } a < w \le b, \\ -\frac{1}{2\alpha}(w-b-\alpha)^2 + (b-a+\alpha)/2 & \text{for } b < w \le b+\alpha, \\ (b-a+\alpha)/2 & \text{for } w > b+\alpha. \end{cases}$$

Then $f'$ is the continuous function given by

$$f'(w) = \begin{cases} 1 & \text{for } a \le w \le b, \\ 0 & \text{for } w \le a-\alpha \text{ or } w \ge b+\alpha, \\ \text{linear} & \text{for } a-\alpha \le w \le a \text{ or } b \le w \le b+\alpha. \end{cases} \tag{9.9}$$

Clearly $|f(w)| \le (b-a+\alpha)/2$. With this choice of $f$, and $\eta_i$, $\hat{K}(t)$ and $K(t)$ as defined in (9.3), by the Cauchy–Schwarz inequality, $EW^2 = 1$ and (9.4),

$$\begin{aligned} (b-a+\alpha)/2 \ge EWf(W) &= E\int_{-\infty}^{\infty} f'(W+t)\hat{K}(t)\,dt \\ &= Ef'(W)\int_{|t|\le 1} K(t)\,dt + E\int_{|t|\le 1}\big(f'(W+t) - f'(W)\big)K(t)\,dt \\ &\quad + E\int_{|t|>1} f'(W+t)\hat{K}(t)\,dt \\ &\quad + E\int_{|t|\le 1} f'(W+t)\big(\hat{K}(t) - K(t)\big)dt \\ &:= H_1 + H_2 + H_3 + H_4. \end{aligned} \tag{9.10}$$

From (9.5), (9.9) and Lemma 9.1 we obtain

$$H_1 \ge Ef'(W)(1-r_1) \ge P(a \le W \le b) - r_1 \quad \text{and} \quad |H_3| \le r_1. \tag{9.11}$$

Moving on to $H_4$, we have

$$|H_4| \leq (1/8)E\int_{|t|\leq 1}(f'(W+t))^2 dt + 2E\int_{|t|\leq 1}(\hat{K}(t) - K(t))^2 dt$$
$$\leq (b - a + 2\alpha)/8 + 2r_3. \tag{9.12}$$

Lastly to bound $H_2$, let

$$L(\alpha) = \sup_{x\in\mathbb{R}} P(x \leq W \leq x + \alpha).$$

Then, noting that $f''(w) = \alpha^{-1}(\mathbf{1}_{[a-\alpha,a]}(w) - \mathbf{1}_{[b,b+\alpha]}(w))$ a.s., write

$$H_2 = E\int_0^1\int_0^t f''(W+s)ds K(t)dt - E\int_{-1}^0\int_t^0 f''(W+s)ds K(t)dt$$

as

$$\alpha^{-1}\int_0^1\int_0^t \{P(a-\alpha \leq W+s \leq a) - P(b \leq W+s \leq b+\alpha)\} ds K(t)\, dt$$
$$-\alpha^{-1}\int_{-1}^0\int_t^0 \{P(a-\alpha \leq W+s \leq a) - P(b \leq W+s \leq b+\alpha)\} ds K(t)\, dt.$$

Now, by Lemma 9.1 and that $\alpha = r_2$,

$$|H_2| \leq \alpha^{-1}\int_0^1\int_0^t L(\alpha) ds |K(t)|\, dt + \alpha^{-1}\int_{-1}^0\int_t^0 L(\alpha) ds |K(t)|\, dt$$
$$= \alpha^{-1}L(\alpha)\int_{|t|\leq 1}|tK(t)|\, dt$$
$$\leq \frac{1}{2}\alpha^{-1}r_2 L(\alpha) = \frac{1}{2}L(\alpha). \tag{9.13}$$

It follows from (9.10)–(9.13) that for all $a < b$

$$P(a \leq W \leq b) \leq 0.625(b-a) + 0.75\alpha + 2r_1 + 2r_3 + 0.5L(\alpha). \tag{9.14}$$

Substituting $a = x$ and $b = x + \alpha$ in (9.14) and taking supremum over $x$ we obtain

$$L(\alpha) \leq 1.375\alpha + 2r_1 + 2r_3 + 0.5L(\alpha),$$

and hence

$$L(\alpha) \leq 2.75\alpha + 4r_1 + 4r_3. \tag{9.15}$$

Finally combining (9.14) and (9.15), and again recalling $\alpha = r_2$, we obtain (9.8). □

Using Proposition 9.1 we prove the following Berry–Esseen bound for random fields satisfying (LD1), which enables one to derive Theorem 9.1. We leave details to the reader.

**Theorem 9.5** *Under* (LD1) *we have*

$$\sup_{z\in\mathbb{R}}\left|P(W\le z)-\Phi(z)\right|\le 3.9r_1+5.8r_2+4.6r_3+r_4+0.5r_5+1.5r_6$$

*where* $r_1, r_2$ *and* $r_3$ *are defined in* (9.6), *and*

$$r_4=E\left|\sum_{i\in\mathcal{J}}(\xi_i\eta_i-E\xi_i\eta_i)\right|,\quad r_5=\sum_{i\in\mathcal{J}}E\left\{|W\xi_i|\left(\eta_i^2\wedge 1\right)\right\}\quad \textit{and}$$

$$r_6=\left(\int_{|t|\le 1}|t|\operatorname{Var}\left(\hat{K}(t)\right)dt\right)^{1/2}.$$

*Proof* For $z\in\mathbb{R}$ and $\alpha>0$ let $f$ be the solution of Stein equation (2.4) for the smoothed indicator function $h_{z,\alpha}(w)$ given in (2.14). Substituting $f$ into identity (9.4) and using (9.5) we obtain

$$\begin{aligned}
&E\left\{f'(W)-Wf(W)\right\}\\
&=E\int_{-\infty}^{\infty}f'(W)\left(K(t)-\hat{K}(t)\right)dt+E\int_{|t|>1}\left(f'(W)-f'(W+t)\right)\hat{K}(t)dt\\
&\quad+E\int_{|t|\le 1}\left(f'(W)-f'(W+t)\right)\left(\hat{K}(t)-K(t)\right)dt\\
&\quad+E\int_{|t|\le 1}\left(f'(W)-f'(W+t)\right)K(t)dt\\
&:=R_1+R_2+R_3+R_4.
\end{aligned}$$

By calculating as in (9.5), and applying the second inequality in (2.15) of Lemma 2.5 we obtain

$$|R_1|=\left|Ef'(W)\sum_{i\in\mathcal{J}}(\xi_i\eta_i-E\xi_i\eta_i)\right|\le r_4,$$

and by the final inequality in (2.15), and Lemma 9.1, we have

$$|R_2|\le E\int_{|t|>1}\left|f'(W)-f'(W+t)\right|\left|\hat{K}(t)\right|dt\le\int_{|t|>1}E\left|\hat{K}(t)\right|dt\le r_1.$$

Applying the simple change of variable $u=rt$ to the bound (2.16) of Lemma 2.5 on the smoothed indicator solution, we have

$$\begin{aligned}
&\left|f'(w+t)-f'(w)\right|\\
&\le|t|\left(1+|w|+\frac{1}{\alpha}\int_0^1\mathbf{1}_{[z,z+\alpha]}(w+rt)dr\right)\\
&=\left(1+|w|\right)|t|+\frac{1}{\alpha}\left|\int_0^t\mathbf{1}(z\le w+u\le z+\alpha)du\right| &(9.16)\\
&\le\left(1+|w|\right)|t|+\mathbf{1}(z-0\vee t\le w\le z-0\wedge t+\alpha). &(9.17)
\end{aligned}$$

For $R_3$, the bound (9.17) will produce two terms. For the first,

$$E\int_{|t|\le 1}(1+|W|)|t|\big|\hat{K}(t)-K(t)\big|dt$$
$$=E\int_{|t|\le 1}|t|\big|\hat{K}(t)-K(t)\big|dt+E|W|\int_{|t|\le 1}|t|\big|\hat{K}(t)-K(t)\big|dt.$$

Applying the triangle inequality and the bounds from Lemma 9.1, the first term above is bounded by $r_2$. Similarly, the second term may be bounded by $0.5r_5+0.5r_2$. Hence

$$|R_3|\le 1.5r_2+0.5r_5+R_{3,1}+R_{3,2},$$

where

$$R_{3,1}=E\int_0^1 \mathbf{1}(z-t\le W\le z+\alpha)\big|\hat{K}(t)-K(t)\big|dt \quad \text{and}$$
$$R_{3,2}=E\int_{-1}^0 \mathbf{1}(z\le W\le z-t+\alpha)\big|\hat{K}(t)-K(t)\big|dt.$$

Let $\delta=0.625\alpha+4r_1+2.125r_2+4r_3$. Then by Proposition 9.1,

$$P(z-t\le W\le z+\alpha)\le \delta+0.625t \tag{9.18}$$

for $t\ge 0$. Hence,

$$R_{3,1}\le E\bigg\{\int_0^1\big(0.5\alpha(\delta+0.625t)^{-1}\mathbf{1}(z-t\le W\le z+\alpha)$$
$$+0.5\alpha^{-1}(\delta+0.625t)\big|\hat{K}(t)-K(t)\big|^2\big)dt\bigg\}$$
$$\le 0.5\alpha+0.5\alpha^{-1}\delta\int_0^1 \mathrm{Var}\big(\hat{K}(t)\big)dt+0.32\alpha^{-1}\int_0^1 t\,\mathrm{Var}\big(\hat{K}(t)\big)dt.$$

As a corresponding upper bound holds for $R_{3,2}$, we arrive at

$$|R_3|\le \alpha+0.5\alpha^{-1}\delta r_3+0.32\alpha^{-1}r_6^2+1.5r_2+0.5r_5.$$

By (9.16), (9.18) with $t=0$, and Lemma 9.1 we have

$$|R_4|\le E\int_{|t|\le 1}(1+|W|)\big|tK(t)\big|dt$$
$$+\alpha^{-1}\int_{|t|\le 1}\bigg|\int_0^t P(z\le W+u\le z+\alpha)du\bigg|\big|K(t)\big|dt$$
$$\le r_2+\alpha^{-1}\int_{|t|\le 1}\delta\big|tK(t)\big|dt\le r_2+0.5\alpha^{-1}\delta r_2. \tag{9.19}$$

Combining the above inequalities yields

$$\big|Eh_{z,\alpha}(W)-Nh_{z,\alpha}\big|$$
$$\le r_4+r_1+2.5r_2+0.5r_5+\alpha+\alpha^{-1}\big\{\delta(0.5r_3+0.5r_2)+0.32r_6^2\big\}$$
$$\le r_4+r_1+2.82r_2+0.5r_5+0.32r_3+\alpha$$
$$+\alpha^{-1}\big\{(4r_1+2.125r_2+4r_3)(0.5r_3+0.5r_2)+0.32r_6^2\big\}.$$

Using the fact that $Eh_{z-\alpha,\alpha}(W) \le P(W \le z) \le Eh_{z,\alpha}(W)$ and that $|\Phi(z+\alpha) - \Phi(z)| \le (2\pi)^{-1/2}\alpha$, we have

$$\sup_{z\in\mathbb{R}}\big|P(W \le z) - \Phi(z)\big| \le \sup_{z\in\mathbb{R}}\big|Eh_{z,\alpha}(W) - Nh_{z,\alpha}\big| + 0.5\alpha.$$

Letting

$$\alpha = \big((4r_1 + 2.125r_2 + 4r_3)(0.5r_3 + 0.5r_2) + 0.32r_6^2\big)^{1/2}$$

and applying the inequality $(a+b)^{1/2} \le a^{1/2} + b^{1/2}$ yields

$$\begin{aligned}
&\sup_{z\in\mathbb{R}}\big|P(W \le z) - \Phi(z)\big| \\
&\quad \le r_4 + r_1 + 2.82r_2 + 0.5r_5 + 0.32r_3 \\
&\qquad + 2.5\big((4r_1 + 2.125r_2 + 4r_3)(0.5r_3 + 0.5r_2) + 0.32r_6^2\big)^{1/2} \\
&\quad \le r_4 + r_1 + 2.82r_2 + 0.5r_5 + 0.32r_3 + 1.5r_6 \\
&\qquad + 2\big((4r_1 + 2.125r_2 + 4r_3)(r_3 + r_2)\big)^{1/2}.
\end{aligned}$$

Now, applying inequality (9.7) on the last term, we obtain

$$\begin{aligned}
&\sup_{z\in\mathbb{R}}\big|P(W \le z) - \Phi(z)\big| \\
&\quad \le r_4 + r_1 + 2.82r_2 + 0.5r_5 + 0.32r_3 + 1.5r_6 \\
&\qquad + \sqrt{2}\big(0.5(4r_1 + 2.125r_2 + 4r_3) + 2(0.5r_3 + 0.5r_2)\big) \\
&\quad \le r_4 + 3.9r_1 + 5.8r_2 + 0.5r_5 + 4.6r_3 + 1.5r_6,
\end{aligned}$$

completing the proof of Theorem 9.5. □

We remark that if we use the Stein solution for the indicator $h_z(w) = \mathbf{1}_{(-\infty,z]}(w)$ instead of the one for the smoothed indicator $h_{z,\alpha}(w)$, then the final integral in (9.19) can be no more than $\delta\int_{|t|\le 1}|K(t)|dt$, a term which is not clearly bounded by $\int_{|t|\le 1}|K(t)|dt$, though $\int_{|t|\le 1}K(t)dt \le 1 + r_1$.

Under (LD2), letting $\tau_i = \sum_{j\in B_i}\xi_j$, the proof of Theorem 9.2 is based on a conditional concentration inequality for $P(a_{\tau_i} \le W \le b_{\tau_i}|\tau_i)$, where $\tau_i = (\xi, \eta_i, \zeta_i)$, $\zeta_i = \sum_{j\in B_i}\xi_j$ and $a_{\tau_i} \le b_{\tau_i}$ are measurable functions of $\tau_i$, while the proof of Theorem 9.3 relies on a non-uniform concentration inequality for $E((1+W)^3\mathbf{1}_{\{a_{\tau_i}\le W\le b_{\tau_i}\}}|\tau_i)$. We refer to Chen and Shao (2004) for details.

## 9.3 Applications

The following three applications of our local dependence results were considered in Chen and Shao (2004).

*Example 9.1* (Dependency Graphs) This example was discussed in Baldi and Rinott (1989) and Rinott (1994) where some results on uniform bound were obtained.

Consider a set of random variables $\{X_i, i \in \mathcal{V}\}$ indexed by the vertices of a graph $\mathcal{G} = (\mathcal{V}, \mathcal{E})$. $\mathcal{G}$ is said to be a dependency graph if for any pair of disjoint sets $\Gamma_1$ and $\Gamma_2$ in $\mathcal{V}$ such that no edge in $\mathcal{E}$ has one endpoint in $\Gamma_1$ and the other in $\Gamma_2$, the sets of random variables $\{X_i, i \in \Gamma_1\}$ and $\{X_i, i \in \Gamma_2\}$ are independent. Let $D$ denote the maximal degree of $G$, i.e., the maximal number of edges incident to a single vertex. Let $A_i = \{j \in \mathcal{V}\colon$ there is an edge connecting $j$ and $i\}$, $B_i = \bigcup_{j \in A_i} A_j$, $C_i = \bigcup_{j \in B_i} A_j$, $B_i^* = \bigcup_{j \in A_i} B_j$, $C_i^* = \bigcup_{j \in B_i^*} B_j$ and $D_i^* = \bigcup_{j \in C_i^*} B_j$. Noting that

$$|A_i| \le D, \quad |B_i| \le D^2, \quad |C_i| \le D^3, \quad \left|B_i^*\right| \le D^3, \quad \left|C_i^*\right| \le D^5$$
$$\text{and} \quad \left|D_i^*\right| \le D^7,$$

we have that

$$\kappa_1 = \left|\{j \in \mathcal{J}\colon C_i \cap B_j \neq \emptyset\}\right| \le D^5 \quad \text{and}$$
$$\kappa_2 = \max_{i \in \mathcal{J}} \left\{\left|D_i^*\right|, \left|\{j\colon i \in D_j^*\}\right|\right\} \le D^7.$$

Hence, applying Theorem 9.2 with $\kappa = \kappa_1$, and Theorem 9.3 with $\kappa = \kappa_2$, yields the following theorem.

**Theorem 9.6** *Let $\{X_i, i \in \mathcal{V}\}$ be random variables indexed by the vertices of a dependency graph. Put $W = \sum_{i \in \mathcal{V}} X_i$. Assume that $EW^2 = 1$, $EX_i = 0$ and $E|X_i|^p \le \theta^p$ for $i \in \mathcal{V}$ and for some $\theta > 0$.*

$$\sup_z \left|P(W \le z) - \Phi(z)\right| \le 75 D^{5(p-1)} |\mathcal{V}| \theta^p \tag{9.20}$$

*and for $z \in R$,*

$$\left|P(W \le z) - \Phi(z)\right| \le C(1 + |z|)^{-p} D^{7p} |\mathcal{V}| \theta^p.$$

The bound (9.20) compares favorably with those of Baldi and Rinott (1989).

*Example 9.2* (Exceedances of the $m$-scans process) Let $X_1, X_2, \ldots,$ be i.i.d. random variables and let $R_i = \sum_{k=0}^{m-1} X_{i+k}$, $i = 1, 2, \ldots, n$ be the $m$-scans process. For $a \in \mathbb{R}$ consider the number of exceedances of $a$ by $\{R_i\colon i = 1, \ldots, n\}$,

$$Y = \sum_{i=1}^{n} \mathbf{1}\{R_i > a\}.$$

Assessing the statistical significance of exceedances of scan statistics in one and higher dimensions plays a key role in many areas of applied statistics, and is a well studied problem, see, for example Glaz et al. (2001) and Naus (1982). Scan statistics have been used, for example, for the evaluation of the significance of observed inhomogeneities in the distribution of markers along the length of long DNA sequences, see Dembo and Karlin (1992), and Karlub and Brede (1992). Dembo and Rinott

(1996) obtain a uniform Berry–Esseen bound for $Y$, of the best possible order, as $n \to \infty$.

Let $p = P(R_1 > a)$ and $\sigma^2 = \mathrm{Var}(Y)$. From Dembo and Rinott (1996) we have $\sigma^2 \ge np(1-p)$, and that $\{\mathbf{1}\{R_i > a\},\ 1 \le i \le n\}$ are $m$-dependent. Let

$$W = \frac{Y - np}{\sigma} = \sum_{i=1}^{n} \xi_i \quad \text{where } \xi_i = \big(\mathbf{1}(R_i > a) - p\big)/\sigma.$$

Since $\sigma^2 \ge np(1-p)$, we have

$$\sum_{i=1}^{n} E\left|\xi_i^3\right| = n\frac{p(1-p)^3 + p^3(1-p)}{\sigma^3} \le \frac{np(1-p)}{\sigma^3} \le \frac{1}{\sqrt{np(1-p)}}.$$

Hence the following non-uniform bound is a consequence of Theorem 9.4.

**Theorem 9.7** *There exists a universal constant $C$ such that for all $z \in \mathbb{R}$,*

$$\left|P(W \le z) - \Phi(z)\right| \le \frac{Cm^3}{(1+|z|)^3\sqrt{np(1-p)}}.$$

# Chapter 10
# Uniform and Non-uniform Bounds for Non-linear Statistics

In this chapter we consider uniform and non-uniform Berry–Esseen bounds for non-linear statistics that can be written as a linear statistic plus an error term. We apply our results to $U$-statistics, multi-sample $U$-statistics, $L$-statistics, random sums, and functions of non-linear statistics, obtaining bounds with optimal asymptotic rates. The main tools are uniform and non-uniform randomized concentration inequalities. The work of Chen and Shao (2007) forms the basis of this chapter.

## 10.1 Introduction and Main Results

Let $X_1, X_2, \ldots, X_n$ be independent random variables and let $T := T(X_1, \ldots, X_n)$ be a general sampling statistic. In many cases of interest $T$ can be written as a linear statistic plus a manageable error term, that is, as $T = W + \Delta$ where

$$W = \sum_{i=1}^{n} g_{n,i}(X_i), \quad \text{and} \quad \Delta := \Delta(X_1, \ldots, X_n) = T - W,$$

for some functions $g_{n,i}$. Let $\xi_i = g_{n,i}(X_i)$. We assume that

$$E\xi_i = 0 \quad \text{for } i = 1, 2, \ldots, n, \quad \text{and} \quad \sum_{i=1}^{n} \mathrm{Var}(\xi_i) = 1, \tag{10.1}$$

and also that $\Delta$ depends on $X_i$ only through $g_{n,i}(X_i)$, that is, with slight abuse of notation,

$$\Delta = \Delta(\xi_1, \ldots, \xi_n).$$

It is clear that if $\Delta \to 0$ in probability as $n \to \infty$ then the central limit theorem holds for $W$ provided the Lindeberg condition is satisfied. If in addition, $E|\Delta|^p < \infty$ for some $p > 0$, then by the Chebyshev inequality followed by a simple minimization, one can obtain the following uniform bound,

$$\sup_{z\in\mathbb{R}}\bigl|P(T \le z) - \Phi(z)\bigr| \le \sup_{z\in\mathbb{R}}\bigl|P(W \le z) - \Phi(z)\bigr| + 2\bigl(E|\Delta|^p\bigr)^{1/(1+p)}, \tag{10.2}$$

L.H.Y. Chen et al., *Normal Approximation by Stein's Method*,
Probability and Its Applications,
DOI 10.1007/978-3-642-15007-4_10, © Springer-Verlag Berlin Heidelberg 2011

where the first term on the right hand side of (10.2) may be readily estimated by the Berry–Esseen inequality. However, after the addition of the second term the resulting bound will not generally be sharp for many commonly used statistics. Taking a different approach, by developing randomized versions of the concentration inequalities in Sects. 3.4.1 and 8.1, we can establish uniform and non-uniform Berry–Esseen bounds for $T$ with optimal asymptotic rates.

Let $\delta > 0$ satisfy

$$\sum_{i=1}^{n} E|\xi_i| \min(\delta, |\xi_i|) \geq 1/2 \tag{10.3}$$

and recall that

$$\beta_2 = \sum_{i=1}^{n} E\xi_i^2 \mathbf{1}_{\{|\xi_i|>1\}} \quad \text{and} \quad \beta_3 = \sum_{i=1}^{n} E|\xi_i|^3 \mathbf{1}_{\{|\xi_i|\leq 1\}}. \tag{10.4}$$

The following approximation of $T$ by $W$ provides our uniform Berry–Esseen bound for $T$.

**Theorem 10.1** *Let $\xi_1, \ldots, \xi_n$ be independent random variables satisfying* (10.1), *$W = \sum_{i=1}^{n} \xi_i$ and $T = W + \Delta$. For each $i = 1, \ldots, n$, let $\Delta_i$ be a random variable such that $\xi_i$ and $(W - \xi_i, \Delta_i)$ are independent. Then for any $\delta$ satisfying* (10.3),

$$\sup_{z\in\mathbb{R}} \left| P(T \leq z) - P(W \leq z) \right| \leq 4\delta + E|W\Delta| + \sum_{i=1}^{n} E\left|\xi_i(\Delta - \Delta_i)\right|. \tag{10.5}$$

*In particular,*

$$\sup_{z\in\mathbb{R}} \left| P(T \leq z) - P(W \leq z) \right| \leq 2(\beta_2 + \beta_3) + E|W\Delta| + \sum_{i=1}^{n} E\left|\xi_i(\Delta - \Delta_i)\right| \tag{10.6}$$

*and*

$$\sup_{z\in\mathbb{R}} \left| P(T \leq z) - \Phi(z) \right| \leq 6.1(\beta_2 + \beta_3) + E|W\Delta| + \sum_{i=1}^{n} E\left|\xi_i(\Delta - \Delta_i)\right|. \tag{10.7}$$

With $\|X\|_2$ denoting the $L^2$ norm $\|X\|_2 = (EX^2)^{1/2}$ of a random variable $X$, we now provide a corresponding non-uniform bound.

**Theorem 10.2** *Let $\xi_1, \ldots, \xi_n$ be independent random variables satisfying* (10.1), *$W = \sum_{i=1}^{n} \xi_i$ and $T = W + \Delta$. For each $1 \leq i \leq n$, let $\Delta_i$ be a random variable such that $\xi_i$ and $(W - \xi_i, \Delta_i)$ are independent. Then for $\delta$ satisfying* (10.3), *and any $p \geq 2$,*

$$\left| P(T \leq z) - P(W \leq z) \right| \leq \gamma_{z,p} + e^{-|z|/3}\tau \quad \text{for all } z \in \mathbb{R}, \tag{10.8}$$

*where*

$$\gamma_{z,p} = P\big(|\Delta| > \big(|z|+1\big)/3\big) + 2\sum_{i=1}^{n} P\big(|\xi_i| > \big(|z|+1\big)/(6p)\big)$$
$$+ e^p\big(1+z^2/(36p)\big)^{-p}\beta_2 \quad and \tag{10.9}$$
$$\tau = 22\delta + 8.6\|\Delta\|_2 + 3.6\sum_{i=1}^{n}\|\xi_i\|_2\|\Delta - \Delta_i\|_2.$$

*If $E|\xi_i|^p < \infty$ for some $p > 2$, then for some constant $C_p$ depending on $p$ only,*

$$\begin{aligned}&\big|P(T \le z) - \Phi(z)\big|\\ &\le P\big(|\Delta| > \big(|z|+1\big)/3\big)\\ &\quad + \frac{C_p}{(|z|+1)^p}\left(\|\Delta\|_2 + \sum_{i=1}^{n}\|\xi_i\|_2\|\Delta-\Delta_i\|_2 + \sum_{i=1}^{n}\big\{E|\xi_i|^p + E|\xi_i|^{3\wedge p}\big\}\right).\end{aligned} \tag{10.10}$$

The following remark shows how to choose $\delta$ so that (10.3) is satisfied.

*Remark 10.1*

(i) When $E|\xi_i|^p < \infty$ for $p > 2$ then one may verify that

$$\delta = \left(\frac{2(p-2)^{p-2}}{(p-1)^{p-1}}\sum_{i=1}^{n} E|\xi_i|^p\right)^{1/(p-2)} \tag{10.11}$$

satisfies (10.3) using the inequality

$$\min(x,y) \ge y - \frac{(p-2)^{p-2}y^{p-1}}{(p-1)^{p-1}x^{p-2}} \quad \text{for } x > 0,\ y \ge 0. \tag{10.12}$$

Inequality (10.12) is trivial when $y \le x$. For $y > x$ the inequality follows by replacing $x$ and $y$ by $x/(p-1)$ and $y/(p-2)$, respectively, resulting in the inequality

$$1 \le \frac{p-2}{p-1}\left(\frac{x}{y}\right) + \frac{1}{p-1}\left(\frac{y}{x}\right)^{p-2},$$

which holds as the function

$$\frac{p-2}{p-1}a + \frac{1}{p-1}a^{2-p} \quad \text{for } a > 0$$

has a minimum of 1 at $a = 1$.

(ii) If $\beta_2 + \beta_3 \le 1/2$, then (10.3) holds with $\delta = (\beta_2+\beta_3)/2$. In fact, as (10.12) for $p = 3$ yields $\min(x,y) \ge y - y^2/(4x)$, we have

$$\sum_{i=1}^{n} E|\xi_i|\min\big(\delta, |\xi_i|\big) \ge \sum_{i=1}^{n} E|\xi_i|\mathbf{1}_{\{|\xi_i|\le 1\}}\min\big(\delta, |\xi_i|\big)$$

$$\geq \sum_{i=1}^{n} \left\{ E\xi_i^2 \mathbf{1}_{\{|\xi_i| \leq 1\}} - \frac{E|\xi_i|^3 \mathbf{1}_{\{|\xi_i| \leq 1\}}}{4\delta} \right\}$$

$$= 1 - \frac{4\delta\beta_2 + \beta_3}{4\delta} \geq 1 - \frac{\beta_2 + \beta_3}{4\delta} = 1/2.$$

(iii) Recalling (10.1), we see that if $\delta > 0$ satisfies

$$\sum_{i=1}^{n} E\xi_i^2 \mathbf{1}_{\{|\xi_i| \geq \delta\}} < 1/2,$$

then (10.3) holds. In particular, when $\xi_i, 1 \leq i \leq n$ are standardized i.i.d. random variables, then $\delta$ may be taken to be of the order $1/\sqrt{n}$, which may be much smaller than $\beta_2 + \beta_3$.

We turn now to our applications, deferring the proofs of Theorems 10.1 and 10.2 to Sect. 10.3.

## 10.2 Applications

Theorems 10.1 and 10.2 can be applied to a wide range of different statistics, providing bounds of the best possible order in many instances. To illustrate the usefulness and generality of these results we present the following five applications.

### *10.2.1 U-statistics*

Let $X_1, X_2, \ldots, X_n$ be a sequence of i.i.d. random variables, and for some $m \geq 2$ let $h(x_1, \ldots, x_m)$ be a symmetric, real-valued function, where $m < n/2$ may depend on $n$. Introduced by Hoeffding (1948), the class of $U$-statistics are those random variables that can be written as

$$U_n = \binom{n}{m}^{-1} \sum_{1 \leq i_1 < \cdots < i_m \leq n} h(X_{i_1}, \ldots, X_{i_m}). \tag{10.13}$$

Special cases include (i) the sample mean, when $h(x_1, x_2) = \frac{1}{2}(x_1 + x_2)$, (ii) the sample variance, where $h(x_1, x_2) = \frac{1}{2}(x_1 - x_2)^2$, and (iii) the one-sample Wilcoxon statistic, when $h(x_1, x_2) = \mathbf{1}(x_1 + x_2 \leq 0)$. We refer the reader to Koroljuk and Borovskich (1994) for a systematic treatment of $U$-statistics, and note that Rinott and Rotar (1997) also handle a variety of $U$-statistics using Stein's method.

The Hoeffding decomposition (see (10.19) below) allows us to write $U$ as a linear statistic plus an error term, allowing for the application of Theorems 10.1 and 10.2, yielding the following result.

**Theorem 10.3** *Let $X_1, \ldots, X_n$ be i.i.d. random variables and let $U$ be given by (10.13) with $Eh(X_1, \ldots, X_m) = 0$, $\sigma^2 = Eh^2(X_1, \ldots, X_m) < \infty$ and $\sigma_1^2 = Eg^2(X_1) > 0$ where $g(x) = E(h(X_1, X_2, \ldots, X_m)|X_1 = x)$. Then*

$$\sup_{z\in\mathbb{R}}\left|P\left(\frac{\sqrt{n}}{m\sigma_1}U_n \le z\right) - P\left(\frac{1}{\sqrt{n}\sigma_1}\sum_{i=1}^{n} g(X_i) \le z\right)\right| \le \frac{4c_0}{\sqrt{n}} + \frac{(1+\sqrt{2})(m-1)\sigma}{(m(n-m+1))^{1/2}\sigma_1}, \tag{10.14}$$

*where $c_0$ is any constant such that $Eg^2(X_1)\mathbf{1}(|g(X_1)| > c_0\sigma_1) \le \sigma_1^2/2$. If in addition $E|g(X_1)|^p < \infty$ for some $2 < p \le 3$, then*

$$\sup_{z\in\mathbb{R}}\left|P\left(\frac{\sqrt{n}}{m\sigma_1}U_n \le z\right) - \Phi(z)\right| \le \frac{6.1E|g(X_1)|^p}{n^{(p-2)/2}\sigma_1^p} + \frac{(1+\sqrt{2})(m-1)\sigma}{(m(n-m+1))^{1/2}\sigma_1}, \tag{10.15}$$

*and there exists a universal constant $C$ such that for all $z \in \mathbb{R}$*

$$\left|P\left(\frac{\sqrt{n}}{m\sigma_1}U_n \le z\right) - \Phi(z)\right| \le \frac{9m\sigma^2}{(|z|+1)^2(n-m+1)\sigma_1^2} + \frac{C}{(|z|+1)^p}\left(\frac{m^{1/2}\sigma}{(n-m+1)^{1/2}\sigma_1} + \frac{E|g(X_1)|^p}{n^{(p-2)/2}\sigma_1^p}\right). \tag{10.16}$$

Note that for the error bound in (10.14) to be of order $O(n^{-1/2})$ it is necessary that $\sigma^2$, the second moment of $h$, be finite. However, requiring $\sigma^2 < \infty$ is not the weakest assumption under which the uniform bound at this rate is known to hold; Friedrich (1989) obtained the order $O(n^{-1/2})$ when $E|h|^{5/3} < \infty$. It would be interesting to use Stein's method to obtain this same result. We refer to Benkus et al. (1994) and Jing and Zhou (2005) for a discussion regarding the necessity of the moment condition.

For $1 \le k \le m$, let

$$h_k(x_1, \ldots, x_k) = E\big(h(X_1, \ldots, X_m)|X_1 = x_1, \ldots, X_k = x_k\big),$$

$$\bar{h}_k(x_1, \ldots, x_k) = h_k(x_1, \ldots, x_k) - \sum_{i=1}^{k} g(x_i),$$

$$\Delta = \frac{\sqrt{n}}{m\sigma_1}\binom{n}{m}^{-1}\sum_{1\le i_1<\cdots<i_m\le n}\bar{h}_m(X_{i_1}, \ldots, X_{i_m}), \tag{10.17}$$

and for $l \in \{1, \ldots, n\}$,

$$\Delta_l = \frac{\sqrt{n}}{m\sigma_1}\binom{n}{m}^{-1}\sum_{1\le i_1<\cdots<i_m\le n,\, i_j\ne l \text{ for all } j}\bar{h}_m(X_{i_1}, \ldots, X_{i_m}). \tag{10.18}$$

We now prove Theorem 10.3 by applying Theorems 10.1 and 10.2.

*Proof* Observing that

$$U_n = \frac{m}{n}\sum_{i=1}^{n} g(X_i) + \binom{n}{m}^{-1} \sum_{1\le i_1<\cdots<i_m\le n} \bar{h}_m(X_{i_1}, \ldots, X_{i_m}), \tag{10.19}$$

we have

$$\frac{\sqrt{n}}{m\sigma_1} U_n = W + \Delta,$$

where

$$W = \sum_{i=1}^{n} \xi_i \quad \text{with } \xi_i = \frac{1}{\sqrt{n}\sigma_1} g(X_i).$$

For each $l \in 1, \ldots, n$ the random variables $W - \xi_l$ and $\Delta_l$ are functions of $\xi_j$, $j \neq l$, and therefore $\xi_i$ is independent of $(W - \xi_i, \Delta_i)$. Hence, by Theorems 10.1 and 10.2, and (iii) of Remark 10.1, the result follows by Lemma 10.1. □

**Lemma 10.1** *Let* $\Delta$ *and* $\Delta_l$ *for* $l = 1, \ldots, n$ *be as in* (10.17) *and* (10.18), *respectively. Then*

$$E\Delta^2 \le \frac{(m-1)^2\sigma^2}{m(n-m+1)\sigma_1^2} \tag{10.20}$$

*and*

$$E(\Delta - \Delta_l)^2 \le \frac{2(m-1)^2\sigma^2}{nm(n-m+1)\sigma_1^2}. \tag{10.21}$$

The reader is encouraged to prove Lemma 10.1 for the case $m = 2$, and refer to the Appendix for the general case.

### *10.2.2 Multi-sample U-statistics*

Consider $k$ independent sequences, $X_{j1}, \ldots, X_{jn_j}$, $j = 1, \ldots, k$ of i.i.d. random variables, of lengths $n_1, \ldots, n_k$. With $m_j \ge 1$ for $j = 1, \ldots, k$, let $h(x_{jl}, l = 1, \ldots, m_j; j = 1, \ldots, k)$ be a function which is symmetric with respect to the $m_j$ arguments of the $j$-th set, that is, invariant under permutations of $x_{jl}, l = 1, \ldots, m_j$. Let

$$\theta = Eh(X_{jl}, j = 1, \ldots, k, l = 1, \ldots, m_j).$$

The multi-sample $U$-statistic is defined as

$$U_{\mathbf{n}}=\left\{\prod_{j=1}^{k}\binom{n_j}{m_j}^{-1}\right\}\sum h(X_{jl}, j=1,\ldots,k, l=i_{j1},\ldots,i_{jm_j}), \quad (10.22)$$

where $\mathbf{n}=(n_1,\ldots,n_k)$ and the summation is carried out over all indices satisfying $1\le i_{j1}<\cdots<i_{jm_j}\le n_j$. Clearly, $U_{\mathbf{n}}$ is an unbiased estimate of $\theta$. The two-sample Wilcoxon statistic, where

$$h(x_{11};x_{21})=\mathbf{1}_{\{x_{11}<x_{21}\}},$$

and the two-sample $\omega^2$-statistic, where

$$h(x_{11},x_{12};x_{21},x_{22})=\begin{cases}1/3 & \text{if } \max(x_{11},x_{12})<\min(x_{21},x_{22}) \\ & \text{or } \min(x_{11},x_{12})<\max(x_{21},x_{22}),\\ -1/6 & \text{otherwise}\end{cases}$$

are both special cases of the general multi-sample $U$-statistic; see Koroljuk and Borovskich (1994), pp. 36–37, for additional examples.

For multi-sample $U$-statistics of the form (10.22), Helmers and van Zwet (1982) and Borovskich (1983) (see Koroljuk and Borovskich 1994, pp. 304–311) obtain a uniform Berry–Esseen bound of order $O((\min_{1\le j\le k} n_j)^{-1/2})$. Theorem 10.4 not only refines their results but also gives an optimal non-uniform bound.

To state the theorem, first note that we may assume $\theta=0$ without loss of generality. Next, let

$$\sigma^2=Eh^2(X_{11},\ldots,X_{1m_1};\ldots;X_{k1},\ldots,X_{km_k}),$$

for $j=1,\ldots,k$, define

$$h_j(x)=Eh(X_{11},\ldots,X_{1m_1};\ldots;X_{k1},\ldots,X_{km_k})|X_{j1}=x),$$

let $\sigma_j^2=Eh_j^2(X_{j1})$, and set

$$\sigma_{\mathbf{n}}^2=\sum_{j=1}^{k}\frac{m_j^2}{n_j}\sigma_j^2.$$

**Theorem 10.4** *Assume that* $\theta=0$, $\sigma^2<\infty$, $\max_{1\le j\le k}\sigma_j^2>0$ *and let* $n_j\ge 2m_j$ *for all* $j=1,\ldots,k$. *Then for* $2<p\le 3$

$$\sup_{z\in\mathbb{R}}\left|P\left(\sigma_{\mathbf{n}}^{-1}U_{\mathbf{n}}\le z\right)-\Phi(z)\right| \le \frac{6.1}{\sigma_{\mathbf{n}}^p}\sum_{j=1}^{k}\frac{m_j^p}{n_j^{p-1}}E\left|h_j(X_{j1})\right|^p+\frac{(1+\sqrt{2})\sigma}{\sigma_{\mathbf{n}}}\sum_{j=1}^{k}\frac{m_j^2}{n_j}, \quad (10.23)$$

*and for* $z\in\mathbb{R}$

$$\left|P\left(\sigma_{\mathbf{n}}^{-1}U_n\le z\right)-\Phi(z)\right| \le \frac{9\sigma^2}{(1+|z|)^2\sigma_{\mathbf{n}}^2}\left(\sum_{j=1}^{k}\frac{m_j^2}{n_j}\right)^2$$

$$+\frac{C}{(1+|z|)^p}\left(\frac{\sigma}{\sigma_{\mathbf{n}}}\sum_{j=1}^{k}\frac{m_j^2}{n_j}+\frac{1}{\sigma_{\mathbf{n}}^p}\sum_{j=1}^{k}\frac{m_j^p E|h_j(X_{j1})|^p}{n_j^{p-1}}\right). \tag{10.24}$$

*Proof* We follow an argument similar to the one used for the proof of Theorem 10.3. For $1 \le j \le k$, let $\boldsymbol{X}_j = (X_{j1}, \ldots, X_{jm_j})$, $\boldsymbol{x}_j = (x_{j1}, \ldots, x_{jm_j})$ and define

$$\bar{h}(\boldsymbol{x}_1, \ldots, \boldsymbol{x}_k) = h(\boldsymbol{x}_1, \ldots, \boldsymbol{x}_k) - \sum_{j=1}^{k}\sum_{l=1}^{m_j} h_j(x_{jl}). \tag{10.25}$$

For the given $U$-statistic $U_{\mathbf{n}}$, define the projection

$$\hat{U}_{\mathbf{n}} = \sum_{j=1}^{k}\sum_{l=1}^{n_j} E(U_{\mathbf{n}}|X_{jl}).$$

Since

$$m_j/n_j = \binom{n_j-1}{m_j-1}\Big/\binom{n_j}{m_j},$$

we have

$$\hat{U}_{\mathbf{n}} = \sum_{j=1}^{k}\sum_{l=1}^{n_j}\frac{m_j}{n_j}h_j(X_{jl}),$$

and the difference $U_{\mathbf{n}} - \hat{U}_{\mathbf{n}}$ can be expressed as

$$U_{\mathbf{n}} - \hat{U}_{\mathbf{n}} = \left\{\prod_{j=1}^{k}\binom{n_j}{m_j}^{-1}\right\}\sum \bar{h}(\boldsymbol{X}_{1i_1}, \ldots, \boldsymbol{X}_{ki_k}),$$

where $\boldsymbol{X}_{ji_j} = (X_{ji_{j1}}, \ldots, X_{ji_{jm_j}})$ and the summation is carried out over all indices $1 \le i_{j1} < i_{j2} < \cdots < i_{jm_j} \le n_j$, $j = 1, 2, \ldots, k$. Thus, we obtain

$$\sigma_{\mathbf{n}}^{-1}U_{\mathbf{n}} = W + \Delta$$

with

$$W = \sigma_{\mathbf{n}}^{-1}\sum_{j=1}^{k}\sum_{l=1}^{n_j}\frac{m_j}{n_j}h_j(X_{jl}) \quad \text{and}$$

$$\Delta = \sigma_{\mathbf{n}}^{-1}\left\{\prod_{j=1}^{k}\binom{n_j}{m_j}^{-1}\right\}\sum \bar{h}(\boldsymbol{X}_{1i_1}, \ldots, \boldsymbol{X}_{ki_k}).$$

To apply Theorems 10.1 and 10.2, let $\xi_{jl} = \sigma_{\mathbf{n}}^{-1}\frac{m_j}{n_j}h_j(X_{jl})$ and

$$\Delta_{jl} = \sigma_{\mathbf{n}}^{-1}\left\{\prod_{v=1}^{k}\binom{n_v}{m_v}^{-1}\right\}\sum{}^{(jl)}\bar{h}(\boldsymbol{X}_{1i_1}, \ldots, \boldsymbol{X}_{ki_k}),$$

where the sum $\sum^{(jl)}$ excludes the variable $X_{jl}$, that is, the summation is carried out over all indices $1 \le i_{v1} < i_{v2} < \cdots < i_{vm_v} \le n_v$, $1 \le v \le k$, $v \ne j$ and $1 \le i_{j1} < i_{j2} < \cdots < i_{jm_j} \le n_j$ with $i_{js} \ne l$ for $1 \le s \le m_j$.

Clearly, $\xi_{jl}$ and $(W - \xi_{jl}, \Delta_{jl})$ are independent. Theorem 10.4 now follows from Theorems 10.1 and 10.2 and the lemma below. □

**Lemma 10.2** *We have*

$$E\Delta^2 \le \frac{\sigma^2}{\sigma_{\mathbf{n}}^2}\left(\sum_{j=1}^{k}\frac{m_j^2}{n_j}\right)^2 \tag{10.26}$$

*and*

$$E(\Delta - \Delta_{jl})^2 \le \frac{2m_j^2\sigma^2}{n_j^2\sigma_{\mathbf{n}}^2}\sum_{v=1}^{k}\frac{m_v^2}{n_v}. \tag{10.27}$$

We refer the reader to the Appendix for a proof.

### 10.2.3 L-statistics

Suppose that one wishes to estimate a measure of spread of an unknown distribution $F$ based on a sample $X_1, \ldots, X_n$ of independent observations. Under a second moment assumption, one could form the unbiased estimator

$$\widehat{\sigma^2} = \frac{1}{n-1}\sum_{i=1}^{n}(X_i - \overline{X})^2 \quad \text{where } \overline{X} = \frac{1}{n}\sum_{i=1}^{n} X_i$$

of the variance of the underlying $F$. A little algebra shows that the variance estimator $\widehat{\sigma^2}$ may be written more 'symmetrically' as

$$\widehat{\sigma^2} = \frac{1}{n(n-1)}\sum_{i<j}(X_i - X_j)^2.$$

Indeed, from this second formula, the unbiasedness of $\widehat{\sigma^2}$ is clearly seen. Without the need for assuming the existence of second moments, one could instead compute the estimator

$$T = \binom{n}{2}^{-1}\sum_{i<j}|X_i - X_j| \tag{10.28}$$

to obtain an idea of the underlying spread, in this case unbiasedly estimating Gini's mean difference $E|X_1 - X_2|$.

The estimate $T$ of Gini's mean difference can actually be written as a linear combination

$$T = \sum_{i=1}^{n} c_{ni}X_{ni} \tag{10.29}$$

of the order statistics $X_{n1} \le X_{n2} \le \cdots \le X_{nn}$ of the sample, as follows. To begin, as $|x-y| = |\max(x,y) - \min(x,y)|$, we have

$$\begin{aligned}
\frac{2}{n(n-1)} \sum_{i<j} |X_i - X_j| &= \frac{2}{n(n-1)} \sum_{i<j} |X_{nj} - X_{ni}| \\
&= \frac{2}{n(n-1)} \sum_{i=1}^{n-1} \sum_{j=i+1}^{n} (X_{nj} - X_{ni}) \\
&= \frac{2}{n(n-1)} \left( \sum_{j=2}^{n} (j-1) X_{nj} - \sum_{i=1}^{n-1} (n-i) X_{ni} \right) \\
&= \frac{2}{n(n-1)} \sum_{i=1}^{n} (2i - n - i) X_{ni},
\end{aligned}$$

and hence is of the form (10.29).

A subclass of estimators of this form that includes many typical applications may be written using the empirical distribution function

$$F_n(x) = n^{-1} \sum_{i=1}^{n} \mathbf{1}(X_i \le x)$$

and a real-valued function $J(t)$ on $[0,1]$. Indeed, letting

$$T(G) = \int_{-\infty}^{\infty} x J\big(G(x)\big) dG(x)$$

for non-decreasing functions $G$, we have that

$$T(F_n) = \frac{1}{n} \sum_{i=1}^{n} J(i/n) X_{ni},$$

a linear combination of the order statistics of the sample, with coefficients, or weights, determined for all sample sizes by the function $J$. Estimators of the form $T(F_n)$ are known as an $L$-statistic, referring to their formation as linear combinations of order statistics; the trimmed mean and smoothly trimmed mean are two special cases. We refer to Serfling (1980), Chap. 8 for additional examples and some asymptotic properties of $L$-statistics.

As $T(F_n)$ estimates some parameter of interested of the underlying, unknown distribution $F$, a natural question is to determine its variation about its asymptotic value $T(F)$ as a function of $n$. Uniform Berry–Esseen bounds for $L$-statistics for smooth functions $J$ were given by Helmers (1977), and Helmers et al. (1990). In order to apply Theorems 10.1 and 10.2 to yield uniform and non-uniform bounds for $L$-statistics, let

$$g(x) = \int_{-\infty}^{\infty} \big(\mathbf{1}(x \le s) - F(s)\big) J\big(F(s)\big) ds \tag{10.30}$$

and

$$\sigma^2 = \int_{-\infty}^{\infty}\int_{-\infty}^{\infty} J\big(F(s)\big)J\big(F(t)\big)F\big(\min(s,t)\big)\big(1 - F\big(\max(s,t)\big)\big)dsdt,$$

which is easily checked to be the variance of $g(X)$ when $X$ has distribution function $F(x)$.

**Theorem 10.5** *Let $n \geq 2$ and assume that $EX_1^2 < \infty$ and $E|g(X_1)|^p < \infty$ for some $2 < p \leq 3$. If the weight function $J(t)$ is Lipschitz of order* 1 *on* $[0, 1]$, *that is, there exists a constant $c_0$ such that*

$$\big|J(t) - J(s)\big| \leq c_0|t - s| \quad \text{for } 0 \leq s, t \leq 1, \tag{10.31}$$

*then*

$$\sup_{z\in\mathbb{R}}\big|P\big(\sqrt{n}\sigma^{-1}\big(T(F_n) - T(F)\big) \leq z\big) - \Phi(z)\big| \leq \frac{6.1E|g(X_1)|^p}{n^{(p-2)/2}\sigma^p} + \frac{(1+\sqrt{2})c_0(EX_1^2)^{1/2}}{\sqrt{n}\sigma} \tag{10.32}$$

*and for all $z \in \mathbb{R}$,*

$$\big|P\big(\sqrt{n}\sigma^{-1}\big(T(F_n) - T(F)\big) \leq z\big) - \Phi(z)\big| \leq \frac{9c_0^2 EX_1^2}{(|z|+1)^2 n\sigma^2} + \frac{C}{(1+|z|)^p}\bigg(\frac{c_0(EX_1^2)^{1/2}}{\sqrt{n}\sigma} + \frac{E|g(X_1)|^p}{n^{(p-2)/2}\sigma^p}\bigg). \tag{10.33}$$

*Proof* Let $\psi(t) = \int_0^t J(s)ds$. Using integration by parts (see, e.g., Serfling 1980, p. 265), we have

$$T(F_n) - T(F) = -\int_{-\infty}^{\infty}\big[\psi\big(F_n(x)\big) - \psi\big(F(x)\big)\big]dx.$$

Therefore, letting

$$g_{n,i}(X_i) = -\frac{1}{\sqrt{n}\sigma}\int_{-\infty}^{\infty}\big(\mathbf{1}(X_i \leq x) - F(x)\big)J\big(F(x)\big)dx$$

we may write

$$\sqrt{n}\sigma^{-1}\big(T(F_n) - T(F)\big) = W + \Delta$$

with

$$W = \sum_{i=1}^{n} g_{n,i}(X_i) \quad \text{and}$$

$$\Delta = -\sqrt{n}\sigma^{-1}\int_{-\infty}^{\infty}\big[\psi\big(F_n(x)\big) - \psi\big(F(x)\big) - \big(F_n(x) - F(x)\big)J\big(F(x)\big)\big]dx.$$

As for every $i = 1, \ldots, n$, the variable $g_{n,i}(X_i)$ is independent of $W - g_{n,i}(X_i)$ and of

$$\Delta_i = -\sqrt{n}\sigma^{-1} \int_{-\infty}^{\infty} \left[\psi\big(F_{n,i}(x)\big) - \psi\big(F(x)\big) - \big(F_{n,i}(x) - F(x)\big)J\big(F(x)\big)\right]dx,$$

where

$$F_{n,i}(x) = \frac{1}{n}\left\{F(x) + \sum_{1 \le j \le n,\, j \ne i} \mathbf{1}(X_j \le x)\right\},$$

we may apply Theorems 10.1 and 10.2. Hence, to prove Theorem 10.5 it suffices to show

$$\sigma^2 E\Delta^2 \le c_0^2 n^{-1} E X_1^2 \tag{10.34}$$

and

$$\sigma^2 E|\Delta - \Delta_i|^2 \le 2c_0^2 n^{-2} E X_1^2. \tag{10.35}$$

Observe that the Lipschitz condition (10.31) implies

$$\big|\psi(t) - \psi(s) - (t-s)J(s)\big| = \left|\int_0^{t-s} \big(J(u+s) - J(s)\big)du\right| \le 0.5c_0(t-s)^2 \tag{10.36}$$

for $0 \le s \le t \le 1$. Hence with $\eta_i(x) = \mathbf{1}(X_i \le x) - F(x)$, we have

$$\begin{aligned}
&\sigma^2 E\Delta^2 \\
&\le 0.25c_0^2 n E\left(\int_{-\infty}^{\infty} \big(F_n(x) - F(x)\big)^2 dx\right)^2 \\
&= 0.25c_0^2 n^{-3} \int_{-\infty}^{\infty}\int_{-\infty}^{\infty} E\left(\sum_{i=1}^{n}\sum_{j=1}^{n} \eta_i(x)\eta_j(y)\right)^2 dxdy \\
&\le 0.25c_0^2 n^{-3} \int_{-\infty}^{\infty}\int_{-\infty}^{\infty} \big(3n^2 E\eta_1^2(x) E\eta_1^2(y) + nE\{\eta_1^2(x)\eta_1^2(y)\}\big)dxdy.
\end{aligned}$$

For the first term in the integral, we observe that

$$\begin{aligned}
&\int_{-\infty}^{\infty}\int_{-\infty}^{\infty} E\eta_1^2(x) E\eta_1^2(y) dxdy \\
&\quad = \left(\int_{-\infty}^{\infty} F(x)\big(1 - F(x)\big)dx\right)^2 \le \big(E|X_1|\big)^2 \le EX_1^2,
\end{aligned}$$

while for the second term,

$$\begin{aligned}
&\int_{-\infty}^{\infty}\int_{-\infty}^{\infty} E\{\eta_1^2(x)\eta_1^2(y)\}dxdy \\
&\quad = 2\int\int_{x \le y} E\{\eta_1^2(x)\eta_1^2(y)\}dxdy
\end{aligned}$$

$$
\begin{aligned}
&= 2\int\int_{x\le y}\{(1-F(x))^2(1-F(y))^2F(x)\\
&\quad + F^2(x)(1-F(y))^2(F(y)-F(x))\\
&\quad + F^2(x)F^2(y)(1-F(y))\}dxdy\\
&\le 2\int\int_{x\le y}F(x)(1-F(y))dxdy\\
&= 2\left\{\int\int_{x\le y\le 0}+\int\int_{0<x\le y}+\int\int_{x\le 0,y>0}\right\}F(x)(1-F(y))dxdy\\
&\le 2\left\{\int_{x\le 0}|x|F(x)dx+\int_{y\ge 0}y(1-F(y))dy\right.\\
&\quad \left.+\int_{x\le 0}F(x)dx\int_{y>0}(1-F(y))dy\right\}\\
&\le E(X_1^-)^2+E(X_1^+)^2+2EX_1^-EX_1^+\\
&\le 2EX_1^2.
\end{aligned}
\tag{10.37}
$$

Recalling $n \ge 2$, the proof of inequality (10.34) is complete.

To prove (10.35), first observe that

$$
\begin{aligned}
&\frac{\sigma}{\sqrt{n}}|\Delta-\Delta_i|\\
&= \left|\int_{-\infty}^{\infty}[\psi(F_n(x))-\psi(F_{n,i}(x))-(F_n(x)-F_{n,i}(x))J(F_{n,i}(x))]dx\right.\\
&\quad \left.+\int_{-\infty}^{\infty}(F_n(x)-F_{n,i}(x))[J(F_{n,i}(x))-J(F(x))]dx\right|\\
&\le 0.5c_0\int_{-\infty}^{\infty}(F_n(x)-F_{n,i}(x))^2dx\\
&\quad + c_0\int_{-\infty}^{\infty}|F_n(x)-F_{n,i}(x)||F_{n,i}(x)-F(x)|dx\\
&= 0.5c_0n^{-2}\int_{-\infty}^{\infty}(\mathbf{1}(X_i\le x)-F(x))^2dx\\
&\quad + c_0n^{-2}\int_{-\infty}^{\infty}|\mathbf{1}(X_i\le x)-F(x)|\left|\sum_{j\ne i}\{\mathbf{1}(X_j\le x)-F(x)\}\right|dx\\
&= 0.5c_0n^{-2}\int_{-\infty}^{\infty}\eta_i^2(x)dx + c_0n^{-2}\int_{-\infty}^{\infty}|\eta_i(x)|\left|\sum_{j\ne i}\eta_j(x)\right|dx.
\end{aligned}
$$

Now, to handle the first term above, from (10.37) we have

$$
E\left(\int_{-\infty}^{\infty}\eta_i^2(x)dx\right)^2 = \int_{-\infty}^{\infty}\int_{-\infty}^{\infty}E\eta_1^2(x)\eta_1^2(y)dxdy \le 4EX_1^2,
$$

while for the second term,

$$
\begin{aligned}
&E\Big(\int_{-\infty}^{\infty}|\eta_i(x)|\Big|\sum_{j\neq i}\eta_j(x)\Big|dx\Big)^2\\
&=\int_{-\infty}^{\infty}\int_{-\infty}^{\infty}E\Big\{|\eta_i(x)|\Big|\sum_{j\neq i}\eta_j(x)\Big||\eta_i(y)|\Big|\sum_{j\neq i}\eta_j(y)\Big|\Big\}dxdy\\
&=\int_{-\infty}^{\infty}\int_{-\infty}^{\infty}E|\eta_i(x)\eta_i(y)|E\Big\{\Big|\sum_{j\neq i}\eta_j(x)\Big|\Big|\sum_{j\neq i}\eta_j(y)\Big|\Big\}dxdy\\
&\le\int_{-\infty}^{\infty}\int_{-\infty}^{\infty}\|\eta_i(x)\|_2\|\eta_i(y)\|_2\Big\|\sum_{j\neq i}\eta_j(x)\Big\|_2\Big\|\sum_{j\neq i}\eta_j(y)\Big\|_2dxdy\\
&=(n-1)\int_{-\infty}^{\infty}\int_{-\infty}^{\infty}\|\eta_i(x)\|_2^2\|\eta_i(y)\|_2^2dxdy\\
&\le(n-1)\big(E|X_1|\big)^2\le(n-1)EX_1^2.
\end{aligned}
$$

Hence, applying (9.7) and recalling $n\ge 4$ we obtain

$$
\begin{aligned}
&\sigma^2E|\Delta-\Delta_i|^2\\
&\le n^{-3}c_0^2E\Big(0.5\int_{-\infty}^{\infty}\eta_i^2(x)dx+\int_{-\infty}^{\infty}|\eta_i(x)|\Big|\sum_{j\neq i}\eta_j(x)\Big|dx\Big)^2\\
&\le n^{-3}c_0^2\Big\{0.75E\Big(\int_{-\infty}^{\infty}\eta_i^2(x)dx\Big)^2+1.5E\Big(\int_{-\infty}^{\infty}|\eta_i(x)|\Big|\sum_{j\neq i}\eta_j(x)\Big|dx\Big)^2\Big\}\\
&\le n^{-3}c_0^2\{3EX_1^2+1.5(n-1)EX_1^2\}\\
&\le 2n^{-2}c_0^2EX_1^2.
\end{aligned}
$$

This proves (10.35), and hence the theorem. □

### 10.2.4 Random Sums of Independent Random Variables with Non-random Centering

Let $\{X_i, i\ge 1\}$ be i.i.d. random variables with $EX_i=\mu$ and $\mathrm{Var}(X_i)=\sigma^2$, and let $\{N_n, n\ge 1\}$ be a sequence of non-negative integer-valued random variables that are independent of $\{X_i, i\ge 1\}$. Assume for each $n=1,2,\ldots$ that $EN_n^2<\infty$ and

$$\frac{N_n-EN_n}{\sqrt{\mathrm{Var}(N_n)}}\xrightarrow{d}\mathcal{N}(0,1).$$

Then, by Robbins (1948),

$$\frac{\sum_{i=1}^{N_n}X_i-(EN_n)\mu}{\sqrt{\sigma^2EN_n+\mu^2\,\mathrm{Var}(N_n)}}\xrightarrow{d}\mathcal{N}(0,1). \tag{10.38}$$

This result is a special case of limit theorems for random sums with non-random centering. Such problems arise, for example, in the study of Galton–Watson branching processes. We refer to Finkelstein et al. (1994), and references therein, for recent developments in this area.

As another application of Theorem 10.1, we give the following uniform bound for the convergence in (10.38) when $N_n$ is the sum of i.i.d. random variables.

**Theorem 10.6** *Let $\{X_i, i \geq 1\}$ be i.i.d. random variables with $EX_i = \mu$, $\mathrm{Var}(X_i) = \sigma^2$ and $E|X_i|^3 < \infty$, and $\{Y_i, i \geq 1\}$ i.i.d. non-negative integer-valued random variables with $EY_i = \nu$, $\mathrm{Var}(Y_i) = \tau^2$, $EY_1^3 < \infty$ with $\{X_i, i \geq 1\}$ and $\{Y_i, i \geq 1\}$ independent. If $N_n = \sum_{i=1}^n Y_i$, then there exists a universal constant $C$ such that*

$$\sup_{z\in\mathbb{R}}\left|P\left(\frac{\sum_{i=1}^{N_n} X_i - n\mu\nu}{\sqrt{n(\nu\sigma^2+\tau^2\mu^2)}} \leq z\right) - \Phi(z)\right| \leq Cn^{-1/2}\left(\frac{\tau^2}{\nu^2} + \frac{EY_1^3}{\tau^3} + \frac{E|X_1|^3}{\nu^{1/2}\sigma^3} + \frac{\sigma}{\mu\sqrt{\nu}}\right). \tag{10.39}$$

*Proof* Let $Z_1$ and $Z_2$ be independent standard normal random variables that are independent of $\{X_i, i \geq 1\}$ and $\{Y_i, i \geq 1\}$. Put

$$b = \sqrt{\nu\sigma^2 + \tau^2\mu^2},$$

$$T_n = \frac{\sum_{i=1}^{N_n} X_i - n\mu\nu}{\sqrt{n}b} \quad \text{and} \quad H_n = \frac{\sum_{i=1}^{N_n} X_i - N_n\mu}{\sqrt{N_n}\sigma},$$

and write

$$T_n = \frac{\sqrt{N_n}\sigma}{\sqrt{n}b}H_n + \frac{(N_n - n\nu)\mu}{\sqrt{n}b}$$

and

$$T_n(Z_1) = \frac{\sqrt{N_n}\sigma}{\sqrt{n}b}Z_1 + \frac{(N_n - n\nu)\mu}{\sqrt{n}b}.$$

Applying the Berry–Esseen inequality (3.27) to $H_n$ by first conditioning on $N_n$ yields, with $C$ not necessarily the same at each occurrence, that

$$\begin{aligned}
&\sup_{z\in\mathbb{R}}\left|P(T_n \leq z) - P\big(T_n(Z_1) \leq z\big)\right| \\
&\quad \leq P\big(|N_n - n\nu| > n\nu/2\big) + CE\left(\frac{|X_1|^3}{\sqrt{N_n}\sigma^3}\mathbf{1}\{|N_n - n\nu| \leq n\nu/2\}\right) \\
&\quad \leq \frac{4\tau^2}{n\nu^2} + Cn^{-1/2}\frac{E|X_1|^3}{\nu^{1/2}\sigma^3}.
\end{aligned} \tag{10.40}$$

Now, letting the truncation $\overline{x}$ for any $x \in \mathbb{R}$ be given by

$$\overline{x} = \begin{cases} n\nu/2 & \text{for } x < n\nu/2, \\ x & \text{for } n\nu/2 \leq x \leq 3n\nu/2, \\ 3n\nu/2 & \text{for } x > 3n\nu/2, \end{cases}$$

we may write

$$\widehat{T}_n(Z_1) = \frac{\sqrt{N_n}\sigma}{\sqrt{n}b} Z_1 + \frac{(N_n - n\nu)\mu}{\sqrt{n}b} = \frac{\tau\mu}{b}\left(W + \Delta + \frac{\sigma\sqrt{\nu}}{\tau\mu} Z_1\right),$$

where

$$W = \frac{N_n - n\nu}{\sqrt{n\tau}} \quad \text{and} \quad \Delta = \frac{(\sqrt{N_n} - \sqrt{n\nu})\sigma Z_1}{\sqrt{n\tau\mu}}.$$

As $Y_i$ is independent of $N_n - Y_i$ for all $i = 1, \ldots, n$, we may apply Theorem 10.1 to $W + \Delta$ setting

$$\Delta_i = \frac{(\sqrt{N_n - Y_i + \nu} - \sqrt{n\nu})\sigma Z_1}{\sqrt{n\tau\mu}} \quad \text{for } i = 1, \ldots, n. \tag{10.41}$$

To evaluate the second term in the bound (10.7), condition on $Z_1$ and apply the identity

$$\sqrt{x} - \sqrt{y} = \frac{x - y}{\sqrt{x} + \sqrt{y}}$$

to obtain

$$E\big(|W\Delta|\,|Z_1\big) \le \frac{\sigma|Z_1|}{\sqrt{n\tau\mu}} E\left(\frac{|W(N_n - n\nu)|}{\sqrt{n\nu}}\right) \le \frac{\sigma|Z_1|}{\sqrt{n}\mu\sqrt{\nu}},$$

while for the third term, apply

$$\frac{1}{\sqrt{n\tau}} E\big(|(Y_i - \nu)(\Delta - \Delta_i)|\,|\,Z_1\big) \le \frac{\sqrt{2}\sigma|Z_1|}{n^{3/2}\tau^2\mu\sqrt{\nu}} E(Y_i - \nu)^2 = \frac{\sqrt{2}\sigma|Z_1|}{n^{3/2}\mu\sqrt{\nu}}.$$

Now letting

$$T_n(Z_1, Z_2) = \frac{\tau\mu}{b}\left(Z_2 + \frac{\sigma\sqrt{\nu}}{\tau\mu} Z_1\right)$$

and applying Theorem 10.1 for given $Z_1$ yields

$$\begin{aligned}
&\sup_{z\in\mathbb{R}}\big|P\big(T_n(Z_1) \le z\big) - P\big(T_n(Z_1, Z_2) \le z\big)\big| \\
&\quad\le P\big(|N_n - n\nu| > 0.5n\nu\big) + \sup_{z\in\mathbb{R}}\big|P\big(\widehat{T}_n(Z_1) \le z\big) - P\big(T_n(Z_1, Z_2) \le z\big)\big| \\
&\quad\le \frac{4\tau^2}{n\nu^2} + C\left(\frac{EY_1^3}{n^{1/2}\tau^3} + \frac{\sigma E|Z_1|}{n^{1/2}\mu\sqrt{\nu}}\right) \\
&\quad\le Cn^{-1/2}\left(\frac{\tau^2}{\nu^2} + \frac{EY_1^3}{\tau^3} + \frac{\sigma}{\mu\sqrt{\nu}}\right).
\end{aligned} \tag{10.42}$$

As $T_n(Z_1, Z_2)$ has a standard normal distribution, (10.39) follows from (10.40) and (10.42). □

### 10.2.5 Functions of Non-linear Statistics

Let $X_1, X_2, \ldots, X_n$ be a random sample and $\hat{\theta}_n = \hat{\theta}_n(X_1, \ldots, X_n)$ a weakly consistent estimator of an unknown parameter $\theta$. Assume that $\hat{\theta}_n$ can be written as

$$\hat{\theta}_n = \theta + \frac{1}{\sqrt{n}}\left(\sum_{i=1}^{n} \xi_i + \Delta\right) \tag{10.43}$$

where $\xi_i = g_{n,i}(X_i)$ are functions of $X_i$ satisfying $E\xi_i = 0$ and $\sum_{i=1}^n E\xi_i^2 = 1$, and $\Delta := \Delta_n(X_1, \ldots, X_n) \to 0$ in probability. Under these conditions,

$$\sqrt{n}(\hat{\theta}_n - \theta) \to_d \mathcal{N}(0,1) \quad \text{as } n \to \infty.$$

The class of $U$-statistics, multi-sample $U$-statistics and $L$-statistics discussed in previous subsections fit into this setting. The so called 'Delta Method' in statistics (see Theorem 7 of Ferguson 1996, for instance) allows us to determine the asymptotic distribution of functions of the estimator $\hat{\theta}_n$. In particular, if $h$ is differentiable in a neighborhood of $\theta$ with $h'$ continuous at $\theta$ with $h'(\theta) \neq 0$, then

$$\frac{\sqrt{n}(h(\hat{\theta}_n) - h(\theta))}{h'(\theta)} \to_d \mathcal{N}(0,1). \tag{10.44}$$

Of course, results that give some idea as to the accuracy of the Delta Method are of interest. When $\hat{\theta}_n$ is the sample mean, the Berry–Esseen bound and Edgeworth expansion have been well studied (see Bhattacharya and Ghosh 1978). The next theorem shows that the results in Sect. 10.1 can be extended to functions of non-linear statistics.

**Theorem 10.7** *Suppose the statistic $\hat{\theta}_n$ may be expressed in the form* (10.43) *where $\xi_1, \ldots, \xi_n$ are independent random variables with mean zero and satisfy* $\mathrm{Var}(W) = 1$ *where $W = \sum_{i=1}^n \xi_i$. Assume that $h'(\theta) \neq 0$ and $\delta(c_0) = \sup_{|x-\theta| \le c_0} |h''(x)| < \infty$ for some $c_0 > 0$. Then for all $2 < p \le 3$,*

$$\begin{aligned}
&\sup_{z\in\mathbb{R}} \left| P\left(\frac{\sqrt{n}(h(\hat{\theta}_n) - h(\theta))}{h'(\theta)} \le z\right) - \Phi(z)\right| \\
&\quad\le \left(1 + \frac{2c_0\delta(c_0)}{|h'(\theta)|}\right)\left(E|W\Delta| + \sum_{i=1}^n E\left|\xi_i(\Delta - \Delta_i)\right|\right) \\
&\qquad + 6.1\sum_{i=1}^n E|\xi_i|^p + \frac{4}{c_0^2 n} + \frac{2E|\Delta|}{c_0 n^{1/2}} + \frac{3.4c_0^{3-p}\delta(c_0)}{|h'(\theta)|n^{(p-2)/2}} \\
&\qquad + \frac{n^{-1/2}\delta(c_0)}{|h'(\theta)|}\sum_{i=1}^n E\xi_i^2 E|\Delta_i|,
\end{aligned} \tag{10.45}$$

*for any $\Delta_i, i = 1, \ldots, n$ such that $\xi_i$ and $(W - \xi_i, \Delta_i)$ are independent.*

Naturally, Theorem 10.7 may be applied as well to functions of linear statistics, and before turning to the proof we present the following simple example. Suppose one is making inference based on a random sample $X_1, \ldots, X_n$ from the Poisson distribution with unknown parameter $\lambda > 0$. Letting

$$\overline{X}_n = \frac{1}{n}\sum_{i=1}^{n} X_i$$

be the sample mean, the central limit theorem yields

$$\sqrt{n}(\overline{X}_n - \lambda) \to_d \mathcal{N}(0, \lambda). \tag{10.46}$$

As the limiting distribution depends on the very parameter one is trying to estimate, a confidence interval based directly on (10.46) would depend on the estimate $\overline{X}_n$ of $\lambda$ not only, naturally, for its centering, but also for its length, thus contributing some additional, unwanted, uncertainty. However, here, as in other important cases of interest, there exists a variance stabilizing transformation, that is, a function $g$ such that the standardized asymptotic distribution of $g(\overline{X}_n)$ does not depend on $\lambda$. For the case at hand, a direct application of the Delta method with $g(x) = 2\sqrt{x}$ yields

$$\sqrt{n}(2\sqrt{\overline{X}_n} - 2\sqrt{\lambda}) \to_d \mathcal{N}(0, 1). \tag{10.47}$$

In (10.47) the limiting distribution is known even though $\lambda$ is not, and the resulting confidence interval, for $2\sqrt{\lambda}$, will use an estimate of $\lambda$ only for centering.

To apply Theorem 10.7 to calculate a bound on the error in the normal approximation justified by (10.47), we note that (10.43) holds with $\Delta = 0$ upon letting

$$\hat{\theta}_n = \frac{\overline{X}_n}{\sqrt{\lambda}}, \quad \theta = \sqrt{\lambda} \quad \text{and} \quad \xi_i = \frac{X_i - \lambda}{\sqrt{\lambda n}}.$$

With these choices (10.47) is equivalent to (10.44) when $h(x) = 2\lambda^{1/4}\sqrt{x}$; in particular, note that $h'(\sqrt{\lambda}) = 1$. With $\Delta_i = 0$, Theorem 10.7 yields for, say, $p = 3$ and $c_0 = \sqrt{\lambda}/2$, that

$$\sup_{z \in \mathbb{R}} \left| P\left(\sqrt{n}(2\sqrt{\overline{X}_n} - 2\sqrt{\lambda}) \le z\right) - \Phi(z)\right|$$
$$\le \frac{6.1E|X_1 - \lambda|^3}{\lambda^{3/2}\sqrt{n}} + \frac{4.9}{\sqrt{\lambda n}} + \frac{16}{\lambda n}.$$

We now proceed to the proof of the theorem.

*Proof* Let $p \in (2, 3]$. Since (10.45) is trivial if $\sum_{i=1}^{n} E|\xi_i|^p > 1/6$, we assume

$$\sum_{i=1}^{n} E|\xi_i|^p \le 1/6. \tag{10.48}$$

Similar to the proof of Theorem 10.6, let

$$\overline{x} = \begin{cases} -c_0/2 & \text{for } x < -c_0/2, \\ x & \text{for } -c_0/2 \le x \le c_0/2, \\ c_0/2 & \text{for } x > c_0/2. \end{cases}$$

Observe that

$$\begin{aligned} &\frac{\sqrt{n}(h(\hat{\theta}_n) - h(\theta))}{h'(\theta)} \\ &= \frac{\sqrt{n}}{h'(\theta)}\left(h'(\theta)(\hat{\theta}_n - \theta) + \int_0^{\hat{\theta}_n - \theta} \left[h'(\theta + t) - h'(\theta)\right]dt\right) \\ &= W + \Delta + \frac{\sqrt{n}}{h'(\theta)} \int_0^{n^{-1/2}(W+\Delta)} \left[h'(\theta + t) - h'(\theta)\right]dt \\ &:= W + \Lambda + R, \end{aligned}$$

where

$$\Lambda = \Delta + \frac{\sqrt{n}}{h'(\theta)} \int_0^{\overline{n^{-1/2}W} + \overline{n^{-1/2}\Delta}} \left[h'(\theta + t) - h'(\theta)\right]dt \quad \text{and}$$

$$R = \frac{\sqrt{n}}{h'(\theta)} \int_{\overline{n^{-1/2}W} + \overline{n^{-1/2}\Delta}}^{n^{-1/2}(W+\Delta)} \left[h'(\theta + t) - h'(\theta)\right]dt.$$

We will apply Theorem 10.1 with $\Lambda$ and $\Lambda_i$, defined in (10.53), playing the role of $\Delta$ and $\Delta_i$, respectively. But first, in order to handle the remainder term $R$, note that if $|n^{-1/2}W| \le c_0/2$ and $|n^{-1/2}\Delta| \le c_0/2$ then $R = 0$. Hence

$$\begin{aligned} P\big(|R| > 0\big) &\le P\big(|W| > c_0 n^{1/2}/2\big) + P\big(|\Delta| > c_0 n^{1/2}/2\big) \\ &\le 4/\big(c_0^2 n\big) + 2E|\Delta|/\big(c_0 n^{1/2}\big). \end{aligned} \tag{10.49}$$

Recall $W = \sum_{i=1}^n \xi_i$. We prove in the Appendix that under (10.48), for all $2 < p \le 3$,

$$E|W|^p \le 2\big(EW^2\big)^{p/2} + \sum_{i=1}^n E|\xi_i|^p \le 2.2. \tag{10.50}$$

With $W^{(i)}$ denoting $W - \xi_i$ as usual, we have

$$\begin{aligned} &\left|\int_0^{\overline{n^{-1/2}W} + \overline{n^{-1/2}\Delta}} \left[h'(\theta + t) - h'(\theta)\right]dt\right| \\ &\le 0.5\delta(c_0)\big(\overline{n^{-1/2}W} + \overline{n^{-1/2}\Delta}\big)^2 \\ &\le \delta(c_0)\big(\big(\overline{n^{-1/2}W}\big)^2 + \big(\overline{n^{-1/2}\Delta}\big)^2\big) \\ &\le \delta(c_0)\big((c_0/2)^{3-p}\big(n^{-1/2}|W|\big)^{p-1} + (c_0/2)n^{-1/2}|\Delta|\big), \end{aligned} \tag{10.51}$$

and therefore

$$
\begin{aligned}
E|W\Lambda| &\le E|W\Delta| + \frac{(c_0/2)^{3-p}\delta(c_0)}{|h'(\theta)|n^{(p-2)/2}} E|W|^p + \frac{c_0\delta(c_0)}{|h'(\theta)|} E|W\Delta| \\
&\le \left(1 + \frac{c_0\delta(c_0)}{|h'(\theta)|}\right) E|W\Delta| + \frac{2.2c_0^{3-p}\delta(c_0)}{|h'(\theta)|n^{(p-2)/2}},
\end{aligned} \tag{10.52}
$$

where for the last term we have applied inequality (10.50).

Now introducing

$$
\Lambda_i = \Delta_i + \frac{\sqrt{n}}{h'(\theta)} \int_0^{\overline{n^{-1/2}W^{(i)}} + \overline{n^{-1/2}\Delta_i}} \left[h'(\theta+t) - h'(\theta)\right] dt, \tag{10.53}
$$

the difference $\Lambda - \Lambda_i$ will equal $\Delta - \Delta_i$ plus $\sqrt{n}/h'(\theta)$ times the term in the absolute value of (10.54), which we bound in a manner similar to (10.51). In particular, applying the bound

$$
\left| \int_a^b \left[h'(\theta+t) - h'(\theta)\right] dt \right| \le \frac{\delta(c_0)}{2} |b-a| \big(|a| + |b|\big) \quad \text{for } a, b \in [-c_0, c_0],
$$

we obtain

$$
\begin{aligned}
&\left| \int_{\overline{n^{-1/2}W^{(i)}} + \overline{n^{-1/2}\Delta_i}}^{\overline{n^{-1/2}W} + \overline{n^{-1/2}\Delta}} \left[h'(\theta+t) - h'(\theta)\right] dt \right| \\
&\le \frac{\delta(c_0)}{2} \Big( \big|\overline{n^{-1/2}W} - \overline{n^{-1/2}W^{(i)}}\big| \Big( \big|\overline{n^{-1/2}W}\big| + \big|\overline{n^{-1/2}W^{(i)}}\big| \\
&\quad + \big|\overline{n^{-1/2}\Delta}\big| + \big|\overline{n^{-1/2}\Delta_i}\big| \Big) + 2c_0 \big|\overline{n^{-1/2}\Delta} - \overline{n^{-1/2}\Delta_i}\big| \Big) \\
&\le \frac{\delta(c_0)}{2} \Big( \big|\overline{n^{-1/2}W} - \overline{n^{-1/2}W^{(i)}}\big| \Big( c_0^{3-p} |n^{-1/2}W|^{p-2} + c_0^{3-p} |n^{-1/2}W^{(i)}|^{p-2} \\
&\quad + n^{-1/2}|\Delta| + n^{-1/2}|\Delta_i| \Big) + 2c_0 n^{-1/2} |\Delta - \Delta_i| \Big) \\
&\le \delta(c_0) \Big( c_0^{3-p} n^{-(p-1)/2} |\xi_i| \big( |W^{(i)}|^{p-2} + |\xi_i|^{p-2} \big) \\
&\quad + n^{-1} |\xi_i| |\Delta_i| + 2c_0 n^{-1/2} |\Delta - \Delta_i| \Big).
\end{aligned} \tag{10.54}
$$

Now, to attend to the final term in the bound (10.7), where, again, $\Lambda$ and $\Lambda_i$ are playing the role of $\Delta$ and $\Delta_i$, from the inequality above we obtain

$$
\begin{aligned}
&\sum_{i=1}^n E\big|\xi_i(\Lambda - \Lambda_i)\big| \\
&\quad \le \sum_{i=1}^n E\big|\xi_i(\Delta - \Delta_i)\big| \\
&\qquad + \frac{\sqrt{n}\delta(c_0)}{|h'(\theta)|} \Bigg\{ c_0^{3-p} n^{-(p-1)/2} \sum_{i=1}^n E\Big(|\xi_i|^2 \big( |W^{(i)}|^{p-2} + |\xi_i|^{p-2} \big)\Big) \\
&\qquad + n^{-1} \sum_{i=1}^n E\xi_i^2 |\Delta_i| + 2c_0 n^{-1/2} \sum_{i=1}^n E\big|\xi_i(\Delta - \Delta_i)\big| \Bigg\}
\end{aligned}
$$

$$\begin{aligned}
&\le \left(1+\frac{2c_0\delta(c_0)}{|h'(\theta)|}\right)\sum_{i=1}^n E\big|\xi_i(\Delta-\Delta_i)\big| + \frac{c_0^{3-p}\delta(c_0)}{|h'(\theta)|n^{(p-2)/2}}\sum_{i=1}^n \big(E\xi_i^2+E|\xi_i|^p\big)\\
&\quad+\frac{n^{-1/2}\delta(c_0)}{|h'(\theta)|}\sum_{i=1}^n E\xi_i^2 E|\Delta_i|\\
&\le \left(1+\frac{2c_0\delta(c_0)}{|h'(\theta)|}\right)\sum_{i=1}^n E\big|\xi_i(\Delta-\Delta_i)\big|\\
&\quad+\frac{1.2c_0^{3-p}\delta(c_0)}{|h'(\theta)|n^{(p-2)/2}}+\frac{n^{-1/2}\delta(c_0)}{|h'(\theta)|}\sum_{i=1}^n E\xi_i^2 E|\Delta_i|, \qquad (10.55)
\end{aligned}$$

recalling (10.48) for the last inequality. The theorem now follows by combining (10.7), (10.49), (10.52) and (10.55). □

## 10.3 Uniform and Non-uniform Randomized Concentration Inequalities

As the previous chapters have demonstrated, the concentration inequality approach is a powerful tool for deriving sharp Berry–Esseen bounds for independent random variables. In this section we develop uniform and non-uniform randomized concentration inequalities which we will use to prove Theorems 10.1 and 10.2.

Let $\xi_1,\ldots,\xi_n$ be independent random variables satisfying (10.1), $W=\sum_{i=1}^n \xi_i$ and $T=W+\Delta$. The simple inequality

$$\begin{aligned}
-P\big(z-|\Delta|\le W\le z\big)\le P(T\le z)-P(W\le z)&\\
\le P\big(z\le W\le z+|\Delta|\big)& \qquad (10.56)
\end{aligned}$$

provides lower and upper bounds for the difference between the distribution functions of $T$ and its approximation $W$, and involves the probability that $W$ lies in an interval of random length. Hence, we are led to consider concentration inequalities that bound quantities of the form $P(\Delta_1\le W\le\Delta_2)$.

**Proposition 10.1** *Let $\delta>0$ satisfy* (10.3). *Then*

$$\begin{aligned}
P(\Delta_1\le W\le\Delta_2)&\le 4\delta+E\big|W(\Delta_2-\Delta_1)\big|\\
&\quad+\sum_{i=1}^n\big\{E\big|\xi_i(\Delta_1-\Delta_{1,i})\big|+E\big|\xi_i(\Delta_2-\Delta_{2,i})\big|\big\}, \qquad (10.57)
\end{aligned}$$

*whenever $\xi_i$ is independent of $(W-\xi_i,\Delta_{1,i},\Delta_{2,i})$ for all $i=1,\ldots,n$.*

When both $\Delta_1$ and $\Delta_2$ are not random, say, $\Delta_1=a$ and $\Delta_2=b$ with $a\le b$, then, by (ii) of Remark 10.1, whenever $\beta_1+\beta_2\le 1/2$ Proposition 10.1 recovers (3.38) by letting $\Delta_{1,i}=a$ and $\Delta_{i,2}=b$ for each $i=1,\ldots,n$.

*Proof* As the probability $P(\Delta_1 \le W \le \Delta_2)$ is zero if $\Delta_1 > \Delta_2$ we may assume without loss of generality that $\Delta_1 \le \Delta_2$ a.s. We follow the proof of (3.28). For $a \le b$ let

$$f_{a,b}(w) = \begin{cases} -\frac{1}{2}(b-a) - \delta & \text{for } w < a - \delta, \\ w - \frac{1}{2}(a+b) & \text{for } a - \delta \le w \le b + \delta, \\ \frac{1}{2}(b-a) + \delta & \text{for } w > b + \delta, \end{cases}$$

and set

$$\hat{K}_i(t) = \xi_i\{\mathbf{1}(-\xi_i \le t \le 0) - \mathbf{1}(0 < t \le -\xi_i)\} \quad \text{and} \quad \hat{K}(t) = \sum_{i=1}^n \hat{K}_i(t).$$

Since $\xi_i$ and $f_{\Delta_{1,i},\Delta_{2,i}}(W - \xi_i)$ are independent for $1 \le i \le n$ and $E\xi_i = 0$, we have

$$\begin{aligned} EWf_{\Delta_1,\Delta_2}(W) &= \sum_{i=1}^n E\{\xi_i(f_{\Delta_1,\Delta_2}(W) - f_{\Delta_1,\Delta_2}(W - \xi_i))\} \\ &\quad + \sum_{i=1}^n E\{\xi_i(f_{\Delta_1,\Delta_2}(W - \xi_i) - f_{\Delta_{1,i},\Delta_{2,i}}(W - \xi_i))\} \\ &:= H_1 + H_2. \end{aligned} \tag{10.58}$$

Using the fact that $\hat{K}(t) \ge 0$ and $f'_{\Delta_1,\Delta_2}(w) \ge 0$, we have

$$\begin{aligned} H_1 &= \sum_{i=1}^n E\left\{\xi_i \int_{-\xi_i}^0 f'_{\Delta_1,\Delta_2}(W + t)dt\right\} \\ &= \sum_{i=1}^n E\left\{\int_{-\infty}^{\infty} f'_{\Delta_1,\Delta_2}(W + t)\hat{K}_i(t)dt\right\} \\ &= E\left\{\int_{-\infty}^{\infty} f'_{\Delta_1,\Delta_2}(W + t)\hat{K}(t)dt\right\} \\ &\ge E\left\{\int_{|t|\le\delta} f'_{\Delta_1,\Delta_2}(W + t)\hat{K}(t)dt\right\} \\ &\ge E\left\{\mathbf{1}_{\{\Delta_1 \le W \le \Delta_2\}} \int_{|t|\le\delta} \hat{K}(t)dt\right\} \\ &= E\left\{\mathbf{1}_{\{\Delta_1 \le W \le \Delta_2\}} \sum_{i=1}^n |\xi_i| \min(\delta, |\xi_i|)\right\} \\ &\ge H_{1,1} - H_{1,2}, \end{aligned} \tag{10.59}$$

where

$$H_{1,1} = P(\Delta_1 \le W \le \Delta_2) \sum_{i=1}^n E|\xi_i| \min(\delta, |\xi_i|) \ge \frac{1}{2} P(\Delta_1 \le W \le \Delta_2), \tag{10.60}$$

by (10.3), and

$$H_{1,2} = E\left|\sum_{i=1}^{n}\left\{|\xi_i|\min(\delta, |\xi_i|) - E|\xi_i|\min(\delta, |\xi_i|)\right\}\right|$$
$$\le \left(\mathrm{Var}\left(\sum_{i=1}^{n}|\xi_i|\min(\delta, |\xi_i|)\right)\right)^{1/2} \le \delta. \tag{10.61}$$

As to $H_2$, first, one verifies

$$\left|f_{\Delta_1,\Delta_2}(w) - f_{\Delta_{1,i},\Delta_{2,i}}(w)\right| \le |\Delta_1 - \Delta_{1,i}|/2 + |\Delta_2 - \Delta_{2,i}|/2,$$

which then yields

$$|H_2| \le \frac{1}{2}\sum_{i=1}^{n}\left\{E\left|\xi_i(\Delta_1 - \Delta_{1,i})\right| + E\left|\xi_i(\Delta_2 - \Delta_{2,i})\right|\right\}. \tag{10.62}$$

It follows from the definition of $f_{\Delta_1,\Delta_2}$ that

$$\left|f_{\Delta_1,\Delta_2}(w)\right| \le \frac{1}{2}(\Delta_2 - \Delta_1) + \delta.$$

Hence, by (10.58)–(10.62)

$$\begin{aligned}
&P(\Delta_1 \le W \le \Delta_2)\\
&\quad\le 2EWf_{\Delta_1,\Delta_2}(W) + 2\delta + \sum_{i=1}^{n}\left\{E\left|\xi_i(\Delta_1 - \Delta_{1,i})\right| + E\left|\xi_i(\Delta_2 - \Delta_{2,i})\right|\right\}\\
&\quad\le E\left|W(\Delta_2 - \Delta_1)\right|\\
&\qquad + 2\delta E|W| + 2\delta + \sum_{i=1}^{n}\left\{E\left|\xi_i(\Delta_1 - \Delta_{1,i})\right| + E\left|\xi_i(\Delta_2 - \Delta_{2,i})\right|\right\}\\
&\quad\le E\left|W(\Delta_2 - \Delta_1)\right| + 4\delta + \sum_{i=1}^{n}\left\{E\left|\xi_i(\Delta_1 - \Delta_{1,i})\right| + E\left|\xi_i(\Delta_2 - \Delta_{2,i})\right|\right\},
\end{aligned}$$

as desired. □

*Proof of Theorem 10.1* Claim (10.5) follows from applying (10.56) and Proposition 10.1 with

$$(\Delta_1, \Delta_2, \Delta_{i,1}, \Delta_{i,2}) = \begin{cases} (z+\Delta, z, \Delta_i, z) & \Delta < 0,\\ (z, z+\Delta, z, \Delta_i) & \Delta \ge 0.\end{cases}$$

Next, (10.6) is trivial when $\beta_1 + \beta_2 > 1/2$, and otherwise follows from (10.5) and (ii) of Remark 10.1. Lastly, (10.7) is a direct corollary of (10.6) and (3.31). □

Theorem 10.2 is based on the following non-uniform randomized concentration inequality.

**Proposition 10.2** *Let $\delta > 0$ satisfy* (10.3). *If $\xi_i$ is independent of $(W - \xi_i, \Delta_{1,i}, \Delta_{2,i})$ for all $i = 1, \ldots, n$, then for all $a \in \mathbb{R}$ and $p \ge 2$,*

$$P(\Delta_1 \le W \le \Delta_2,\ \Delta_1 \ge a)$$
$$\le 2 \sum_{1\le i\le n} P\big(|\xi_i| > (1 \vee a)/(2p)\big) + e^p\big(1 + a^2/(4p)\big)^{-p}\beta_2 + e^{-a/2}\tau_1, \quad (10.63)$$

*where* $\beta_2$ *is given in* (10.4) *and*

$$\tau_1 = 18\delta + 7.2\|\Delta_2 - \Delta_1\|_2 + 3\sum_{i=1}^{n} \|\xi_i\|_2\big(\|\Delta_1 - \Delta_{1,i}\|_2 + \|\Delta_2 - \Delta_{2,i}\|_2\big). \quad (10.64)$$

*Proof* When $a \le 2$, (10.63) follows from Proposition 10.1. For $a > 2$, without loss of generality assume that

$$a \le \Delta_1 \le \Delta_2, \quad (10.65)$$

as otherwise we may consider $\overline{\Delta}_1 = \max(a, \Delta_1)$ and $\overline{\Delta}_2 = \max(a, \Delta_1, \Delta_2)$ and use the fact that $|\overline{\Delta}_2 - \overline{\Delta}_1| \le |\Delta_2 - \Delta_1|$. We follow the lines of argument in the proofs of Propositions 8.1 and 10.1. Let

$$\bar{xi}_i = \xi_i \mathbf{1}_{\{\xi_i \le 1\}}, \quad \bar{W} = \sum_{i=1}^{n} \bar{xi}_i, \quad \text{and} \quad \bar{W}^{(i)} = \bar{W} - \bar{xi}_i.$$

As in (8.20), we have

$$\{\Delta_1 \le W \le \Delta_2\} \subset \{\Delta_1 \le \bar{W} \le \Delta_2\} \cup \Big\{\Delta_1 \le W \le \Delta_2,\ \max_{1\le i\le n} \xi_i > 1\Big\}$$
$$\subset \{\Delta_1 \le \bar{W} \le \Delta_2\} \cup \Big\{W \ge a,\ \max_{1\le i\le n} \xi_i > 1\Big\}$$

by (10.65).

Invoking Lemma 8.3 for the second term above, it only remains to show

$$P(\Delta_1 \le \bar{W} \le \Delta_2) \le e^{-a/2}\tau_1. \quad (10.66)$$

We can assume that $\delta \le 0.065$ since otherwise, by (8.5) of Lemma 8.1 with $\alpha = 1$,

$$P(\Delta_1 \le \bar{W} \le \Delta_2) \le P(\bar{W} \ge a) \le e^{-a/2} E e^{\bar{W}/2}$$
$$\le e^{-a/2}\exp\big(e^{0.5} - 1.5\big) \le 1.17e^{-a/2} \le 18\delta e^{-a/2},$$

implying (10.66).

For $\alpha, \beta \in \mathbb{R}$ let

$$f_{\alpha,\beta}(w) = \begin{cases} 0 & \text{for } w < \alpha - \delta, \\ e^{w/2}(w - \alpha + \delta) & \text{for } \alpha - \delta \le w \le \beta + \delta, \\ e^{w/2}(\beta - \alpha + 2\delta) & \text{for } w > \beta + \delta, \end{cases} \quad (10.67)$$

and set

$$\bar{M}_i(t) = \xi_i(\mathbf{1}_{\{-\bar{xi}_i \le t \le 0\}} - \mathbf{1}_{\{0 < t \le -\bar{xi}_i\}}) \quad \text{and} \quad \bar{M}(t) = \sum_{i=1}^{n} \bar{M}_i(t).$$

Similarly to (10.58), we may write

$$EWf_{\Delta_1,\Delta_2}(\bar{W}) = H_3 + H_4, \tag{10.68}$$

where

$$H_3 = E\int_{-\infty}^{\infty} f'_{\Delta_1,\Delta_2}(\bar{W}+t)\bar{M}(t)dt \quad \text{and}$$
$$H_4 = \sum_{i=1}^{n} E\{\xi_i\big(f_{\Delta_1,\Delta_2}(\bar{W}-\bar{x}i_i) - f_{\Delta_{1,i},\Delta_{2,i}}(\bar{W}-\bar{x}i_i)\big)\}.$$

It follows from the fact that $\bar{M}(t) \ge 0$, $f'_{\Delta_1,\Delta_2}(w) \ge e^{w/2}$ for $\Delta_1 - \delta \le w \le \Delta_2 + \delta$ and $f'_{\Delta_1,\Delta_2}(w) \ge 0$ for all $w$ that

$$\begin{aligned}
H_3 &\ge E\left\{\int_{|t|\le\delta} f'_{\Delta_1,\Delta_2}(\bar{W}+t)\bar{M}(t)dt\right\} \\
&\ge E\left\{e^{(\bar{W}-\delta)/2}\mathbf{1}_{\{\Delta_1\le\bar{W}\le\Delta_2\}}\int_{|t|\le\delta}\bar{M}(t)dt\right\} \\
&= E\left\{e^{(\bar{W}-\delta)/2}\mathbf{1}_{\{\Delta_1\le\bar{W}\le\Delta_2\}}\sum_{i=1}^{n}|\xi_i|\min\big(\delta,|\bar{x}i_i|\big)\right\} \\
&\ge Ee^{(\bar{W}-\delta)/2}\mathbf{1}_{\{\Delta_1\le\bar{W}\le\Delta_2\}}\sum_{i=1}^{n}E|\xi_i|\min\big(\delta,|\bar{x}i_i|\big) \\
&\quad - Ee^{(\bar{W}-\delta)/2}\left|\sum_{i=1}^{n}\{|\xi_i|\min\big(\delta,|\bar{x}i_i|\big) - E|\xi_i|\min\big(\delta,|\bar{x}i_i|\big)\}\right| \\
&\ge H_{3,1} - H_{3,2},
\end{aligned} \tag{10.69}$$

where, as in the proof of Proposition 10.1,

$$H_{3,1} \ge e^{(a-\delta)/2}P(\Delta_1 \le \bar{W} \le \Delta_2)\sum_{i=1}^{n}E|\xi_i|\min\big(\delta,|\bar{x}i_i|\big) \quad \text{and}$$
$$H_{3,2} \le \big(Ee^{\bar{W}}\big)^{1/2}\left(\mathrm{Var}\left(\sum_{i=1}^{n}|\xi_i|\min\big(\delta,|\bar{x}i_i|\big)\right)\right)^{1/2}.$$

Since $\delta$ satisfies (10.3),

$$\sum_{i=1}^{n}E|\xi_i|\min\big(\delta,|\bar{x}i_i|\big) = \sum_{i=1}^{n}E|\xi_i|\min\big(\delta,|\xi_i|\big) \ge 1/2$$

and now $\delta \le 0.065$ yields

$$H_{3,1} \ge 0.48e^{a/2}P(\Delta_1 \le \bar{W} \le \Delta_2). \tag{10.70}$$

As in (8.15), using (8.5) to obtain $Ee^{\bar{W}^{(i)}} \le \exp(e-2) \le 2.06$ we have

$$H_{3,2} \le 1.44\delta. \tag{10.71}$$

As to $H_4$, it is easy to see that

$$\left|f_{\Delta_1,\Delta_2}(w) - f_{\Delta_{1,i},\Delta_{2,i}}(w)\right| \le e^{w/2}\big(|\Delta_1 - \Delta_{1,i}| + |\Delta_2 - \Delta_{2,i}|\big).$$

Hence, by the Hölder inequality and the independence of $\xi_i$ and $\bar{W} - \bar{x}i_i$,

$$\begin{aligned}
|H_4| &\le \sum_{i=1}^n E|\xi_i| e^{(\bar{W}-\bar{x}i_i)/2}\big(|\Delta_1 - \Delta_{1,i}| + |\Delta_2 - \Delta_{2,i}|\big) \\
&\le \sum_{i=1}^n \big(E\xi_i^2 e^{\bar{W}-\bar{x}i_i}\big)^{1/2}\big(\|\Delta_1 - \Delta_{1,i}\|_2 + \|\Delta_2 - \Delta_{2,i}\|_2\big) \\
&= \sum_{i=1}^n \big(E\xi_i^2 E e^{\bar{W}-\bar{x}i_i}\big)^{1/2}\big(\|\Delta_1 - \Delta_{1,i}\|_2 + \|\Delta_2 - \Delta_{2,i}\|_2\big) \\
&\le 1.44 \sum_{i=1}^n \|\xi_i\|_2\big(\|\Delta_1 - \Delta_{1,i}\|_2 + \|\Delta_2 - \Delta_{2,i}\|_2\big). \qquad (10.72)
\end{aligned}$$

Now, recalling $\bar{x}i_i \le 1$ for all $i$, we have

$$\begin{aligned}
&E W^2 e^{\bar{W}} \\
&\quad = \sum_{i=1}^n E\xi_i^2 e^{\bar{x}i_i} E e^{\bar{W}-\bar{x}i_i} + \sum_{1\le i\ne j\le n} E\xi_i\big(e^{\bar{x}i_i} - 1\big) E\xi_j\big(e^{\bar{x}i_j} - 1\big) E e^{\bar{W}-\bar{x}i_i-\bar{x}i_j} \\
&\quad \le 2.06e \sum_{i=1}^n E\xi_i^2 + 2.06(e-1)^2 \sum_{1\le i\ne j\le n} E\xi_i^2 E\xi_j^2 \\
&\quad \le 2.06e + 2.06(e-1)^2 < 3.42^2.
\end{aligned}$$

Thus, we obtain

$$\begin{aligned}
E W f_{\Delta_1,\Delta_2}(\bar{W}) &\le E|W| e^{\bar{W}/2}\big(|\Delta_2 - \Delta_1| + 2\delta\big) \\
&\le \big(E\big(W^2 e^{\bar{W}}\big)\big)^{1/2}\big(\|\Delta_2 - \Delta_1\|_2 + 2\delta\big) \\
&\le 3.42\big(\|\Delta_2 - \Delta_1\|_2 + 2\delta\big). \qquad (10.73)
\end{aligned}$$

Combining (10.68)–(10.73) yields

$$\begin{aligned}
&P(\Delta_1 \le \bar{W} \le \Delta_2) \\
&\quad \le e^{-a/2}(0.48)^{-1}\Bigg(8.28\delta + 3.42\|\Delta_2 - \Delta_1\|_2 \\
&\qquad + 1.44 \sum_{i=1}^n \|\xi_i\|_2\big(\|\Delta_1 - \Delta_{1,i}\|_2 + \|\Delta_2 - \Delta_{2,i}\|_2\big)\Bigg),
\end{aligned}$$

and collecting terms completes the verification of (10.66). □

*Proof of Theorem 10.2* Without loss of generality, assume that $z \ge 0$. Since for $0 \le z \le 2$ inequality (10.8) follows from (10.5), we may assume $z > 2$.

Applying Proposition 10.2 with

$$(\Delta_1, \Delta_2, \Delta_{1,i}, \Delta_{2,i}) = \big(z - |\Delta|, z, z - |\Delta_i|, z\big)$$

and $a = (2z-1)/3$ yields

$$\begin{aligned}
&P\big(z - |\Delta| \le W \le z, |\Delta| \le (z+1)/3\big) \\
&\quad \le 2 \sum_{1\le i\le n} P\big(|\xi_i| > \big(1 \vee (2z-1)/3\big)/(2p)\big) + e^p\big(1 + (2z-1)^2/(36p)\big)^{-p}\beta_2 \\
&\qquad + e^{-(2z-1)/6}\left(18\delta + 7.2\|\Delta\|_2 + 3\sum_{i=1}^n \|\xi_i\|_2\|\Delta - \Delta_i\|_2\right) \\
&\quad \le 2 \sum_{1\le i\le n} P\big(|\xi_i| > (z+1)/(6p)\big) + e^p\big(1 + z^2/(36p)\big)^{-p}\beta_2 + e^{-z/3}\tau.
\end{aligned}$$

Now combining the bound above with (10.56) and the inequality

$$\begin{aligned}
P\big(z - |\Delta| \le W \le z\big) &\le P\big(|\Delta| > (z+1)/3\big) \\
&\quad + P\big(z - |\Delta| \le W \le z, |\Delta| \le (z+1)/3\big)
\end{aligned}$$

yields

$$-\gamma_{z,p} - e^{-z/3}\tau \le P(T \le z) - P(W \le z).$$

Similarly showing the corresponding upper bound completes the proof of (10.8).

When $\beta_1 + \beta_2 \le 1/2$, in light of (ii) of Remark 10.1, choosing $\delta = (\beta_2 + \beta_3)/2$ and noting that $\beta_2 \le \sum_{i=1}^n E|\xi_i|^p$, $\beta_3 \le \sum_{i=1}^n E|\xi_i|^{3\wedge p}$ and

$$\sum_{i=1}^n P\left(|\xi_i| > \frac{|z|+1}{6p}\right) \le \frac{(6p)^p}{(|z|+1)^p}\sum_{i=1}^n E|\xi_i|^p,$$

we see (10.10) holds by (10.8) and Theorem 8.1.

If $\beta_2 + \beta_3 > 1/2$, then

$$\sum_{i=1}^n \big(E|\xi_i|^p + E|\xi_i|^{3\wedge p}\big) \ge 1/2$$

and

$$\begin{aligned}
P(T \ge z) &\le P\big(W \ge (2z-1)/3\big) + P\big(|\Delta| > (z+1)/3\big) \\
&\le \frac{C_p}{(1+z)^p}\left(1 + \sum_{i=1}^n E|\xi_i|^p\right) + P\big(|\Delta| > (z+1)/3\big)
\end{aligned}$$

by (8.10). Therefore (10.10) remains valid. □

# Appendix

*Proof of Lemma 10.1* It is known (see, e.g., Koroljuk and Borovskich 1994, p. 271) that

$$E\Big(\sum_{1\le i_1<\cdots<i_m\le n} \bar{h}_m(X_{i_1},\ldots,X_{i_m})\Big)^2 = \binom{n}{m}\sum_{j=2}^{m}\binom{m}{j}\binom{n-m}{m-j}E\bar{h}_j^2(X_1,\ldots,X_j). \tag{10.74}$$

Using that the variables are i.i.d., the symmetry of $h(x_1,\ldots,x_m)$, and that $Eh(X_1,\ldots,X_m)=0$ implies $Eg(X_i)=0$, we have

$$\begin{aligned}
&E\bar{h}_j^2(X_1,\ldots,X_j)\\
&= Eh_j^2(X_1,\ldots,X_j) - 2\sum_{i=1}^{j}E\big[g(X_i)h_j(X_1,\ldots,X_j)\big] + E\Big(\sum_{i=1}^{j}g(X_i)\Big)^2\\
&= Eh_j^2(X_1,\ldots,X_j) - 2jE\big[g(X_1)E\big(h(X_1,\ldots,X_m)|X_1,\ldots,X_j\big)\big]\\
&\quad + jEg^2(X_1)\\
&= Eh_j^2(X_1,\ldots,X_j) - 2jE\big[g(X_1)h(X_1,\ldots,X_m)\big] + jEg^2(X_1)\\
&= Eh_j^2(X_1,\ldots,X_j) - 2jEg^2(X_1) + jEg^2(X_1)\\
&= Eh_j^2(X_1,\ldots,X_j) - jEg_1^2(X_1),
\end{aligned} \tag{10.75}$$

so in particular

$$E\bar{h}_j^2(X_1,\ldots,X_j) \le Eh_j^2(X_1,\ldots,X_j). \tag{10.76}$$

We next prove by induction that for $2\le j\le m$

$$Eh_{j-1}^2(X_1,\ldots,X_{j-1}) \le \frac{j-1}{j}Eh_j^2(X_1,\ldots,X_j). \tag{10.77}$$

Since $E\bar{h}_2^2(X_1,X_2)\ge 0$ and $g(x)=h_1(x)$, (10.77) holds for $j=2$ by (10.75). Assume that (10.77) is true for $j\ge 2$. Then

$$\begin{aligned}
&E\big(h_{j+1}(X_1,\ldots,X_{j+1}) - h_j(X_1,\ldots,X_j) - h_j(X_2,\ldots,X_{j+1})\big)^2\\
&\quad = Eh_{j+1}^2(X_1,\ldots,X_{j+1}) - 4E\big[h_{j+1}(X_1,\ldots,X_{j+1})h_j(X_1,\ldots,X_j)\big]\\
&\qquad + 2Eh_j^2(X_1,\ldots,X_j) + 2Eh_j(X_1,\ldots,X_j)h_j(X_2,\ldots,X_{j+1})\\
&\quad = Eh_{j+1}^2(X_1,\ldots,X_{j+1}) - 2Eh_j^2(X_1,\ldots,X_j)\\
&\qquad + 2E\big(E\big(h_j(X_1,\ldots,X_j)h_j(X_2,\ldots,X_{j+1})\,|\,X_2,\ldots,X_j\big)\big)\\
&\quad = Eh_{j+1}^2(X_1,\ldots,X_{j+1}) - 2Eh_j^2(X_1,\ldots,X_j) + 2Eh_{j-1}^2(X_1,\ldots,X_{j-1}).
\end{aligned} \tag{10.78}$$

On the other hand, by (4.143)

$$\begin{aligned}
&E\big(h_{j+1}(X_1,\ldots,X_{j+1}) - h_j(X_1,\ldots,X_j) - h_j(X_2,\ldots,X_{j+1})\big)^2 \\
&\quad \ge E\big(E\big(h_{j+1}(X_1,\ldots,X_{j+1}) - h_j(X_1,\ldots,X_j) \\
&\qquad - h_j(X_2,\ldots,X_{j+1}) \,|\, X_1,\ldots,X_j)\big)^2 \\
&\quad = Eh_{j-1}^2(X_1,\ldots,X_{j-1}). \qquad (10.79)
\end{aligned}$$

Now (10.78), (10.79) and the induction hypothesis yield

$$\begin{aligned}
2Eh_j^2(X_1,\ldots,X_j) &\le Eh_{j+1}^2(X_1,\ldots,X_{j+1}) + Eh_{j-1}^2(X_1,\ldots,X_{j-1}) \\
&\le Eh_{j+1}^2(X_1,\ldots,X_{j+1}) + \frac{j-1}{j}Eh_j^2(X_1,\ldots,X_j),
\end{aligned}$$

which simplifies to (10.77) with $j$ replaced by $j+1$, completing the inductive step. Now iterating (10.77) we obtain

$$Eh_j^2(X_1,\ldots,X_j) \le \frac{j}{m}Eh_m^2(X_1,\ldots,X_m) = \frac{j}{m}\sigma^2. \qquad (10.80)$$

In order to demonstrate the bounds (10.20) and (10.21), respectively, we prove that

$$\sum_{j=2}^{m}\binom{m}{j}\binom{n-m}{m-j}\frac{j}{m} \le \frac{m(m-1)^2}{n(n-m+1)}\binom{n}{m} \qquad (10.81)$$

and

$$\sum_{j=1}^{m-1}\binom{m-1}{j}\binom{n-m}{m-1-j}\frac{j+1}{m} \le \frac{2(m-1)^2}{n(n-m+1)}\binom{n}{m} \qquad (10.82)$$

hold for all $n/2 > m \ge 2$. Regarding (10.81),

$$\begin{aligned}
\sum_{j=2}^{m}\binom{m}{j}\binom{n-m}{m-j}\frac{j}{m} &= \sum_{j=2}^{m}\binom{m-1}{j-1}\binom{n-m}{m-j} \\
&= \sum_{j=0}^{m-1}\binom{m-1}{j}\binom{n-m}{m-1-j} - \binom{n-m}{m-1} \\
&= \binom{n-1}{m-1} - \binom{n-m}{m-1} \\
&= \binom{n-1}{m-1}\left\{1 - \frac{(n-m)!/(n-2m+1)!}{(n-1)!/(n-m)!}\right\} \\
&= \binom{n-1}{m-1}\left\{1 - \prod_{j=n-m+1}^{n-1}\left(1-\frac{m-1}{j}\right)\right\} \\
&\le \binom{n-1}{m-1}\sum_{j=n-m+1}^{n-1}\frac{m-1}{j}
\end{aligned}$$

$$
\begin{aligned}
&\le \binom{n-1}{m-1}\frac{(m-1)^2}{n-m+1}\\
&= \frac{m(m-1)^2}{n(n-m+1)}\binom{n}{m}.
\end{aligned} \tag{10.83}
$$

As for (10.82),

$$
\begin{aligned}
&\sum_{j=1}^{m-1}\binom{m-1}{j}\binom{n-m}{m-1-j}\frac{j+1}{m}\\
&= \frac{m-1}{m}\sum_{j=1}^{m-1}\binom{m-1}{j}\binom{n-m}{m-1-j}\frac{j}{m-1}\\
&\quad + \frac{1}{m}\sum_{j=1}^{m-1}\binom{m-1}{j}\binom{n-m}{m-1-j}\\
&= \frac{m-1}{m}\sum_{j=0}^{m-2}\binom{m-2}{j}\binom{n-m}{m-2-j} + \frac{1}{m}\sum_{j=1}^{m-1}\binom{m-1}{j}\binom{n-m}{m-1-j}\\
&= \frac{m-1}{m}\binom{n-2}{m-2} + \frac{1}{m}\left\{\binom{n-1}{m-1} - \binom{n-m}{m-1}\right\}\\
&\le \binom{n-1}{m-1}\left(\frac{(m-1)^2}{m(n-1)} + \frac{(m-1)^2}{m(n-m+1)}\right)\\
&\le \frac{2(m-1)^2}{n(n-m+1)}\binom{n}{m},
\end{aligned}
$$

where in the second to last inequality we have applied (10.83).

From (10.17), (10.74), (10.76), (10.80) and (10.81) we obtain that

$$
\begin{aligned}
E\Delta^2 &= \frac{n}{m^2\sigma_1^2}\binom{n}{m}^{-2} E\left\{\sum_{1\le i_1<\cdots<i_m\le n} \bar{h}_m(X_1,\ldots,X_m)\right\}^2\\
&\le \frac{n}{m^2\sigma_1^2}\binom{n}{m}^{-1}\sum_{j=2}^{m}\binom{m}{j}\binom{n-m}{m-j} E\bar{h}_j^2(X_1,\ldots,X_j)\\
&\le \frac{n}{m^2\sigma_1^2}\binom{n}{m}^{-1}\sum_{j=2}^{m}\binom{m}{j}\binom{n-m}{m-j} Eh_j^2(X_1,\ldots,X_j)\\
&\le \frac{n\sigma^2}{m^2\sigma_1^2}\binom{n}{m}^{-1}\sum_{j=2}^{m}\binom{m}{j}\binom{n-m}{m-j}\frac{j}{m}\\
&\le \frac{(m-1)^2\sigma^2}{m(n-m+1)\sigma_1^2}.
\end{aligned} \tag{10.84}
$$

This proves (10.20).

Similarly, using (10.18), a slightly modified form of (10.74), (10.76), (10.80) and (10.82), we obtain

$$
\begin{aligned}
&E(\Delta - \Delta_l)^2 \\
&= \frac{n}{m^2\sigma_1^2}\binom{n}{m}^{-2} \\
&\quad \times E\left(\left\{\sum_{1\le i_1<\cdots<i_m\le n} - \sum_{1\le i_1<\cdots<i_m\le n, i_j\ne l \text{ for all } j}\right\}\bar{h}_m(X_{i_1},\ldots,X_{i_m})\right)^2 \\
&= \frac{n}{m^2\sigma_1^2}\binom{n}{m}^{-2} E\left(\sum_{1\le i_1<\cdots<i_{m-1}\le n-1} \bar{h}_m(X_{i_1},\ldots,X_{i_{m-1}},X_n)\right)^2 \\
&= \frac{n}{m^2\sigma_1^2}\binom{n}{m}^{-2}\binom{n-1}{m-1}\sum_{j=1}^{m-1}\binom{m-1}{j}\binom{n-m}{m-1-j} E\bar{h}_{j+1}^2(X_1,\ldots,X_j,X_n) \\
&\le \frac{n}{m^2\sigma_1^2}\binom{n}{m}^{-2}\binom{n-1}{m-1}\sum_{j=1}^{m-1}\binom{m-1}{j}\binom{n-m}{m-1-j} Eh_{j+1}^2(X_1,\ldots,X_j,X_n) \\
&\le \frac{\sigma^2}{m\sigma_1^2}\binom{n}{m}^{-1}\sum_{j=1}^{m-1}\binom{m-1}{j}\binom{n-m}{m-1-j}\frac{j+1}{m} \\
&\le \frac{2(m-1)^2\sigma^2}{nm(n-m+1)\sigma_1^2}.
\end{aligned} \tag{10.85}
$$

This proves (10.21) and hence completes the proof of Lemma 10.1. □

*Proof of Lemma 10.2* For $0 \le d_j \le m_j$, $1 \le j \le k$ and $\bar{h}$ as in (10.25), let

$$
\begin{aligned}
&Y_{d_1,\ldots,d_k}(x_{j1},\ldots,x_{jd_j}, 1\le j\le k) \\
&\quad = E\bar{h}(x_{j1},\ldots,x_{jd_j},X_{jd_j+1},\ldots,X_{jm_j}, 1\le j\le k)
\end{aligned}
$$

and

$$
y_{d_1,\ldots,d_k} = EY_{d_1,\ldots,d_k}^2(X_{j1},\ldots,X_{jd_j}, 1\le j\le k).
$$

Noting that

$$
E\big(\bar{h}(\boldsymbol{X}_{1i_1},\ldots,\boldsymbol{X}_{ki_k})|X_{jl}\big) = 0 \quad \text{for every } 1\le l\le m_j,\ 1\le j\le k,
$$

using (4.5.8) in Koroljuk and Borovskich (1994) for the first equality and the fact that $y_{d_1,\ldots,d_k} \le \sigma^2$, we have

$$
\begin{aligned}
&E(U_{\mathbf{n}} - \hat{U}_{\mathbf{n}})^2 \\
&= \left\{\prod_{j=1}^{k}\binom{n_j}{m_j}^{-1}\right\} \sum_{\substack{d_1+\cdots+d_k\ge 2 \\ 0\le d_j\le m_j, 1\le j\le k}} \prod_{j=1}^{k}\left\{\binom{m_j}{d_j}\binom{n_j-m_j}{m_j-d_j}\right\} y_{d_1,\ldots,d_k}
\end{aligned}
$$

$$\le \sigma^2 \left\{ \prod_{j=1}^{k} \binom{n_j}{m_j}^{-1} \right\} \sum_{\substack{d_1+\cdots+d_k\ge 2\\ 0\le d_j\le m_j, 1\le j\le k}} \prod_{j=1}^{k} \left\{ \binom{m_j}{d_j}\binom{n_j-m_j}{m_j-d_j} \right\}. \tag{10.86}$$

The inequality

$$\sum_{\substack{d_1+\cdots+d_k\ge 2\\ 0\le d_j\le m_j, 1\le j\le k}} \prod_{j=1}^{k} \left\{ \binom{m_j}{d_j}\binom{n_j-m_j}{m_j-d_j} \right\} \le \left( \sum_{j=1}^{k} \frac{m_j^2}{n_j} \right)^2 \prod_{j=1}^{k} \binom{n_j}{m_j}, \tag{10.87}$$

which we prove later, now completes the proof of (10.26) by noting that $\Delta = (U_n - \hat{U}_n)\sigma_{\mathbf{n}}^{-1}$.

As to (10.27) it suffices to consider the case $j = 1$. By analogy with (10.85) and (10.86), setting $\overline{X_{1i_1}} = (X_{1i_{1,1}}, \ldots, X_{1i_{1,m_1-1}}, X_{1,n_1})$ we have

$$\begin{aligned}
&\sigma_{\mathbf{n}}^2 E(\Delta - \Delta_{1l})^2 \\
&= \left\{ \prod_{v=1}^{k} \binom{n_v}{m_v} \right\}^{-2} \\
&\quad \times E\Bigg( \sum_{\substack{1\le i_{v1}<i_{v2}<\cdots<i_{vm_v}\le n_v, 2\le v\le k\\ 1\le i_{1,1}<i_{1,2}<\cdots<i_{1,m_1-1}\le n_1-1}} \bar{h}(\overline{X_{1i_1}}, X_{2i_2}, \ldots, X_{ki_k}) \Bigg)^2 \\
&= \sigma^2 \left\{ \prod_{v=1}^{k} \binom{n_v}{m_v} \right\}^{-2} \binom{n_1-1}{m_1-1} \prod_{v=2}^{k} \binom{n_v}{m_v} \\
&\quad \times \sum_{\substack{d_1+\cdots+d_k\ge 1\\ 0\le d_1\le m_1-1, 0\le d_v\le m_v, 2\le v\le k}} \binom{m_1-1}{d_1}\binom{n_1-m_1}{m_1-1-d_1} \\
&\quad \times \prod_{v=2}^{k} \binom{m_v}{d_v}\binom{n_v-m_v}{m_v-d_v} \\
&= \frac{\sigma^2 m_1}{n_1} \left\{ \prod_{v=1}^{k} \binom{n_v}{m_v} \right\}^{-1} \left\{ \sum_{0\le d_1\le m_1-1, 0\le d_v\le m_v, 2\le v\le k} - \sum_{d_j=0, 1\le j\le k} \right\} \\
&\quad \times \binom{m_1-1}{d_1}\binom{n_1-m_1}{m_1-1-d_1} \prod_{v=2}^{k} \binom{m_v}{d_v}\binom{n_v-m_v}{m_v-d_v} \\
&= \frac{\sigma^2 m_1}{n_1} \left\{ \prod_{v=1}^{k} \binom{n_v}{m_v} \right\}^{-1} \\
&\quad \times \left\{ \binom{n_1-1}{m_1-1} \prod_{v=2}^{k} \binom{n_v}{m_v} - \binom{n_1-m_1}{m_1-1} \prod_{v=2}^{k} \binom{n_v-m_v}{m_v} \right\}
\end{aligned}$$

$$= \frac{\sigma^2 m_1}{n_1}\left\{\prod_{v=1}^{k}\binom{n_v}{m_v}\right\}^{-1}\left\{\left(\binom{n_1-1}{m_1-1}-\binom{n_1-m_1}{m_1-1}\right)\prod_{v=2}^{k}\binom{n_v}{m_v}\right.$$
$$\left.+\binom{n_1-m_1}{m_1-1}\prod_{v=2}^{k}\binom{n_v}{m_v}-\prod_{v=2}^{k}\binom{n_v-m_v}{m_v}\right\}.$$

Now, using inequality (10.83) for the difference of binomial coefficients, and induction to prove that for $a_j \geq b_j \geq 1$,

$$\prod_{j=1}^{k} a_j - \prod_{j=1}^{k} b_j \leq \sum_{v=1}^{k}(a_v - b_v)\prod_{j=1}^{k} a_j,$$

we obtain

$$\sigma_{\mathbf{n}}^2 E(\Delta - \Delta_{1l})^2$$
$$\leq \frac{\sigma^2 m_1}{n_1}\left\{\prod_{v=1}^{k}\binom{n_v}{m_v}\right\}^{-1}\left\{\frac{(m_1-1)^2}{n_1-m_1+1}\binom{n_1-1}{m_1-1}\prod_{v=2}^{k}\binom{n_v}{m_v}\right.$$
$$\left.+\sum_{2\leq v\leq k}\frac{(m_v-1)^2}{n_v-m_v+1}\binom{n_1-m_1}{m_1-1}\prod_{j=2}^{k}\binom{n_j}{m_j}\right\}$$
$$\leq \frac{\sigma^2 m_1^2}{n_1^2}\sum_{1\leq v\leq k}\frac{m_v^2}{n_v-m_v+1} \leq \frac{2\sigma^2 m_1^2}{n_1^2}\sum_{1\leq v\leq k}\frac{m_v^2}{n_v}$$

for $n_1 \geq 2m_1$, proving (10.27).

Now we prove (10.87). Consider two cases in the summation:

*Case 1* At least one of $d_j \geq 2$, say $d_1 \geq 2$. In this case, we have

$$\sum_{\substack{2\leq d_1\leq m_1 \\ 1\leq d_l\leq m_l,1\leq l\leq k}}\prod_{j=1}^{k}\left\{\binom{m_j}{d_j}\binom{n_j-m_j}{m_j-d_j}\right\}$$
$$\leq \left\{\prod_{j=2}^{k}\binom{n_j}{m_j}\right\}\sum_{2\leq d_1\leq m_1}\binom{m_1}{d_1}\binom{n_1-m_1}{m_1-d_1}$$
$$\leq \left\{\prod_{j=2}^{k}\binom{n_j}{m_j}\right\}\frac{m_1}{2}\sum_{2\leq d_1\leq m_1}\binom{m_1}{d_1}\binom{n_1-m_1}{m_1-d_1}\frac{d_1}{m_1}$$
$$\leq \frac{m_1^2(m_1-1)^2}{2n_1(n_1-m_1+1)}\prod_{j=1}^{k}\binom{n_j}{m_j} \quad \text{[by (10.81)]}$$
$$\leq \frac{m_1^4}{n_1^2}\prod_{j=1}^{k}\binom{n_j}{m_j}$$

for $n_1 \geq 2m_1$.

*Case 2* At least two of $\{d_j\}$ are equal to 1, say $d_1 = d_2 = 1$. Then

$$\begin{aligned}
&\sum_{\substack{d_1=d_2=1\\ 1\le d_l\le m_l,3\le l\le k}} \prod_{j=1}^{k}\left\{\binom{m_j}{d_j}\binom{n_j-m_j}{m_j-d_j}\right\}\\
&\le m_1 m_2 \binom{n_1-m_1}{m_1-1}\binom{n_2-m_2}{m_2-1}\prod_{j=3}^{k}\binom{n_j}{m_j}\\
&\le \frac{m_1^2 m_2^2}{n_1,n_2}\prod_{j=1}^{k}\binom{n_j}{m_j}.
\end{aligned}$$

Thus, we have

$$\begin{aligned}
&\sum_{\substack{d_1+\cdots+d_k\ge 2\\ 1\le d_j\le m_j,1\le j\le k}} \prod_{j=1}^{k}\left\{\binom{m_j}{d_j}\binom{n_j-m_j}{m_j-d_j}\right\}\\
&\le \left(\sum_{j=1}^{k}\frac{m_j^4}{n_j^2} + \sum_{1\le i\ne j\le k}\frac{m_i^2 m_j^2}{n_i n_j}\right)\prod_{j=1}^{k}\binom{n_j}{m_j}\\
&= \left(\sum_{j=1}^{k}\frac{m_j^2}{n_j}\right)^2\prod_{j=1}^{k}\binom{n_j}{m_j}.
\end{aligned}$$

This proves (10.87) and completes the proof of Lemma 10.2. □

*Proof of (10.50)* We prove the stronger inequality

$$E|W|^p \le (p-1)B_n^p + \sum_{i=1}^{n} E|\xi_i|^p \tag{10.88}$$

for $2 < p \le 3$, where $B_n^2 = EW^2$ is not necessarily equal to 1. Let $W^{(i)} = W - \xi_i$. Then

$$\begin{aligned}
E|W|^p &= \sum_{i=1}^{n} E\xi_i W|W|^{p-2}\\
&= \sum_{i=1}^{n} E\xi_i\big(W|W|^{p-2} - W^{(i)}|W|^{p-2}\big)\\
&\quad + \sum_{i=1}^{n} E\xi_i\big(W^{(i)}|W|^{p-2} - W^{(i)}|W^{(i)}|^{p-2}\big),
\end{aligned}$$

because $\xi_i$ and $W^{(i)}$ are independent, and $E\xi_i = 0$. Thus we have

$$E|W|^p \le \sum_{i=1}^n E\xi_i^2 |W|^{p-2} + \sum_{i=1}^n E|\xi_i| \left|W^{(i)}\right| \left\{ \left( \left|W^{(i)}\right| + |\xi_i| \right)^{p-2} - \left|W^{(i)}\right|^{p-2} \right\}$$
$$\le \sum_{i=1}^n E\xi_i^2 \left( |\xi_i|^{p-2} + \left|W^{(i)}\right|^{p-2} \right)$$
$$+ \sum_{i=1}^n E|\xi_i| \left|W^{(i)}\right|^{p-1} \left\{ \left( 1 + |\xi_i| / \left|W^{(i)}\right| \right)^{p-2} - 1 \right\}.$$

Since $(1+x)^{p-2} - 1 \le (p-2)x$ for $x \ge 0$, we have

$$E|W|^p \le \sum_{i=1}^n E|\xi_i|^p + \sum_{i=1}^n E\xi_i^2 E\left|W^{(i)}\right|^{p-2}$$
$$+ \sum_{i=1}^n E|\xi_i| \left|W^{(i)}\right|^{p-1} (p-2)|\xi_i| / \left|W^{(i)}\right|$$
$$= \sum_{i=1}^n E|\xi_i|^p + (p-1) \sum_{i=1}^n E\xi_i^2 E\left|W^{(i)}\right|^{p-2},$$

and Hölder's inequality now gives

$$E|W|^p \le \sum_{i=1}^n E|\xi_i|^p + (p-1) \sum_{i=1}^n E\xi_i^2 \left( E\left|W^{(i)}\right|^2 \right)^{(p-2)/2}$$
$$\le \sum_{i=1}^n E|\xi_i|^p + (p-1) B_n^p,$$

as desired. □

# Chapter 11
# Moderate Deviations

The Berry–Esseen inequality, which gives a bound on the absolute error in the normal approximation, may not be very informative when $\Phi(x)$ is close to 0 or 1. Cramér's theory of moderate deviations, on the other hand, provides relative errors. Let $X_1, X_2, \ldots$ be i.i.d. random variables with $E(X_1) = 0$, $EX_1^2 = 1$ and $Ee^{t_0|X_1|} < \infty$ for some $t_0 > 0$ and set $W = \sum_{i=1}^n X_i/\sqrt{n}$. Then, see Petrov (1995) for instance, for $z \ge 0$ with $z = o(n^{1/2})$,

$$\frac{P(W \ge z)}{1 - \Phi(z)} = \exp\left\{z^2 \lambda\left(\frac{z}{\sqrt{n}}\right)\right\}\left(1 + O\left(\frac{1+z}{\sqrt{n}}\right)\right),$$

for a function $\lambda(t)$ known as the Cramér series. In particular,

$$\frac{P(W \ge z)}{1 - \Phi(z)} = 1 + O(1)\frac{(1+z^3)E|X_1|^3}{\sqrt{n}} \tag{11.1}$$

for $0 \le z \le n^{1/6}/(E|X_1|^3)^{1/3}$, where $O(1)$ is a sequence of real numbers bounded by a universal constant for $n \in \mathbb{N}$. As a consequence of (11.1),

$$\frac{P(W \ge z)}{1 - \Phi(z)} \to 1 \tag{11.2}$$

as $n \to \infty$, uniformly in $z \in [0, o(n^{1/6}))$. It is known in general that $o(n^{1/6})$ is the largest possible value for the range of $z$ such that (11.2) holds.

In this chapter, following Chen et al. (2009), we first establish a Cramér type moderate deviation theorem in the mold of (11.1) by use of the Stein identity (2.42) for approximate exchangeable pairs, and then apply the result to four examples: the combinatorial central limit theorem, the anti-voter model on complete graphs, the binary expansion of a random integer, and the Curie–Weiss model. Related ideas appear in Raič (2007); see also Chatterjee (2007).

## 11.1 A Cramér Type Moderate Deviation Theorem

The following theorem gives moderate deviation bounds of Cramér type for $W$ in the context of the Stein identity (2.42).

L.H.Y. Chen et al., *Normal Approximation by Stein's Method*,
Probability and Its Applications,
DOI 10.1007/978-3-642-15007-4_11, 

**Theorem 11.1** *Suppose that for a given random variable $W$ there exist a constant $\delta > 0$, a random function $\hat{K}(t)$ and a random variable $R$ such that*

$$EWf(W) = E\int_{|t|\le\delta} f'(W+t)\hat{K}(t)dt + E\big(Rf(W)\big) \tag{11.3}$$

*for all absolutely continuous function $f$ for which the expectations on either side exists, and let*

$$\hat{K}_1 = \int_{|t|\le\delta} \hat{K}(t)dt. \tag{11.4}$$

*If $\hat{K}(t) \ge 0$ and there exist constants $\delta_1, \delta_2$ and $d_0 \ge 1$ such that*

$$\big|E(\hat{K}_1|W) - 1\big| \le \delta_1\big(1+W^2\big), \tag{11.5}$$

$$\big|E(R|W)\big| \le \delta_2\big(1+|W|\big) \tag{11.6}$$

*and*

$$E(\hat{K}_1|W) \le d_0, \tag{11.7}$$

*then*

$$\frac{P(W\ge z)}{1-\Phi(z)} = 1 + O(1)\big(d_0\big(1+z^3\big)\delta + \big(1+z^4\big)\delta_1 + \big(1+z^2\big)\delta_2\big) \tag{11.8}$$

*for $0 \le z \le d_0^{-1}\min(\delta^{-1/3}, \delta_1^{-1/4}, \delta_2^{-1/3})$, where $O(1)$ is bounded by a universal constant.*

Additionally, a moderate deviation bound holds when a bounded zero bias coupling exists.

**Theorem 11.2** *Let $W$ and $W^*$ be defined on the same space, with $W^*$ having the $W$-zero biased distribution, such that $|W^* - W| \le \delta$. Then*

$$\frac{P(W\ge z)}{1-\Phi(z)} = 1 + O(1)\big(1+z^3\big)\delta$$

*for $0 \le z \le \delta^{-1/3}$, where $O(1)$ is bounded by a universal constant.*

The following remarks may be useful.

*Remark 11.1* Let $(W, W')$ be an exchangeable pair satisfying

$$E(W - W'|W) = \lambda\big(W - R(W)\big) \tag{11.9}$$

and $|W - W'| \le \delta$. Then with (11.3) is satisfied with $\hat{K}_1 = \Delta^2/(2\lambda)$ where $\Delta = W - W'$; see Sect. 2.3.2, and (2.41) and (2.42) in particular.

*Remark 11.2* One can show that if condition (11.5) in Theorem 11.1 is replaced by

$$\left|E(\hat{K}_1|W) - 1\right| \le \delta_1\left(1 + |W|\right), \tag{11.10}$$

then

$$\frac{P(W \ge z)}{1 - \Phi(z)} = 1 + O(1)d_0\left(1 + z^3\right)(\delta + \delta_1) + O(1)\left(1 + z^2\right)\delta_2$$

for $0 \le z \le d_0^{-1}\min(\delta^{-1/3}, \delta_1^{-1/3}, \delta_2^{-1/3})$.

## 11.2 Applications

We apply Theorems 11.1 and 11.2 to four cases, all involving dependent random variables; namely, the combinatorial central limit theorem, the binary expansion of a random integer, the anti-voter model on a complete graph, and the Curie–Weiss model.

*Example 11.1* (Combinatorial central limit theorem) With $n \ge 3$ let $\{a_{ij}\}_{i,j=1}^n$ be an array of real numbers and $\pi$ a random permutation with the uniform distribution over $\mathcal{S}_n$, the symmetric group on $\{1, \ldots, n\}$ and

$$Y = \sum_{i=1}^n a_{i\pi(i)}.$$

Derivation of $L^1$ and Berry–Esseen bounds for the standardized $W = (Y - EY)/\sigma$ are obtained in Sects. 4.4 and 6.1, respectively, where $\sigma^2$ is given in (4.105). With

$$c_0 = \max_{1 \le i,j \le n} |a_{ij} - a_{i\bullet} - a_{\bullet j} - a_{\bullet\bullet}|,$$

the construction of $W^*$ with the $W$-zero biased distribution in Lemma 4.6, following the proof of Theorem 6.1, satisfies

$$|W^* - W| \le 8c_0/\sigma.$$

Therefore, by Theorem 11.2

$$\frac{P(W \ge z)}{1 - \Phi(z)} = 1 + O(1)\left(1 + z^3\right)c_0/\sigma$$

for $0 \le z \le (\sigma/c_0)^{1/3}$.

Similar remarks apply to the combinatorial central limit theorem when $\pi$ is chosen with a distribution uniform, or constant, on cycle type; the bounded coupling for the former instance was applied to prove Theorem 6.3.

*Example 11.2* (The anti-voter model) The discrete time anti-voter model was described in Sect. 6.4, where a bound to the normal was shown using the method of exchangeable pairs; some references and history may also be found there. We recall that in this model each vertex of a graph is in state $+1$ or $-1$, and at each time step a randomly chosen vertex adopts the opposite state of one of its randomly chosen neighbors. Adopting notation from Sect. 6.4, here we prove a moderate deviation result for $W$ given by

$$W = U/\sigma \quad \text{where } U = \sum_{v \in V} X_v \text{ with } \sigma^2 = \mathrm{Var}(U),$$

the net standardized sign of the stationary distribution of the anti-voter chain, run on the complete graph with $n \geq 3$ vertices.

Theorem 6.6 yields that if the process is run in equilibrium for a single time step to obtain $U'$, then $(U, U')$, a $2/n$-Stein pair which satisfies

$$|U' - U| \leq 2 \quad \text{and} \quad E\big[(U' - U)^2 | \mathbf{X}\big] = 8(a + b)/\big(n(n-1)\big),$$

where $a$ and $b$ are the number of edges that are incident on vertices both of which are in state $+1$, or $-1$, respectively. In particular, according to Remark 11.1, identity (11.3) is satisfied with $\delta = 2/\sigma$ and, since $R = 0$, with $\delta_2 = 0$ in (11.6).

Given $U$, there are $(n + U)/2$ vertices in state $+1$ and $(n - U)/2$ vertices in state $-1$. Since the graph is complete, we see that $a = (n + U)(n + U - 2)/8$ and $b = (n - U)(n - U - 2)/8$. Therefore

$$\begin{aligned} E\big[(W - W')^2 | X\big] &= \frac{1}{\sigma^2} E\big[(U' - U)^2 | X\big] = \frac{8(a+b)}{\sigma^2 n(n-1)} \\ &= \frac{2U^2 + 2n^2 - 4n}{\sigma^2 n(n-1)} = \frac{2\sigma^2 W^2 + 2n^2 - 4n}{\sigma^2 n(n-1)}, \end{aligned}$$

and so

$$\begin{aligned} E(\hat{K}_1 | W) - 1 &= \frac{n}{4} E\big((W' - W)^2 | W\big) - 1 \\ &= \frac{W^2}{2(n-1)} - \frac{2\sigma^2(n-1) - (n^2 - 2n)}{2\sigma^2(n-1)}. \end{aligned}$$

As $E(E(\hat{K}_1|W) - 1) = 0$ and $EW^2 = 1$, we conclude that $\sigma^2 = (n^2 - 2n)/(2n - 3)$. Hence,

$$E(\hat{K}_1 | W) - 1 = \frac{W^2}{2(n-1)} - \frac{1}{2(n-1)}, \tag{11.11}$$

implying that (11.5) is satisfied with $\delta_1 = 1/(2(n-1))$. Additionally, as $\sigma^2 \geq n/2$ and $|U| \leq n$, we find that $W^2 \leq 2n$ and the quantity in (11.11) can be at most 2; hence we may take $d_0 = 3$ in (11.7).

Hence, Theorem 11.1 yields the moderate deviation result

$$\frac{P(W \geq z)}{1 - \Phi(z)} = 1 + O(1)\big(1 + z^3\big)/\sqrt{n}$$

for $0 \leq z \leq n^{1/6}/(3\sqrt{2})$, noting that $z^4/n$ is dominated by $z^3/\sqrt{n}$ over the given range.

*Example 11.3* (Binary expansion of a random integer) Following the conventions in Sect. 6.5, let $X$ be a random variable uniformly distributed over the set $\{0, 1, \ldots, n-1\}$ for some $n \geq 2$, and with $m = [\log_2(n-1)] + 1$, let $S = X_1 + \cdots + X_m$, the number of ones in the binary expansion of $X$. Set

$$W = \frac{S - m/2}{\sqrt{m/4}} \tag{11.12}$$

and recall that for $(S, S')$ and $(W, W')$, the exchangeable pairs constructed there, $|S - S'| \leq 1$, so

$$|W - W'| \leq \frac{2}{\sqrt{m}} \tag{11.13}$$

and equalities (6.138) and (6.139) hold, that is,

$$E(W - W'|W) = \lambda\left(\frac{W + E(Q|W)}{\sqrt{m}}\right) \tag{11.14}$$

and

$$\frac{1}{2\lambda} E\big((W - W')^2|W\big) = 1 - \frac{E(Q|W)}{m}, \tag{11.15}$$

where $\lambda = 2/m$ and $Q = Q(X, n)$, as defined in (6.134).

We prove a moderate deviation bound for $W$ in (11.12) with the help of the following result.

**Lemma 11.1** *There exists a constant $C$ such that with $W$ and $Q = Q(X, n)$ defined in* (11.12) *and* (6.134), *respectively,*

$$E(Q|W) \leq C\big(1 + |W|\big).$$

As the proof of this lemma is quite involved we defer it to the Appendix. By (11.13) we may take $\delta = 2/\sqrt{m}$. Equality (11.14), (2.41) and Lemma 11.1 yield

$$\big|E(R|W)\big| = \big|E(Q|W)\big|/\sqrt{m} \leq (C/\sqrt{m})\big(1 + |W|\big),$$

so we may take $\delta_2 = C/\sqrt{m}$ to satisfy (11.10) in Remark 11.2. Similarly, (11.15) and Lemma 11.1 yield that (11.5) is satisfied with $\delta_1 = C/m$. Lastly, as $Q$ is non-negative, by (11.15) we may take $d_0 = 1$ in (11.7). Hence, by Remark 11.2,

$$\frac{P(W \geq z)}{1 - \Phi(z)} = 1 + O(1)\big(1 + z^3\big)/\sqrt{m}$$

for $0 \leq z \leq m^{1/6}/(\max(2, C))^{1/3}$.

*Example 11.4* (The Curie–Weiss model) The Curie–Weiss model is a simple statistical mechanical model of ferromagnetic interaction, where for $n \in \mathbb{N}$, a vector $\boldsymbol{\sigma} = (\sigma_1, \ldots, \sigma_n)$ of 'spins' in $\{-1, 1\}^n$ has joint probability mass function

$$p(\boldsymbol{\sigma}) = C_\beta \exp\left(\frac{\beta}{n} \sum_{i<j} \sigma_i \sigma_j\right) \tag{11.16}$$

where $C_\beta$ is a normalizing constant and $\beta > 0$ is known as the inverse temperature.

Let $\sigma^2$ be the variance of the total sum of spins $\sum_{i=1}^n \sigma_i$, and $W = \sum_{i=1}^n \sigma_i/\sigma$, the normalized sum. It is known that $W$ converges to a standard normal distribution as $n \to \infty$ when $0 < \beta < 1$, see, e.g., Ellis and Newman (1978a, 1978b). However, when $\beta = 1$ the limiting distribution is non-normal; see Sect. 13.3 for a Berry–Esseen type bound in this instance.

Here, focusing on the case $\beta \in (0, 1)$, we prove

$$\frac{P(W \geq z)}{1 - \Phi(z)} = 1 + O(1)\big(1 + z^3\big)/\sqrt{n} \tag{11.17}$$

for $0 \leq z \leq n^{1/6}$, and $O(1)$ a constant that may depend on $\beta$. The proof is postponed to Sect. 11.4.

## 11.3 Preliminary Lemmas

To prove Theorem 11.1 we first develop two preliminary lemmas. Our first lemma gives a bound for the moment generating function of $W$.

**Lemma 11.2** *Let $W$ be a random variable satisfying the hypotheses of Theorem* 11.1 *for some $\delta > 0$, $\delta_1 \geq 0, 0 \leq \delta_2 \leq 1/4$ and $d_0 \geq 1$. Then for all $0 < t \leq 1/(2\delta)$ satisfying*

$$8td_0(t\delta_1 + 2\delta_2) \leq 1 \tag{11.18}$$

*we have*

$$Ee^{tW} \leq \exp\big(t^2/2 + c_0(t)\big) \tag{11.19}$$

*where*

$$c_0(t) = 30d_0\big\{\delta_2 t + \delta_1 t^2 + (\delta_2 + \delta)t^3 + \delta_1 t^4\big\}. \tag{11.20}$$

*Proof* Fix $a > 0$, $t \in (0, 1/(2\delta)]$ and $s \in (0, t]$, and let $f(w) = e^{s(w\wedge a)}$. Letting $h(s) = Ee^{s(W\wedge a)}$, firstly we prove that $h'(s)$ can be bounded by an expression in $h(s)$ and $EW^2 f(W)$. Applying the bounded convergence theorem to differentiate under the expectation, and (11.3),

$$\begin{aligned}
h'(s) &= E(W \wedge a)e^{s(W\wedge a)} \leq E\big(Wf(W)\big) \\
&= E\int_{|u|\leq\delta} f'(W+u)\hat{K}(u)du + E\big(Rf(W)\big) \\
&= sE\int_{|u|\leq\delta} e^{s(W+u)}\mathbf{1}(W+u \leq a)\hat{K}(u)du + E\big(e^{s(W\wedge a)}E(R|W)\big) \\
&\leq sE\int_{|u|\leq\delta} e^{s[(W+u)\wedge a]}\hat{K}(u)du + E\big(e^{s(W\wedge a)}E(R|W)\big) \\
&\leq sE\int_{|u|\leq\delta} e^{s(W\wedge a+\delta)}\hat{K}(u)du + E\big(e^{s(W\wedge a)}E(R|W)\big)
\end{aligned}$$

$$
\begin{aligned}
&= sE\int_{|u|\le\delta} e^{s(W\wedge a)}\hat{K}(u)du \\
&\quad + sE\int_{|u|\le\delta} e^{s(W\wedge a)}\big(e^{s\delta}-1\big)\hat{K}(u)du + E\big(e^{s(W\wedge a)}E(R|W)\big) \\
&\le sEe^{s(W\wedge a)}\hat{K}_1 + sEe^{s(W\wedge a)}\big|e^{s\delta}-1\big|\hat{K}_1 + \delta_2 E\big(\big(1+|W|\big)e^{s(W\wedge a)}\big),
\end{aligned}
$$

where we have applied (11.4) and (11.6) to obtain the last inequality. Now, applying the simple inequality

$$
\big|e^x - 1\big| \le 2|x| \quad \text{for } |x|\le 1
$$

followed by (11.7), the fact that $1+|w|\le 2(1+w^2)$ and then (11.5), we find that

$$
\begin{aligned}
h'(s) &\le sEe^{s(W\wedge a)}\hat{K}_1 + sEe^{s(W\wedge a)}2s\delta\hat{K}_1 + \delta_2 E\big(\big(1+|W|\big)e^{s(W\wedge a)}\big) \\
&\le sE\big(e^{s(W\wedge a)}E(\hat{K}_1|W)\big) + 2s^2 d_0\delta Ee^{s(W\wedge a)} \\
&\quad + 2\delta_2 E\big(\big(1+W^2\big)e^{s(W\wedge a)}\big) \\
&= sEe^{s(W\wedge a)} + sEe^{s(W\wedge a)}\big[E(\hat{K}_1|W)-1\big] + 2s^2 d_0\delta Ee^{s(W\wedge a)} \\
&\quad + 2\delta_2 E\big(\big(1+W^2\big)e^{s(W\wedge a)}\big) \\
&\le sEe^{s(W\wedge a)} + s\delta_1 Ee^{s(W\wedge a)}\big(1+W^2\big) \\
&\quad + 2s^2 d_0\delta Ee^{s(W\wedge a)} + 2\delta_2 E\big(\big(1+W^2\big)e^{s(W\wedge a)}\big). \qquad (11.21)
\end{aligned}
$$

Collecting terms and recalling $0<s\le t$ we obtain

$$
h'(s) \le \big(s(1+\delta_1+2td_0\delta)+2\delta_2\big)h(s) + (s\delta_1+2\delta_2)EW^2 f(W). \qquad (11.22)
$$

Secondly, we show that $EW^2 f(W)$ can be bounded by a function of $h(s)$ and $h'(s)$. Letting $g(w)=we^{s(w\wedge a)}$, and then arguing as for (11.22) and recalling $0< s\delta\le t\delta\le 1/2$,

$$
\begin{aligned}
EW^2 f(W) &= EWg(W) \\
&= E\int_{|u|\le\delta}\big(e^{s[(W+u)\wedge a]} + s(W+u)e^{s[(W+u)\wedge a]}\mathbf{1}(W+u\le a)\big)\hat{K}(u)du \\
&\quad + E\big(RWf(W)\big) \\
&\le E\int_{|u|\le\delta}\big(e^{s(W\wedge a)}e^{s\delta} + s\big[(W+u)\wedge a\big]e^{s(W\wedge a)}e^{s\delta}\big)\hat{K}(u)du \\
&\quad + E\big(RWf(W)\big) \\
&\le e^{s\delta}E\big(f(W)+sf(W)\big((W\wedge a)+\delta\big)\big)\hat{K}_1 + \delta_2 Ef(W)\big(1+|W|\big)|W| \\
&\le d_0 e^{0.5}(1+0.5)Ef(W) + sd_0 e^{0.5}E(W\wedge a)f(W) \\
&\quad + \delta_2 Ef(W)\big(1+2W^2\big) \\
&\le 3d_0 h(s) + 2sd_0 h'(s) + \delta_2 h(s) + 2\delta_2 EW^2 f(W). \qquad (11.23)
\end{aligned}
$$

Thus, recalling $\delta_2\le 1/4$, we have

$$
EW^2 f(W) \le (6d_0+2\delta_2)h(s) + 4sd_0 h'(s). \qquad (11.24)
$$

We are now ready to prove (11.19). Substituting (11.24) into (11.22) yields

$$\begin{aligned}
h'(s) &\le \big(s(1+\delta_1+2td_0\delta)+2\delta_2\big)h(s) \\
&\quad + (s\delta_1+2\delta_2)\big((6d_0+2\delta_2)h(s)+4sd_0h'(s)\big) \\
&= \big(s\big(1+\delta_1(1+6d_0+2\delta_2)+2td_0\delta\big)+2\delta_2(1+6d_0+2\delta_2)\big)h(s) \\
&\quad + 4sd_0(s\delta_1+2\delta_2)h'(s) \\
&\le \big(s\big(1+\delta_1(1+6d_0+2\delta_2)+2td_0\delta\big)+2\delta_2(1+6d_0+2\delta_2)\big)h(s) \\
&\quad + 4td_0(s\delta_1+2\delta_2)h'(s).
\end{aligned}$$

Solving for $h'(s)$, we obtain

$$h'(s) \le \big(sc_1(t)+c_2(t)\big)h(s), \tag{11.25}$$

where

$$\begin{aligned}
c_1(t) &= \frac{1+\delta_1(1+6d_0+2\delta_2)+2td_0\delta}{1-c_3(t)}, \\
c_2(t) &= \frac{2\delta_2(1+6d_0+2\delta_2)}{1-c_3(t)} \quad \text{with} \\
c_3(t) &= 4td_0(t\delta_1+2\delta_2).
\end{aligned} \tag{11.26}$$

Taking $t$ to satisfy (11.18) yields $c_3(t) \le 1/2$.

Solving (11.25) we obtain

$$h(t) \le \exp\left(\frac{t^2}{2}c_1(t)+tc_2(t)\right). \tag{11.27}$$

Now, using $c_3(t) \le 1/2$, (11.26), $\delta_2 \le 1/4$, $d_0 \ge 1$ and (11.20),

$$\begin{aligned}
&\frac{t^2}{2}\big(c_1(t)-1\big)+tc_2(t) \\
&\quad = \frac{t^2}{2}\left(\frac{\delta_1(1+6d_0+2\delta_2)+2td_0\delta+c_3(t)}{1-c_3(t)}\right)+\frac{2t\delta_2(1+6d_0+2\delta_2)}{1-c_3(t)} \\
&\quad \le t^2\big(\delta_1(1+6d_0+2\delta_2)+2td_0\delta+4td_0(t\delta_1+2\delta_2)\big)+4t\delta_2(1+6d_0+2\delta_2) \\
&\quad \le c_0(t)
\end{aligned}$$

and hence

$$\frac{t^2}{2}c_1(t)+tc_2(t) \le \frac{t^2}{2}+c_0(t).$$

Hence letting $a \to \infty$ in (11.27) we obtain (11.19), as desired. □

**Lemma 11.3** *Let $W$ be a random variable satisfying the hypotheses of Theorem 11.1 for some nonnegative $\delta$, $\delta_1$ and $\delta_2$ with $\max(\delta,\delta_1,\delta_2) \le 1/256$ and $d_0 \ge 1$. Then for all*

$$t \in \big[0, d_0^{-1}\min\big(\delta^{-1/3}, \delta_1^{-1/4}, \delta_2^{-1/3}\big)\big] \tag{11.28}$$

*and integers $k \geq 1$, there exists a finite universal constant $C$ such that*

$$\int_0^t u^k e^{u^2/2} P(W \geq u) du \leq C\big(1 + t^k\big).$$

*Proof* Recalling (11.20), for $t$ satisfying (11.28) it is easy to see that $c_0(t) \leq 93$ and (11.18) is satisfied. Hence, by Lemma 11.2, inequality (11.19) holds.

Write

$$\begin{aligned}
&\int_0^t u^k e^{u^2/2} P(W \geq u) du \\
&\quad = \int_0^{[t]} u^k e^{u^2/2} P(W \geq u) du + \int_{[t]}^t u^k e^{u^2/2} P(W \geq u) du.
\end{aligned}$$

For the first integral, noting that for $j \geq 1$ we have $\sup_{j-1 \leq u \leq j} e^{u^2/2 - ju} = e^{(j-1)^2/2 - j(j-1)}$,

$$\begin{aligned}
\int_0^{[t]} u^k e^{u^2/2} P(W \geq u) du &\leq \sum_{j=1}^{[t]} j^k \int_{j-1}^j e^{u^2/2 - ju} e^{ju} P(W \geq u) du \\
&\leq \sum_{j=1}^{[t]} j^k e^{(j-1)^2/2 - j(j-1)} \int_{j-1}^j e^{ju} P(W \geq u) du \\
&\leq 2 \sum_{j=1}^{[t]} j^k e^{-j^2/2} \int_{-\infty}^{\infty} e^{ju} P(W \geq u) du \\
&= 2 \sum_{j=1}^{[t]} j^k e^{-j^2/2} (1/j) E e^{jW} \\
&\leq 2 \sum_{j=1}^{[t]} j^{k-1} \exp\big(-j^2/2 + j^2/2 + c_0(j)\big) \\
&\leq 2 e^{c_0(t)} \sum_{j=1}^{[t]} j^{k-1} \\
&\leq C\big(1 + t^k\big).
\end{aligned}$$

Similarly, we have

$$\begin{aligned}
\int_{[t]}^t u^k e^{u^2/2} P(W \geq u) du &\leq t^k \int_{[t]}^t e^{u^2/2 - tu} e^{tu} P(W \geq u) du \\
&\leq t^k e^{[t]^2/2 - t[t]} \int_{[t]}^t e^{tu} P(W \geq u) du \\
&\leq 2 t^k e^{-t^2/2} \int_{-\infty}^{\infty} e^{tu} P(W \geq u) du \\
&\leq C\big(1 + t^k\big),
\end{aligned}$$

completing the proof. □

## 11.4 Proofs of Main Results

*Proof of Theorem 11.1* For $z \ge 0$ the factor $d_0(1+z^3)\delta + (1+z^4)\delta_1 + (1+z^2)\delta_2$ in the error term of (11.8) is bounded below by $d_0\delta$. Note also that $1/(1-\Phi(z)) \le 1/(1-\Phi(z_0))$ for $0 \le z \le z_0$. Therefore, (11.8) is trivial if the range of $z$ is bounded, by, say 7. Hence, we can assume

$$d_0^{-1} \min\bigl(\delta^{-1/3}, \delta_1^{-1/4}, \delta_2^{-1/3}\bigr) \ge 7. \tag{11.29}$$

Let $f = f_z$ be the solution of the Stein equation (2.2) for $0 \le z \le d_0^{-1} \min(\delta^{-1/3}, \delta_1^{-1/4}, \delta_2^{-1/3})$. By (11.3), (2.2) and (11.4),

$$\begin{aligned}
&EWf(W) - ERf(W) \\
&\quad = E\int_{|t|\le\delta} f'(W+t)\hat{K}(t)dt \\
&\quad = E\int_{|t|\le\delta} \bigl\{(W+t)f(W+t) + 1 - \Phi(z) - \mathbf{1}(W+t>z)\bigr\}\hat{K}(t)dt \\
&\quad = E\int_{|t|\le\delta} \bigl((W+t)f(W+t) - Wf(W)\bigr)\hat{K}(t)dt + EWf(W)\hat{K}_1 \\
&\qquad + E\int_{|t|\le\delta} \bigl\{1 - \Phi(z) - \mathbf{1}(W+t>z)\bigr\}\hat{K}(t)dt \\
&\quad \le E\int_{|t|\le\delta} \bigl((W+\delta)f(W+\delta) - Wf(W)\bigr)\hat{K}(t)dt + EWf(W)\hat{K}_1 \\
&\qquad + E\int_{|t|\le\delta} \bigl\{1 - \Phi(z) - \mathbf{1}(W>z+\delta)\bigr\}\hat{K}(t)dt,
\end{aligned}$$

where, in the final inequality, we have applied (2.6), that is, the monotonicity of $wf(w)$, and the assumption that $\hat{K}(t)$ is non-negative. Again applying (11.4), the expression above can be written

$$\begin{aligned}
&E\bigl((W+\delta)f(W+\delta) - Wf(W)\bigr)\hat{K}_1 \\
&\qquad + EWf(W)\hat{K}_1 + E\bigl\{1 - \Phi(z) - \mathbf{1}(W>z+\delta)\bigr\}\hat{K}_1 \\
&\quad = 1 - \Phi(z) - P(W>z+\delta) \\
&\qquad + E\bigl((W+\delta)f(W+\delta) - Wf(W)\bigr)\hat{K}_1 + EWf(W)\hat{K}_1 \\
&\qquad + E\bigl\{1 - \Phi(z) - \mathbf{1}(W>z+\delta)\bigr\}(\hat{K}_1 - 1).
\end{aligned}$$

Therefore, we have

$$\begin{aligned}
&P(W>z+\delta) - \bigl(1-\Phi(z)\bigr) \\
&\quad \le E\bigl((W+\delta)f(W+\delta) - Wf(W)\bigr)\hat{K}_1 + EWf(W)(\hat{K}_1 - 1) \\
&\qquad + E\bigl\{1 - \Phi(z) - \mathbf{1}(W>z+\delta)\bigr\}(\hat{K}_1 - 1) + ERf(W) \\
&\quad \le d_0 E\bigl((W+\delta)f(W+\delta) - Wf(W)\bigr) + \delta_1 E\bigl(|W|(1+W^2)f(W)\bigr) \\
&\qquad + \delta_1 E\bigl|1 - \Phi(z) - \mathbf{1}(W>z+\delta)\bigr|\bigl(1+W^2\bigr) + \delta_2 E\bigl(1+|W|\bigr)f(W)
\end{aligned}$$

where we have again applied the monotonicity of $wf(w)$, inequality (2.9) giving that $f(w) \geq 0$, as well as (11.5), (11.6) and (11.7). Rewriting, we have that

$$P(W > z+\delta) - \big(1-\Phi(z)\big) \leq d_0 I_1 + \delta_1 I_2 + \delta_1 I_3 + \delta_2 I_4, \tag{11.30}$$

where

$$\begin{aligned}
I_1 &= E\big((W+\delta)f(W+\delta) - Wf(W)\big),\\
I_2 &= E\big(|W|\big(1+W^2\big)f(W)\big),\\
I_3 &= E\big|1-\Phi(z) - \mathbf{1}(W > z+\delta)\big|\big(1+W^2\big) \quad \text{and}\\
I_4 &= E\big(1+|W|\big)f(W).
\end{aligned}$$

We will consider $I_2$ first, and again using $f(w) \geq 0$, apply

$$|w|\big(1+w^2\big)f(w) \leq 2\big(1+|w|^3\big)f(w). \tag{11.31}$$

Recalling inequality (2.11),

$$e^{z^2/2}\big(1-\Phi(z)\big) \leq \min\left(\frac{1}{2}, \frac{1}{z\sqrt{2\pi}}\right) \quad \text{for } z > 0, \tag{11.32}$$

and the form (2.3) of the solution $f = f_z$ from Lemma 2.2, to bound the first term arising from the expectation of (11.31) we have

$$\begin{aligned}
Ef(W) &\leq \sqrt{\pi/2}P(W > z) + \sqrt{\pi/2}\big(1-\Phi(z)\big)P(W \leq 0)\\
&\quad + \sqrt{2\pi}\big(1-\Phi(z)\big)Ee^{W^2/2}\mathbf{1}(0 < W \leq z)\\
&\leq \sqrt{\pi/2}P(W > z) + \sqrt{\pi/2}\big(1-\Phi(z)\big)\\
&\quad + \sqrt{2\pi}\big(1-\Phi(z)\big)Ee^{W^2/2}\mathbf{1}(0 < W \leq z).
\end{aligned} \tag{11.33}$$

Note that (11.29) implies $\max(\delta, \delta_1, \delta_2) \leq 1/256$. Hence the hypotheses of Lemma 11.2 are satisfied, and therefore also the conclusion of Lemma 11.3.

Now note that since $c_0$ in (11.20) is bounded over the given range of $z$, it follows from Lemma 11.2 that

$$Ee^{zW} \leq Ce^{z^2/2} \quad \text{and hence} \quad P(W > z) \leq e^{-z^2}Ee^{zW} \leq Ce^{-z^2/2}, \tag{11.34}$$

where $C$ denotes an absolute constant, not necessarily the same at each occurrence. This last inequality handles the first term in (11.33).

We will apply the identities, for any absolutely continuous function $g$, that

$$\begin{aligned}
&\int_0^z g'(y)P(W > y)dy\\
&\quad = g(z)P(W > z) - g(0)P(W > 0) + Eg(W)\mathbf{1}(0 < W \leq z),
\end{aligned}$$

and

$$\int_z^\infty g'(y)P(W > y)dy = -g(z)P(W > z) + Eg(W)\mathbf{1}(W > z). \tag{11.35}$$

Now, to handle the last term in (11.33), by Lemma 11.3,

$$\begin{aligned}Ee^{W^2/2}\mathbf{1}(0<W\le z) &\le P(0<W\le z)+\int_0^z ye^{y^2/2}P(W>y)dy\\ &\le C(1+z).\end{aligned}$$

For the second term in (11.31), similarly, by (2.7), (11.32) and (2.3),

$$\begin{aligned}&E|W|^3f(W)\\ &\quad\le EW^2\mathbf{1}(W>z)+\big(1-\Phi(z)\big)EW^2\mathbf{1}(W<0)\\ &\qquad+\sqrt{2\pi}\big(1-\Phi(z)\big)EW^3e^{W^2/2}\mathbf{1}(0<W\le z).\end{aligned}$$

The second term is clearly bounded by $2(1-\Phi(z))$, and we may bound the last expectation as

$$\begin{aligned}EW^3e^{W^2/2}\mathbf{1}(0<W\le z) &\le \int_0^z \big(y^4+3y^2\big)e^{y^2/2}P(W>y)dy\\ &\le C\big(1+z^4\big),\end{aligned}\tag{11.36}$$

applying Lemma 11.3 again.

As to $EW^2\mathbf{1}(W>z)$, first, using (11.34),

$$\begin{aligned}\int_z^\infty yP(W>y)dy &\le Ee^{zW}\int_z^\infty ye^{-zy}dy\\ &=\big(Ee^{zW}\big)z^{-2}\big(1+z^2\big)e^{-z^2}\le Ce^{-z^2/2}z^{-2}\big(1+z^2\big)\\ &\le Ce^{-z^2/2}\end{aligned}$$

for $z>1$. Thus, for all such $z$, by (11.35) and (11.34),

$$\begin{aligned}EW^2\mathbf{1}(W>z) &= z^2P(W>z)+\int_z^\infty 2yP(W>y)dy\\ &\le C\big(1+z^2\big)e^{-z^2/2}\le C\big(1+z^3\big)\big(1-\Phi(z)\big).\end{aligned}\tag{11.37}$$

Now, by (11.3) with $f(w)=w$ and (11.6) and (11.5), we have

$$\begin{aligned}EW^2 &= E\int_{|t|\le\delta}\hat{K}(t)dt+E(RW)\\ &\le E(\hat{K}_1)+\delta_2E\big(|W|+W^2\big)\\ &\le E(\hat{K}_1)+\delta_2E\big(1+2W^2\big)\\ &\le (1+\delta_1+\delta_2)+(\delta_1+2\delta_2)EW^2\\ &\le 5/4+EW^2/4,\end{aligned}$$

yielding $EW^2\le 2$. Hence (11.37) remains valid for $0\le z\le 1$ since $EW^2\mathbf{1}(W>z)\le EW^2\le 2$. Summarizing, we have

$$I_2\le C\big(1+z^4\big)\big(1-\Phi(z)\big),$$

and in a similar fashion one may demonstrate

$$I_4\le C\big(1+z^2\big)\big(1-\Phi(z)\big)\tag{11.38}$$

and

$$I_3 \le 3\big(1-\Phi(z)\big) + E\mathbf{1}(W \ge \delta + z)\big(1+W^2\big) \le C\big(1+z^3\big)\big(1-\Phi(z)\big).$$

Lastly, to handle $I_1$ letting $g(w) = (wf(w))'$ and recalling (2.81),

$$g(w) = \begin{cases} (\sqrt{2\pi}(1+w^2)e^{w^2/2}(1-\Phi(w)) - w)\Phi(z), & w > z, \\ (\sqrt{2\pi}(1+w^2)e^{w^2/2}\Phi(w) + w)(1-\Phi(z)), & w < z \end{cases}$$

and the inequality

$$0 \le \sqrt{2\pi}\big(1+w^2\big)e^{w^2/2}\big(1-\Phi(w)\big) - w \le \frac{2}{1+w^3} \quad \text{for } w \ge 0$$

from (5.4) of Chen and Shao (2001), we have for $0 \le t \le \delta$,

$$\begin{aligned} &Eg(W+t) \\ &\quad = Eg(W+t)\mathbf{1}\{W+t \ge z\} + Eg(W+t)\mathbf{1}\{W+t \le 0\} \\ &\qquad + Eg(W+t)\mathbf{1}\{0 < W+t < z\} \\ &\quad \le \frac{2}{1+z^3}P(W+t \ge z) + 2\big(1-\Phi(z)\big)P(W+t \le 0) \\ &\qquad + \sqrt{2\pi}\big(1-\Phi(z)\big)E\big(\big(1+(W+t)^2+(W+t)\big)e^{(W+t)^2/2}\big) \\ &\qquad \times \mathbf{1}\{0 < W+t < z\} \\ &\quad \le C\big(1+z^3\big)\big(1-\Phi(z)\big), \end{aligned}$$

by arguing as in (11.36) for the final term. Now writing $I_1 = \int_0^\delta Eg(W+t)dt$, putting everything together and using the continuity of the right hand side in $z$ to replace the strict inequality in (11.30) by a non-strict one, we obtain

$$\begin{aligned} &P(W \ge z+\delta) - \big(1-\Phi(z)\big) \\ &\quad \le C\big(1-\Phi(z)\big)\big(d_0\big(1+z^3\big)\delta + \big(1+z^4\big)\delta_1 + \big(1+z^2\big)\delta_2\big). \end{aligned} \tag{11.39}$$

Now note that for $\delta z \le 1$ and $z \ge 0$,

$$\begin{aligned} &1-\Phi(z-\delta) - \big(1-\Phi(z)\big) \\ &\quad = \frac{1}{\sqrt{2\pi}}\int_{z-\delta}^{z} e^{-t^2/2}dt \\ &\quad \le \frac{1}{\sqrt{2\pi}}\delta e^{-(z-\delta)^2/2} \\ &\quad \le \frac{1}{\sqrt{2\pi}}\delta e^{-z^2/2+z\delta} \\ &\quad \le \frac{1}{\sqrt{2\pi}}\delta e^{-z^2/2+1} \\ &\quad \le e\delta(1+z)\big(1-\Phi(z)\big) \\ &\quad \le 3(1+z)\delta\big(1-\Phi(z)\big) \le 6\big(1+z^3\big)\delta\big(1-\Phi(z)\big). \end{aligned}$$

For the third to last inequality we have used the fact that $g(z) \geq 0$ for all $z \geq 0$, where

$$g(z) = 1 - \Phi(z) - \frac{1}{\sqrt{2\pi}(1+z)} e^{-z^2/2},$$

which can be shown by verifying $g'(z) \leq 0$ for all $z \geq 0$, and $\lim_{z\to\infty} g(z) = 0$. Hence

$$\begin{aligned}
&P(W \geq z) - \bigl(1 - \Phi(z)\bigr) \\
&\quad = P(W \geq z) - \bigl(1 - \Phi(z-\delta)\bigr) + \bigl(1 - \Phi(z-\delta)\bigr) - \bigl(1 - \Phi(z)\bigr) \\
&\quad \leq P(W \geq z) - \bigl(1 - \Phi(z-\delta)\bigr) + 6\bigl(1+z^3\bigr)\delta\bigl(1 - \Phi(z)\bigr).
\end{aligned}$$

Now, from (11.39), with $C$ not necessarily the same at the occurrence,

$$\begin{aligned}
&P(W \geq z) - \bigl(1 - \Phi(z)\bigr) \\
&\quad \leq C\bigl(1 - \Phi(z)\bigr)\bigl(d_0\bigl(1+z^3\bigr)\delta + \bigl(1+z^4\bigr)\delta_1 + \bigl(1+z^2\bigr)\delta_2\bigr).
\end{aligned}$$

As a corresponding lower bound may be shown in the same manner, the proof of Theorem 11.1 is complete. □

The proof of Theorem 11.2 follows the lines same as the proof of Theorem 11.1, with $\hat{K}_1 = 1$, $\delta_1 = \delta_2 = 0$ and $d_0 = 1$; we omit the details.

We now prove our moderate deviation result for the Curie–Weiss model.

*Proof of (11.17)* For each $i \in \{1, \ldots, n\}$ let $\sigma_i'$ be a random sample from the conditional distribution of $\sigma_i$ given $\{\sigma_j, j \neq i, 1 \leq j \leq n\}$. Let $I$ be a random index uniformly distributed over $\{1, \ldots, n\}$ independent of $\{\sigma_i, \sigma_i'\colon 1 \leq i \leq n\}$. Recalling that $\sigma^2$ is the variance of the total spin $\sum_{i=1}^n \sigma_i$, and that $W = \sum_{i=1}^n \sigma_i/\sigma$, define $W' = W - (\sigma_I - \sigma_I')/\sigma$. Then $(W, W')$ is an exchangeable pair. Let

$$A(w) = \frac{\exp(-\beta\sigma w/n + \beta/n)}{\exp(\beta\sigma w/n - \beta/n) + \exp(-\beta\sigma w/n + \beta/n)},$$

and

$$B(w) = \frac{\exp(\beta\sigma w/n + \beta/n)}{\exp(\beta\sigma w/n + \beta/n) + \exp(-\beta\sigma w/n - \beta/n)}.$$

With $\boldsymbol{\sigma} = (\sigma_1, \ldots, \sigma_n)$, from (11.16) we obtain

$$\begin{aligned}
E(W - W'|\boldsymbol{\sigma}) &= \frac{1}{n\sigma}\sum_{i=1}^n E\bigl(\sigma_i - \sigma_i'|\boldsymbol{\sigma}\bigr) \\
&= \frac{1}{n\sigma}\Bigl\{\sum_{i\colon \sigma_i=1} 2P\bigl(\sigma_i' = -1|\boldsymbol{\sigma}\bigr) + \sum_{i\colon \sigma_i=-1} (-2)P\bigl(\sigma_i' = 1|\boldsymbol{\sigma}\bigr)\Bigr\} \\
&= \frac{1}{n\sigma}\bigl\{(n + \sigma W)A(W) - (n - \sigma W)B(W)\bigr\} \\
&= \biggl(\frac{A(W) + B(W)}{n}\biggr)W + \frac{A(W) - B(W)}{\sigma},
\end{aligned}$$

and hence

$$E(W - W'|W) = \left(\frac{A(W) + B(W)}{n}\right)W + \frac{A(W) - B(W)}{\sigma}.$$

Similarly,

$$\begin{aligned}
E\big((W - W')^2|W\big) &= E\big(E\big((W - W')^2|\boldsymbol{\sigma}\big)|W\big) \\
&= E\left(\frac{1}{n\sigma^2}\sum_{i=1}^{n} E\big((\sigma_i - \sigma_i')^2|\boldsymbol{\sigma}\big)\bigg|W\right) \\
&= \frac{1}{n\sigma^2}\big\{(n + \sigma W)2A(W) + (n - \sigma W)2B(W)\big\} \\
&= \frac{2(A(W) + B(W))}{\sigma^2} + \frac{2(A(W) - B(W))}{n\sigma}W.
\end{aligned}$$

It is easy to see that

$$\begin{aligned}
\frac{e^{-\beta\sigma w/n}}{e^{-\beta\sigma w/n} + e^{\beta\sigma w/n}} \le A(w) &= \frac{1}{1 + \exp(2\beta\sigma w/n - 2\beta/n)} \\
&\le \frac{e^{2\beta/n}}{1 + \exp(2\beta\sigma w/n)} \\
&= \frac{e^{-\beta\sigma w/n}e^{2\beta/n}}{e^{-\beta\sigma w/n} + e^{\beta\sigma w/n}}
\end{aligned}$$

and similarly,

$$\begin{aligned}
\frac{e^{\beta\sigma w/n}}{e^{-\beta\sigma w/n} + e^{\beta\sigma w/n}} \le B(w) &= \frac{1}{1 + \exp(-2\beta\sigma w/n - 2\beta/n)} \\
&\le \frac{e^{\beta\sigma w/n}e^{2\beta/n}}{e^{-\beta\sigma w/n} + e^{\beta\sigma w/n}}.
\end{aligned}$$

Therefore

$$A(W) + B(W) = 1 + O(1)\frac{1}{n}$$

and

$$A(W) - B(W) = -\tanh(\beta\sigma W/n) + O(1)\frac{1}{n}.$$

Hence we have

$$\begin{aligned}
&E(W - W'|W) \\
&= \left(1 + O(1)\frac{1}{n}\right)\frac{W}{n} + \frac{1}{\sigma}\left(-\tanh(\beta\sigma W/n) + O(1)\frac{1}{n}\right) \\
&= \frac{W}{n} + O(1)\frac{W}{n^2} + O(1)\frac{1}{n\sigma} + \frac{1}{\sigma}\left(-\beta\sigma W/n - \frac{\tanh''(\xi)}{2}(\beta\sigma W/n)^2\right) \\
&= \frac{1-\beta}{n}W - \frac{\tanh''(\xi)\beta^2\sigma W^2}{2n^2} + O(1)\frac{1}{n\sigma},
\end{aligned} \tag{11.40}$$

using the fact that $|\sigma W| \le n$, and likewise

$$\begin{aligned}
&E\big((W-W')^2|W\big)\\
&= \frac{2}{\sigma^2} + O(1)\frac{1}{n\sigma^2} + O(1)\frac{W}{n^2\sigma} + \frac{2W}{n\sigma}\big(-\tanh'(\eta)\beta\sigma W/n\big)\\
&= \frac{2}{\sigma^2} - \frac{2\tanh'(\eta)\beta W^2}{n^2} + O(1)\frac{1}{n\sigma^2},
\end{aligned} \tag{11.41}$$

where $\xi$ and $\eta$ lie between 0 and $\beta\sigma W/n$.

From (11.40) and Remark 11.1, $W$ satisfies (11.3) with $\lambda = (1-\beta)/n$, $\hat{K}_1 = (W-W')^2/2\lambda$ and

$$R = \frac{\tanh''(\xi)\beta^2\sigma W^2}{2n(1-\beta)} + O(1)\frac{1}{\sigma}. \tag{11.42}$$

Further, from (11.41),

$$\begin{aligned}
&E[\hat{K}_1|W] - 1\\
&= \frac{1}{2\lambda}E\big((W-W')^2|W\big) - 1\\
&= \frac{n}{(1-\beta)\sigma^2} - 1 - \frac{\tanh'(\eta)\beta W^2}{n(1-\beta)} + O(1)\frac{1}{\sigma^2}.
\end{aligned} \tag{11.43}$$

Since (11.9) holds, the expected value of the left hand side of (11.43) is $-E[RW]$. Hence, using that $EW = 0$, making the second term in (11.42) vanish after multiplying by $W$ and taking expectation, we obtain

$$\begin{aligned}
&\frac{n}{(1-\beta)\sigma^2} - 1 - \frac{E(\tanh'(\eta)\beta W^2)}{n(1-\beta)} + O(1)\frac{1}{\sigma^2}\\
&= -E\left(\frac{\tanh''(\xi)\beta^2\sigma W^3}{2n(1-\beta)}\right).
\end{aligned} \tag{11.44}$$

On the left hand side, since $\tanh'(x)$ is bounded on $\mathbb{R}$ and $EW^2 = 1$, the third term is $O(1/n)$, and the last term is of smaller order than the first.

On the right hand side, as $\tanh''(x)$ has sign opposite that of $x$, we conclude $\tanh''(\xi)W^3 \le 0$, as $\xi$ lies between 0 and $\beta\sigma W/n$. Hence the right hand side above is nonnegative. As $\tanh''(x)$ is bounded on $\mathbb{R}$, $|W^3| \le nW^2/\sigma$ and $EW^2 = 1$, the right hand side is also bounded. Hence $n/((1-\beta)\sigma^2)$ is of order 1, and $\sigma/\sqrt{n}$ is bounded away from 0 and infinity.

Note now that from (11.44) that if $E|W^3| \le C$ then

$$\frac{n}{(1-\beta)\sigma^2} - 1 = O(1/\sqrt{n}),$$

implying, by (11.43), that

$$\left|\frac{1}{2\lambda}E\big((W-W')^2|W\big) - 1\right| \le Cn^{-1/2}\big(1+|W|\big). \tag{11.45}$$

Next we prove $E|W^3| \le C$. Letting $f(w) = w|w|$, for which $f'(w) = 2|w|$, substitution into (11.3), and (11.43) and (11.42) yield

$$\begin{aligned}
&E|W^3| \\
&= EWf(W) = E\int_{|t|\le\delta} 2|W|\hat{K}(t)dt + E\big(Rf(W)\big) \\
&= 2E|W| + 2E\big(|W|\big(E[\hat{K}_1|W] - 1\big)\big) + E\big(Rf(W)\big) \\
&= 2E|W| + 2E|W|\left(\frac{n}{(1-\beta)\sigma^2} - 1 - \frac{\tanh'(\eta)\beta W^2}{n(1-\beta)} + O(1)\frac{1}{\sigma^2}\right) \\
&\quad + E\left(\frac{\tanh''(\xi)\beta^2\sigma W^2}{2n(1-\beta)} + O(1)\frac{1}{\sigma}\right)f(W) \\
&= 2E|W| + 2E|W|\left(\frac{n}{(1-\beta)\sigma^2} - 1\right) + O\left(\frac{1}{n}\right)E|W|^3 + O(1)\frac{1}{\sigma^2}E|W| \\
&\quad + E\left(\frac{\tanh''(\xi)\beta^2\sigma W^3}{2n(1-\beta)}|W|\right) + O(1)\frac{1}{\sigma}Ef(W).
\end{aligned}$$

As $\tanh''(\xi)W^3 \le 0$, and $n/((1-\beta)\sigma^2) - 1 = O(1)$, the right hand side is $O(1) + O(1/n)E|W^3|$, hence $E|W^3| = O(1)$, as desired.

By (11.42) and the fact that $\sigma/\sqrt{n}$ is bounded away from zero and infinity, in place of Condition (11.6) we have instead that

$$\big|E(R|W)\big| \le \delta_2\big(1 + W^2\big) \quad \text{where } \delta_2 = \frac{C}{\sqrt{n}}. \tag{11.46}$$

However, simple modifications can be made in the proofs of Lemma 11.2 and Theorem 11.1 so that (11.17) holds. First, note that the inequality $(1 + |w|) \le 2(1 + w^2)$ is used in (11.21) to bound the first application of (11.6) in Lemma 11.2. Next, since $\xi$ is between 0 and $\beta\sigma W/n$, the terms $\tanh''(\xi)$ and $W$ have opposite signs. Hence, in the display (11.23) in Lemma 11.2, for the first term of the remainder $R$ in (11.42) we have

$$E\left(\frac{\tanh''(\xi)\beta^2\sigma W^2}{2n(1-\beta)} We^{s(W\wedge a)}\right) \le 0,$$

while the second term, of order $1/\sigma$, that is, order $1/\sqrt{n}$, can be absorbed after the indicated multiplication by $W$ in the existing term $\delta_2 Ef(W)(1 + |W|)|W|$, with $\delta_2$ of order $1/\sqrt{n}$. Hence (11.24), and Lemma 11.2 remain valid.

In the proof of Theorem 11.1, the present case can be handled by replacing $I_4 = (1 + |W|)f(W)$ by $I_4 = (1 + W^2)f(W)$, resulting in the bound

$$I_4 \le C\big(1 + z^3\big)\big(1 - \Phi(z)\big)$$

in place of (11.38). By (11.43) we may take $d_0 = O(1)$, and since $|W' - W| = |\sigma_I' - \sigma_I|/\sigma$, we have $\delta = O(1/\sqrt{n})$. Likewise, by (11.46) and (11.45) we may take $\delta_2$ and $\delta_1$ respectively, both of order $O(1/\sqrt{n})$. Hence, in view of (11.45) and Remark 11.2, we have the following moderate deviation result for $W$

$$\frac{P(W \geq z)}{1 - \Phi(z)} = 1 + O(1)d_0(1 + z^3)\delta + O(1)(1 + z^3)\delta_1 + O(1)(1 + z^3)\delta_2.$$

This completes the proof of (11.17). □

## Appendix

*Proof of Lemma 11.1* Write

$$n - 1 = \sum_{i \geq 1} 2^{m - p_i},$$

with $1 = p_1 < p_2 < \cdots \leq m_1$ the positions of the ones in the binary expansion of $n - 1$, where $m_1 \leq m$. Recall that $X$ is uniformly distributed over $\{0, 1, \ldots, n - 1\}$, and that

$$X = \sum_{i=1}^{m} X_i 2^{m-i},$$

with exactly $S$ of the indicator variables $X_1, \ldots, X_m$ equal to 1.

We say that $X$ falls in category $i$, $i = 1, \ldots, m_1$, when

$$X_{p_1} = 1, \quad X_{p_2} = 1, \quad \ldots, \quad X_{p_{i-1}} = 1 \quad \text{and} \quad X_{p_i} = 0, \tag{11.47}$$

and in category $m_1 + 1$ if $X = n - 1$. This last category is nonempty only when $S = m_1$ and in this case, $Q = m - m_1$, which gives the last term in (11.48).

Note that if $X$ is in category $i$ for $i \leq m_1$, then, since $X$ can be no greater than $n - 1$, the digits of $X$ and $n - 1$ match up to the $p_i^{th}$, except for the digit in place $p_i$, where $n - 1$ has a one, and $X$ a zero. Further, up to this digit, $n - 1$ has $p_i - i$ zeros, and so $X$ has $a_i = p_i - i + 1$ zeros. Changing any of these $a_i$ zeros of $X$, except the zero in position $p_i$, to one results in a number $n - 1$ or greater, while changing any other zeros, since digit $p_i$ of $n - 1$ is one and that same digit of $X$ is zero, does not. Hence $Q$ is at most $a_i$ when $X$ falls in category $i$. Since $X$ has $S$ ones in its expansion, $i - 1$ of which are accounted for by (11.47), conditional on $S$ the remaining $S - (i - 1)$ ones are uniformly distributed over the $m - p_i = m - (i - 1) - a_i$ remaining digits $\{X_{p_i+1}, \ldots, X_m\}$. Thus, we have the inequality

$$E(Q|S) \leq \frac{1}{A} \sum_{i \geq 1} \binom{m - (i-1) - a_i}{S - (i-1)} a_i + \frac{I(S = m_1)}{A}(m - m_1) \tag{11.48}$$

where

$$A = \sum_{i \geq 1} \binom{m - (i-1) - a_i}{S - (i-1)} + I(S = m_1),$$

and $1 = a_1 \leq a_2 \leq a_3 \leq \cdots$.

Note that if $m_1 = m$, the last term of (11.48) equals 0. When $m_1 < m$, we have

$$\frac{I(S = m_1)}{A}(m - m_1) \leq \binom{m-1}{m_1}^{-1} (m - m_1) \leq 1,$$

so we may consider only the remaining terms of (11.48) in the following argument. We consider two cases; constants $C$ may not necessarily be the same at each occurrence.

*Case 1* $S \geq m/2$.

As $a_i \geq 1$ for all $i$, there are at most $m+1$ nonzero terms in the sum (11.48). Divide the summands into two groups, those for which $a_i \leq 2\log_2 m$ and those with $a_i > 2\log_2 m$. The first group can sum to no more than $2\log_2 m$, as the sum is a weighted average of the $a_i$ terms, with weights summing to less than 1.

For the second group, note that

$$\begin{aligned}
&\binom{m-(i-1)-a_i}{S-(i-1)} \Big/ A \\
&\leq \binom{m-(i-1)-a_i}{S-(i-1)} \Big/ \binom{m-1}{S} \\
&= \prod_{j=1}^{a_i-1} \left(\frac{m-S-j}{m-j}\right) \prod_{j=0}^{i-2} \left(\frac{S-j}{m-(a_i-1)-1-j}\right) \\
&\leq \frac{1}{2^{a_i-1}} \leq \frac{2}{m^2},
\end{aligned} \tag{11.49}$$

where the second to last inequality follows from $S \geq m/2$ and the fact that the term considered is nonzero only when $S \leq m-a_i$, and the last from $a_i > 2\log_2 m$. As $a_i \leq m$ and there are at most $m+1$ terms in the sum, the terms in the second group can sum to no more than 4.

*Case 2* $S < m/2$. Divide the sum in (11.48) into two groups according as to whether $i > 2\log_2 m$ or $i \leq 2\log_2 m$. Reordering the product in (11.49),

$$\begin{aligned}
\binom{m-(i-1)-a_i}{S-(i-1)} \Big/ A &\leq \prod_{j=0}^{i-2} \left(\frac{S-j}{m-1-j}\right) \prod_{j=1}^{a_i-1} \left(\frac{m-S-j}{m-(i-1)-j}\right) \\
&\leq 1/2^{i-1}
\end{aligned}$$

using the assumption $S < m/2$, and noting that the term considered is zero unless $S \geq i-1$. The above inequality is true for all $i$, so in particular the summation over $i$ satisfying $i > 2\log_2 m$ is bounded by 4.

Next consider $i \leq 2\log_2 m$. For $a_i \geq 2$ the inequality

$$S \geq m\left(\frac{\log a_i}{a_i-1}\right) + 2\log_2 m \tag{11.50}$$

implies $S \geq (1 - e^{-\frac{\log a_i}{a_i-1}})m - 1 + e^{-\frac{\log a_i}{a_i-1}} i$, which is equivalent to $a_i(\frac{m-S-1}{m-(i-1)-1})^{a_i-1} \leq 1$, which clearly holds also for $a_i = 1$. Hence,

$$\begin{aligned}
&a_i \binom{m-(i-1)-a_i}{S-(i-1)} \Big/ A \\
&\leq a_i \binom{m-(i-1)-a_i}{S-(i-1)} \Big/ \binom{m-1}{S} \\
&= a_i \prod_{j=0}^{i-2} \left( \frac{S-j}{m-1-j} \right) \prod_{j=1}^{a_i-1} \left( \frac{m-S-j}{m-(i-1)-j} \right) \\
&\leq \frac{1}{2^{i-1}} a_i \left( \frac{m-S-1}{m-(i-1)-1} \right)^{a_i-1} \leq \frac{1}{2^{i-1}}
\end{aligned}$$

using the fact that $a_i (\frac{m-S-1}{m-(i-1)-1})^{a_i-1} \leq 1$.

On the other hand, if $S < m(\frac{\log a_i}{a_i-1}) + 2\log_2 m$ then $a_i S/(m-1) \leq C \log_2 m$, which implies

$$\begin{aligned}
&a_i \binom{m-(i-1)-a_i}{S-(i-1)} \Big/ A \\
&\leq \frac{a_i S}{m-1} \prod_{j=1}^{i-2} \left( \frac{S-j}{m-1-j} \right) \prod_{j=1}^{a_i-1} \left( \frac{m-S-j}{m-(i-1)-j} \right) \\
&\leq C \log_2 m / 2^{i-2}.
\end{aligned}$$

Hence the sum over $i$ is bounded by some constant time $\log_2 m$. Combining the two cases we have that the right hand side of (11.48), and therefore $E(Q|S)$, is bounded by $C \log_2 m$.

To complete the proof of the lemma, that is, to prove $E(Q|W) \leq C(1+|W|)$, we only need to show

$$E(Q|S) \leq C \quad \text{when } |W| \leq \log_2 m, \tag{11.51}$$

as when $|W| > \log_2 m$ we already have $E(Q|W) \leq C \log_2 m \leq C|W|$.

In case 1 we have shown $E(Q|S)$ is bounded, and in case 2 that the contribution of the summands where $i > 2\log_2 m$ is bounded. Hence we need only consider summands where $i \leq 2\log_2 m$. Note that $|W| \leq \log_2 m$ implies $S \geq m/2 - \sqrt{m/4}\log_2 m$. When $a_i, m$ are bigger than some universal constant, $m/2 - \sqrt{m/4}\log_2 m \geq m(\frac{\log a_i}{a_i-1}) + 2\log_2 m$, which implies that (11.50) holds. Hence, as in case 2, we have that $(\frac{m-S-1}{m-(i-1)-1})^{a_i-1} \times a_i \leq 1$ and $\binom{m-(i-1)-a_i}{S-(i-1)} \times a_i/A \leq 1/2^{i-1}$. Summing, we see the contribution from the remaining terms are also bounded, completing the proof of (11.51), and the lemma. □

# Chapter 12
# Multivariate Normal Approximation

In this chapter we consider multivariate normal approximation. We begin with the extension of the ideas in Sect. 4.8 on bounds for smooth functions, using the results in Sect. 2.3.4 which may be applied in the multivariate setting. The first goal is to develop smooth function bounds in $\mathbb{R}^p$. In Sect. 12.1 we obtain such bounds using multivariate size bias couplings, and in Sect. 12.3 by multivariate exchangeable pairs. In Sect. 12.4 we turn to local dependence, and bounds in the Kolmogorov distance. We consider applications of these results to questions in random graphs.

Generalizing notions from Sect. 4.8, for

$$\mathbf{k} = (k_1, \ldots, k_p) \in \mathbb{N}_0^p \quad \text{let} \quad |\mathbf{k}| = \sum_{i=1}^{p} k_i,$$

and for functions $h : \mathbb{R}^p \to \mathbb{R}$ whose partial derivatives

$$h^{\mathbf{k}}(\mathbf{x}) = \frac{\partial^{k_1+\cdots+k_p} h}{\partial^{k_1} x_1 \cdots \partial^{k_p} x_p} \quad \text{exists for all } 0 \le |\mathbf{k}| \le m,$$

and $\|\cdot\|$ the supremum norm, recall that $L_m^\infty(\mathbb{R}^p)$ is the collection of all functions $h : \mathbb{R}^p \to \mathbb{R}$ with

$$\|h\|_{L_m^\infty(\mathbb{R}^p)} = \max_{0 \le |\mathbf{k}| \le m} \left\| h^{(\mathbf{k})} \right\| \tag{12.1}$$

finite. Now, for random vectors $\mathbf{X}$ and $\mathbf{Y}$ in $\mathbb{R}^p$, letting

$$\mathcal{H}_{m,\infty,p} = \left\{ h \in L_m^\infty(\mathbb{R}^p) \colon \|h\|_{L_m^\infty(\mathbb{R}^p)} \le 1 \right\} \tag{12.2}$$

define

$$\left\| \mathcal{L}(\mathbf{X}) - \mathcal{L}(\mathbf{Y}) \right\|_{\mathcal{H}_{m,\infty,p}} = \sup_{h \in \mathcal{H}_{m,\infty,p}} \left| Eh(\mathbf{X}) - Eh(\mathbf{Y}) \right|.$$

For a vector, matrix, or more generally, any array $A = (a_\alpha)_{\alpha \in \mathcal{A}}$ with $\mathcal{A}$ finite, let

$$\|A\| = \max_{\alpha \in \mathcal{A}} |a_\alpha|. \tag{12.3}$$

L.H.Y. Chen et al., *Normal Approximation by Stein's Method*,
Probability and Its Applications,
DOI 10.1007/978-3-642-15007-4_12, 

## 12.1 Multivariate Normal Approximation via Size Bias Couplings

The following theorem gives a smooth function bound via multivariate size bias couplings.

**Theorem 12.1** *Let* $\mathbf{Y}$ *be a random vector in* $\mathbb{R}^p$ *with nonnegative components, mean* $\boldsymbol{\mu} = E\mathbf{Y}$, *and invertible covariance matrix* $\operatorname{Var}(\mathbf{Y}) = \boldsymbol{\Sigma}$. *For each* $i = 1, \ldots, p$ *let* $(\mathbf{Y}, \mathbf{Y}^i)$ *be random vectors defined on a joint probability space such that* $\mathbf{Y}^i$ *has the* $\mathbf{Y}$*-size biased distribution in direction* $i$, *as in* (2.68). *Then, with* $\mathbf{Z}$ *a mean zero, covariance* $I$ *normal vector in* $\mathbb{R}^p$,

$$\begin{aligned}
&\big\|\mathcal{L}(\boldsymbol{\Sigma}^{-1/2}(\mathbf{Y}-\boldsymbol{\mu}) - \mathcal{L}(\mathbf{Z})\big\|_{\mathcal{H}_{3,\infty,p}} \\
&\quad\le \frac{p^2}{2}\big\|\Sigma^{-1/2}\big\|^2 \sum_{i=1}^p\sum_{j=1}^p \mu_i \sqrt{\operatorname{Var} E\big[Y_j^i - Y_j \,|\, \mathbf{Y}\big]} \\
&\qquad + \frac{1}{2}\frac{p^3}{3}\big\|\Sigma^{-1/2}\big\|^3 \sum_{i=1}^p\sum_{j=1}^p\sum_{k=1}^p \mu_i E\big|\big(Y_j^i - Y_j\big)\big(Y_k^i - Y_k\big)\big|.
\end{aligned} \tag{12.4}$$

Note that the theorem does not require the joint construction of $(\mathbf{Y}^1, \ldots, \mathbf{Y}^p)$.

*Proof* Given $h$ with $\|h\|_{L_3^\infty(\mathbb{R}^p)} \le 1$, let $f$ be the solution of (2.22) given by (2.21) and (2.20). Writing out the expressions in (2.22),

$$\begin{aligned}
&E\big\{h\big(\Sigma^{-1/2}(\mathbf{Y}-\boldsymbol{\mu})\big) - Nh\big\} \\
&\quad = E\left\{\sum_{i=1}^p\sum_{j=1}^p \sigma_{ij}\frac{\partial^2}{\partial y_i\partial y_j} f(\mathbf{Y}) - \sum_{i=1}^p (Y_i - \mu_i)\frac{\partial}{\partial y_i} f(\mathbf{Y})\right\}.
\end{aligned} \tag{12.5}$$

Recall from (2.68) that $\mathbf{Y}^i$ is characterized by the fact that

$$EY_i G(\mathbf{Y}) = \mu_i EG\big(\mathbf{Y}^i\big) \tag{12.6}$$

for all functions $G : \mathbb{R}^p \to \mathbb{R}$ for which the expectations exist. For the coordinate function $G(\mathbf{y}) = y_j$, (12.6) gives

$$\sigma_{ij} = \operatorname{Cov}(Y_i, Y_j) = EY_iY_j - \mu_i\mu_j = E\mu_i\big(Y_j^i - Y_j\big). \tag{12.7}$$

Subtracting $\mu_i EG(\mathbf{Y})$ from both sides of (12.6), we obtain

$$E(Y_i - \mu_i)G(\mathbf{Y}) = \mu_i E\big[G\big(\mathbf{Y}^i\big) - G(\mathbf{Y})\big]. \tag{12.8}$$

Equation (12.5), and (12.8) with $G = (\partial/\partial y_i) f$, yield

$$\begin{aligned}
&E\big\{h\big(\Sigma^{-1/2}(\mathbf{Y}-\boldsymbol{\mu})\big) - Nh\big\} \\
&\quad = E\left\{\sum_{i=1}^p\sum_{j=1}^p \sigma_{ij}\frac{\partial^2}{\partial y_i\partial y_j} f(\mathbf{Y}) - \sum_{i=1}^p \mu_i\left[\frac{\partial}{\partial y_i} f\big(\mathbf{Y}^i\big) - \frac{\partial}{\partial y_i} f(\mathbf{Y})\right]\right\}.
\end{aligned} \tag{12.9}$$

Taylor expanding $(\partial/\partial y_i)f(\mathbf{Y}^i)$ about $\mathbf{Y}$, with remainder in integral form, and simple calculations show that the right hand side of (12.9) equals

$$-E\sum_{i=1}^{p}\sum_{j=1}^{p}\left[\mu_i(Y_j^i - Y_j) - \sigma_{ij}\right]\frac{\partial^2}{\partial y_i \partial y_j}f(\mathbf{Y}) - E\sum_{i=1}^{p}\sum_{j=1}^{p}\sum_{k=1}^{p}\mu_i$$
$$\times \int_0^1 (1-t)\frac{\partial^3}{\partial y_i \partial y_j \partial y_k} f\big(\mathbf{Y} + t(\mathbf{Y}^i - \mathbf{Y})\big)\big(Y_j^i - Y_j\big)\big(Y_k^i - Y_k\big)dt. \quad (12.10)$$

In the first term, we condition on $\mathbf{Y}$, apply the Cauchy–Schwarz inequality and use (12.7), and then apply the bound (2.23) with $k = 2$ to obtain the first term in (12.4). The second term in (12.10) gives the second term in (12.4) by applying (2.23) with $k = 3$. □

## 12.2 Degrees of Random Graphs

In the classical Erdös and Rényi (1959b) random graph model (see also Bollobás 1985) for $n \in \mathbb{N}$ and $\varrho \in (0,1)$, $K = K_{n,\varrho}$ is the random graph on the vertex set $\mathcal{V} = \{1, \ldots, n\}$ with random edge set $\mathcal{E}$ where each pair of vertices has probability $\varrho$ of being connected, independently of all other such pairs. For $v \in \mathcal{V}$ let

$$D(v) = \sum_{w \in \mathcal{V}} \mathbf{1}_{\{v,w\} \in \mathcal{E}},$$

the degree of vertex $v$, and for $d \in \{0, 1, 2, \ldots\}$ let

$$Y = \sum_{v \in \mathcal{V}} X_v \quad \text{where } X_v = \mathbf{1}_{\{D(v)=d\}},$$

the number of vertices with degree $d$.

Karoński and Ruciński (1987) proved asymptotic normality of $Y$ when $\varrho n^{(d+1)/d} \to \infty$ and $\varrho n \to 0$, or $\varrho n \to \infty$ and $\varrho n - \log n - d \log\log n \to -\infty$; see also Palka (1984) and Bollobás (1985). Asymptotic normality when $\varrho n \to c > 0$, was obtained by Barbour et al. (1989); see also Kordecki (1990) for the case $d = 0$, for nonsmooth $h$. Goldstein (2010b) gives a Berry–Esseen theorem for $Y$ for all $d$ by applying the size bias coupling in Bolthausen's (1984) inductive method. Other univariate results on asymptotic normality of counts on random graphs, including counts of the type discussed in Theorems 12.2, are given in Janson and Nowicki (1991), and references therein.

Based on the work of Goldstein and Rinott (1996) we consider the joint asymptotic normality of a vector of degree counts. For $p \in \mathbb{N}$ let $d_i$ for $i = 1, \ldots, p$ be distinct, fixed nonnegative integers, and let $\mathbf{Y} \in \mathbb{R}^p$ have $i$th coordinate

$$Y_i = \sum_{v \in \mathcal{V}} X_{vi} \quad \text{where } X_{vi} = \mathbf{1}_{\{D(v)=d_i\}},$$

the number of vertices of the graph with degree $d_i$. For simplicity we assume $0 < \varrho = c/(n-1) < 1$ in what follows, though the results below can be weakened to cover the case $n\varrho_n \to c > 0$ as $n \to \infty$. To keep track of asymptotic constants, for a sequence $a_n$ and a sequence of positive numbers $b_n$ write

$$a_n = \Omega(b_n)$$

if $\limsup_{n\to\infty} |a_n|/b_n \le 1$.

**Theorem 12.2** *If $\varrho = \varrho_n = c/(n-1)$ for some $c > 0$ and $\mathbf{Z} \in \mathbb{R}^p$ is a mean zero normal vector with identity covariance matrix, then*

$$\left\| \mathcal{L}(\boldsymbol{\Sigma}^{-1/2}(\mathbf{Y} - \boldsymbol{\mu}) - \mathcal{L}(\mathbf{Z}) \right\|_{\mathcal{H}_{3,\infty,p}} \le n^{-1/2}(r_1 + r_2), \tag{12.11}$$

*where*

$$r_1 = \frac{p^3 b}{2} \sum_{i=1}^{p} \beta_i \Omega\left(\sqrt{24c + 48c^2 + 144c^3 + 48d_i^2 + 144cd_i^2 + 12}\right) \quad \text{and}$$

$$r_2 = \frac{p^5 b^{3/2}}{3} \sum_{i=1}^{p} \beta_i \left(c + c^2 + (d_i + 1)^2\right),$$

*where the components $\mu_i, \sigma_{ij}, i, j = 1, \ldots, n$ of the mean vector $\boldsymbol{\mu} = E\mathbf{Y}$ and covariance matrix $\boldsymbol{\Sigma} = \mathrm{Var}(\mathbf{Y})$ respectively, are given by*

$$\begin{aligned} \mu_i &= n\beta_i \quad \text{and} \\ \sigma_{ij} &= n\beta_i\beta_j \left[ \frac{(d_i - c)(d_j - c)}{c(1 - c/(n-1))} - 1 \right] + \mathbf{1}_{\{i=j\}} n\beta_i, \end{aligned} \tag{12.12}$$

*and*

$$\begin{aligned} \beta_i &= \binom{n-1}{d_i} \varrho^{d_i} (1-\varrho)^{n-1-d_i} \quad \text{and} \\ b &= \left[ \frac{1}{\min_j \beta_j (1 - \sum_{i=1}^{p} \beta_i)} \right]. \end{aligned} \tag{12.13}$$

Note that $\sum_{i=1}^{p} \beta_i < 1$ when $\{d_1, \ldots, d_p\} \ne \{0, 1, \ldots, n-1\}$, and then the quantities $r_1$ and $r_2$ are both of order $O(1)$.

*Proof* As for any $v \in \mathcal{V}$ the degree $D(v)$ is the sum of $n-1$ independent Bernoulli variables with success probability $\varrho$, we have $D(v) \sim \mathrm{Bin}(n-1, \varrho)$. In particular, $\beta_i$ in (12.13) equals

$$P\big(D(v) = d_i\big) = EX_{vi},$$

yielding the expression for $\mu_i$ in (12.12).

To calculate the covariance $\sigma_{ij}$ for $i \ne j$, with $v \ne u$ write

$$\begin{aligned} &EX_{vi}X_{uj} \\ &\quad = E\big(X_{vi}X_{uj} | \{v, u\} \in \mathcal{E}\big)\varrho + E\big(X_{vi}X_{uj} | \{v, u\} \notin \mathcal{E}\big)(1 - \varrho). \end{aligned} \tag{12.14}$$

Given that there is an edge connecting $v$ and $u$, $X_{vi}X_{uj}=1$ if and only if $v$ is connected to $d_i-1$ vertices in $\mathcal{V}\setminus\{u\}$, and $u$ to $d_j-1$ vertices in $\mathcal{V}\setminus\{v\}$, which are functions of independent Bernoulli variables. Hence

$$\begin{aligned}E\big(X_{vi}X_{uj}|\{v,u\}\in\mathcal{E}\big)&=\binom{n-2}{d_i-1}\binom{n-2}{d_j-1}\varrho^{d_i+d_j-2}(1-\varrho)^{2n-2-d_i-d_j}\\&=\beta_i\beta_j\frac{d_id_j}{\varrho^2(n-1)^2}\\&=\beta_i\beta_j\frac{d_id_j}{c^2}.\end{aligned}$$

Likewise, given that there is no edge between $v$ and $u$, $X_{vi}X_{uj}=1$ if and only if $v$ is connected to $d_i$ vertices in $\mathcal{V}\setminus\{u\}$, and $u$ to $d_j$ vertices in $\mathcal{V}\setminus\{v\}$, and so

$$\begin{aligned}E\big(X_{vi}X_{uj}|\{v,u\}\notin\mathcal{E}\big)&=\binom{n-2}{d_i}\binom{n-2}{d_j}\varrho^{d_i+d_j}(1-\varrho)^{2n-4-d_i-d_j}\\&=\beta_i\beta_j\frac{(n-1-d_i)(n-1-d_j)}{(1-\varrho)^2(n-1)^2}\\&=\beta_i\beta_j\frac{(n-1-d_i)(n-1-d_j)}{(n-1-c)^2}.\end{aligned}$$

Adding these expressions according to (12.14) yields

$$EX_{vi}X_{uj}=\beta_i\beta_j\left(\frac{d_id_j+c(n-1)-cd_i-cd_j}{c(n-1-c)}\right).$$

Now, multiplying by $n^2-n$, as $X_{vi}X_{vj}=0$ for $d_i\neq d_j$, we have

$$EY_iY_j=n\beta_i\beta_j\left(\frac{d_id_j+c(n-1)-cd_i-cd_j}{c(1-c/(n-1))}\right),$$

and subtracting $n^2\beta_i\beta_j$ yields (12.12) for $i\neq j$. When $i=j$ the calculation is the same, but for the addition in the second moment of the expectation of $n$ diagonal terms of the form $X_{vi}^2=X_{vi}$.

We may write the covariance matrix $\mathbf{\Sigma}$ more compactly as follows. Let

$$\mathbf{b}=\big(\beta_1^{1/2},\ldots,\beta_p^{1/2}\big)^\mathsf{T},$$

$$\mathbf{g}=\left(\frac{\beta_1^{1/2}(d_1-c)}{\sqrt{c(1-c/(n-1))}},\ldots,\frac{\beta_p^{1/2}(d_p-c)}{\sqrt{c(1-c/(n-1))}}\right)^\mathsf{T},$$

and

$$D=\operatorname{diag}\big(\beta_1^{1/2},\ldots,\beta_p^{1/2}\big),$$

that is, the diagonal matrix whose diagonal elements are the components of $\mathbf{b}$. Then it is not difficult to see that

$$n^{-1}\mathbf{\Sigma}=D\big(I+\mathbf{g}\mathbf{g}^\mathsf{T}-\mathbf{b}\mathbf{b}^\mathsf{T}\big)D.$$

For nonnegative definite matrices $A$ and $B$, write

$$A \preceq B \quad \text{when } \mathbf{x}^\mathsf{T} A\mathbf{x} \le \mathbf{x}^\mathsf{T} B\mathbf{x} \text{ for all } \mathbf{x}. \tag{12.15}$$

It is clear that

$$D\big(I - \mathbf{b}\mathbf{b}^\mathsf{T}\big)D \preceq n^{-1}\boldsymbol{\Sigma}.$$

Letting $\lambda_1(A) \le \cdots \le \lambda_p(A)$ be the eigenvalues of $A$ in non-decreasing order, then, see e.g. Horn and Johnson (1985),

$$\lambda_k\big(D\big(I - \mathbf{b}\mathbf{b}^\mathsf{T}\big)D\big) \le \lambda_k\big(n^{-1}\boldsymbol{\Sigma}\big).$$

It is simple to verify that the eigenvalues of $B = I - \mathbf{b}\mathbf{b}^\mathsf{T}$ are 1, with multiplicity $p-1$, and, corresponding to the eigenvector $\mathbf{b}$, $\lambda_1(B) = 1 - \mathbf{b}^\mathsf{T}\mathbf{b}$. Now, by the Rayleigh-Ritz characterization of eigenvalues we obtain

$$\lambda_1(DBD) = \min_{\mathbf{x}\in\mathbb{R}^p, \mathbf{x}\ne 0} \frac{\mathbf{x}^\mathsf{T} DBD\mathbf{x}}{\mathbf{x}^\mathsf{T}\mathbf{x}} = \min_{\mathbf{y}\in\mathbb{R}^p, \mathbf{y}\ne 0} \frac{\mathbf{y}^\mathsf{T} B\mathbf{y}}{\mathbf{y}^\mathsf{T} D^{-2}\mathbf{y}} \ge \frac{\lambda_1(B)}{\lambda_p(D^{-2})}.$$

Hence

$$\lambda_1\big(n^{-1}\boldsymbol{\Sigma}\big) \ge \min_j \beta_j \left(1 - \sum_{i=1}^p \beta_i\right) = b_1^{-1},$$

and

$$\big\|\boldsymbol{\Sigma}^{-1/2}\big\| \le \lambda_p\big(\boldsymbol{\Sigma}^{-1/2}\big) = \frac{n^{-1/2}}{\lambda_1((n^{-1}\boldsymbol{\Sigma})^{1/2})} \le n^{-1/2} b_1^{1/2}. \tag{12.16}$$

To apply Theorem 12.1, for all $i \in \{1, \ldots, p\}$ we need to couple $\mathbf{Y}$ to a vector $\mathbf{Y}^i$ having the size bias distribution of $\mathbf{Y}$ in direction $i$. Let $\mathcal{A} = \{vi, v \in \mathcal{V}, i = 1, \ldots, p\}$ so that $\mathbf{X} = \{X_{vi}, v \in \mathcal{V}, i = 1, \ldots, p\} = \{X_\alpha, \alpha \in \mathcal{A}\}$. We will apply Proposition 2.2 to yield $\mathbf{Y}^i$ from $\mathbf{X}^\alpha$ for $\alpha \in \mathcal{A}$.

To achieve $\mathbf{X}^\alpha$ for $\alpha \in \mathcal{A}$, we follow the outline given after Proposition 2.2. First we generate $X_\alpha^\alpha$ from the $X_\alpha$-size bias distribution. Since $X_\alpha$ is a nontrivial Bernoulli variable, we have $X_\alpha^\alpha = 1$. Then we must generate the remaining variables with distribution $\mathcal{L}(X_\beta^\alpha | X_\alpha^\alpha = 1)$. That is, for $\alpha = vi$, say, we need to have $D(v) = d_i$, the degree of $v$ equal to $d_i$, and the remaining variables so conditioned. We can achieve such variables as follows. If $D(v) > d_i$ let $K^{vi}$ be the graph obtained by removing $D(v) - d_i$ edges from $K$, selected uniformly from the $D(v)$ edges of $v$. If $D(v) < d_i$ let $K^{vi}$ be the graph obtained by adding $d_i - D(v)$ edges of the form $\{v, u\}$ to $K$, where the vertices $u$ are selected uniformly from the $n - 1 - D(v)$ vertices not connected to $v$. If $D(v) = d_i$ let $K^{vi} = K$. Using exchangeability, it is easy to see that the distribution of the graph $K^{vi}$ is the conditional distribution of $K$ given that the degree of $v$ is $d_i$.

Now, for $j = 1, \ldots, p$ letting

$$B_j = \{vj \colon v \in \mathcal{V}\} \quad \text{we may write} \quad Y_j = \sum_{\alpha\in B_j} X_\alpha.$$

By Proposition 2.2, to construct $Y_j^i$, we first choose a summand of $Y_j$ according to the distribution given in (2.71), that is, proportional to its expectation. As $EX_{vj}$ is constant and $|B_j| = n$, we set $P(V = v) = 1/n$, so that $V$ uniform over $\mathcal{V}$, with $V$ independent of $K$. Then letting $X_{vj}^{Vi}$ be the indicator that vertex $v$ has degree $d_j$ in $K^{Vi}$, Proposition 2.2 yields that the vector $\mathbf{Y}^i$ with components

$$Y_j^i = \sum_{v=1}^n X_{vj}^{Vi}, \quad j = 1, \ldots, p$$

has the $\mathbf{Y}$-size biased distribution in direction $i$. In other words, for the given $i$, one vertex of $K$ is chosen uniformly to have edges added or removed as necessary in order for it to have degree $d_i$, and then $Y_j^i$ counts the number of vertices of degree $d_j$ in the graph that results.

We now proceed to obtain a bound for the last term in (12.4) of Theorem 12.1. Note that since exactly $|D(V) - d_i|$ edges are either added or removed from $K$ to form $K^{Vi}$, and that the vertex degrees can only change on vertices incident to these edges and on vertex $V$ itself, we have

$$|Y_j^i - Y_j| \le |D(V) - d_i| + 1.$$

This upper bound is achieved, for example, when $i \ne j, d_i < d_j$ and the degree of $V$ and the degrees of all the $D(V)$ vertices connected to $V$ have degree $d_j$. Hence, as $D(V) \sim \mathrm{Bin}(n-1, \varrho)$, and $\varrho = c/(n-1)$,

$$\begin{aligned} E\left|(Y_j^i - Y_j)(Y_k^i - Y_k)\right| &\le E\left(|D(V) - d_i| + 1\right)^2 \\ &\le 2E\left(D^2(V) + (d_i + 1)^2\right) \\ &\le 2\left((n-1)\varrho + (n-1)^2\varrho^2 + (d_i + 1)^2\right) \\ &= 2\left(c + c^2 + (d_i + 1)^2\right). \end{aligned}$$

Now, considering the last term in (12.4), since the bound above depends only on $i$, applying (12.16) and that $\mu_i = n\beta_i$ from (12.12), we obtain

$$\begin{aligned} &\frac{1}{2}\frac{p^3}{3}\left\|\Sigma^{-1/2}\right\|^3 \sum_{i=1}^p \sum_{j=1}^p \sum_{k=1}^p \mu_i E\left|(Y_j^i - Y_j)(Y_k^i - Y_k)\right| \\ &\quad \le \frac{p^5}{3} n^{-1/2} b^{3/2} \sum_{i=1}^p \beta_i\left(c + c^2 + (d_i + 1)^2\right), \end{aligned}$$

yielding the term $r_2$ in the bound (12.11).

Since $\mathbf{Y}$ is measurable with respect to $K$, following (4.143) we obtain the upper bound

$$\operatorname{Var} E\left[Y_j^i - Y_j|\mathbf{Y}\right] \le \operatorname{Var} E\left[Y_j^i - Y_j|K\right],$$

and will demonstrate

$$\begin{aligned} &\operatorname{Var} E\left[Y_j^i - Y_j|K\right] \\ &\quad = n^{-1}\Omega\left(24c + 48c^2 + 144c^3 + 48d_i^2 + 144cd_i^2 + 12\right) \end{aligned} \tag{12.17}$$

Then, for the first term in (12.4), again applying (12.12) to give $\mu_i = n\beta_i$, and (12.16), we obtain

$$\frac{p^2}{2}\|\Sigma^{-1/2}\|^2 \sum_{i=1}^{p}\sum_{j=1}^{p}\mu_i\sqrt{\operatorname{Var} E[Y_j^i - Y_j \mid \mathbf{Y}]}$$
$$\le n^{-1/2}\frac{p^3 b}{2}\sum_{i=1}^{n}\beta_i \Omega\left(\sqrt{24c + 48c^2 + 144c^3 + 48d_i^2 + 144cd_i^2 + 12}\right),$$

yielding $r_1$.

To obtain (12.17) we first condition on $V = v$. Recalling $V$ is uniform and letting $|\cdot|$ denote cardinality, in this way we obtain

$$\begin{aligned} &E[Y_j^i - Y_j | K] \\ &\quad = \frac{1}{n}\sum_{\substack{u:\,\{u,v\}\in\mathcal{E}\\ v:\,D(v)>d_i}}\left[\left|\{D(u) = d_j + 1\}\right| - \left|\{D(u) = d_j\}\right|\right]\frac{D(v) - d_i}{D(v)} \\ &\qquad + \frac{1}{n}\sum_{u\ne v,\,\{u,v\}\notin\mathcal{E},\,D(v)<d_i}\left[\left|\{D(u) = d_j - 1\}\right| - \left|\{D(u) = d_j\}\right|\right] \\ &\qquad \times \frac{d_i - D(v)}{n - 1 - D(v)} \\ &\qquad + \frac{1}{n}\left|\{v\colon D(v) \ne d_i\}\right|\delta_{i,j} - \frac{1}{n}\left|\{v\colon D(v) = d_j\}\right|(1 - \delta_{i,j}). \end{aligned} \tag{12.18}$$

To understand the first term, for example, note that if $V = v$ and $D(v) > d_i$, then $X_{uj}^i - X_{uj} = 1$ if $\{u, v\} \in \mathcal{E}$, $D(u) = d_j + 1$, and $\{u, v\}$ is one of the $d_i - D(v)$ edges removed at $v$ at random, chosen with probability $(D(v) - d_i)/D(v)$.

Note that the factor $1/n$ multiplies all terms in (12.18), which provides a factor of $1/n^2$ in the variance. Breaking the two sums into two separate sums, so that six terms result, we will bound the variance of each term separately and then apply the bound

$$\operatorname{Var}\left(\sum_{j=1}^{k} U_j\right) \le k\sum_{j=1}^{k}\operatorname{Var}(U_j) \tag{12.19}$$

for $k = 6$.

The first term of (12.18) yields two sums, both of the form

$$\sum_{u,v}\mathbf{1}_A(u, v)\frac{D(v) - d_i}{D(v)}$$
$$\text{for } A = \left\{(u, v)\colon \{u, v\} \in \mathcal{E}\colon D(v) > d_i, D(u) = d_j + a\right\}, \tag{12.20}$$

the first with $a = 1$, and the second with $a = 0$. We show

$$\operatorname{Var}\left(\sum_{u,v}\mathbf{1}_A(u, v)\frac{D(v) - d_i}{D(v)}\right) = \Omega\left(2cn + 4c^2 n + 12c^3 n\right). \tag{12.21}$$

To calculate this variance requires the consideration of terms all of the form

$$\mathrm{Cov}\left(\mathbf{1}_A(u,v)\frac{D(v)-d_i}{D(v)}, \mathbf{1}_A(u',v')\frac{D(v')-d_i}{D(v')}\right). \tag{12.22}$$

Let $N$ be the number of distinct vertices among $u, u', v, v'$. From the definition of $A$ in (12.20), and that no edge connects a vertex to itself, we see that we need only consider cases where $u \neq v$ and $u' \neq v'$. Hence $N$ may only take on the values 2, 3 and 4, leading to the three terms in (12.21).

There are two cases for $N = 2$. The $n(n-1)$ diagonal variance terms with $(u, v) = (u', v')$ can be bounded by their second moments as

$$\begin{aligned}
\mathrm{Var}\left(\mathbf{1}_A(u,v)\frac{D(v)-d_i}{D(v)}\right) &\le E\left(\mathbf{1}_A(u,v)\frac{D(v)-d_i}{D(v)}\right)^2 \\
&\le P(\{u,v\} \in \mathcal{E}) \\
&= \frac{c}{n-1},
\end{aligned}$$

leading to a factor of $n\Omega(c)$. Handling the case $(u, v) = (v', u')$ in the same manner gives an overall contribution of $2n\Omega(c)$ for the case $N = 2$, and the first term in (12.21).

For $N = 3$ there are four subcases, all of which may be handled in a similar way. Consider, for example, the case $u = u'$, $v \neq v'$. Using the inequality $\mathrm{Cov}(X, Y) \le EXY$, valid for nonnegative $X$ and $Y$, we obtain

$$\begin{aligned}
&\mathrm{Cov}\left(\mathbf{1}_A(u,v)\frac{D(v)-d_i}{D(v)}, \mathbf{1}_A(u,v')\frac{D(v')-d_i}{D(v')}\right) \\
&\le P(\{u,v\} \in \mathcal{E}, \{u,v'\} \in \mathcal{E}) \\
&= c^2/(n-1)^2.
\end{aligned}$$

Handling the three other cases similarly and noting that the total number of $N = 3$ terms is no more than $4n^3$ leads to a contribution of $4n\Omega(c^2)$ from the case $N = 3$ and the second term in (12.21).

In the case $N = 4$ the vertices $u, u', v, v'$ are distinct, and we have

$$\begin{aligned}
&\mathrm{Cov}\left(\mathbf{1}_A(u,v)\frac{D(v)-d_i}{D(v)}, \mathbf{1}_A(u',v')\frac{D(v')-d_i}{D(v')}\right) \\
&= E\left(\mathbf{1}_A(u,v)\frac{D(v)-d_i}{D(v)}\mathbf{1}_A(u',v')\frac{D(v')-d_i}{D(v')}\right) - \beta^2,
\end{aligned} \tag{12.23}$$

where

$$\begin{aligned}
\beta &= E\left(\mathbf{1}_A(u,v)\frac{D(v)-d_i}{D(v)}\right) \\
&= E\left(\mathbf{1}_{\{\{u,v\}\in\mathcal{E}\}}\mathbf{1}_{\{D(u)=d_j+a\}}\frac{(D(v)-d_i)_+}{D(v)}\right).
\end{aligned}$$

With $C$ the event that

$$\big\{\{u,u'\},\{v,v'\},\{u,v'\},\{u',v\}\big\}\cap\mathcal{E}=\emptyset \tag{12.24}$$

we have $P(C)=(1-c/(n-1))^4=1-4\Omega(c/n)$. This estimate implies, noting that the events $\{u,v\},\{u',v'\}\in\mathcal{E}$ each have probability $c/(n-1)$ and are independent of $C$, that

$$\begin{aligned}
&E\left(\mathbf{1}_A(u,v)\frac{D(v)-d_i}{D(v)}\mathbf{1}_A(u',v')\frac{D(v')-d_i}{D(v')}\right)\\
&\quad=E\left(\mathbf{1}_A(u,v)\frac{D(v)-d_i}{D(v)}\mathbf{1}_A(u',v')\frac{D(v')-d_i}{D(v')}\bigg|C\right)P(C)+4\Omega\left(\frac{c^3}{n^3}\right)\\
&\quad\le\alpha^2+4\Omega\left(\frac{c^3}{n^3}\right),
\end{aligned} \tag{12.25}$$

with

$$\alpha=E\left(\mathbf{1}_{\{\{u,v\}\in\mathcal{E}\}}\mathbf{1}_{\{D(u)=d_j+a\}}\frac{(D(v)-d_i)_+}{D(v)}\bigg|C\right),$$

where in the last inequality we used the conditional independence given $C$ of the events indicated, for $(u,v)$ and $(u',v')$.

Bounding both $\alpha$ and $\beta$ by the probability that $\{u,v\}\in\mathcal{E}$, an event independent of $C$, we bound the covariance term (12.23) as

$$\begin{aligned}
\left|\alpha^2+4\Omega\left(\frac{c^3}{n^3}\right)-\beta^2\right|&=\left|(\alpha+\beta)(\alpha-\beta)+4\Omega\left(\frac{c^3}{n^3}\right)\right|\\
&\le2|\alpha-\beta|\Omega\left(\frac{c}{n}\right)+4\Omega\left(\frac{c^3}{n^3}\right).
\end{aligned} \tag{12.26}$$

To handle $\alpha-\beta$, letting

$$R=\big\{\{u,v\}\in\mathcal{E}\big\},\quad S=\mathbf{1}_{\{D(u)=d_j+a\}}\quad\text{and}\quad T=\frac{(D(v)-d_i)_+}{D(v)},$$

we have

$$\alpha-\beta=E[\mathbf{1}_RST|C]-E[\mathbf{1}_RST]=E[ST|CR]\frac{P(RC)}{P(C)}-E[ST|R]P(R).$$

As $R$ and $C$ are independent and $P(R)=c/(n-1)$,

$$|\alpha-\beta|=\big|E[ST|CR]-E[ST|R]\big|\Omega(c/n).$$

Since $S$ and $T$ are conditionally independent given $R$ or given $CR$, we have

$$|\alpha-\beta|=\big|E[S|CR]E[T|CR]-E[S|R]E[T|R]\big|\Omega(c/n).$$

Let $X,Y\sim\text{Binomial}(n-4,\varrho)$ and $X',Y'\sim\text{Binomial}(n-2,\varrho)$, all independent. In $\alpha$, conditioning on $CR$, $D(u)-1$ and $D(v)-1$ are equal in distribution to $X$ and $Y$ respectively; in $\beta$, conditioning on $R$, the same variables are distributed as $X',Y'$. Hence,

$$|\alpha - \beta| = \left| E\mathbf{1}_{\{X=d_j+a-1\}} E\left(\frac{(Y+1-d_i)_+}{Y+1}\right) - E\mathbf{1}_{\{X'=d_j+a-1\}} E\left(\frac{(Y'+1-d_i)_+}{Y'+1}\right)\right| \Omega\left(\frac{c}{n}\right). \qquad (12.27)$$

Next, note

$$|E\mathbf{1}_{\{X=d_j+a-1\}} - E\mathbf{1}_{\{X'=d_j+a-1\}}| = 2\Omega(c/n) \quad \text{and}$$
$$\left|E\left(\frac{(Y+1-d_i)_+}{Y+1}\right) - E\left(\frac{(Y'+1-d_i)_+}{Y'+1}\right)\right| = 2\Omega(c/n),$$

which can be easily understood by defining $X$ and $X'$ jointly, with $X' = X + \xi$, with $\xi \sim \text{Binomial}(2, \varrho)$, independently of $X$, so that $P(X \neq X') = 2\Omega(c/n)$, and constructing $Y$ and $Y'$ similarly. Hence, by (12.27),

$$|\alpha - \beta| = 4\Omega\left(\frac{c^2}{n^2}\right),$$

and the $N = 4$ covariance term (12.26) is $12\Omega(\frac{c^3}{n^3})$. As there are no more than $n^4$ where $u$, $u'$, $v$, $v'$ are all distinct, their total contribution is $\Omega(12c^3 n)$, yielding the final term in (12.21).

We apply a similar argument to the third and fourth terms arising from (12.18), both of the form

$$\sum_{u\neq v} \mathbf{1}_D(u, v)\frac{d_i - D(v)}{n-1-D(v)}$$
$$\text{for } D = \{(u, v)\colon \{u, v\} \notin \mathcal{E}\colon D(v) < d_i, D(u) = d_j - a\}, \qquad (12.28)$$

the first with $a = 1$, the second with $a = 0$, and show

$$\operatorname{Var}\left(\sum_{u\neq v} \mathbf{1}_D(u, v)\frac{d_i - D(v)}{n-1-D(v)}\right) = \Omega\left(2d_i^2 + 4d_i^2 n + 12cd_i^2 n\right). \qquad (12.29)$$

With $N$ again counting the number of distinct indices among $u$, $u'$, $v$, $v'$, for the cases $N = 2$ and $N = 3$ it will suffice to apply the inequality

$$\mathbf{1}_{\{D(v)<d_i\}}\left(\frac{d_i - D(v)}{n-1-D(v)}\right) \leq \frac{d_i}{n-2-d_i}.$$

In particular, for the $n(n-1)$ variance terms where $(u, v) = (u', v')$ we obtain

$$\operatorname{Var}\left(\mathbf{1}_D(u, v)\left(\frac{d_i - D(v)}{n-1-D(v)}\right)\right) \leq \Omega\left(d_i^2/n^2\right),$$

and handling the case $(u, v) = (v', u')$ in the same manner gives an overall contribution of $2\Omega(d_i^2)$ for the case $N = 2$.

For $N = 3$, considering again the instance $u = u', v \neq v'$,

$$\operatorname{Cov}\left(\mathbf{1}_D(u, v)\frac{d_i - D(v)}{n-1-D(v)}, \mathbf{1}_D(u, v')\frac{d_i - D(v')}{n-1-D(v')}\right) \leq \Omega\left(d_i^2/n^2\right);$$

handling the other cases similarly and, again, noting that the total number of $N=3$ terms is no more than $4n^3$ leads to a contribution of $4\Omega(d_i^2 n)$.

For $N=4$, write

$$\begin{aligned}
&\mathrm{Cov}\left(\mathbf{1}_D(u,v)\frac{d_i-D(v)}{n-1-D(v)}, \mathbf{1}_D(u',v')\frac{d_i-D(v')}{n-1-D(v')}\right)\\
&= E\left(\mathbf{1}_D(u,v)\frac{d_i-D(v)}{n-1-D(v)}\mathbf{1}_D(u',v')\frac{d_i-D(v')}{n-1-D(v')}\right)-\delta^2, \quad (12.30)
\end{aligned}$$

where

$$\delta = E\left(\mathbf{1}_{\{\{u,v\}\notin\mathcal{E}, D(u)=d_j-a\}}\frac{(d_i-D(v))_+}{n-1-D(v)}\right).$$

With $C$ as in (12.24), for the first term in (12.30), as for (12.25), we have

$$\begin{aligned}
&E\left(\mathbf{1}_D(u,v)\frac{d_i-D(v)}{n-1-D(v)}\mathbf{1}_D(u',v')\frac{d_i-D(v')}{n-1-D(v')}\middle| C\right)P(C)+4\Omega\left(\frac{cd_i^2}{n^3}\right)\\
&\le \gamma^2+4\Omega\left(\frac{cd_i^2}{n^3}\right)
\end{aligned}$$

where

$$\gamma = E\left(\mathbf{1}_{\{\{u,v\}\notin\mathcal{E},\, D(u)=d_j-a\}}\frac{(d_i-D(v))_+}{n-1-D(v)}\middle| C\right).$$

Now we may bound the covariance (12.30) as

$$\begin{aligned}
\left|\gamma^2+4\Omega\left(\frac{cd_i^2}{n^3}\right)-\delta^2\right| &\le |\gamma-\delta||\gamma+\delta|+4\Omega\left(\frac{cd_i^2}{n^3}\right)\\
&= 2|\gamma-\delta|\Omega\left(\frac{d_i}{n}\right)+4\Omega\left(\frac{cd_i^2}{n^3}\right). \quad (12.31)
\end{aligned}$$

Let $X, Y \sim \text{Binomial}(n-4, \varrho)$ and $X', Y' \sim \text{Binomial}(n-2, \varrho)$, all independent. By arguments similar to those which yield (12.27), with $\overline{R}$ denoting $R$ complement, noting that in $\gamma$, conditioning on $C\overline{R}$, $D(u)$, $D(v)$ are equal in distribution to $X$, $Y$, and in $\delta$, conditioning on $\overline{R}$, these same variables are distributed as $X'$, $Y'$, we have

$$|\gamma-\delta| \le \left|E\mathbf{1}_{\{X=d_j-a\}}E\left(\frac{(d_i-Y)_+}{n-1-Y}\right)-E\mathbf{1}_{\{X'=d_j-a\}}E\left(\frac{(d_i-Y')_+}{n-1-Y'}\right)\right|, \quad (12.32)$$

with

$$\begin{aligned}
&|E\mathbf{1}_{\{X=d_j-a\}}-E\mathbf{1}_{\{X'=d_j-a\}}| = 2\Omega(c/n) \quad \text{and}\\
&\left|E\left(\frac{(d_i-Y)_+}{n-1-Y}\right)-E\left(\frac{(d_i-Y')_+}{n-1-Y'}\right)\right| = 2\Omega\left(\frac{cd_i}{n^2}\right),
\end{aligned}$$

yielding

$$|\gamma-\delta| = 4\Omega\left(\frac{cd_i}{n^2}\right).$$

Hence the $N=4$ covariance term (12.30), by (12.31), is $12\Omega(cd_i^2/n^3)$. As there are at most $n^4$ such terms the $N=4$ contribution is $\Omega(12cd_i^2 n)$. Summing this amount to the results of the $N=2$ and $N=3$ computations yields (12.29).

For the last two terms in (12.18), note that

$$\left|v\colon D(v)\neq d_i\right| = n - Y_i \quad \text{and} \quad \left|v\colon D(v)=d_j\right| = Y_j,$$

which respectively have variances $\sigma_{ii}$ and $\sigma_{jj}$, given in (12.12). Now, for instance,

$$\begin{aligned}\sigma_{ii} &\le \frac{n}{c(1-c/(n-1))}\sum_{k=1}^{p}\beta_k^2(d_k-c)^2 \\ &\le n\Omega(1/c)\sum_{k=1}^{p}\beta_k(d_k-c)^2 \\ &\le n\Omega(1/c)(n-1)\varrho(1-\varrho)=\Omega(n),\end{aligned} \tag{12.33}$$

with the same result holding, clearly, for $\sigma_{jj}$. Adding together (12.21), (12.29) and (12.33), and then multiplying by 2 to match each term of the first given type to the following one of the same type yields the sum of the six variance terms. Applying (12.19), which mandates an additional factor of 6, now yields (12.17). □

## 12.3 Multivariate Exchangeable Pairs

To construct a Stein pair for a given application in order to apply, say, Theorem 5.4, one must create mean zero, variance 1, exchangeable variables $W$, $W'$, which satisfy the linearity condition

$$E(W'|W)=(1-\lambda)W \quad \text{for some } \lambda\in(0,1). \tag{12.34}$$

Typically it is easy to construct an exchangeable pair, by, say, sampling some variables from their conditional distribution given the others, but the linearity condition (12.34) is never guaranteed to hold. However, even when (12.34) fails to hold there are at least two remedies available, each of which may allow a bound to the normal to be computed. One remedy, already mentioned, is to work with an approximate version of (12.34), such as (2.41), which has a remainder term, and bounds may then be computed using Theorem 3.5; see also the work of Rinott and Rotar (1997). In this section, we introduce a second technique, due to Reinert and Röllin (2009), which may be used when the linearity condition fails. The success of this method depends on being able to 'embed' the given collection of random variables in a larger one so that linearity holds in some multivariate sense. The price to pay is the extra complication of the higher dimensional setting, and often an accompanying loss in rates.

The one dimensional linearity condition (12.34) may be rephrased as saying that the conditional expectation agrees with the linear regression of $W'$ on $W$. Extending to the multivariate case, we recall that when $\mathbf{Y}'$, $\mathbf{Y}$ are random vectors with covariance matrix

$$\Sigma = \begin{bmatrix} \Sigma_{11} & \Sigma_{12} \\ \Sigma_{21} & \Sigma_{22} \end{bmatrix}$$

with $\Sigma_{22}$ invertible, centralizing by letting

$$\mathbf{W} = \mathbf{Y} - E\mathbf{Y} \quad \text{and} \quad \mathbf{W}' = \mathbf{Y}' - E\mathbf{W}', \tag{12.35}$$

the linear regression $\mathbf{L}$ of $\mathbf{W}'$ on $\mathbf{W}$ is given by

$$\mathbf{L}(\mathbf{W}'|\mathbf{W}) = \Sigma_{12}\Sigma_{22}^{-1}\mathbf{W}. \tag{12.36}$$

Driven by (12.34) and these considerations, we are led to search for an exchangeable pair of vectors $\mathbf{W}, \mathbf{W}'$ such that $E(\mathbf{W}'|\mathbf{W}) = (I - \Lambda)\mathbf{W}$, or

$$E(\mathbf{W}' - \mathbf{W}|\mathbf{W}) = -\Lambda\mathbf{W} \tag{12.37}$$

for some matrix $\Lambda$. Multivariate exchangeable pairs were first studied by Chatterjee and Meckes (2008), under a condition somewhat more restrictive than (12.37).

Following Reinert and Röllin (2010), we show how in some cases a random quantity may be embedded in a higher dimensional vector so that (12.37) is satisfied, even if linearity does not hold for the originally given problem. We take again as our example some characteristic of $K = K_{n,\varrho}$, the random graph considered in Sect. 12.2. In this case, say the quantity of interest is $T$, the total number of triangles. Let the vertices of $K$ be labeled by $\{1, 2, \ldots, n\}$, and for distinct $i, j \in \{1, \ldots, n\}$ let $\mathbf{1}_{i,j}$ be the indicator of the event that $\{i, j\}$ is in the edge set of $K$. We assume $n \geq 4$. With these conventions, we may write the number of triangles $T$ as

$$T = \sum_{i<j<k} \mathbf{1}_{i,j}\mathbf{1}_{j,k}\mathbf{1}_{k,i}.$$

To construct an exchangeable pair select vertices $1 \leq I < J \leq n$ uniformly from all such pairs, that is, with probability

$$P(I = i, J = j) = \binom{n}{2}^{-1} \quad \text{for all } 1 \leq i < j \leq n,$$

independently of $K$. Form a new graph $K'$ with edge indicators $\mathbf{1}'_{i,j} = \mathbf{1}_{i,j}$ for $\{i, j\} \neq \{I, J\}$, and $\mathbf{1}_{I,J}$ replaced by an independent copy $\mathbf{1}'_{I,J}$.

The change in the number of triangles is given by

$$T' - T = \sum_{k:\, k \notin \{I,J\}} \left(\mathbf{1}'_{I,J}\mathbf{1}_{J,k}\mathbf{1}_{k,I} - \mathbf{1}_{I,J}\mathbf{1}_{J,k}\mathbf{1}_{k,I}\right).$$

Calculating the expectation of the change $T' - T$, conditional on $T$, by averaging over the $\binom{n}{2}$ choices of $i$ and $j$ we have

$$\begin{aligned} &E(T' - T|T) \\ &\quad = \binom{n}{2}^{-1} \sum_{i<j} E\left(\sum_{k:\, k \notin \{i,j\}} (\mathbf{1}'_{i,j}\mathbf{1}_{j,k}\mathbf{1}_{i,k} - \mathbf{1}_{i,j}\mathbf{1}_{j,k}\mathbf{1}_{i,k}) \Big| T\right) \\ &\quad = \binom{n}{2}^{-1} \left\{\varrho E\left(\sum_{i<j,\, k \notin \{i,j\}} \mathbf{1}_{j,k}\mathbf{1}_{i,k} \Big| T\right) - 3T\right\}, \end{aligned}$$

where for the first term we have used that $\mathbf{1}'_{i,j}$ is an indicator independent of $K$ with success probability $\varrho$, while for the second term the factor of three accounts for the number of ways $k$ can appear relative to $i < j$. Hence we see that the conditional expectation depends on the number of 2-stars $S$, given by

$$S = \sum_{i<j:\, k \notin \{i,j\}} \mathbf{1}_{i,k}\mathbf{1}_{j,k},$$

and hence by conditioning in addition on $S$ we obtain

$$E(T' - T|S, T) = \binom{n}{2}^{-1} (\varrho S - 3T). \tag{12.38}$$

However, if we are to form an enlarged vector by appending $S$ to $T$ we must now also verify the linearity condition for $S$ as well. For the difference in the number of two stars we have

$$S' - S = \sum_{k:\, k \notin \{I,J\}} (\mathbf{1}'_{I,J} - \mathbf{1}_{I,J})(\mathbf{1}_{J,k} + \mathbf{1}_{I,k}),$$

and hence

$$\begin{aligned}
&E(S' - S|S, T) \\
&\quad = \binom{n}{2}^{-1} \sum_{i<j} E\left( \sum_{k:\, k \notin \{i,j\}} (\mathbf{1}'_{i,j} - \mathbf{1}_{i,j})(\mathbf{1}_{j,k} + \mathbf{1}_{i,k}) \middle| S, T \right) \\
&\quad = \frac{\varrho}{\binom{n}{2}} E\left( \sum_{i<j,\, k \notin \{i,j\}} (\mathbf{1}_{j,k} + \mathbf{1}_{i,k}) \middle| S, T \right) \\
&\qquad - \frac{1}{\binom{n}{2}} E\left( \sum_{i<j,\, k \notin \{i,j\}} \mathbf{1}_{i,j}\mathbf{1}_{j,k} + \mathbf{1}_{i,j}\mathbf{1}_{i,k} \middle| S, T \right) \\
&\quad = 2\binom{n}{2}^{-1} \big(\varrho(n-2)E(\mathtt{E}|S, T) - S\big)
\end{aligned} \tag{12.39}$$

where $\mathtt{E}$ is the total number of edges in $K_{n,\varrho}$,

$$\mathtt{E} = \sum_{i<j} \mathbf{1}_{i,j}.$$

Continuing the process, so now appending $\mathtt{E}$ to the vector $S$, $T$, taking the difference,

$$\mathtt{E}' - \mathtt{E} = \mathbf{1}_{I,J} - \mathbf{1}'_{I,J}$$

we find

$$\begin{aligned}
&E(\mathtt{E}' - \mathtt{E}|\mathtt{E}) \\
&\quad = \binom{n}{2}^{-1} \sum_{i<j} E(\mathbf{1}'_{i,j} - \mathbf{1}_{i,j}|\mathtt{E}) = \binom{n}{2}^{-1} \left(\binom{n}{2}\varrho - \mathtt{E}\right),
\end{aligned} \tag{12.40}$$

a linear function of $\mathtt{E}$.

Hence, letting $\mathbf{Y} = (\mathrm{E}, S, T)^\mathsf{T}$, the conditional expectation $E(\mathbf{Y}'|\mathbf{Y})$ will be a linear function of $\mathbf{Y}$. In particular, centralizing by letting $\mathbf{W}$ and $\mathbf{W}'$ be given by (12.35), from (12.38), (12.39) and (12.40) with the additional conditioning on $\mathrm{E}$, we conclude that (12.37) holds with

$$E(\mathbf{W}' - \mathbf{W}|\mathbf{W}) = -\binom{n}{2}^{-1} \begin{bmatrix} 1 & 0 & 0 \\ -2(n-2)\varrho & 2 & 0 \\ 0 & -\varrho & 3 \end{bmatrix} \mathbf{W}, \tag{12.41}$$

thus achieving the multivariate version of (12.34).

In this particular example, the different powers of $n$ in the matrix in equality (12.41) indicates the presence of a scaling issue. In particular, from Reinert and Röllin (2010), and as one can easily confirm,

$$E\mathrm{E} = \binom{n}{2}\varrho, \quad ES = 3\binom{n}{3}\varrho^2 \quad \text{and} \quad ET = \binom{n}{3}\varrho^3,$$

and in addition they show

$$\mathrm{Var}(\mathrm{E}) = 3\binom{n}{3}\frac{1}{n-2}\varrho(1-\varrho),$$
$$\mathrm{Var}(S) = 3\binom{n}{3}\varrho^2(1-\varrho)\big(1-\varrho+4(n-2)\varrho\big)$$

and

$$\mathrm{Var}(T) = \binom{n}{3}\varrho^3(1-\varrho)\big((1-\varrho)^2 + 3\varrho(1-\varrho) + 3(n-2)\varrho^2\big).$$

In particular, we see that the variance of the number of edges grows at the rate $n^2$, while the variance of the number of 2-stars and triangles grow like $n^4$. Considering, then, the standardized variables

$$\mathrm{E}_1 = \frac{n-2}{n^2}\mathrm{E}, \quad S_1 = \frac{1}{n^2}S \quad \text{and} \quad T_1 = \frac{1}{n^2}T,$$

which have limiting variances, letting

$$\mathbf{W}_1 = (\mathrm{E}_1 - E\mathrm{E}_1, S_1 - ES_1, T_1 - ET_1), \tag{12.42}$$

and defining $\mathbf{W}_1'$ likewise, one obtains $E(\mathbf{W}_1'|\mathbf{W}_1) = (I - \Lambda)\mathbf{W}_1$ with

$$\Lambda = \binom{n}{2}^{-1} \begin{bmatrix} 1 & 0 & 0 \\ -2\varrho & 2 & 0 \\ 0 & -\varrho & 3 \end{bmatrix} \mathbf{W}_1.$$

Calculating the remaining covariance terms, Reinert and Röllin (2010) find that the covariance matrix $\Sigma_1$ of $\mathbf{W}_1$ is given by

$$\Sigma_1 = 3\frac{(n-2)\binom{n}{3}}{n^4}\varrho(1-\varrho) \begin{bmatrix} 1 & 2\varrho & \varrho^2 \\ 2\varrho & 4\varrho^2 + \frac{\varrho(1-\varrho)}{n-2} & 2\varrho^3 + \frac{\varrho^2(1-\varrho)}{n-2} \\ \varrho^2 & 2\varrho^3 + \frac{\varrho^2(1-\varrho)}{n-2} & \varrho^4 + \frac{\varrho^2(1+\varrho-2\varrho^2)}{3(n-2)} \end{bmatrix}. \tag{12.43}$$

As the variables have been scaled so that their covariance matrix $\Sigma_1$ has a non-trivial limit, convergence in distribution of $\mathbf{W}_1$ to $\Sigma_1^{1/2}\mathbf{Z}$ is a consequence of the following result.

**Theorem 12.3** *Let* $\mathbf{W}_1$ *and* $\Sigma_1$ *be given by* (12.42) *and* (12.43) *respectively. Then with* $\mathbf{Z}$ *a standard normal vector in* $\mathbb{R}^3$,

$$\left\|\mathcal{L}(\mathbf{W}_1)-\mathcal{L}\left(\Sigma_1^{1/2}\mathbf{Z}\right)\right\|_{\mathcal{H}_{3,\infty,3}} \le \frac{1}{n}\left(\frac{35}{4}+9n^{-1}\right)+\frac{8}{3n}\left(1+n^{-1}+n^{-2}\right).$$

Reinert and Röllin (2010) prove Theorem 12.3 by applying Theorem 12.4 for multivariate exchangeable pairs, from Reinert and Röllin (2009), and calculating the quantities that appear in the bounds below; we refer the reader there for these latter details. In these two works they also consider identity (12.37) generalized to have remainder,

$$E(\mathbf{W}'-\mathbf{W}|\mathbf{W})=-\Lambda\mathbf{W}+\mathbf{R}, \tag{12.44}$$

and supply further applications of their embedding technique to runs on the line, complete $U$-statistics, and doubly indexed permutation statistics.

In Theorem 12.4, rather than compare the distribution of a random vector with covariance matrix $I$ to that of a standard normal $\mathbf{Z}$, as in Theorem 12.2, one compares the distribution of a vector with covariance $\Sigma$ to that of $\Sigma^{1/2}\mathbf{Z}$. In this case, one considers the variation on the multivariate Stein equation (2.22) given by

$$\operatorname{Tr}\Sigma D^2 f(\mathbf{w})-\mathbf{w}^{\mathsf{T}}\nabla f(\mathbf{w})=h(\mathbf{w})-Eh\left(\Sigma^{1/2}\mathbf{Z}\right), \tag{12.45}$$

where $\mathbf{Z}$ is a standard normal vector in $\mathbb{R}^p$. Equation (12.45) is (2.22) for $\boldsymbol{\mu}=0$, with the function $h(\mathbf{w})$ evaluated at $\Sigma^{1/2}\mathbf{w}$. Using the change of variable arguments in Lemma 2.6 it is straightforward to construct a solution $f$ to (12.45) and show that it satisfies

$$\left|f_{i_1\cdots i_k}(\mathbf{w})\right| \le \frac{1}{k}\left|h_{i_1\cdots i_k}(\mathbf{w})\right|, \tag{12.46}$$

where $f_{i_1\cdots i_k}(\mathbf{w})$ denotes the partial derivative of $f$ with respect to $w_{i_1},\ldots,w_{i_k}$, and likewise for $h$, whenever the partial of $h$ on the right hand side exists. As in Sect. 12.2, we adopt the supremum norm (12.3) for matrices.

**Theorem 12.4** *Let* $(\mathbf{W},\mathbf{W}')$ *be an exchangeable pair of* $\mathbb{R}^d$ *valued random vectors satisfying*

$$E\mathbf{W}=0,\quad E\mathbf{W}\mathbf{W}^{\mathsf{T}}=\Sigma \tag{12.47}$$

*with* $\Sigma$ *positive definite. Suppose further that* (12.44) *holds with* $\Lambda$ *an invertible matrix and* $\mathbf{R}$ *some* $\mathbb{R}^p$ *valued random vector. Then, if* $\mathbf{Z}$ *has the standard* $p$*-dimensional normal distribution,*

$$\left\|\mathcal{L}(\mathbf{W})-\mathcal{L}\left(\Sigma^{1/2}\mathbf{Z}\right)\right\|_{\mathcal{H}_{3,\infty,p}} \le \frac{A}{4}+\frac{B}{12}+\left(1+\frac{p}{2}\left\|\Sigma^{1/2}\right\|\right)C,$$

*where, with* $\gamma_i = \sum_{m=1}^p |(\Lambda^{-1})_{m,i}|$,

$$A = \sum_{i,j=1}^p \gamma_i \sqrt{\mathrm{Var}(E((W_i' - W_i)(W_j' - W_j)|\mathbf{W}))},$$

$$B = \sum_{i,j,k=1}^p \gamma_i E|(W_i' - W_i)(W_j' - W_j)(W_k' - W_k)| \quad \textit{and}$$

$$C = \sum_{i=1}^p \gamma_i \sqrt{\mathrm{Var}(R_i)}.$$

*Proof* Recalling (12.2), let $f$ be the solution to the multivariate Stein equation (12.45) for a given $h \in \mathcal{H}_{3,\infty,p}$. Note that the function $F : \mathbb{R}^p \times \mathbb{R}^p \to \mathbb{R}$ given by

$$F(\mathbf{w}', \mathbf{w}) = \frac{1}{2}(\mathbf{w}' - \mathbf{w})^\mathsf{T} \Lambda^{-\mathsf{T}} \big(\nabla f(\mathbf{w}') + \nabla f(\mathbf{w})\big)$$

is anti-symmetric, and therefore, by exchangeability, $EF(\mathbf{W}', \mathbf{W}) = 0$. Thus

$$\begin{aligned} 0 &= \frac{1}{2} E(\mathbf{W}' - \mathbf{W})^\mathsf{T} \Lambda^{-\mathsf{T}} \big(\nabla f(\mathbf{W}') + \nabla f(\mathbf{W})\big) \\ &= E\big\{(\mathbf{W}' - \mathbf{W})^\mathsf{T} \Lambda^{-\mathsf{T}} \nabla f(\mathbf{W})\big\} \\ &\quad + \frac{1}{2} E\big\{(\mathbf{W}' - \mathbf{W})^\mathsf{T} \Lambda^{-\mathsf{T}} \big(\nabla f(\mathbf{W}') - \nabla f(\mathbf{W})\big)\big\} \\ &= E\big\{\mathbf{R}^\mathsf{T} \Lambda^{-\mathsf{T}} \nabla f(\mathbf{W})\big\} - E\big\{\mathbf{W}^\mathsf{T} \nabla f(\mathbf{W})\big\} \\ &\quad + \frac{1}{2} E\big\{(\mathbf{W}' - \mathbf{W})^\mathsf{T} \Lambda^{-\mathsf{T}} \big(\nabla f(\mathbf{W}') - \nabla f(\mathbf{W})\big)\big\}, \end{aligned} \tag{12.48}$$

where we have applied (12.44).

Focusing on the final expression in (12.48), Taylor expansion yields

$$\begin{aligned} &E\big\{(\mathbf{W}' - \mathbf{W})^\mathsf{T} \Lambda^{-\mathsf{T}} \big(\nabla f(\mathbf{W}') - \nabla f(\mathbf{W})\big)\big\} \\ &\quad = \sum_{m,i,j} (\Lambda^{-1})_{m,i} E\big\{(W_i' - W_i)(W_j' - W_j) f_{m,j}(\mathbf{W})\big\} \\ &\qquad + \sum_{m,i,j,k} (\Lambda^{-1})_{m,i} E\big\{(W_i' - W_i)(W_j' - W_j)(W_k' - W_k) \tilde{R}_{mjk}\big\} \end{aligned} \tag{12.49}$$

where, by (12.46),

$$|\tilde{R}_{mjk}| \le \frac{1}{2} \sup_{\mathbf{w} \in \mathbb{R}^p} \big|f_{m,j,k}(\mathbf{w})\big| \le \frac{1}{6}. \tag{12.50}$$

By exchangeability, (12.44) and (12.47),

$$\begin{aligned} &E(\mathbf{W}' - \mathbf{W})(\mathbf{W}' - \mathbf{W})^\mathsf{T} \\ &\quad = E\big\{\mathbf{W}(\mathbf{W} - \mathbf{W}')^\mathsf{T}\big\} + E\big\{\mathbf{W}(\mathbf{W} - \mathbf{W}')^\mathsf{T}\big\} \\ &\quad = 2E\big\{\mathbf{W}(\Lambda \mathbf{W} - \mathbf{R})^\mathsf{T}\big\} = 2\Sigma \Lambda^\mathsf{T} - 2E\big(\mathbf{W}\mathbf{R}^\mathsf{T}\big) := Q, \quad \text{say.} \end{aligned}$$

Solving for $\Sigma$ we obtain

$$\Sigma = \frac{1}{2} Q\Lambda^{-\mathsf{T}} + E\big(\mathbf{W}\mathbf{R}^{\mathsf{T}}\big)\Lambda^{-\mathsf{T}},$$

and therefore

$$\begin{aligned}
&\operatorname{Tr}\big(\Sigma D^2 f(\mathbf{W})\big)\\
&= \frac{1}{2}\operatorname{Tr} Q\Lambda^{-\mathsf{T}} D^2 f(\mathbf{W}) + E\operatorname{Tr}\big(\mathbf{W}\mathbf{R}^{\mathsf{T}}\big)\Lambda^{-\mathsf{T}} D^2 f(\mathbf{W})\\
&= \frac{1}{2}\sum_{m,i,j}\big(\Lambda^{-1}\big)_{m,i} Q_{j,i} f_{m,j}(\mathbf{W}) + \sum_{m,i,j}\big(\Lambda^{-1}\big)_{m,i} E(W_j R_i) f_{m,j}(\mathbf{W}).
\end{aligned}$$

Combining this identity with (12.48) and (12.49), we obtain

$$\begin{aligned}
&\big|E\big\{\operatorname{Tr}\big(\Sigma D^2 f(\mathbf{W})\big) - \mathbf{W}^{\mathsf{T}}\nabla f(\mathbf{W})\big\}\big|\\
&\le \frac{1}{2}\left|\sum_{m,i,j,k} E\big(\Lambda^{-1}\big)_{m,i}\big[Q_{j,i} - E\big(\big(W_i' - W_i\big)\big(W_j' - W_j\big)|\mathbf{W}\big)\big] f_{m,j}(\mathbf{W})\right|\\
&\quad + \frac{1}{2}\left|\sum_{m,i,j,k} E\big(\Lambda^{-1}\big)_{m,i}\big[\big(W_i' - W_i\big)\big(W_j' - W_j\big)\big(W_k' - W_k\big)\big]\tilde{R}_{mjk}(\mathbf{W})\right|\\
&\quad + \left|\sum_{i,m}\big(\Lambda^{-1}\big)_{m,i} E R_i f_m(\mathbf{W})\right| + \left|\sum_{m,i,j}\big(\Lambda^{-1}\big)_{m,i} E(W_j R_i) E f_{m,j}(\mathbf{W})\right|\\
&\le \frac{1}{4}\sum_{i,j}\lambda^{(i)} E\big|Q_{j,i} - E\big(\big(W_i' - W_i\big)\big(W_j' - W_j\big)|\mathbf{W}\big)\big| + \frac{1}{12}B\\
&\quad + \sum_i \lambda^{(i)} E|R_i| + \frac{1}{2}\sum_{i,j}\lambda^{(i)} E|W_j R_i|,
\end{aligned} \tag{12.51}$$

where we have applied (12.46) and (12.50) to obtain the last inequality. The Cauchy–Schwarz inequality yields $E|R_j| \le \sqrt{ER_j^2}$ and

$$E|W_j R_i| \le \sqrt{EW_j^2 E R_i^2} \le \|\Sigma\|^{1/2}\sqrt{ER_i^2}.$$

The term $C$ in the bound of the theorem now follows from the last two terms of (12.51). Lastly, recalling that $E(\mathbf{W}' - \mathbf{W})(\mathbf{W}' - \mathbf{W})^{\mathsf{T}} = Q$, the first term of the bound, $A/4$, arises from the first term of (12.51). □

## 12.4 Local Dependence, and Bounds in Kolmogorov Distance

We now consider multivariate results in distances which include the Kolmogorov metric, due to Rinott and Rotar (1996), that apply to sums

$$\mathbf{W}=\sum_{i=1}^{n}\mathbf{X}_i \tag{12.52}$$

of bounded, locally dependent random variables. In this section, for an array $A$, let $|A|$ denote the sum of the absolute values of the components of $A$. Since in finite dimension all norms are equivalent, and, in addition, as constants in the results of this section are not given explicitly, this convention is only a matter of convenience.

Let $\mathbf{X}_1,\ldots,\mathbf{X}_n$ be random vectors in $\mathbb{R}^p$ satisfying $|\mathbf{X}_i|\le B$ for all $i=1,\ldots,n$; the constant $B$ is allowed to depend on $n$. For each $i=1,\ldots,n$, let $S_i\subset\mathcal{N}_i$ be dependency neighborhoods for $\mathbf{X}_i$, such that with $D_1\le D_2$ we have

$$\max\bigl\{|S_i|,i=1,\ldots,n\bigr\}\le D_1 \quad\text{and}\quad \max\bigl\{|\mathcal{N}_i|,i=1,\ldots,n\bigr\}\le D_2,$$

where for a finite set we also use $|\cdot|$ to denote cardinality. The sets $S_i$ and $\mathcal{N}_i$ may be random. For $i=1,\ldots,n$, let

$$\begin{aligned}\mathbf{U}_i&=\sum_{k\in S_i}\mathbf{X}_k, \quad \mathbf{V}_i=\mathbf{W}-\mathbf{U}_i,\\ \mathbf{R}_i&=\sum_{k\in\mathcal{N}_i}\mathbf{X}_k \quad\text{and}\quad \mathbf{T}_i=\mathbf{W}-\mathbf{R}_i,\end{aligned} \tag{12.53}$$

and

$$\begin{aligned}\chi_1&=\sum_{i=1}^{n}E\bigl|E(\mathbf{X}_i|\mathbf{V}_i)\bigr|,\\ \chi_2&=\sum_{i=1}^{n}E\bigl|E\bigl(\mathbf{X}_i\mathbf{U}_i^{\mathsf{T}}\bigr)-E\bigl(\mathbf{X}_i\mathbf{U}_i^{\mathsf{T}}|\mathbf{T}_i\bigr)\bigr| \quad\text{and}\quad \chi_3=\left|I-\sum_{i=1}^{n}E\bigl(\mathbf{X}_i\mathbf{U}_i^{\mathsf{T}}\bigr)\right|.\end{aligned} \tag{12.54}$$

As in Sect. 5.4, we consider a collection $\mathcal{H}$ of functions satisfying Condition 5.1, and recall that $a$ is a constant that measures the roughness of the collection with respect to the normal distribution. Specifically, with $\mathbf{Z}$ a standard normal vector in $\mathbb{R}^p$, we assume the collection $\mathcal{H}$ satisfies

$$E\tilde{h}_\epsilon(\mathbf{Z})\le a\epsilon \quad\text{for all } \epsilon>0, \tag{12.55}$$

where $\tilde{h}_\epsilon$ is defined in (iii) of Condition 5.1. The class of indicators of convex sets is one example of a collection $\mathcal{H}$ that satisfies Condition 5.1, see Sazonov (1968), and Bhattacharya and Rao (1986). As in the one dimensional case, given a function class $\mathcal{H}$ and random vectors $\mathbf{X}$ and $\mathbf{Y}$, we let

$$\bigl\|\mathcal{L}(\mathbf{X})-\mathcal{L}(\mathbf{Y})\bigr\|_{\mathcal{H}}=\sup_{h\in\mathcal{H}}\bigl|Eh(\mathbf{X})-Eh(\mathbf{Y})\bigr|.$$

Letting $\mathcal{H}$ be the collection of indicators of 'lower quadrants' the distance above specializes to the Kolmogorov distance.

As our interest in this section is in multivariate results, we consider only the general $p\ge 1$ case of the results of Rinott and Rotar (1996), and refer the reader there for improvements of Theorems 12.5 and 12.8 in the case $p=1$.

**Theorem 12.5** *For $p \geq 1$, there exists a constant $C$ depending only on the dimension $p$, such that*

$$\begin{aligned}\|\mathcal{L}(\mathbf{W}) - \mathcal{L}(\mathbf{Z})\|_{\mathcal{H}} \leq C\{&aD_2B + naD_1D_2B^3(|\log B| + \log n)\\ &+ \chi_1 + (|\log B| + \log n)(\chi_2 + \chi_3)\}. \end{aligned} \tag{12.56}$$

The $\log n$ terms in the bound will typically preclude the $n^{-1/2}$ rate possible in the $p = 1$ case. Though it is not assumed that $\mathbf{W}$ have mean zero and identity covariance, typically, as below, the theorem is invoked in the standardized case. Theorem 12.5 follows from Theorem 12.8, which proceeds by the way of smoothing inequalities, as in the proof of Theorem 5.8. As in Rinott and Rotar (1996), we apply Theorem 12.5 to two questions in random graphs.

First, let $\mathcal{G}$ be a fixed regular graph with $n$ vertices, each of degree $m$. Let each vertex be independently assigned color $i = 1, \ldots, p$ with probability $\pi_i \geq 0$ and $\sum_{i=1}^p \pi_i = 1$. We are interested in counting the number of edges that connect vertices having the same color. Indexing the $N = nm/2$ edges of $\mathcal{G}$ by $\{1, \ldots, N\}$, for $i = 1, \ldots, p$ we may write $Y_i$, the number of edges whose vertices both have color $i$, as the sum of indicators

$$Y_i = \sum_{k=1}^{N} X_{ki},$$

where $X_{ki}$ indicates if edge $k$ connects vertices both of which have color $i$. Letting $\mathbf{Y} = (Y_1, \ldots, Y_p)$, clearly, the mean $\boldsymbol{\mu} = E\mathbf{Y}$ is given by

$$\boldsymbol{\mu} = N(\pi_1^2, \ldots, \pi_p^2). \tag{12.57}$$

To calculate the variance of $Y_i$, note first that there will be $N$ variance terms of the form $\mathrm{Var}(X_{ki}) = \pi_i^2(1 - \pi_i^2)$. Next, for a give edge $k$, the covariance between $X_{ki}$ and $X_{li}$ for $l \neq k$ will be nonzero only for the $2(m-1)$ edges $l \neq k$ that share a vertex with $k$, and in this case

$$\mathrm{Cov}(X_{ki}, X_{li}) = \pi_i^3 - \pi_i^4.$$

Hence, letting $\Sigma = (\sigma_{ij})$ be the covariance matrix of $\mathbf{Y}$,

$$\sigma_{ii} = N\pi_i^2(1 - \pi_i^2) + 2N(m-1)(\pi_i^3 - \pi_i^4).$$

For $i \neq j$ note that $\mathrm{Cov}(X_{ki}, X_{lj})$ is zero if edges $k$ and $l$ have no vertex in common, while otherwise, as $X_{ki}X_{lj} = 0$, this covariance is $-\pi_i^2\pi_j^2$. Hence

$$\sigma_{ij} = -N(2m-1)\pi_i^2\pi_j^2 \quad \text{for } i \neq j.$$

It is not difficult to see that we may write $\Sigma$ a bit more compactly as

$$\Sigma = N(2m-1)\left[A - bb^{\mathsf{T}}\right] + NH, \tag{12.58}$$

where $A$ and $H$ are diagonal matrices with $i$th diagonal entries $\pi_i^3$ and $\pi_i^2 - \pi_i^3$, respectively, and $b$ the column vector with $i$th component $\pi_i^2$. Using this representation and recalling (12.15), we now show

$$NH \preceq \Sigma. \tag{12.59}$$

Let $D$ be the diagonal matrix with diagonal entries $\pi_i^{3/2}$ and $g$ the column vector with entries $\pi_i^{1/2}$. Then $A - bb^\mathsf{T} = D(I - gg^\mathsf{T})D$. Since $\sum \pi_i = 1$, its is easy to see that the smallest eigenvalue of $I - gg^\mathsf{T}$ is 0. Hence $A - gg^\mathsf{T}$ is nonnegative definite, and (12.59) follows. It follows that, with

$$L = \left[\min_{1 \le i \le p} \left\{\pi_i^2(1-\pi_i)\right\}\right]^{-1/2}, \tag{12.60}$$

and $\|\cdot\|$ denoting the maximal absolute value of the entries of an array as in (12.3),

$$\left\|\Sigma^{-1/2}\right\| \le N^{-1/2} L.$$

We apply Theorem 12.5 to the normalized counts $\mathbf{W} = \Sigma^{-1/2}(\mathbf{Y} - \boldsymbol{\mu})$ with the bound $B = pN^{-1/2}L$ on the standardized summands

$$\mathbf{X}_k = \Sigma^{-1/2}\left(X_{k1} - \pi_1^2, \ldots, X_{kp} - \pi_k^2\right)$$

in (12.52). For any edge $j$ we choose $S_j$ to be the collection of all edges that share a vertex with $j$, and $\mathcal{N}_i = \bigcup_{j \in S_i} S_j$, yielding the bounds $D_1 = 2m - 1$ and $D_2 = (2m-1)^2$ on the cardinalities of $S_j$ and $\mathcal{N}_i$, respectively. These choices yield $\chi_1 = \chi_2 = \chi_3 = 0$. Observing that $m \le n$ and $L \ge 1$, we have obtained the following result.

**Theorem 12.6** *For $i = 1, \ldots, p$, let $Y_i$ count the number of edges, of a regular degree $m$ graph with $n$ vertices, that connect vertices both of color $i$, where colors $1, \ldots, p$ are assigned independently with probabilities $\pi_1, \ldots, \pi_p$. Then there exists a constant $C$, depending only on $p$, such that*

$$\left\|\mathcal{L}(\mathbf{W}) - \mathcal{L}(\mathbf{Z})\right\|_{\mathcal{H}} \le Cam^{3/2}L^3\left(|\log L| + \log n\right)n^{-1/2}, \tag{12.61}$$

*where $\mathbf{W} = \Sigma^{-1/2}(\mathbf{Y} - \boldsymbol{\mu})$ with the mean $\boldsymbol{\mu}$ and variance $\Sigma$ of $\mathbf{Y}$ given by* (12.57) *and* (12.58), *respectively, $L$ is given by* (12.60) *and the constant $a$ depends on the class $\mathcal{H}$ through* (12.55).

In this same case, a bound of rate $n^{-1/2}$ was obtained by Goldstein and Rinott (1996), but only for smooth functions.

The following example illustrates an advantage of Theorem 12.5 in allowing the neighborhoods of dependence $S_i$ and $\mathcal{N}_i$ to be random. Consider a sample of $n$ points chosen independently in $\mathbb{R}^k$ according to some absolutely continuous distribution. Similar to the earlier example, color each point with color $i$ with probability $\pi_i, i = 1, \ldots, p$, independently of the sample and of the colors assigned to other points. Now form the nearest neighbor graph by making a directed edge from each point in the collection to its nearest neighbor. Let $X_{ji}$ be the indicator that vertex $j$ and its nearest neighbor both have color $i$, so that

$$Y_i = \sum_{j=1}^n X_{ji}$$

counts the number of times a vertex and its nearest neighbor share color $i$, with mutual nearest neighbors counted twice.

The vector $\mathbf{Y} = (Y_1, \ldots, Y_p)$ is of interest in multivariate tests of equality of distribution. In particular, consider observations that are drawn from $p = 2$ distributions, say $F_1$ and $F_2$, with probabilities $\pi_1$ and $\pi_2 = 1 - \pi_1$, respectively. When $F_1 \neq F_2$ one would expect that there would be a certain degree of clustering of the points drawn from the same distribution, and that the nearest neighbor of a point drawn from $F_1$, say, would more likely be a point from this same population, rather than one from $F_2$. Clustering of this type could be detected by computing functions of $\mathbf{Y}$ for a given sample, and testing against the null hypotheses that $F_1 = F_2$, that is, the case where the colors are assigned independently of the sample. Hence, the normal approximation of $\mathbf{Y}$ for this instance gives bounds on how well the null distribution of a test statistic can be approximated by the same function of a multivariate normal vector. The use of $\mathbf{Y}$ for such test was considered by Schilling (1986) and Henze (1988); the latter proves asymptotic normality, without rates, of certain test statistics.

Clearly, the mean $\boldsymbol{\mu}$ of $\mathbf{Y}$ is given by

$$\boldsymbol{\mu} = n\big(\pi_1^2, \ldots, \pi_p^2\big). \tag{12.62}$$

Referring further details to Rinott and Rotar (1996), letting $\alpha$ be the probability that vertex $j$ is the nearest neighbor of its own nearest neighbor, $\beta = ED^2(j)$ for $D(j)$ the degree of $j$, $H$, $J$ and $K$ the $p \times p$ diagonal matrices having $i$th diagonal elements $\pi_i^3$, $\pi_i^2 - \pi_i^4$ and $\pi_i^2 - \pi_i^3$ respectively, and $b$ the column vector with $i$th component $\pi_i^2$, we may write the covariance matrix $\Sigma$ of $\mathbf{Y}$ as

$$\Sigma = \frac{1}{2} n(\beta - 2)\big[H - bb^{\mathsf{T}}\big] + nJ + n\alpha K. \tag{12.63}$$

By use of this representation, one may derive

$$\big\|\Sigma^{-1/2}\big\| \le n^{-1/2} M \quad \text{where } M = \Big[\min_{1 \le i \le p}\big\{\pi_i^2 - \pi_i^4\big\}\Big]^{-1/2}. \tag{12.64}$$

Hence, as in the previous example, we may take the bound $B$ on the standardized summand variables to equal $pn^{-1/2}M$.

Since the dependency neighborhoods may be random they can, in particular, depend on the nearest neighbor graph $\mathcal{G}$ itself. In particular, let $S_j$ consist of vertex $j$ and all vertices that are connected to $j$ by an edge. Then for any set $A$ we have

$$\begin{aligned} &P\big(\{X_{ji}, 1 \le i \le p\} \in A\big) \\ &\quad = P\big(\{X_{ji}, 1 \le i \le p\} \in A \,|\, \mathcal{G}, \{X_{li}, 1 \le i \le p, l \notin S_j\}\big). \end{aligned}$$

Taking expectations conditioned on $\{X_{li}, 1 \le i \le p, l \notin S_j\}$ we obtain the independence of $\{X_{ji}, 1 \le i \le p\}$ and $\{X_{li}, 1 \le i \le p, l \notin S_j\}$. With similar arguments for $\mathcal{N}_i = \bigcup_{j \in S_i} S_j$, one may conclude that $\chi_1 = \chi_2 = \chi_3 = 0$.

It is easy to see that in $\mathbb{R}^k$ the degree of the nearest neighbor graph is bounded by the kissing number $\kappa_k$, the maximum number of spheres of radius 1 that can simultaneously touch the unit sphere at the origin. The kissing number was discussed in

Sect. 4.6, where further references, some to bounds, were provided. For our choices of dependency neighborhoods, we may take $D_1 = \kappa_k$ and $D_2 = \kappa_k^2$ as bounds on the cardinalities of $S_i$ and $\mathcal{N}_i$, respectively.

Applying Theorem 12.5, as for the bound (12.61), one obtains the following result.

**Theorem 12.7** *Let $n$ i.i.d. points from an absolutely continuous distribution in $\mathbb{R}^k$ be assigned colors $i = 1, \ldots, p$ independently with probabilities $\pi_1, \ldots, \pi_p$, and let the component $Y_i$ of $\mathbf{Y} = (Y_1, \ldots, Y_p)$ count the number of vertices which share color $i$ with its nearest neighbor. Then, with $\mathbf{W} = \Sigma^{-1/2}(\mathbf{Y} - \boldsymbol{\mu})$, where the mean $\boldsymbol{\mu}$ and the variance $\Sigma$ of $\mathbf{Y}$ are given by* (12.62) *and* (12.64), *respectively, there exists a constant $C$, depending only on $p$, such that*

$$\left\|\mathcal{L}(\mathbf{W}) - \mathcal{L}(\mathbf{Z})\right\|_{\mathcal{H}} \le C a \kappa_k^3 M^3 \left(|\log M| + \log n\right) n^{-1/2},$$

*where $\kappa_k$ is the kissing number in dimension $k$, $M$ is given in* (12.64), *and the constant $a$ depends on the class $\mathcal{H}$ through* (12.55).

The quantities $\alpha$ and $\beta$ in (12.63) converge to finite limits as $n$ tends to infinity, and therefore $n^{-1}\Sigma$ tends to a limiting matrix. When the sample of points have a continuous distribution $F$, these limits do not depend on $F$, yielding, asymptotically, a non-parametric test for the equality of multivariate distributions. See Schilling (1986), Henze (1988) and Rinott and Rotar (1996) for further details.

We now proceed to the proof of Theorem 12.5, which is a consequence of the following result.

**Theorem 12.8** *For each $i = 1, \ldots, n$ assume that we have two representations of $\mathbf{W}$, $\mathbf{W} = \mathbf{U}_i + \mathbf{V}_i$, and $\mathbf{W} = \mathbf{R}_i + \mathbf{T}_i$, such that $|\mathbf{U}_i| \le A_1$ and $|\mathbf{R}_i| \le A_2$ for constants $A_1 \le A_2$. Then for $p \ge 1$ there exists a constant $C$ depending only on the dimension $p$ such that*

$$\begin{aligned}\left\|\mathcal{L}(\mathbf{W}) - \mathcal{L}(\mathbf{Z})\right\|_{\mathcal{H}} \le C\big\{&aA_2 + naA_1A_2B\left(|\log A_2 B| + \log n\right) \\ &+ \chi_1 + \left(|\log A_2 B| + \log n\right)(\chi_2 + \chi_3)\big\},\end{aligned} \tag{12.65}$$

*where the constant $a$ depends on the class $\mathcal{H}$ through* (12.55), *$|\mathbf{X}_i| \le B$ for all $i = 1, \ldots, n$ and $\chi_1$, $\chi_2$ and $\chi_3$ are specified in* (12.54).

Theorem 12.5 follows by observing that the quantities defined in (12.53) satisfy the assumptions of Theorem 12.8 with $A_1 = D_1 B$ and $A_2 = D_2 B$. We note that the $\log A_2$ term that appears in (12.65) does not give rise to a $\log D_2$ term in (12.56) as $D_2 \le n$.

Throughout the proof of Theorem 12.8 we write $C$ for universal constants, not necessarily the same at each occurrence. For a given $h \in \mathcal{H}$ and $s \in (0, 1)$ we work with the smoothed version of $h$ given by

$$h_s(x) = \int_{\mathbb{R}^p} h(\sqrt{s}z + \sqrt{1-s}x)\phi(z)dz \tag{12.66}$$

where $\phi(z)$ denotes the standard normal density in $\mathbb{R}^p$. It is straightforward to verify that $Nh_s = Nh$ for all $s \in [0, 1]$.

We require the following smoothing result, Lemma 2.11 of Götze (1991), from Bhattacharya and Rao (1986). Lemma 12.1 is essentially the same as Lemma 5.3, though there the smoothing is at scale $t$, and here at scale $\sqrt{t}$. Here we let $\Phi$ denote the standard normal distribution in $\mathbb{R}^p$.

**Lemma 12.1** *Let $Q$ be a probability measure on $\mathbb{R}^p$. Then there exists a constant $C > 0$, which depends only on $p$, such that for all $t \in (0, 1)$*

$$\sup\left\{\left|\int_{\mathbb{R}^p} hd(Q - \Phi)\right| : h \in \mathcal{H}\right\} \le C\left[\sup\left\{\left|\int_{\mathbb{R}^p} hd(Q - \Phi)_t\right| : h \in \mathcal{H}\right\} + a\sqrt{t}\right],$$

*where $a$ is any constant that satisfies* (12.55).

For $f : \mathbb{R}^p \to \mathbb{R}$, let $\nabla f$ and $D^2 f$ denote the Hessian matrix and gradient of $f$, respectively. Consider the multivariate Stein equation (2.22) with $\boldsymbol{\mu} = 0$ and $\Sigma = I$, that is,

$$\operatorname{Tr} D^2 f(\mathbf{w}) - \mathbf{w} \cdot \nabla f(\mathbf{w}) = h(\mathbf{w}) - Nh. \tag{12.67}$$

By Götze (1991), the function

$$f_t(x) = -\frac{1}{2}\int_t^1 \left[h_s(x) - Nh\right]\frac{ds}{1-s} \tag{12.68}$$

solves the Stein equation (12.67) for $h_t$. Again by Götze (1991), when $|h| \le 1$, there exists a constant $C$ such that

$$\|\nabla f_t\| \le C \quad \text{and} \quad \left\|D^2 f_t\right\| \le C\log\left(t^{-1}\right). \tag{12.69}$$

Setting $K_i = \mathbf{X}_i \mathbf{U}_i^{\mathsf{T}}$, by (12.67),

$$Eh_t(\mathbf{W}) - Nh = E\left[\operatorname{Tr} D^2 f_t(\mathbf{W}) - \mathbf{W} \cdot \nabla f_t(\mathbf{W})\right] = \mathcal{A} - \mathcal{B} - \mathcal{C} + \mathcal{D},$$

where

$$\begin{aligned}
\mathcal{A} &= E\operatorname{Tr}\left[D^2 f_t(\mathbf{W})\left(I - \sum_{i=1}^n K_i\right)\right], \\
\mathcal{B} &= \sum_{i=1}^n E\left[\mathbf{X}_i \cdot \nabla f_t(\mathbf{V}_i)\right], \\
\mathcal{C} &= \sum_{i=1}^n E\left\{\mathbf{X}_i \cdot \left[\nabla f_t(\mathbf{W}) - \nabla f_t(\mathbf{V}_i) - D^2 f_t(\mathbf{V}_i)\mathbf{U}_i^{\mathsf{T}}\right]\right\} \quad \text{and} \\
\mathcal{D} &= \sum_{i=1}^n E\operatorname{Tr}\left\{K_i\left[D^2 f_t(\mathbf{W}) - D^2 f_t(\mathbf{V}_i)\right]\right\}.
\end{aligned} \tag{12.70}$$

The next lemma is used to bound Taylor series remainder terms arising from the decomposition of terms (12.70). We will let $f_{t(jkl)}^{(3)}(x)$ denote the third derivative of $f_t$ at $x$, with respect to $x_j, x_k, x_l$, with a similar notation for the partial derivatives of the normal density $\phi(z)$, and derivatives of lower orders.

**Lemma 12.2** *Let* $\mathbf{W}, \mathbf{V}$ *and* $\mathbf{U}$ *be any random vectors in* $\mathbb{R}^p$ *satisfying* $\mathbf{W} = \mathbf{V} + \mathbf{U}$, *and let* $Y$ *be any random variable. Suppose there exists constants* $C_1$ *and* $C_2$ *such that* $|\mathbf{U}| \le C_1$ *and* $|Y| \le C_2$. *Set*

$$\kappa = \sup\{|Eh(\mathbf{W}) - Nh| : h \in \mathcal{H}\}. \tag{12.71}$$

*Then there exists a constant* $C$, *depending only on* $p$, *such that for all* $\tau \in [0, 1]$ *and* $h \in \mathcal{H}$,

$$|EYf_{t(jkl)}^{(3)}(\mathbf{V} + \tau\mathbf{U})| \le CC_2(\kappa/\sqrt{t} + aC_1/\sqrt{t} + a|\log t|),$$

*where* $a$ *satisfies* (12.55).

*Proof* By replacing $h_t$ by $h_t - Nh$ we may assume $Nh_t = 0$. Differentiation of (12.66), using a change of variable, and (12.68), yield

$$f_{t(jkl)}^{(3)}(x) = \frac{1}{2}\int_t^1 \frac{(1-s)^{1/2}}{s^{3/2}} ds \int_{\mathbb{R}^p} h(\sqrt{s}z + \sqrt{1-s}x)\phi_{jkl}^{(3)}(z)dz.$$

Observe that

$$\int_{\mathbb{R}^p} \phi_{jkl}^{(3)}(z)dz = \frac{\partial^3}{\partial_j\partial_k\partial_l}\int_{\mathbb{R}^p} \phi(z+x)dz\bigg|_{x=0} = \frac{\partial^3}{\partial_j\partial_k\partial_l}1\bigg|_{x=0} = 0. \tag{12.72}$$

Now,

$$\begin{aligned}
&|EYf_{t(jkl)}^{(3)}(\mathbf{V} + \tau\mathbf{U})| \\
&\quad = \left|\frac{1}{2}\int_t^1 \frac{(1-s)^{1/2}}{s^{3/2}} ds \int_{\mathbb{R}^p} EYh(\sqrt{s}z + \sqrt{1-s}(\mathbf{V} + \tau\mathbf{U}))\phi_{jkl}^{(3)}(z)dz\right| \\
&\quad = \left|\frac{1}{2}\int_t^1 \frac{(1-s)^{1/2}}{s^{3/2}} ds \right. \\
&\qquad \left. \times \int_{\mathbb{R}^p} EYh(\sqrt{1-s}\mathbf{W} - \sqrt{1-s}(1-\tau)\mathbf{U} + \sqrt{s}z)\phi_{jkl}^{(3)}(z)dz\right| \\
&\quad = \left|\frac{1}{2}\int_t^1 \frac{(1-s)^{1/2}}{s^{3/2}} ds \int_{\mathbb{R}^p} EY\{h(\sqrt{1-s}\mathbf{W} - \sqrt{1-s}(1-\tau)\mathbf{U} + \sqrt{s}z)\right. \\
&\qquad \left. - h(\sqrt{1-s}\mathbf{W} - \sqrt{1-s}(1-\tau)\mathbf{U})\}\phi_{jkl}^{(3)}(z)dz\right| \qquad (12.73) \\
&\quad \le \frac{1}{2}\int_t^1 \frac{1}{s^{3/2}} ds\, C_2 \int_{\mathbb{R}^p} E\Big\{\sup_{|u| \le C_1 + \sqrt{s}|z|} h(\sqrt{1-s}\mathbf{W} + u). \\
&\qquad - \inf_{|u| \le C_1 + \sqrt{s}|z|} h(\sqrt{1-s}\mathbf{W} + u)\Big\}|\phi_{jkl}^{(3)}(z)|dz \\
&\quad = \frac{C_2}{2}\int_t^1 \frac{1}{s^{3/2}} ds \int_{\mathbb{R}^p} E\tilde{h}(\sqrt{1-s}\mathbf{W}; C_1 + \sqrt{s}|z|)|\phi_{jkl}^{(3)}(z)|dz, \qquad (12.74)
\end{aligned}$$

where we have used (12.72) to obtain (12.73), and recalled the definition of $\tilde{h}$ in (5.28).

Let $\mathbf{Z}$ denote an independent standard normal vector in $\mathbb{R}^p$. Adding and subtracting, the quantity (12.74) equals

$$\frac{C_2}{2}\int_t^1 \frac{1}{s^{3/2}}\int_{\mathbb{R}^p} E\{\tilde{h}(\sqrt{1-s}\mathbf{W}; C_1+\sqrt{s}|z|) - \tilde{h}(\sqrt{1-s}\mathbf{Z}; C_1+\sqrt{s}|z|) + \tilde{h}(\sqrt{1-s}\mathbf{Z}; C_1+\sqrt{s}|z|)\}|\phi_{jkl}^{(3)}(z)|dz. \tag{12.75}$$

Again in view of definition (5.28) of $\tilde{h}$, for any $\epsilon > 0$,

$$\begin{aligned} &|E\{\tilde{h}(\sqrt{1-s}\mathbf{W}; \epsilon) - \tilde{h}(\sqrt{1-s}\mathbf{Z}; \epsilon)\}| \\ &\le |E\{h_\epsilon^+(\sqrt{1-s}\mathbf{W}; \epsilon) - h_\epsilon^+(\sqrt{1-s}\mathbf{Z}; \epsilon)\}| \\ &\quad + |E\{h_\epsilon^-(\sqrt{1-s}\mathbf{W}; \epsilon) - h_\epsilon^-(\sqrt{1-s}\mathbf{Z}; \epsilon)\}|. \end{aligned} \tag{12.76}$$

By the closure conditions on the class $\mathcal{H}$ and the definition (12.71) of $\kappa$, we see that for any $\epsilon > 0$ the expression (12.76) is bounded by $2\kappa$.

As

$$\int_t^1 \frac{1}{s^{3/2}}ds \le \frac{C}{\sqrt{t}},$$

we conclude, for some $C$,

$$\begin{aligned} &\int_t^1 \frac{1}{s^{3/2}}ds \int_{\mathbb{R}^p} |E\{\tilde{h}(\sqrt{1-s}\mathbf{W}; C_1+\sqrt{s}|z|) \\ &\quad - \tilde{h}((\sqrt{1-s}\mathbf{Z}; C_1+\sqrt{s}|z|))\}\phi_{jkl}^{(3)}(z)|dz \le C\kappa/\sqrt{t}. \end{aligned} \tag{12.77}$$

Turning now to the last term of (12.75), by (12.55),

$$E\tilde{h}(\sqrt{1-s}\mathbf{Z}; C_1+\sqrt{s}|z|) \le a(C_1+\sqrt{s}|z|).$$

Hence

$$\begin{aligned} &\int_t^1 \frac{1}{s^{3/2}}ds \int_{\mathbb{R}^p} E\tilde{h}(\sqrt{1-s}\mathbf{Z}; C_1+\sqrt{s}|z|)|\phi_{jkl}^{(3)}(z)|dz \\ &\le a\int_t^1 \frac{1}{s^{3/2}}ds \int_{\mathbb{R}^p} (C_1+\sqrt{s}|z|)|\phi_{jkl}^{(3)}(z)|dz \\ &\le Ca(C_1/\sqrt{t}+|\log t|). \end{aligned}$$

Lemma 12.2 now follows by collecting terms. □

We are now ready for the proof of Theorem 12.8.

*Proof* Consider the decomposition in (12.70), starting with the term $\mathcal{C}$. Let $X_{ij}$ and $U_{ij}$ denote the $j$th components of $\mathbf{X}_i$ and $\mathbf{U}_i$, respectively. For $i = 1, \ldots, n$, Taylor expansion of $\nabla f_t(\mathbf{W})$ about $\mathbf{V}_i$ shows that $\mathcal{C}$ equals

$$\sum_{i=1}^n E\int_0^1 (1-\tau)\sum_{j=1}^p\sum_{k=1}^p\sum_{l=1}^p X_{ij}U_{ik}U_{il}f_{t(jkl)}^{(3)}(\mathbf{V}_i+\tau\mathbf{U}_i)d\tau. \tag{12.78}$$

Applying Lemma 12.2 for each $i$, with $\mathbf{U}=\mathbf{U}_i$ and $Y=X_{ij}U_{ik}U_{il}$, recalling $|\mathbf{X}_i|\le B$ and $|\mathbf{U}_i|\le A_1$, we obtain

$$|\mathcal{C}|\le CnA_1^2B\big(\kappa/\sqrt{t}+aA_1/\sqrt{t}+a|\log t|\big). \tag{12.79}$$

Next consider the term $\mathcal{D}$ in (12.70). A first order Taylor expansion yields

$$f_{t(jk)}^{(2)}(\mathbf{W})-f_{t(jk)}^{(2)}(\mathbf{V}_i)=\sum_{l=1}^{p}\int_0^1 f_{t(jkl)}^{(3)}(\mathbf{V}_i+\tau\mathbf{U}_i)U_{il}d\tau. \tag{12.80}$$

The term $\mathcal{D}$ is obtained by multiplying (12.80) by the entries of $K_i$, and it is easy to see from their definition that this leads to a term which is similar to (12.78), allowing us to conclude that $|\mathcal{D}|$ is bounded by the right hand side of (12.79), with a possibly different constant.

Next, note that we may write $\mathcal{B}$ of (12.70) as

$$\mathcal{B}=\sum_{i=1}^{n}E\big[\nabla f_t(\mathbf{V}_i)\cdot E(\mathbf{X}_i|\mathbf{V}_i)\big].$$

By the bound (12.69) on the solution to (12.67), the components of $\nabla f_t(\mathbf{V}_i)$ are uniformly bounded, implying that for some positive constant $C$

$$|\mathcal{B}|\le C\sum_{i=1}^{n}\sum_{j=1}^{p}E\big|E(X_{ij}|\mathbf{V}_i)\big|. \tag{12.81}$$

Finally, consider $\mathcal{A}$. With $\delta_{jk}=1$ when $j=k$ and 0 otherwise, we have

$$\begin{aligned}
&\mathrm{Tr}\left[D^2f_t(\mathbf{W})\left(I-\sum_{i=1}^{n}K_i\right)\right]\\
&=\sum_{j=1}^{p}\sum_{k=1}^{p}D^2f_{t(jk)}(\mathbf{W})\left(\delta_{jk}-\sum_{i=1}^{n}X_{ik}U_{ij}\right)\\
&=\sum_{j=1}^{p}\sum_{k=1}^{p}D^2f_{t(jk)}(\mathbf{W})\\
&\quad\times\left(\delta_{jk}-\sum_{i=1}^{n}E(X_{ik}U_{ij})+\sum_{i=1}^{n}E(X_{ik}U_{ij})-\sum_{i=1}^{n}X_{ik}U_{ij}\right).
\end{aligned} \tag{12.82}$$

By the bound (12.69), we have that $|D^2f_{t(jk)}(\mathbf{W})|\le C\log(t^{-1})$ for all $j,k=1,\ldots,p$ and $t\in(0,1)$. Hence, for the expectation of the first two terms of the last line of (12.82), we obtain the bound

$$\begin{aligned}
&E\left|\sum_{j=1}^{p}\sum_{k=1}^{p}D^2f_{t(jk)}(\mathbf{W})\left(\delta_{jk}-\sum_{i=1}^{n}E(X_{ik}U_{ij})\right)\right|\\
&\quad\le C|\log t|\sum_{j=1}^{p}\sum_{k=1}^{p}\left|\delta_{jk}-\sum_{i=1}^{n}E(X_{ik}U_{ij})\right|.
\end{aligned} \tag{12.83}$$

Now write the expression involving the last two terms in the last line of (12.82) in the form

$$\sum_{j=1}^{p}\sum_{k=1}^{p}\{D^2 f_{t(jk)}(\mathbf{W}) - D^2 f_{t(jk)}(\mathbf{T}_i) + D^2 f_{t(jk)}(\mathbf{T}_i)\} \times \big[E(X_{ik}U_{ij}) - X_{ik}U_{ij}\big]. \tag{12.84}$$

Taylor expansion of the difference $D^2 f_{t(jk)}(\mathbf{W}) - D^2 f_{t(jk)}(\mathbf{T}_i)$, and Lemma 12.2 applied for each $i$ with $\mathbf{U} = \mathbf{R}_i$ and $Y = R_{il}(X_{ik}U_{ij} - E(X_{ik}U_{ij}))$, imply

$$\begin{aligned}&\sum_{i=1}^{n}\sum_{j=1}^{p}\sum_{k=1}^{p}\big|E\{[D^2 f_{t(jk)}(\mathbf{W}) - D^2 f_{t(jk)}(\mathbf{T}_i)][E(X_{ik}U_{ij}) - X_{ik}U_{ij}]\}\big| \\ &\quad\le CnA_1A_2B\big(\kappa/\sqrt{t} + aA_2/\sqrt{t} + a|\log t|\big).\end{aligned} \tag{12.85}$$

Returning to (12.84), we apply (12.69) to bound the last term by

$$\begin{aligned}&\sum_{i=1}^{n}\sum_{j=1}^{p}\sum_{k=1}^{p}\big|E\{D^2 f_{t(jk)}(\mathbf{T}_i)[E(X_{ik}U_{ij}) - X_{ik}U_{ij}]\}\big| \\ &\quad\le C|\log t|\sum_{i=1}^{n}\sum_{j=1}^{p}\sum_{k=1}^{p} E\big|E(X_{ik}U_{ij}) - E(X_{ik}U_{ij}|\mathbf{T}_i)\big|.\end{aligned} \tag{12.86}$$

Combining Lemma 12.1, the decomposition (12.70), and the bounds (12.79), (12.81), (12.83), (12.85) and (12.86), noting that since $A_1 \le A_2$ the term (12.79) may be ignored, being of smaller order than (12.85), we obtain

$$\begin{aligned}\kappa &\le CnA_1A_2B\kappa/\sqrt{t} + CnaA_1A_1B\big(A_2/\sqrt{t} + |\log t|\big) \\ &\quad + C\sum_{i=1}^{n}\sum_{j=1}^{p} E\big|E(X_{ij}|\mathbf{V}_i)\big| \\ &\quad + C|\log t|\left\{\sum_{j=1}^{p}\sum_{k=1}^{p}\left|\delta_{jk} - \sum_{i=1}^{n} E(X_{ij}U_{ik})\right|\right. \\ &\quad \left. + \sum_{i=1}^{n}\sum_{j=1}^{p}\sum_{k=1}^{p} E\big|E(X_{ij}U_{ik}) - E(X_{ij}U_{ik}|\mathbf{T}_i)\big|\right\} + Ca\sqrt{t}.\end{aligned} \tag{12.87}$$

Setting $\sqrt{t} = 2CnA_1A_2B$, provided it is less than 1, simple manipulations yield (12.65) after observing that the last term in (12.87) is of lower order than the second term. If $t > 1$ for the choice above, then by enlarging $C$ in (12.65) as necessary, the theorem is trivial. □

# Chapter 13
# Non-normal Approximation

Though the principle theme of this book concerns the normal distribution, in this chapter we explore how Stein's method can be applied to approximations by non-normal distributions as well. There are already many well known distributions other than the normal where Stein's method works, the Poisson case being the most notable. Here we focus on approximation by continuous distributions where an analysis parallel to that for the normal, such as the method of exchangeable pairs, may proceed.

Denoting the random variable of interest as $W = W_n$, it may be the case that appropriate approximating or limiting distributions of $W_n$ are not known a priori. In this chapter, following Chatterjee and Shao (2010) we first develop a method of exchangeable pairs which identifies an appropriate approximating distribution, and which obtains $L^1$ and Berry–Esseen type bounds for that approximation. As applications, in Sect. 13.3 we obtain error bounds of order $1/\sqrt{n}$ in both the $L^1$ and Kolmogorov distance for the non-central limit theorem for the magnetization in the Curie–Weiss model at the critical temperature. In Sect. 13.4, and also using different methods, we derive bounds for approximations by the exponential distribution, and, following Chatterjee et al. (2008), apply the results to the spectrum of the Bernoulli Laplace diffusion model, and following Peköz and Röllin (2009), to first passage times of Markov chains.

## 13.1 Stein's Method via the Density Approach

One way of looking at Stein's characterization for the normal is the following. Since the standard normal density function $\phi(z) = e^{-z^2/2}/\sqrt{2\pi}$ satisfies

$$\frac{\phi'(z)}{\phi(z)} = -z, \quad \text{we have} \quad \phi'(z) + z\phi(z) = 0, \tag{13.1}$$

and integration by parts now yields a kind of 'dual' equation for the distribution of $Z$ having density $\phi$, that is, the Stein characterization

$$E\big[f'(Z) - Zf(Z)\big] = 0$$

L.H.Y. Chen et al., *Normal Approximation by Stein's Method*, Probability and Its Applications,
DOI 10.1007/978-3-642-15007-4_13, © Springer-Verlag Berlin Heidelberg 2011

holding for all functions for which the expectations above exist. We will now see that a number of arguments used for the normal hold more generally for distributions with density $p(y)$ when replacing the ratio $-y$ in (13.1) by $p'(y)/p(y)$.

We will consider approximations by the distribution of $Y$, a random variable with probability density function $p$ satisfying the following condition.

**Condition 13.1** *For some $-\infty \le a < b \le \infty$, the density function $p$ is strictly positive and absolutely continuous over the interval $(a,b)$, zero on $(a,b)^c$, and possesses a right-hand limit $p(a+)$ at $a$ and a left-hand limit $p(b-)$ at $b$. Furthermore, the derivative $p'$ of $p$ satisfies*

$$\int_a^b |p'(y)|dy < \infty. \tag{13.2}$$

The key step for applying the Stein method for approximation by the distribution of $Y$ is the development of a Stein identity and the derivation of bounds on solutions to the Stein equation.

### *13.1.1 The Stein Characterization and Equation*

Let $Y$ have density $p$ satisfying Condition 13.1. Then, letting $f$ be an absolutely continuous function satisfying $f(a+) = f(b-) = 0$, whenever the expectations below exist, on the interval $(a,b)$ we have

$$\begin{aligned} &E\{f'(Y) + f(Y)p'(Y)/p(Y)\} \\ &= E\{(f(Y)p(Y))'/p(Y)\} \\ &= \int_a^b (f(y)p(y))'dy = f(b-)p(b-) - f(a+)p(a+) = 0, \end{aligned} \tag{13.3}$$

that is,

$$E\big(f'(Y) + f(Y)p'(Y)/p(Y)\big) = 0. \tag{13.4}$$

For any measurable function $h$ with $E|h(Y)| < \infty$, let $f = f_h$ be the solution to the Stein equation

$$f'(w) + f(w)p'(w)/p(w) = h(w) - Eh(Y). \tag{13.5}$$

Rewriting (13.5) we have that

$$\big(f(w)p(w)\big)' = \big(h(w) - Eh(Y)\big)p(w)$$

and hence the solution for $w \in (a,b)$ is given by

$$\begin{aligned} f_h(w) &= 1/p(w)\int_a^w \big(h(y) - Eh(Y)\big)p(y)dy \\ &= -1/p(w)\int_w^b \big(h(y) - Eh(Y)\big)p(y)dy. \end{aligned} \tag{13.6}$$

As the Stein equation, and its solution $f$, are valid only over the interval $(a,b)$, we consider, without further mention, approximation of the distributions of random variables $W$ by that of $Y$, having density $p$ on $(a,b)$, only when the support of $W$ is contained in the closure of $(a,b)$.

*Example 13.1* (Exponential Distribution) Let $Y \sim \mathcal{E}(\lambda)$, where $\mathcal{E}(\lambda)$ denotes the exponential distribution with parameter $\lambda$, that is, $Y$ is a random variable with density function $p(y) = \lambda e^{-\lambda y}\mathbf{1}(y > 0)$. Then $p'(y)/p(y) = -\lambda$ and identity (13.4) becomes

$$E\big[f'(Y) - \lambda E f(Y)\big] = 0, \tag{13.7}$$

for any absolutely continuous $f$ for which the expectation above exists, satisfying $f(0) = \lim_{y\to\infty} f(y) = 0$. Similar to the case of the normal, (13.7) is a characterization of the exponential distribution in that if (13.7) holds for all such functions $f$ then $Y \sim \mathcal{E}(\lambda)$.

The exponential distribution is a special case of the Gamma, which we turn to next.

*Example 13.2* (The Gamma and $\chi^2$ distributions) With $\alpha$ and $\beta$ positive numbers, we say $Y$ has the $\Gamma(\alpha,\beta)$ distribution when $Y$ has density

$$p(y) = \frac{y^{\alpha-1}e^{-y/\beta}}{\beta^\alpha\Gamma(\alpha)}\mathbf{1}_{\{y>0\}}.$$

Then $p'(y)/p(y) = \frac{\alpha-1}{y} - \frac{1}{\beta}$ and identity (13.4) becomes

$$E\left(f'(Y) + \left(\frac{\alpha-1}{Y} - \frac{1}{\beta}\right)f(Y)\right) = 0.$$

In the special case where $Y$ has the $\chi_k^2$ distribution, that is, the $\Gamma(k/2,2)$ distribution, the identity specializes to

$$E\left(f'(Y) + \left(\frac{k-2}{2Y} - \frac{1}{2}\right)f(Y)\right) = 0.$$

Approximation by Gamma and $\chi^2$ distributions have been considered by Luk (1994) and Pickett (2004). Gamma approximation of the distribution of stochastic integrals of Weiner processes is handled in Nourdin and Peccati (2009), the normal version of which is explored in Chap. 14.

*Example 13.3* Let $\mathcal{W}(\alpha,\beta)$ denote the density function

$$p(y) = \frac{\alpha e^{-|y|^\alpha/\beta}}{2\beta^{1/\alpha}\Gamma(\frac{1}{\alpha})} \quad \text{for } y \in \mathbb{R}, \text{ with } \alpha > 0,\ \beta > 0. \tag{13.8}$$

For any $\alpha > 0$, by the change of variable $u = y^\alpha$ we have

$$\int_{-\infty}^{\infty} e^{-|y|^\alpha}dy = 2\int_0^\infty e^{-y^\alpha}dy = \frac{2}{\alpha}\int_0^\infty e^{-u}u^{1/\alpha-1}du = \frac{2}{\alpha}\Gamma(1/\alpha),$$

hence, scaling by $\beta^{1/\alpha} > 0$, the family of functions (13.8) are densities on $\mathbb{R}$. Note that the mean zero normal distributions with variance $\sigma^2$ are the special case $\mathcal{W}(2, 2\sigma^2)$.

Of special interest will be the distribution $\mathcal{W}(4, 12)$ with density

$$p(y) = c_1 e^{-y^4/12} \quad \text{for } y \in \mathbb{R}, \text{ where } c_1 = \sqrt{2}/\big(3^{1/4}\Gamma(1/4)\big). \tag{13.9}$$

For $\mathcal{W}(4, 12)$ the ratio of the derivative to the density is given by

$$\frac{p'(y)}{p(y)} = -\frac{y^3}{3}.$$

### 13.1.2 Properties of the Stein Solution

As in the normal case, in order to determine error bounds for approximations by the distribution of $Y$, we need to understand the basic properties of the Stein solution. As we consider the approximation of a random variable $W$ whose support is contained in the closure of $(a, b)$, in this chapter for a function $f$ on $\mathbb{R}$ we take $\|f\|$ to be the supremum of $|f(w)|$ over $w \in (a, b)$.

**Lemma 13.1** *Let $p$ be a density function satisfying Condition* 13.1 *for some* $-\infty \le a < b \le \infty$ *and let*

$$F(y) = \int_a^y p(x)dx$$

*be the associated distribution function. Further, let $h$ be a measurable function and $f_h$ the Stein solution given by* (13.6).

(i) *Suppose there exist $d_1 > 0$ and $d_2 > 0$ such that for all $y \in (a, b)$ we have*

$$\min\big(1 - F(y), F(y)\big) \le d_1 p(y) \tag{13.10}$$

*and*

$$\big|p'(y)\big| \min\big(F(y), 1 - F(y)\big) \le d_2 p^2(y). \tag{13.11}$$

*Then if $h$ is bounded*

$$\|f_h\| \le 2d_1 \|h\|, \tag{13.12}$$

$$\big\|f_h p'/p\big\| \le 2d_2 \|h\| \tag{13.13}$$

*and*

$$\big\|f_h'\big\| \le (2 + 2d_2)\|h\|. \tag{13.14}$$

(ii) *Suppose in addition to* (13.10) *and* (13.11), *there exist $d_3 \ge 0$ such that*

$$\min\big(E|Y|\mathbf{1}_{\{Y \le y\}} + E|Y|F(y), E|Y|\mathbf{1}_{\{Y > y\}} + E|Y|\big(1 - F(y)\big)\big)\left|\left(\frac{p'}{p}\right)'\right| \le d_3 p(y) \tag{13.15}$$

*and $d_4(y)$ such that for all $y \in (a,b)$ we have*

$$\min\bigl(E|Y|\mathbf{1}_{\{Y\le y\}} + E|Y|F(y), E|Y|\mathbf{1}_{\{Y>y\}} + E|Y|\bigl(1-F(y)\bigr)\bigr) \le d_4(y)p(y). \tag{13.16}$$

*Then if $h$ is absolutely continuous with bounded derivative $h'$,*

$$\|f_h''\| \le (1+d_2)(1+d_3)\|h'\|, \tag{13.17}$$

$$|f_h(y)| \le d_4(y)\|h'\| \quad \text{for all } y \in (a,b), \tag{13.18}$$

*and*

$$\|f_h'\| \le (1+d_3)d_1\|h'\|. \tag{13.19}$$

The proof of the lemma is deferred to the Appendix.

## 13.2 $L^1$ and $L^\infty$ Bounds via Exchangeable Pairs

Let $W$ be a random variable of interest and $(W, W')$ an exchangeable pair. Write

$$E(W-W'|W) = g(W) + r(W), \tag{13.20}$$

where we consider $g(W)$ to be the dominant term and $r(W)$ some negligible remainder. When $g(W) = \lambda W$, and $\lambda^{-1}E((W'-W)^2|W)$ is nearly constant, the results in Sect. 5.2 show that the distribution of $W$ can be approximated by the normal, subject to some additional conditions. Here we use the function $g(w)$ to determine an appropriate approximating distribution for $W$, or, more particularly, identify its density function $p$. Once $p$ is determined, we can parallel the development of Stein's method of exchangeable pairs for normal approximation.

As a case in point, the proofs in this section depend on the following exchangeable pair identity, analogous to the one applied in the proof of Lemma 2.7 for the normal. That is, when (13.20) holds, for any absolutely continuous function $f$ for which the expectations below exist, recalling $\Delta = W - W'$, by exchangeability we have

$$\begin{aligned}
0 &= E(W-W')\bigl(f(W') + f(W)\bigr) \\
&= 2Ef(W)(W-W') + E(W-W')\bigl(f(W') - f(W)\bigr) \\
&= 2E\bigl\{f(W)E\bigl((W-W')|W\bigr)\bigr\} - E(W-W')\int_{-\Delta}^{0} f'(W+t)dt \\
&= 2Ef(W)g(W) + 2Ef(W)r(W) - E\int_{-\infty}^{\infty} f'(W+t)\hat{K}(t)dt
\end{aligned} \tag{13.21}$$

where

$$\hat{K}(t) = E\bigl\{\Delta\bigl(\mathbf{1}\{-\Delta \le t \le 0\} - \mathbf{1}\{0 < t \le -\Delta\}\bigr)|W\bigr\}. \tag{13.22}$$

Note that here, similar to (2.39), we have

$$\int_{-\infty}^{\infty} \hat{K}(t)dt = E(\Delta^2|W). \tag{13.23}$$

For a given function $g(y)$ defined on $(a,b)$ let $Y$ be a random variable with density function $p(y)=0$ for $y \notin (a,b)$, and for $y \in (a,b)$,

$$p(y) = c_1 e^{-c_0 G(y)} \quad \text{where } G(y) = \begin{cases} \int_0^y g(s)ds & \text{if } 0 \in [a,b), \\ \int_a^y g(s)ds & \text{if } a > 0, \\ \int_b^y g(s)ds & \text{if } b \le 0 \end{cases} \tag{13.24}$$

with $c_0 > 0$ and

$$c_1^{-1} = \int_a^b e^{-c_0 G(y)} dy < \infty. \tag{13.25}$$

Note that (13.24) implies

$$p'(y) = -c_0 g(y) p(y) \quad \text{for all } y \in (a,b). \tag{13.26}$$

Theorem 13.1 shows that for deriving $L^1$ bounds for approximations by distributions with densities $p$ of the form (13.24), it suffices that there exist a function $b_0(y)$, and constants $b_1$ and $b_2$, such that

$$\begin{aligned} &|f(y)| \le b_0(y) \quad \text{for all } y \in (a,b), \quad \text{and} \\ &\|f'\| \le b_1 \quad \text{and} \quad \|f''\| \le b_2 \end{aligned} \tag{13.27}$$

for all solutions $f$ to the Stein equation (13.5) for absolutely continuous functions $h$ with $\|h'\| \le 1$. For some cases, the following two conditions will help verify the hypotheses of Lemma 13.1 for densities of the form (13.24), thus implying bounds of the form (13.27).

**Condition 13.2** *On the interval* $(a,b)$ *the function* $g(y)$ *is non-decreasing and*

$$yg(y) \ge 0.$$

**Condition 13.3** *On the interval* $(a,b)$ *the function* $g$ *is absolutely continuous, and there exists* $c_2 < \infty$ *such that*

$$\min\left(\frac{1}{c_1}, \frac{1}{|c_0 g(y)|}\right)\left(|y| + \frac{3}{c_1}\right) \max\left(1, c_0|g'(y)|\right) \le c_2.$$

**Lemma 13.2** *Suppose that the density* $p$ *is given by* (13.24) *for some* $c_0 > 0$, *and* $g$ *satisfying Conditions* 13.2 *and* 13.3, *and* $E|g(Y)| < \infty$ *for* $Y$ *having density* $p$. *Then Condition* 13.1 *and all the bounds in Lemma* 13.1 *on the solution* $f$ *and its derivatives hold, with* $d_1 = 1/c_1$, $d_2 = 1$, $d_3 = c_2$ *and* $d_4(y) = c_2$ *for all* $y \in (a,b)$.

We refer the reader to the Appendix for a proof of Lemma 13.2. Equipped with bounds on the solution, we can now provide the following $L^1$ result.

**Theorem 13.1** *Let $(W, W')$ be an exchangeable pair satisfying* (13.20) *and set $\Delta = W - W'$. Let $Y$ have density $p$ of the form* (13.24), *on an interval $(a, b)$, with $c_0 > 0$, and $g$ in* (13.20) *satisfying $E|g(Y)| < \infty$. Suppose that the solution $f$ to the Stein equation* (13.5), *for all absolutely continuous functions $h$ with $\|h'\| \le 1$, satisfies* (13.27) *for some function $b_0(w)$ and constants $b_1$ and $b_2$. Then*

$$\begin{aligned}\|\mathcal{L}(W) - \mathcal{L}(Y)\|_1 &\le b_1 E\left|1 - \frac{c_0}{2}E(\Delta^2|W)\right| \\ &\quad + \frac{c_0 b_2}{4}E|\Delta|^3 + c_0 E\left|r(W)b_0(W)\mathbf{1}_{\{a<W<b\}}\right|. \quad (13.28)\end{aligned}$$

*If Conditions* 13.2 *and* 13.3 *are satisfied, then* (13.28) *holds with $b_0(w) = c_2, b_1 = (1 + c_2)/c_1$ and $b_2 = 2(1 + c_2)$.*

Note that the first term in the bound on the right hand side of (13.28) will be small when $E(\Delta^2|W)$ is nearly constant, that is, nearly equal to its expectation $E(\Delta^2)$, and $c_0$ is chosen close to $2/E(\Delta^2)$.

*Proof* Let $h$ be an absolutely continuous function satisfying $\|h'\| \le 1$ and let $f$ be the solution to the Stein equation (13.5) for $h$. By (13.5) and (13.26) we have

$$\begin{aligned}Eh(W) - Eh(Y) &= E\big(f'(W) + f(W)p'(W)/p(W)\big) \\ &= E\big(f'(W) - c_0 f(W)g(W)\big). \quad (13.29)\end{aligned}$$

As identity (13.21) holds for this $f$, adding and subtracting using (13.23) gives

$$\begin{aligned}&E\big(f'(W) - c_0 f(W)g(W)\big) \\ &= Ef'(W) - (c_0/2)\left\{E\int_{-\infty}^{\infty} f'(W+t)\hat{K}(t)dt - 2Ef(W)r(W)\right\} \\ &= E\big\{f'(W)\big(1 - (c_0/2)E\big(\Delta^2|W\big)\big)\big\} \\ &\quad + \frac{c_0}{2}E\int_{-\infty}^{\infty}\big(f'(W) - f'(W+t)\big)\hat{K}(t)dt + c_0 Ef(W)r(W). \quad (13.30)\end{aligned}$$

Now note that

$$E\int_{-\infty}^{\infty}\big|t\hat{K}(t)\big|dt = \frac{1}{2}E\big|\Delta^3\big|,$$

and that by (13.6) we have $f(a) = f(b) = 0$, so, by the first inequality in (13.27),

$$\big|f(w)\big| \le b_0(w)\mathbf{1}_{\{a<w<b\}}.$$

Now applying the bounds on the solution $f$ to the three terms in (13.30) yields the three terms of (13.28).

The final claim follows by Lemma 13.2, and (13.18), (13.19), and (13.17) of Lemma 13.1. □

When $\Delta$ is bounded, Theorem 13.2 gives a Berry–Esseen type inequality, parallel to Theorem 5.2 for the bounded exchangeable pair coupling with remainder in the normal case.

**Theorem 13.2** *Let $(W, W')$ be an exchangeable pair satisfying* (13.20), *and let $Y$ have density* (13.24) *for some $c_0 > 0$, and $g$ in* (13.20) *satisfying $E|g(Y)| < \infty$ and Conditions* 13.2 *and* 13.3. *If $\Delta = W - W'$ satisfies $|W - W'| \le \delta$ for some constant $\delta$ then*

$$\begin{aligned}&\sup_{z\in\mathbb{R}}\left|P(W \le z) - P(Y \le z)\right| \\ &\quad \le 3E\left|1 - \frac{c_0}{2}E\left(\Delta^2|W\right)\right| + c_1 \max\{1, c_2\}\delta + \frac{2c_0}{c_1}E\left|r(W)\right| \\ &\qquad + \delta^3 c_0\left\{\left(2 + \frac{c_2}{2}\right)E\left|c_0 g(W)\right| + \frac{c_1 c_2}{2}\right\}. \end{aligned} \tag{13.31}$$

*Proof of Theorem 13.2* Since (13.31) is trivial when $c_1c_2\delta > 1$, we assume

$$c_1 c_2 \delta \le 1. \tag{13.32}$$

Let $F$ be the distribution function of $Y$ and for $z \in \mathbb{R}$ let $f = f_z$ be the solution to the Stein equation

$$f'(w) + f(w)p'(w)/p(w) = \mathbf{1}(w \le z) - F(z)$$

or, by (13.26), equivalently, to

$$f'(w) - c_0 f(w)g(w) = \mathbf{1}(w \le z) - F(z). \tag{13.33}$$

By Lemma 13.2, the bound (13.12) of Lemma 13.1 holds, so $\|f\| < \infty$. Letting $\hat{K}(t)$ be given by (13.22), in view of identities (13.21) and (13.23), and that $|W - W'| \le \delta$ and $\hat{K}(t) \ge 0$, we obtain

$$\begin{aligned}&2Ef(W)g(W) + 2Ef(W)r(W) \\ &\quad = E\int_{-\infty}^{\infty} f'(W+t)\hat{K}(t)dt \\ &\quad = E\int_{-\delta}^{\delta}\left\{c_0 f(W+t)g(W+t) + \mathbf{1}(W+t \le z) - F(z)\right\}\hat{K}(t)dt \\ &\quad \ge E\int_{-\delta}^{\delta} c_0 f(W+t)g(W+t)\hat{K}(t)dt + E\mathbf{1}_{\{W\le z-\delta\}}\Delta^2 - F(z)E\Delta^2. \end{aligned}$$

Rewriting, and adding and subtracting using (13.23) again, we have

$$\begin{aligned}&E\mathbf{1}_{\{W\le z-\delta\}}\Delta^2 - F(z)E\Delta^2 \\ &\quad \le 2Ef(W)g(W) + 2Ef(W)r(W) - E\int_{-\delta}^{\delta} c_0 f(W+t)g(W+t)\hat{K}(t)dt \\ &\quad = 2Ef(W)g(W)\left(1 - (c_0/2)E\left(\Delta^2|W\right)\right) + 2Ef(W)r(W) \\ &\qquad + c_0 E\int_{-\delta}^{\delta}\left\{f(W)g(W) - f(W+t)g(W+t)\right\}\hat{K}(t)dt \\ &\quad := J_1 + J_2 + J_3. \end{aligned} \tag{13.34}$$

Lemma 13.2 and (13.12), (13.13), and (13.14) of Lemma 13.1 yield, along with (13.26), that

$$\|f\| \le 2/c_1, \quad \|fg\| \le 2/c_0 \quad \text{and} \quad \|f'\| \le 4. \tag{13.35}$$

Therefore

$$|J_1| \le (4/c_0)E\big|1 - (c_0/2)E\big(\Delta^2|W\big)\big| \tag{13.36}$$

and

$$|J_2| \le (4/c_1)E\big|r(W)\big|. \tag{13.37}$$

To bound $J_3$, we first show that

$$\sup_{|t|\le\delta}\big|g(w+t) - g(w)\big| \le \frac{c_1c_2\delta}{2c_0}\big(c_1 + c_0\big|g(w)\big|\big). \tag{13.38}$$

From Condition 13.3 it follows that

$$\begin{aligned}\big|g'(w)\big| &\le \frac{c_1c_2}{3c_0\min(1/c_1, 1/|c_0g(w)|)}\\ &= \frac{c_1c_2}{3c_0}\max\big(c_1, \big|c_0g(w)\big|\big)\\ &\le \frac{c_1c_2}{3c_0}\big(c_1 + c_0\big|g(w)\big|\big).\end{aligned} \tag{13.39}$$

Thus by the mean value theorem

$$\begin{aligned}\sup_{|t|\le\delta}\big|g(w+t) - g(w)\big| &\le \delta\sup_{|t|\le\delta}\big|g'(w+t)\big|\\ &\le \frac{c_1c_2\delta}{3c_0}\Big(c_1 + c_0\sup_{|t|\le\delta}\big|g(w+t)\big|\Big)\\ &\le \frac{c_1c_2\delta}{3c_0}\Big(c_1 + c_0\big|g(w)\big| + c_0\sup_{|t|\le\delta}\big|g(w+t) - g(w)\big|\Big)\\ &= \frac{c_1c_2\delta}{3c_0}\big(c_1 + c_0\big|g(w)\big|\big) + \frac{c_1c_2\delta}{3}\sup_{|t|\le\delta}\big|g(w+t) - g(w)\big|\\ &\le \frac{c_1c_2\delta}{3c_0}\big(c_1 + c_0\big|g(w)\big|\big) + \frac{1}{3}\sup_{|t|\le\delta}\big|g(w+t) - g(w)\big|,\end{aligned}$$

by (13.32). This proves (13.38).

Now, by (13.35) and (13.38), when $|t| \le \delta$,

$$\begin{aligned}&\big|f(w)g(w) - f(w+t)g(w+t)\big|\\ &\quad\le \big|g(w)\big|\big|f(w+t) - f(w)\big| + \big|f(w+t)\big|\big|g(w+t) - g(w)\big|\\ &\quad\le 4\big|g(w)\big||t| + \frac{2}{c_1}\frac{c_1c_2\delta}{2c_0}\big(c_1 + c_0\big|g(w)\big|\big)\\ &\quad\le (4 + c_2)\delta\big|g(w)\big| + \delta c_1c_2/c_0.\end{aligned}$$

Therefore

$$\begin{aligned}|J_3| &\le c_0(4+c_2)\delta E\big(|g(W)|\Delta^2\big) + \delta c_1 c_2 E\Delta^2 \\ &\le (4+c_2)\delta^3 E|c_0 g(W)| + c_1 c_2 \delta^3. \end{aligned} \tag{13.40}$$

Combining (13.34), (13.36), (13.37) and (13.40) shows that

$$\begin{aligned}E\mathbf{1}_{\{W\le z-\delta\}}\Delta^2 - F(z)E\Delta^2 &\le \frac{4}{c_0}E\left|1-\frac{c_0}{2}E(\Delta^2|W)\right| + \frac{4}{c_1}E|r(W)| \\ &\quad + (4+c_2)\delta^3 E|c_0 g(W)| + c_1 c_2 \delta^3. \end{aligned} \tag{13.41}$$

On the other hand, using $F'(z) = p(z) \le c_1$ by (13.24), we have

$$\begin{aligned}&E\mathbf{1}_{\{W\le z-\delta\}}\Delta^2 - F(z)E\Delta^2 \\ &\quad = \frac{2}{c_0}\big(E\mathbf{1}_{\{W\le z-\delta\}} - F(z-\delta)\big) \\ &\qquad - \frac{2}{c_0}E\left\{\big(\mathbf{1}_{\{W\le z-\delta\}} - F(z)\big)\left(1-\frac{c_0}{2}E(\Delta^2|W)\right)\right\} \\ &\qquad + \frac{2}{c_0}\big(F(z-\delta)-F(z)\big) \\ &\quad \ge \frac{2}{c_0}\big(P(W\le z-\delta) - F(z-\delta)\big) \\ &\qquad - \frac{2}{c_0}E\left|1-\frac{c_0}{2}E(\Delta^2|W)\right| - \frac{2c_1\delta}{c_0}, \end{aligned} \tag{13.42}$$

which together with (13.41) yields

$$\begin{aligned}&P(W\le z-\delta) - F(z-\delta) \\ &\quad \le E\left|1-\frac{c_0}{2}E(\Delta^2|W)\right| + c_1\delta \\ &\qquad + \frac{c_0}{2}\left((4/c_0)E\left|1-\frac{c_0}{2}E(\Delta^2|W) + (4/c_1)E|r(W)|\right|\right. \\ &\qquad \left. + (4+c_2)\delta^3 E|c_0 g(W)| + c_1 c_2\delta^3\right) \\ &\quad = 3E\left|1-\frac{c_0}{2}E(\Delta^2|W)\right| + c_1\delta + 2c_0 E|r(W)|/c_1 \\ &\qquad + \delta^3 c_0\big\{(2+c_2/2)E|c_0 g(W)| + c_1 c_2/2\big\}. \end{aligned} \tag{13.43}$$

Similarly, one can demonstrate

$$\begin{aligned}&F(z+\delta) - P(W\le z+\delta) \\ &\quad \le 3E\left|1-\frac{c_0}{2}E(\Delta^2|W)\right| + c_1\delta + 2c_0 E|r(W)|/c_1 \\ &\qquad + \delta^3 c_0\big\{(2+c_2/2)E|c_0 g(W)| + c_1 c_2/2\big\}. \end{aligned} \tag{13.44}$$

As $c_1\delta \le c_1\max(1,c_2)\delta$, the proof of (13.31) is complete. □

## 13.3 The Curie–Weiss Model

The Curie–Weiss model, or the Ising model on the complete graph, was introduced in Sect. 11.2 and is a simple statistical mechanical model of ferromagnetic interaction. We recall that for $n \in \mathbb{N}$ the vector $\boldsymbol{\sigma} = (\sigma_1, \ldots, \sigma_n)$ of 'spins' in $\{-1, 1\}^n$ are assigned probability

$$p(\boldsymbol{\sigma}) = C_\beta \exp\left(\frac{\beta}{n} \sum_{i<j} \sigma_i \sigma_j\right) \tag{13.45}$$

for a given 'inverse temperature' $\beta > 0$, with $C_\beta$ the appropriate normalizing constant.

For a detailed mathematical treatment of the Curie–Weiss model in general we refer to the book by Ellis (1985). When $0 < \beta < 1$ the total spin $\sum_{i=1}^n \sigma_i$, properly standardized, converges to a standard normal distribution as $n \to \infty$, see, e.g., Ellis and Newman (1978a, 1978b). For $\beta = 1$ it was proved by Ellis and Newman (1978a, 1978b) that as $n \to \infty$, the law of

$$W = n^{-3/4} \sum_{i=1}^n \sigma_i \tag{13.46}$$

converges to the distribution $\mathcal{W}(4, 12)$ of Example 13.3. For various interesting extensions and refinements of their results, see Ellis et al. (1980), and Papangelou (1989).

Here we present the following $L^1$ and Berry–Esseen bounds for the critical $\beta = 1$ non-central limit theorem, obtained via Theorems 13.1 and 13.2, respectively.

**Theorem 13.3** *Let $W$ be the scaled total spin* (13.46) *in the Curie–Weiss model, where the vector $\boldsymbol{\sigma}$ of spins has distribution* (13.45) *at the critical inverse temperature $\beta = 1$, and let $Y$ be a random variable with distribution $\mathcal{W}(4, 12)$ as in* (13.9). *Then there exists a universal constant $C$ such that for all $n \in \mathbb{N}$,*

$$\left\| \mathcal{L}(W) - \mathcal{L}(Y) \right\|_1 \le C n^{-1/2}. \tag{13.47}$$

*and*

$$\sup_{z \in \mathbb{R}} \left| P(W \le z) - P(Y \le z) \right| \le C n^{-1/2}. \tag{13.48}$$

The required exchangeable pair is constructed following Example 2.2. Given $\boldsymbol{\sigma}$ having distribution (13.45), construct $\boldsymbol{\sigma}'$ by choosing $I$ uniformly and independently of $\boldsymbol{\sigma}$, and replacing $\sigma_I$ by $\sigma_I'$, where $\sigma_I'$ is generated from the conditional distribution of $\sigma_I$ given $\{\sigma_j, j \ne I\}$. It is easy to see that $(\boldsymbol{\sigma}, \boldsymbol{\sigma}')$ is an exchangeable pair. Let $W' = W - \sigma_I + \sigma_I'$, the total spin of the configuration when $\sigma_I$ is replaced by $\sigma_I'$. Considering the sequence of distributions (13.45) indexed by $n \in \mathbb{N}$, the key step is to show (13.20), or, more specifically for the case at hand, that

$$E(W - W' | W) = \frac{1}{3} n^{-3/2} W^3 + O\left(n^{-2}\right) \quad \text{as } n \to \infty. \tag{13.49}$$

To explain (13.49), roughly, a simple computation shows that at any inverse temperature,

$$E(W - W'|W) = n^{-3/4}\big(M - \tanh(\beta M)\big) + O\big(n^{-2}\big) \quad \text{as } n \to \infty,$$

where $M = n^{-1/4}W$ is known as the magnetization. Since $M \simeq 0$ with high probability when $\beta \le 1$, and a Taylor expansion about zero yields $\tanh x = x - x^3/3 + O(x^5)$, we see that $M - \tanh(\beta M)$ behaves like $n^{-3/4}M(1-\beta)$ when $\beta < 1$, and like $n^{-3/4}M^3/3$ when $\beta = 1$. This is what distinguishes the high temperature regime $\beta < 1$ from the critical case $\beta = 1$, and how we arrive at (13.49).

Comparing (13.49) with (13.20), we find that if on $(-\infty, \infty)$ we take

$$g(y) = \frac{1}{3}n^{-3/2}y^3 \quad \text{and} \quad c_0 = n^{3/2}, \tag{13.50}$$

then, following (13.24), the density function

$$p(y) = c_1 \exp\left(-c_0 \int_0^y g(s)ds\right) = c_1 e^{-y^4/12}$$

results, that is, the distribution $\mathcal{W}(4, 12)$ of the family considered in Example 13.3. We note that though $c_0$ depends on $n$, the constant $c_1$ given in (13.9) does not.

We now make (13.49) precise, as well as verify the remainder of the hypotheses required in order to invoke Theorem 13.2.

**Lemma 13.3** *Let $W$ be the scaled total spin* (13.46) *in the Curie–Weiss model, where the spins $\boldsymbol{\sigma}$ have distribution given by* (13.45) *at the critical inverse temperature $\beta = 1$, and let $W'$ be the given by* (13.46) *when a uniformly chosen spin from $\boldsymbol{\sigma}$ has been replaced by one having its conditional distribution given the others. Then for all $n \in \mathbb{N}$,*

$$|W - W'| \le 2n^{-3/4}, \tag{13.51}$$

$$E\left|E(W - W'|W) - \frac{1}{3}n^{-3/2}W^3\right| \le 15n^{-2}, \tag{13.52}$$

$$E\left|1 - \frac{n^{3/2}}{2}E\big((W - W')^2|W\big)\right| \le \frac{15}{2}n^{-1/2} \tag{13.53}$$

*and*

$$E|W|^3 \le 15. \tag{13.54}$$

*Proof* As $W$ and $W'$ differ in at most one coordinate, (13.51) is immediate. Next, let $M = n^{-1}\sum_{i=1}^n \sigma_i = n^{-1/4}W$ be the magnetization, and for each $i$, let

$$M_i = n^{-1}\sum_{j \ne i} \sigma_j.$$

It is easy to see that for every $i = 1, 2, \ldots, n$, if one chooses a variables $\sigma_i'$ from the conditional distribution of the $i$th spin given $\sigma_j$, $j \neq i$, independently of $\sigma_i$, then for $\tau \in \{-1, 1\}$,

$$P\left(\sigma_i' = \tau | \sigma_j, j \neq i\right) = P\left(\sigma_i' = \tau | \boldsymbol{\sigma}\right) = \frac{e^{M_i \tau}}{e^{M_i} + e^{-M_i}}, \tag{13.55}$$

and so

$$E\left(\sigma_i' | \boldsymbol{\sigma}\right) = \frac{e^{M_i}}{e^{M_i} + e^{-M_i}} - \frac{e^{-M_i}}{e^{M_i} + e^{-M_i}} = \tanh M_i.$$

Hence

$$\begin{aligned} E(W - W' | \boldsymbol{\sigma}) &= n^{-1} \sum_{i=1}^{n} n^{-3/4} \left(\sigma_i - E\left(\sigma_i' | \boldsymbol{\sigma}\right)\right) \\ &= n^{-3/4} M - n^{-7/4} \sum_{i=1}^{n} \tanh M_i. \end{aligned} \tag{13.56}$$

Now it is easy to verify that the second derivative

$$\frac{d^2}{dx^2} \tanh x = -2(\tanh x)\left(1 - \tanh^2 x\right)$$

has exactly two extrema on the real line, the solutions $x$ to the equation $\tanh^2 x = 1/3$, and is bounded in absolute value by $4/3^{3/2}$. Thus, for all $x, y \in \mathbb{R}$,

$$\left|\tanh x - \tanh y - (x - y)(\cosh y)^{-2}\right| \leq \frac{2(x - y)^2}{3^{3/2}}.$$

It follows that for all $i = 1, \ldots, n$,

$$\left|\sum_{i=1}^{n} \tanh M_i - n \tanh M + n^{-1} (\cosh M)^{-2} \sum_{i=1}^{n} \sigma_i\right| \leq \frac{2n^{-1}}{3^{3/2}},$$

and therefore

$$\left|\sum_{i=1}^{n} \tanh M_i - n \tanh M\right| \leq |M| + \frac{2n^{-1}}{3^{3/2}}.$$

Applying this inequality and the relation $M = n^{-1/4} W$ in (13.56), we obtain

$$\left|E\left(W - W' | \boldsymbol{\sigma}\right) + n^{-3/4} (\tanh M - M)\right| \leq n^{-2} |W| + \frac{2n^{-11/4}}{3^{3/2}}. \tag{13.57}$$

Now consider the function $f(x) = \tanh x - x + x^3/3$. Since $f'(x) = (\cosh x)^{-2} - 1 + x^2 \geq 0$ for all $x$, the function $f$ is increasing, and as $f(0) = 0$ we obtain $f(x) \geq 0$ for all $x \geq 0$. Now, it can be easily verified that the first four derivatives of $f$ vanish at zero, and for all $x \geq 0$,

$$\frac{d^5 f}{dx^5} = \frac{16}{\cosh^2 x} - 120 \frac{\sinh^2 x}{\cosh^4 x} + 120 \frac{\sinh^4 x}{\cosh^6 x} \leq \frac{16}{\cosh^2 x} \leq 16.$$

Thus, for all $x \geq 0$,

$$0 \leq f(x) \leq \frac{16}{5!}x^5 = \frac{2}{15}x^5.$$

Since $f$ is an odd function, we obtain

$$\left|\tanh x - x + \frac{1}{3}x^3\right| \leq \frac{2|x|^5}{15} \quad \text{for all } x \in \mathbb{R}.$$

Using this inequality in (13.57), we arrive at

$$\left|E(W - W'|\sigma) - \frac{1}{3}n^{-3/4}M^3\right| \leq \frac{2n^{-3/4}|M|^5}{15} + n^{-2}|W| + \frac{2n^{-11/4}}{3^{3/2}},$$

or, by the relation $M = n^{-1/4}W$, equivalently,

$$\left|E(W - W'|\sigma) - \frac{1}{3}n^{-3/2}W^3\right| \leq \frac{2n^{-2}|W|^5}{15} + n^{-2}|W| + \frac{2n^{-11/4}}{3^{3/2}}. \tag{13.58}$$

The latter inequality implies, in particular, that

$$\begin{aligned}&\left|E\big((W - W')W^3\big) - \frac{1}{3}n^{-3/2}E\big(W^6\big)\right| \\ &\quad \leq \frac{2n^{-2}E(W^8)}{15} + n^{-2}E\big(W^4\big) + \frac{2n^{-11/4}E|W|^3}{3^{3/2}}. \end{aligned} \tag{13.59}$$

Thus,

$$\begin{aligned}E\big(W^6\big) \leq{}& 3n^{3/2}\big|E\big((W - W')W^3\big)\big| + \frac{2n^{-1/2}E(W^8)}{5} \\ &+ 3n^{-1/2}E\big(W^4\big) + \frac{2n^{-5/4}E|W|^3}{3^{1/2}}. \end{aligned} \tag{13.60}$$

Regarding the first term on the right hand side of (13.60), note that by the exchangeability of $(W, W')$,

$$\begin{aligned}E\big((W - W')W^3\big) &= \frac{1}{2}E\big((W - W')\big(W^3 - W'^3\big)\big) \\ &= -\frac{1}{2}E\big((W - W')^2\big(W'^2 + W'W + W^2\big)\big).\end{aligned}$$

Now, by (13.51) and the Cauchy–Schwarz inequality,

$$\big|E\big((W - W')W^3\big)\big| \leq 6n^{-3/2}E\big(W^2\big). \tag{13.61}$$

For the remaining terms in (13.60), using the crude bound $|W| \leq n^{1/4}$, we obtain

$$\begin{aligned}&\frac{2n^{-1/2}E(W^8)}{5} + 3n^{-1/2}E\big(W^4\big) + \frac{2n^{-5/4}E|W|^3}{3^{1/2}} \\ &\quad \leq \frac{2E(W^6)}{5} + 3E\big(W^2\big) + \frac{2n^{-1}E(W^2)}{3^{1/2}}. \end{aligned} \tag{13.62}$$

Combining (13.60), (13.61), and (13.62), we obtain

$$E(W^6) \le \left(21 + \frac{2n^{-1}}{3^{1/2}}\right) E(W^2) + \frac{2E(W^6)}{5},$$

and therefore, for all $n \in \mathbb{N}$,

$$E(W^6) \le \frac{5}{3}\left(21 + \frac{2n^{-1}}{3^{1/2}}\right) E(W^2) \le 36.925 E(W^2).$$

Since $E(W^2) \le [E(W^6)]^{1/3}$, this gives

$$E(W^6) \le (36.925)^{3/2} \le 224.4 < (15)^2, \tag{13.63}$$

and hence (13.54) holds. Applying the bound (13.63) in (13.58) yields, for all $n \in \mathbb{N}$,

$$\begin{aligned} &E\left|E(W - W'|W) - \frac{1}{3} n^{-3/2} W^3\right| \\ &\le n^{-2}\left(\frac{2(224.4)^{5/6}}{15} + (224.4)^{1/6}\right) + \frac{2n^{-11/4}}{3^{3/2}} \le 15n^{-2}, \end{aligned} \tag{13.64}$$

completing the proof of (13.52).

Lastly, to prove (13.53), by (13.55) we have

$$\begin{aligned} E\big((W - W')^2|\sigma\big) &= n^{-3/2}\frac{1}{n}\sum_{i=1}^{n} 4\frac{e^{-\sigma_i M_i}}{e^{\sigma_i M_i} + e^{-\sigma_i M_i}} \\ &= 2n^{-5/2}\sum_{i=1}^{n}\big(1 - \tanh(\sigma_i M_i)\big) \\ &= 2n^{-3/2} - 2n^{-5/2}\sum_{i=1}^{n} \sigma_i \tanh M_i. \end{aligned}$$

Using $|\tanh M_i - \tanh M| \le |M_i - M| \le n^{-1}$, we obtain

$$\begin{aligned} &\big|E\big((W - W')^2|\sigma\big) - 2n^{-3/2}\big| \\ &\le 2n^{-5/2} + 2n^{-3/2} M \tanh M \\ &\le 2n^{-5/2} + 2n^{-3/2} M^2 \\ &= 2n^{-5/2} + 2n^{-2} W^2. \end{aligned}$$

Using (13.63), we obtain, for all $n \in \mathbb{N}$, that

$$E\big|E\big((W - W')^2|W\big) - 2n^{-3/2}\big| \le 2n^{-5/2} + 2n^{-2}(224.4)^{1/3} \le 15n^{-2}.$$

Now multiplying by $n^{3/2}/2$ completes the proof of (13.53), and the lemma. □

*Proof of Theorem 13.3* We apply Theorems 13.1 and 13.2 with the coupling given in Lemma 13.3. First, inequality (13.52) of Lemma 13.3 shows that the exchangeable pair $(W, W')$ satisfies (13.20) with

$$g(y) = \frac{1}{3} n^{-3/2} y^3 \quad \text{and} \quad \big|r(y)\big| \le 15n^{-2}. \tag{13.65}$$

Recall that the density $p(y)$ of $Y$ on $(-\infty, \infty)$ is given by (13.9), or, equivalently, by (13.24) with $g(y)$ and $c_0 = n^{3/2}$ as in (13.50), and that $c_1$ is a constant not depending on $n$. It is clear that $Y$ has moments of all order, so in particular $E|g(Y)| < \infty$, and that $g(y)$ satisfies Condition 13.2.

As the quantity

$$\min\bigl(1/c_1, 3/\bigl|y^3\bigr|\bigr)(|y| + 3/c_1)\bigl(1 + y^2\bigr)$$

is bounded near zero and has finite limits at plus and minus infinity, Condition 13.3 is satisfied with a constant $c_2$ not depending on $n$.

Regarding the $L^1$ bound (13.47), we have that the hypotheses of Theorem 13.1 are satisfied, and need only verify that all terms in the bound (13.28) of that theorem are of order $O(n^{-1/2})$. Inequality (13.53) shows that the first term in the bound is no more than $(15/2)n^{-1/2}$, inequality (13.51) shows the second term is $O(n^{-9/4})$, and (13.65) shows the last term to be of order $O(n^{-2})$.

Regarding the supremum norm bound (13.48), Lemma 13.3 shows the coupling of $W$ and $W'$ is bounded, and therefore the hypotheses of Theorem 13.2 are satisfied; similarly, we need only verify that all terms in the bound (13.31) are $O(n^{-1/2})$. We have already shown the first term in the bound is of this order. Inequality (13.51) allows us to choose $\delta = 2n^{-3/4}$, showing the second term in the bound is of order $o(n^{-1/2})$. By (13.65) the third term is $O(n^{3/2}n^{-2}) = O(n^{-1/2})$. For the coefficient of the last term, we find $\delta^3 c_0 = O(n^{-9/4}n^{3/2}) = O(n^{-3/4})$. As $c_1$ and $c_2$ do not depend on $n$, and $E|c_0 g(W)| = E|W|^3/3 \le 5$ by (13.54), the final term is $o(n^{-1/2})$. As all terms in the bound are $O(n^{-1/2})$ the claim is shown. □

## 13.4 Exponential Approximation

In this section we focus on approximation by the exponential distribution $\mathcal{E}(\lambda)$ for $\lambda > 0$, that is, the distribution with density $p(x) = \lambda e^{-\lambda x}\mathbf{1}_{\{x>0\}}$ as in Example 13.1. We consider two examples, the spectrum of the Bernoulli–Laplace diffusion model, and first passage time of Markov chains. The first example is handled using exchangeable pairs, and the second by introducing the equilibrium distribution.

### *13.4.1 Spectrum of the Bernoulli–Laplace Markov Chain*

Since for the exponential distribution the ratio $p'(w)/p(w)$ is constant, following (13.20) we hope to construct an exchangeable pair $(W, W')$ satisfying

$$E(W - W'|W) = 1/c_0 + r(W) \tag{13.66}$$

for some positive constant $c_0 > 0$ and small remainder term.

Taking $(a, b) = (0, \infty)$, $g(y) = 1/c_0$ and $G(y) = y/c_0$ in (13.24) yields $p(y) = e^{-y}\mathbf{1}_{\{y>0\}}$, and so $Y \sim \mathcal{E}(1)$, the unit exponential distribution; clearly $c_1 = 1$ and $E|g(Y)| < \infty$. We now obtain bounds on the solution to the Stein equation for the unit exponential.

**Lemma 13.4** *If $f$ is the solution to the Stein equation* (13.5) *with $p(y) = e^{-y}\mathbf{1}(y > 0)$, the unit exponential density, and $h$ any absolutely continuous function with $\|h'\| \le 1$, then the bounds* (13.27) *hold with $b_0(y) = 3.5y$ for $y > 0$, $b_1 = 1$ and $b_2 = 2$.*

*Proof* We verify the hypotheses of Lemma 13.1 are satisfied for $p(y) = e^{-y}$ and $F(y) = 1 - e^{-y}$ for $y > 0$. Clearly (13.10) and (13.10) are satisfied with $d_1 = 1$ and $d_2 = 1$, respectively. As $(p'/p)' = 0$, (13.15) is satisfied with $d_3 = 0$.

Regarding (13.16), we have

$$\begin{aligned} EY\mathbf{1}(Y \le y) + EYP(Y \le y) &= 2\bigl(1 - e^{-y}\bigr) - ye^{-y} \\ &\le 2\bigl(1 - e^{-y}\bigr), \end{aligned}$$

and similarly,

$$EY\mathbf{1}(Y > y) + EYP(Y > y) \le (1 + 2y)e^{-y}.$$

Hence, (13.16) is satisfied with $d_4(y) = 3.5y$ as

$$e^{y}\min\bigl(2\bigl(1 - e^{-y}\bigr), (1 + 2y)e^{-y}\bigr) = \min\bigl(2\bigl(e^{y} - 1\bigr), (1 + 2y)\bigr) \le 3.5y,$$

where the final inequality is shown by considering the cases $0 < y < 2/3$ and $y \ge 2/3$. Invoking Lemma 13.1 with $d_1 = 1, d_2 = 1, d_3 = 0$ and $d_4(y) = 3.5y$ now completes the proof of the claim. □

Theorem 13.1 now immediately yields the following result for approximation by the exponential distribution.

**Theorem 13.4** *Let $(W, W')$ be an exchangeable pair satisfying* (13.66) *for some $c_0 > 0$, let $\Delta = W - W'$, and $Y \sim \mathcal{E}(1)$. Then*

$$\begin{aligned} \bigl\|\mathcal{L}(W) - \mathcal{L}(Y)\bigr\|_1 &\le E\left|1 - \frac{c_0}{2}E\bigl(\Delta^2|W\bigr)\right| + \frac{c_0}{2}E|\Delta|^3 \\ &\quad + 3.5c_0E\bigl|Wr(W)\mathbf{1}_{\{|W|>0\}}\bigr|. \end{aligned}$$

We apply the result above to the Bernoulli–Laplace Markov chain, a simple model of diffusion, following the work of Chatterjee et al. (2008). Two urns contain $n$ balls each. Initially the balls in each urn are all of a single color, with urn 1 containing all white balls, and urn 2 all black. At each stage two balls are picked at random, one from each urn, and interchanged. Let the state of the chain be the number of white balls in the urn 1. Diaconis and Shahshahani (1987) proved that $(n/4)\log(2n) + cn$ steps suffice for this process to reach equilibrium, in the sense that the total variation distance to stationarity is at most $ae^{-dc}$, for some positive universal constants $a$ and $d$. To prove this result they used the fact that the spectrum of the chain consists of the numbers

$$\lambda_i = 1 - \frac{i(2n - i + 1)}{n^2} \quad \text{for } i = 0, \dots, n \tag{13.67}$$

occurring with multiplicities

$$m_i = \binom{2n}{i} - \binom{2n}{i-1} \quad \text{for } i = 0, 1, \dots, n,$$

where we adopt the convention that $\binom{n}{k} = 0$ when $k < 0$, so that the multiplicity $m_0$ of the eigenvalue $\lambda_0 = 1$ is 1.

The Bernoulli–Laplace Markov chain is equivalent to a function of a certain random walk on the Johnson graph $J(2n, n)$. The vertices of the Johnson graph $J(2n, n)$ are all size $n$ subsets of $\{1, 2, \dots, 2n\}$, and two subsets are connected by an edge if they differ by exactly one element. From a given vertex, the random walk moves to a neighbor chosen uniformly at random. Numbering the balls of the Bernoulli–Laplace model 1 through $2n$, with the white balls corresponding to the odd numbers $1, 3, \dots, 2n-1$ and the black balls to the even values $2, 4, \dots, 2n$, the state of the random walk on the Johnson graph is simply the labels of the balls in urn 1.

We apply Stein's method to study an approximation to the distribution of a randomly chosen eigenvalue, that is, to the values $\lambda_i$ given in (13.67), chosen in proportion to their multiplicities $m_i, i = 0, \dots, n$. As the sum $\sum_{i=0}^{n} m_i$ telescopes, this means we choose $\lambda_i$ with probability

$$\pi_i = \frac{\binom{2n}{i} - \binom{2n}{i-1}}{\binom{2n}{n}}, \quad i = 0, 1, \dots, n. \tag{13.68}$$

Letting $I$ have distribution $P(I = i) = \pi_i$, translating and scaling the distribution which chooses $\lambda_i$ with probability $\pi_i$ to be comparable to the unit exponential, we are led to study the random variable

$$W = \mu_I \quad \text{where } \mu_i = \frac{(n-i)(n+1-i)}{n}. \tag{13.69}$$

We construct an exchangeable pair $(W, W')$ using a reversible Markov chain on $\{0, 1, \dots, n\}$, as outlined at the end of Sect. 2.3.2. For the chain to be reversible with respect to $\pi$, its transition kernel $K$ must satisfy the detail balance equation

$$\pi_i K(i, j) = \pi_j K(j, i) \quad \text{for all } i, j \in \{0, \dots, n\}. \tag{13.70}$$

Given such a $K$, one obtains the pair $(W, W')$ by letting $W = \mu_I$ where $I$ is chosen from the equilibrium distribution $\pi$, and $W' = \mu_J$ where $J$ is determined by taking one step from state $I$ according to the transition mechanism $K$.

One can verify that the following transition kernel $K$ satisfies (13.70). For the upward transitions,

$$K(i, i+1) = \frac{2n - i + 1}{4n(n-i)(2n-2i+1)} \quad \text{for } i = 0, \dots, n-1,$$

for the downward transitions,

$$K(i, i-1) = \frac{i}{4n(2n-2i+1)(n-i+1)} \quad \text{for } i = 0, \dots, n,$$

for the probability of returning to the same state,

$$K(i,i) = 1 - K(i,i+1) - K(i,i-1),$$

and $K(i,j) = 0$ otherwise. The following lemma summarizes some of the properties of the exchangeable pair so constructed.

**Lemma 13.5** *Let $W = \mu_I$ as in* (13.69) *with $P(I = i) = \pi_i$, specified in* (13.68), *and let $W' = \mu_J$ where $J$ is obtained from $I$ by taking one step from $I$ according to the transition kernel $K$. Then $(W, W')$ is exchangeable, and, with $\Delta = W - W'$,*

$$E(\Delta|W) = \frac{1}{2n^2} - \frac{n+1}{2n^2}\mathbf{1}_{\{W=0\}}, \tag{13.71}$$

$$E(W) = 1, \qquad E\left(\Delta^2|W\right) = \frac{1}{n^2} \tag{13.72}$$

*and*

$$E|\Delta|^3 \le 4n^{-5/2}. \tag{13.73}$$

*Proof* Since $K$ is reversible with respect to $\pi$, the pair $(I, J)$, and therefore the pair $(\mu_I, \mu_J) = (W, W')$, is exchangeable.

The mapping $i \to \mu_i$ given by (13.69) is strictly decreasing, and is therefore invertible, for $i \in \{0, 1, \dots, n\}$. Hence, conditioning on $W$ is equivalent to conditioning on $I$. For $i \in \{0, 1, \dots, n-1\}$ we have

$$\begin{aligned} E(\Delta|I=i) &= K(i,i+1)(\mu_i - \mu_{i+1}) + K(i,i-1)(\mu_i - \mu_{i-1}) \\ &= \frac{2n-i+1}{4n(n-i)(2n-2i+1)} \frac{(2n-2i)}{n} \\ &\quad + \frac{i}{4n(2n-2i+1)(n-i+1)} \frac{(2i-2n-2)}{n} \\ &= \frac{1}{2n^2}, \end{aligned}$$

and for $i = n$,

$$E(\Delta|I=n) = K(n,n-1)(\mu_n - \mu_{n-1}) = -\frac{1}{2n}.$$

As $\{I = n\} = \{W = 0\}$, the claim (13.71) is shown.

To show that $E(W) = 1$, argue as in the proof of (13.71) to compute that

$$E\left(\Delta^3|W\right) = \frac{2}{n^3}(W-1). \tag{13.74}$$

Since $W$ and $W'$ are exchangeable, $E(\Delta^3) = 0$, and taking expectation in (13.74) yields $E(W) = 1$. Similarly one checks that $E(\Delta^2|W) = 1/n^2$, proving (13.72).

Lastly, to show (13.73), since $\mu_i$ is decreasing in $i$, for $i \in \{0, 1, \dots, n-1\}$ we have

$$
\begin{aligned}
E\big(|\Delta|^3|I=i\big) &= K(i,i+1)(\mu_i-\mu_{i+1})^3 + K(i,i-1)(\mu_{i-1}-\mu_i)^3 \\
&= \frac{2n-i+1}{4n(n-i)(2n-2i+1)}\frac{(2n-2i)^3}{n^3} \\
&\quad + \frac{i}{4n(2n-2i+1)(n-i+1)}\frac{(2n+2-2i)^3}{n^3} \\
&= \frac{2((2n-i+1)(n-i)^2+i(n-i+1)^2)}{n^4(2n-2i+1)} \\
&\le \frac{2((2n-i+1)(n-i)(n-i+1)+i(n-i+1)^2)}{n^4(2n-2i+1)} \\
&= \frac{2(n-i+1)}{n^3},
\end{aligned}
$$

and when $i=n$,

$$
E\big(|\Delta|^3|I=n\big) = K(n,n-1)(\mu_n-\mu_{n-1})^3 = \frac{2}{n^3} = \frac{2(n-i+1)}{n^3}.
$$

Thus,

$$
E\big(|\Delta|^3|W\big) \le \frac{2(n-I+1)}{n^3} \le \frac{2\sqrt{n(W+2)}}{n^3} \le 2n^{-5/2}\sqrt{W+2},
$$

and, by Jensen's inequality,

$$
E|\Delta|^3 \le 2n^{-5/2}\sqrt{E(W+2)} = 2\sqrt{3}n^{-5/2} \le 4n^{-5/2}.
$$

This proves (13.73). □

Applying Theorem 13.4 with $c_0 = 2n^2$ and $r(w) = -(n+1)/(2n^2)\mathbf{1}(w>0)$ now yields the following result.

**Theorem 13.5** *Let $W$ be a scaled, translated randomly chosen eigenvalue of the Bernoulli–Laplace diffusion model given by* (13.69). *Then, with $Y$ having the unit exponential distribution*

$$
\big\|\mathcal{L}(W)-\mathcal{L}(Y)\big\|_1 \le 4n^{-1/2}. \qquad (13.75)
$$

As the difference between $W$ and $W'$ is large when $I$ is close to zero, Theorem 13.2 does not provide a useful bound for the Kolmogorov distance between $W$ and $Y$. However, using a completely different approach and some heavy machinery, Chatterjee et al. (2008) are able to show that

$$
\sup_{z\in\mathbb{R}}\big|P(W\le z)-P(Y\le z)\big| \le Cn^{-1/2} \qquad (13.76)
$$

where $C$ is a universal constant.

### 13.4.2 First Passage Times

We now consider an approach to exponential approximation different from that of the previous sections, following Peköz and Röllin (2009). In the nomenclature of renewal theory, for a non-negative random variable $X$ with finite mean, $X^e$ is said to have the equilibrium distribution with respect to $X$ if for all Lipschitz functions $f$,

$$Ef(X) - f(0) = EXEf'(X^e). \tag{13.77}$$

Clearly (13.77) holds for all Lipschitz functions if and only if it holds for all Lipschitz functions $f$ with $f(0) = 0$.

The equilibrium distribution has a close connection with both the size biased and zero biased distributions. For the first connection, let $X^s$ have the $X$-size bias distribution of $X$, that is,

$$EXf(X) = EXEf(X^s)$$

for all functions $f$ for which these expectations exist. Then, if $U$ has the uniform distribution on [0, 1] independent of $X$, the variable $UX^s$ has the equilibrium distribution $X^e$ with respect to $X$. Indeed, if $f$ is any Lipschitz function then

$$Ef(X) - f(0) = EXf'(UX) = EXEf'(UX^s) = EXEf'(X^e).$$

For the second connection, recall how a random variable with the zero bias distribution can likewise be formed by multiplying a square bias variable by an independent uniform. Note also that for any $a > 0$, parallel to (2.59), we have

$$(aX)^e = U(aX)^s = aUX^s = aX^e. \tag{13.78}$$

Additionally, a simple calculation shows that in general $X^e$ is absolutely continuous with density function $\int_0^x P(X > t)dt/EX$.

As identity (13.7) characterizes the exponential distribution we see that if $X =_d X^e$ then $X \sim \mathcal{E}(\lambda)$, where $\lambda = EX$. Hence the equilibrium distribution of $X$ operates for the exponential distributions in the same way that the zero bias distributions do for the mean zero normals. By analogy then, if the distributions of $X$ and $X^e$ are close then $X$ should be approximately exponential. This intuition is made precise by the following result of Peköz and Röllin (2009).

**Theorem 13.6** *Let $W$ be a non-negative random variable with $EW = 1$ and let $W^e$ have the equilibrium distribution with respect to $W$. Then, for $Y \sim \mathcal{E}(1)$ and any $\beta > 0$,*

$$\sup_{z\in\mathbb{R}} \left|P(W \le z) - P(Y \le z)\right| \le 12\beta + 2P\left(\left|W^e - W\right| > \beta\right) \tag{13.79}$$

*and*

$$\sup_{z\in\mathbb{R}} \left|P\left(W^e \le z\right) - P(Y \le z)\right| \le \beta + P\left(\left|W^e - W\right| > \beta\right). \tag{13.80}$$

*If in addition $W$ has a finite second moment, then*

$$\left\|\mathcal{L}(W) - \mathcal{E}(1)\right\|_1 \le 2E\left|W^e - W\right| \tag{13.81}$$

$$\sup_{z\in\mathbb{R}}\left|P\left(W^e \le z\right) - P(Y \le z)\right| \le E\left|W^e - W\right|. \tag{13.82}$$

The time until the occurrence of a rare event can often be well approximated by an exponential distribution. Aldous (1989) gives a wide survey of some settings where this phenomenon can occur, and Aldous and Fill (1994) summarize many results in the setting of Markov chain hitting times. We consider exponential approximation for first passage times, following the paper by Peköz and Röllin (2009).

If $X_0, X_1, \ldots$ is a Markov chain taking values in a denumerable space $\mathcal{X}$, for $j \in \mathcal{X}$ let

$$T_{\pi,j} = \inf\{t \ge 0\colon X_t = j\}, \tag{13.83}$$

the time of the first visit to state $j$ when the chain is initialized at time 0 with distribution $\pi$, and let

$$T_{i,j} = \inf\{t > 0\colon X_t = j\} \tag{13.84}$$

be the first time the Markov chain started in state $i$ at time 0 next visits state $j$.

**Theorem 13.7** *Let $X_0, X_1, \ldots$ be an ergodic, stationary Markov chain with stationary distribution $\pi$, and $T_{\pi,j}$ and $T_{i,j}$ the first passage times as given in* (13.83) *and* (13.84), *respectively. Then, with $Y \sim \mathcal{E}(1)$, the unit exponential distribution, for every $i \in \mathcal{X}$ we have*

$$\sup_{z\in\mathbb{R}}\left|P(\pi_i T_{\pi,i} \le z) - P(Y \le z)\right| \le 1.5\pi_i + \pi_i E|T_{\pi,i} - T_{i,i}|, \tag{13.85}$$

$$\sup_{z\in\mathbb{R}}\left|P(\pi_i T_{\pi,i} \le z) - P(Y \le z)\right| \le 2\pi_i + P(T_{\pi,i} \ne T_{i,i}) \tag{13.86}$$

*and*

$$\sup_{z\in\mathbb{R}}\left|P(\pi_i T_{\pi,i} \le z) - P(Y \le z)\right| \le 2\pi_i + \sum_{n=1}^{\infty}\left|P_{i,i}^{(n)} - \pi_i\right|, \tag{13.87}$$

*where $P_{i,i}^{(n)} = P(X_n = i|X_0 = i)$.*

*Proof* We first claim that if $U$ is a uniform $[0, 1]$ random variable independent of all else, then the equilibrium distribution of $T_{i,i}$ is given by

$$T_{i,i}^e =_d T_{\pi,i} + U. \tag{13.88}$$

To prove (13.88) we first demonstrate

$$P(T_{\pi,i} = k) = \pi_i P(T_{i,i} > k). \tag{13.89}$$

Consider the renewal–reward process which has renewals at visits to state $i$ and a reward of 1 when times between renewals is greater than $k$. Then, with $(X_1, R_1), (X_2, R_2), \ldots$ the sequence of renewal interarrival times and rewards, the renewal–reward theorem (see, e.g., Grimmett and Stirzaker 2001) yields

$$\lim_{t \to \infty} \frac{ER(t)}{t} = \frac{ER_1}{EX_1}, \tag{13.90}$$

where $R(t)$ is the total reward received by time $t$. As the mean length between renewals is $ET_{i,i} = 1/\pi_i$, the right hand side of (13.90) equals the right hand side of (13.89). On the other hand, there is precisely one time $t$ when the waiting time to the next renewal

$$Y_t = \inf\{s \geq t \colon X_s = i\} - t$$

is exactly $k$ if and only if the cycle length is greater than $k$. Hence $R(t) = \sum_{i=1}^{s} Y_i$, and $R(t)/t \to EY_s$, which is the left hand side of (13.89).

Hence, letting $f$ be a Lipschitz function with $f(0) = 0$, using (13.89) for the first equality, we have

$$\begin{aligned}
Ef'(T_{\pi,i} + U) &= E\big(f(T_{\pi,i} + 1) - f(T_{\pi,i})\big) \\
&= \pi_i \sum_{k=0}^{\infty} P(T_{i,i} > k)\big(f(k+1) - f(k)\big) \\
&= \pi_i \sum_{k=0}^{\infty} \sum_{j=k+1}^{\infty} P(T_{i,i} = j)\big(f(k+1) - f(k)\big) \\
&= \pi_i \sum_{j=0}^{\infty} \sum_{k=0}^{j-1} P(T_{i,i} = j)\big(f(k+1) - f(k)\big) \\
&= \pi_i \sum_{j=0}^{\infty} P(T_{i,i} = j) f(j) \\
&= \pi_i Ef(T_{i,i}).
\end{aligned}$$

The claim (13.88) now follows from definition (13.77).

Writing the $L^\infty$ norm between random variables $\xi$ and $\eta$ more compactly as $\|\mathcal{L}(\xi) - \mathcal{L}(\eta)\|_\infty$, by the triangle inequality, we have

$$\begin{aligned}
&\big\|\mathcal{L}(\pi_i T_{\pi,i}) - \mathcal{E}(1)\big\|_\infty \\
&\quad\leq \big\|\mathcal{L}(\pi_i T_{\pi,i}) - \mathcal{L}\big(\pi_i(T_{\pi,i} + U)\big)\big\|_\infty + \big\|\mathcal{L}\big(\pi_i(T_{\pi,i} + U)\big) - \mathcal{E}(1)\big\|_\infty \\
&\quad\leq \pi_i + \big\|\mathcal{L}\big(\pi_i(T_{\pi,i} + U)\big) - \mathcal{E}(1)\big\|_\infty.
\end{aligned} \tag{13.91}$$

To justify the final inequality (13.91), with $[z]$ denoting the greatest integer not greater than $z$ and $z \geq 0$, note that

$$\begin{aligned}
&P(T_{\pi,i} + U \leq z) \\
&\quad= P\big(T_{\pi,i} \leq [z] - 1\big) + P\big(T_{\pi,i} = [z], U \leq z - [z]\big) \\
&\quad= P\big(T_{\pi,i} \leq [z] - 1\big) + \big(z - [z]\big) P\big(T_{\pi,i} = [z]\big),
\end{aligned}$$

that

$$P(T_{\pi,i} \leq z) = P\big(T_{\pi,i} \leq [z] - 1\big) + P\big(T_{\pi,i} = [z]\big),$$

and with $X_0, X_1, \ldots$ in the equilibrium distribution $\pi$, that

$$P(T_{\pi,i} = [z]) \le P(X_{[z]} = i) = \pi_i.$$

Continuing from (13.91) and applying (13.78), (13.88), and (13.82) of Theorem 13.6, and then (13.88) again, we obtain

$$\begin{aligned}\|\mathcal{L}(\pi_i T_{\pi,i}) - \mathcal{E}(1)\|_\infty &\le \pi_i + \|\mathcal{L}(\pi_i T_{i,i}^e) - \mathcal{E}(1)\|_\infty \\ &\le \pi_i + \|\mathcal{L}((\pi_i T_{i,i})^e) - \mathcal{E}(1)\|_\infty \\ &\le \pi_i + \pi_i E|T_{\pi,i} + U - T_{i,i}| \\ &\le 1.5\pi_i + \pi_i E|T_{\pi,i} - T_{i,i}|, \end{aligned} \tag{13.92}$$

proving (13.85).

Taking $\beta = \pi_i$ in (13.80), from (13.92) and (13.88) we obtain

$$\|\mathcal{L}(\pi_i T_{\pi,i}) - \mathcal{E}(1)\|_\infty \le 2\pi_i + P(|T_{\pi,i} + U - T_{i,i}| > 1),$$

and we now obtain (13.86) by noting that

$$|T_{\pi,i} + U - T_{i,i}| > 1 \quad \text{implies} \quad T_{\pi,i} \ne T_{i,i}.$$

Lastly, to show (13.87), let $X_0, X_1, \ldots$ be the chain in equilibrium and $Y_0, Y_1, \ldots$ a coupled copy started in state $i$ at time 0, according to the maximal coupling, see Griffeath (1974/1975), so that $P(X_n = Y_n = i) = \pi_i \wedge P_{i,i}^{(n)}$. Let $T_{\pi,i}$ and $T_{i,i}$ be the hitting times given by (13.83) and (13.84) defined on the $X$ and $Y$ chain, respectively. Then

$$P(T_{\pi,i} \ne T_{i,i}) \le \sum_{n=0}^{\infty} \big(P(X_n = i, Y_n \ne i) + P(Y_n = i, X_n \ne i)\big),$$

and since

$$\begin{aligned} P(X_n = i, Y_n \ne i) &= \pi_i - P(X_n = i, Y_n = i) \\ &= \pi_i - \pi_i \wedge P_{i,i}^{(n)} = \big(\pi_i - P_{i,i}^{(n)}\big)^+, \end{aligned}$$

and a similar calculation yields

$$P(Y_n = i, X_n \ne i) = \big(P_{i,i}^{(n)} - \pi_i\big)^+,$$

we obtain (13.87) from (13.86). □

From, say, Billingsley (1968) Theorem 8.9, we know that for a finite, irreducible, aperiodic Markov chain, convergence to the unique stationary distribution is exponential, that is, that there exists $A \ge 0$ and $0 \le \rho < 1$ such that

$$\big|P_{i,i}^{(n)} - \pi_i\big| \le A\rho^n \quad \text{for all } n \in \mathbb{N}.$$

In this case, (13.87) of Theorem 13.7 immediately yields the bound

$$\sup_{z\in\mathbb{R}} \big|P(\pi_i T_{\pi,i} \le z) - P(Y \le z)\big| \le \inf_{k\in\mathbb{N}} \big\{(k+1)\pi_i + A\rho^k/(1-\rho)\big\}$$

on the Kolmogorov distance between $\pi_i T_{\pi,i}$ and the unit exponential.

Using the results of Theorem 13.7, Peköz and Röllin (2009) give bounds to the exponential for the times of appearance of patterns in independent coin tosses. In addition, they consider exponential approximations for random sums, and the asymptotic behavior of the scaled population size in a critical Galton–Watson branching process, conditioned on non-extinction.

## Appendix

*Proof of Lemma 13.1* (i) Let $Y'$ be an independent copy of $Y$. Then, for $y \in (a,b)$, we can rewrite $f_h$ in (13.6) as

$$\begin{aligned} f(y) &= \big(1/p(y)\big)E\big(h(Y)-h(Y')\big)\mathbf{1}_{\{Y\le y\}} \\ &= -\big(1/p(y)\big)E\big(h(Y)-h(Y')\big)\mathbf{1}_{\{Y>y\}}, \end{aligned} \tag{13.93}$$

which yields

$$\big|f(y)\big| \le 2\|h\| \min\big(F(y), 1-F(y)\big)/p(y). \tag{13.94}$$

Inequality (13.12) now follows from (13.10) and (13.94). Inequalities (13.94) and (13.11) imply $|f_h p'/p| \le 2d_2\|h\|$, that is, (13.13), and now (13.14) follows from (13.5).

(ii) Let $g_1(y) = p'(y)/p(y)$ for $y \in (a,b)$. Differentiating (13.5) we obtain

$$f'' = h' - f'g_1 - fg_1'. \tag{13.95}$$

To prove (13.17), it suffices to show that

$$\|fg_1'\| \le d_3\|h'\| \tag{13.96}$$

and

$$\|f'g_1\| \le (1+d_3)d_2\|h'\|. \tag{13.97}$$

By (13.93) again, we have

$$\begin{aligned} &\big|f(y)p(y)\big| \\ &\quad\le \|h'\| \min\big(E\big(|Y|+|Y'|\big)\mathbf{1}_{\{Y\le y\}}, E\big(|Y|+|Y'|\big)\mathbf{1}_{\{Y>y\}}\big) \\ &\quad= \|h'\| \min\big(E|Y|\mathbf{1}_{\{Y\le y\}} + E|Y|F(y), E|Y|\mathbf{1}_{\{Y>y\}} + E|Y|\big(1-F(y)\big)\big). \end{aligned} \tag{13.98}$$

This inequality proves (13.96) by assumption (13.15); the claim (13.18) follows in the same way from (13.16).

Rearranging (13.95) and multiplying by $p(y)$ yields

$$\big(h' - fg_1'\big)p = p\big(f'' + f'g_1\big) = f''p + f'p' = (f'p)'.$$

Thus, noting that the boundary terms vanish by (13.6),

$$f'(y)p(y) = \int_a^y \big(h' - fg_1'\big)p\,dx = -\int_y^b \big(h' - fg_1'\big)p\,dx,$$

and hence, using (13.96),

$$\left|f'(y)p(y)\right| \le \|h'\|(1+d_3)\min\bigl(F(y), 1-F(y)\bigr),$$

yielding (13.97), as well as (13.19), by (13.11) and (13.10), respectively. □

*Proof of Lemma 13.2* First, we see that Condition 13.1 is satisfied for densities of the form (13.24) whenever $E|Y| < \infty$ for $Y$ having density $p$, by (13.26). To prove the remaining claims, it suffices to verify that hypotheses (13.10), (13.11), (13.15) and (13.16) of Lemma 13.1 hold with $d_1 = 1/c_1$, $d_2 = 1$, $d_3 = c_2$ and $d_4(y) = c_2$. Let $g_2(y) = c_0 g(y)$ and $F(y) = P(Y \le y)$, the distribution function of $Y$.

Consider the case $0 \in [a, b]$, so that $G(0) = 0$ and $p(0) = c_1$. We first show that (13.10) is satisfied with $d_1 = 1/c_1$. It suffices to show that

$$F(y) \le F(0)p(y)/c_1 \quad \text{for } a < y < 0, \tag{13.99}$$

and

$$1 - F(y) \le \bigl(1 - F(0)\bigr)p(y)/c_1 \quad \text{for } 0 \le y < b. \tag{13.100}$$

Consider the case $a < y < 0$ and let $H(y) = F(y) - F(0)p(y)/c_1$. Differentiating,

$$\begin{aligned} H'(y) &= p(y) - F(0)p'(y)/c_1 \\ &= p(y) + F(0)g_2(y)p(y)/c_1 \\ &= p(y)\bigl(1 + F(0)g_2(y)/c_1\bigr). \end{aligned}$$

Since $g_2(y)$ is non-decreasing by Condition 13.2, if $H'(0) > 0$, then $H'(y)$ has at most one sign change on $(a, 0)$, and therefore $H$ achieves its maximum either are $a$ or at 0. If $H'(0) \le 0$, then $H'(y) \le 0$ for all $y < 0$, and $H$ achieves its maximum at $a$. However, as $H(0) = 0$ and $H(a) = -F(0)p(a)/c_1 \le 0$, we conclude that $H(y) \le 0$ for all $y \le 0$. This proves (13.99). Inequality (13.100) can be shown in a similar fashion.

Next we prove (13.11) holds with $d_2 = 1$. First consider $a < y < 0$. Inequality (13.11) is trivial when $p'(y) = 0$, so consider $y$ such that $p'(y) = -p(y)g_2(y) \ne 0$. By Condition 13.2, we have $g_2(x) \le g_2(y) < 0$ for all $x \le y$, and therefore

$$\begin{aligned} F(y) &= \int_a^y p(x)dx \\ &\le \int_a^y \frac{p(x)g_2(x)}{g_2(y)}dx \\ &= \int_a^y \frac{-p'(x)}{g_2(y)}dx \\ &= \frac{p(y) - p(a)}{-g_2(y)} \le \frac{p(y)}{|g_2(y)|}. \end{aligned} \tag{13.101}$$

Similarly, we have

$$1 - F(y) \le p(y)/g_2(y) \quad \text{for } 0 \le y < b. \tag{13.102}$$

Hence (13.11) is satisfied with $d_2 = 1$.

Note that (13.100) and (13.102) imply that

$$1 - F(y) \le p(y)\min\bigl(1/c_1, 1/g_2(y)\bigr) \quad \text{for } a < y < 0, \tag{13.103}$$

and that likewise we have

$$F(y) \le p(y)\min\bigl(1/c_1, 1/|g_2(y)|\bigr) \quad \text{for } 0 \le y < b. \tag{13.104}$$

To verify (13.15), with $0 \le y < b$ write

$$\begin{aligned}
&E|Y|\mathbf{1}_{\{Y>y\}} \\
&= yP(Y>y) + \int_y^b P(Y>t)dt \\
&\le yp(y)\min\bigl(1/c_1, 1/g_2(y)\bigr) + \int_y^b p(t)\min\bigl(1/c_1, 1/g_2(t)\bigr)dt \\
&\le yp(y)\min\bigl(1/c_1, 1/g_2(y)\bigr) + \min\bigl(1/c_1, 1/g_2(y)\bigr)\int_y^b p(t)dt \\
&= \min\bigl(1/c_1, 1/g_2(y)\bigr)\bigl\{yp(y) + \bigl(1 - F(y)\bigr)\bigr\} \\
&\le \min\bigl(1/c_1, 1/g_2(y)\bigr)\bigl\{yp(y) + p(y)/c_1\bigr\} \\
&= p(y)\min\bigl(1/c_1, 1/g_2(y)\bigr)\{y + 1/c_1\}.
\end{aligned} \tag{13.105}$$

Similarly, for $a < y < 0$ we obtain

$$E|Y|\mathbf{1}_{\{Y \le y\}} \le p(y)\min\bigl(1/c_1, 1/|g_2(y)|\bigr)\bigl\{|y| + 1/c_1\bigr\}. \tag{13.106}$$

Applying inequalities (13.105) and (13.106) at $y = 0$, and noting again that $G(0) = 0$ so that $p(0) = c_1$, gives $E|Y| \le 2/c_1$. Hence, recalling (13.103),

$$\begin{aligned}
&E|Y|\mathbf{1}_{\{Y>y\}} + E|Y|\bigl(1 - F(y)\bigr) \\
&\quad \le p(y)\min\bigl(1/c_1, 1/g_2(y)\bigr)\{y + 3/c_1\} \quad \text{for } 0 \le y < b,
\end{aligned} \tag{13.107}$$

and, using (13.104),

$$\begin{aligned}
&E|Y|\mathbf{1}_{\{Y \le y\}} + E|Y|F(y) \\
&\quad \le p(y)\min\bigl(1/c_1, 1/|g_2(y)|\bigr)\bigl\{|y| + 3/c_1\bigr\} \quad \text{for } a < y < 0.
\end{aligned} \tag{13.108}$$

Thus, (13.15) holds with $d_3 = c_2$ by Condition 13.3. Inequalities (13.107) and (13.108) also show that (13.16) is satisfied $d_4(y) = c_2$, completing the proof of Lemma 13.2 for the case $0 \in [a, b]$. The other cases follow similarly, noting that for $a > 0$ we have $G(a) = 0$ and $p(a) = c_1$, and likewise when $b \le 0$ we have $G(b) = 0$ and $p(b) = c_1$. □

# Chapter 14
# Group Characters and Malliavin Calculus

In this chapter we outline two recent developments in the area of Stein's method, one in the area of algebraic combinatorics, and the other a deep connection to the Malliavin calculus. We provide some background material in these areas in order to help make our presentation more self contained, but refer the reader to more complete sources for a comprehensive picture. The material in Sect. 14.1 is based on Fulman (2009), and that in Sect. 14.2 on Nourdin and Peccati (2009).

## 14.1 Normal Approximation for Group Characters

In the combinatorial central limit theorem studied in Sects. 4.4 and 6.1, one considers

$$Y = \sum_{i=1}^{n} a_{i\pi(i)} \quad \text{where } \pi \sim \mathcal{U}(\mathcal{S}_n), \tag{14.1}$$

that is, the sum of matrix elements, one from each row, where the column index is chosen according to a permutation $\pi$ with the uniform distribution on the symmetric group $\mathcal{S}_n$ of $\{1, \ldots, n\}$. If one were interested, say, in the distribution of the number of fixed points of $\pi$, one could take $a_{ij} = \mathbf{1}_{\{i=j\}}$, that is, let the matrix $(a_{ij})$ be the identity. Or, to look at the matter another way, if $e_1, \ldots, e_n$ are the standard (column) basis vectors in $\mathbb{R}^n$, and if $P_\pi = [e_{\pi(1)}, \ldots, e_{\pi(n)}]$, the permutation matrix associated to $\pi$, then the number of fixed points $Y$ of $\pi$ is the trace $\mathrm{Tr}(P_\pi)$ of $P_\pi$. More generally, we may write $Y$ in (14.1) as $\mathrm{Tr}(AP_\pi)$.

Asking similar questions for other groups leads us fairly directly to the study of the distribution of traces, or group characters on random matrices, and generalizations thereof. One of the earliest results on traces of random matrices is due to Diaconis and Shahshahani (1994), who applied the method of moments to prove joint convergence in distribution for traces of powers in a number of classical compact groups, including the orthogonal group $O(n, \mathbb{R})$. For this case, Stein (1995) showed that the error in the normal approximation decreases faster than the rate $n^{-r}$ for any fixed $r$, and Johansson (1997), working more generally, showed the rate is actually

L.H.Y. Chen et al., *Normal Approximation by Stein's Method*,
Probability and Its Applications,
DOI 10.1007/978-3-642-15007-4_14, 

exponential, validating a conjecture of Diaconis. The theory of representations and group characters being a rich one, below we only provide the most basic relevant definitions, but for omitted details and proofs, refer the reader to Weyl (1997) and Fulton and Harris (1991) for in depth treatments, and in particular to Serre (1997) for finite groups, and Sagan (1991) for the symmetric group.

For $V$ a finite dimensional vector space, let $GL(V)$ denote the set of all invertible linear transformation from $V$ to itself. When taking $V$ to be some subspace of $\mathbb{C}^n$, each such transformation may be considered as an invertible $n \times n$ matrix with complex entries. A representation of a group $G$ is a map

$$\tau : G \to GL(V)$$

which preserves the group structure, that is, which obeys

$$\tau(e) = I \quad \text{and} \quad \tau(gh) = \tau(g)\tau(h),$$

where $e$ is the identity element of $g$, and $I$ the identity matrix. For groups where $G$ itself has a topology we also require $\tau$ to be continuous. The map $\tau(g) = 1$ for all $g \in G$, clearly a representation, is called the trivial representation. We define the dimension $\dim(\tau)$ of the representation $\tau$ to be the dimension of the vector space $V$. The defining representation, for groups such as the ones we consider below which are already presented as matrices, is just the map that sends each matrix to itself.

If $\tau$ is a given representation over the vector space $V$, we say that a subspace $W \subset V$ is invariant if

$$w \in W \quad \text{implies} \quad \tau(g)w \in W \quad \text{for all } g \in G.$$

If $\tau$ has no invariant subspaces other than the trivial ones $W = V$ and $W = 0$, we say $\tau$ is irreducible.

Recall that $Y$ in the motivating example (14.1) could be written in terms of the trace of a matrix. Given a representation $\tau$, the character $\chi^\tau$ of $\tau$, also written simply as $\chi$ when $\tau$ is implicit, is given by

$$\chi(g) = \operatorname{Tr} \tau(g).$$

We let the dimension, or degree, of $\chi$ be the dimension of $\tau$, and say that $\chi$ is irreducible whenever $\tau$ is. The character $\chi$ inherits the properties of the trace, and in particular is the sum of the eigenvalues of $\tau(g)$, counting multiplicities. As a representation always sends the identity element to the identity matrix, the dimension of any representation may be calculated by evaluating its character on the identity.

A few more simple facts regarding characters are in order. For a complex number $z \in \mathbb{C}$, let $\overline{z}$ denote the complex conjugate of $z$. For $\chi$ a character of the representation $\tau$, we have

(i) $\chi(g^{-1}) = \overline{\chi}(g)$ for all $g \in G$.
(ii) $\chi(hgh^{-1}) = \chi(g)$ for all $g, h \in G$.

The first property may be shown by arguing that the transformations $\tau(g)$ may be taken to be unitary without loss of generality. Hence, any eigenvalue of $\tau(g)$ has modulus 1, so its inverse and complex conjugate are equal. Clearly then the same holds for their sum. The second property is simply a consequence of the cyclic invariance of the trace. Recalling that $C$ is a conjugacy class of the group $G$ if

$$g^{-1}hg \in C \quad \text{for all } g \in G \text{ and } h \in C,$$

property (ii) may be stated as the fact that characters are constant on the conjugacy classes of $G$, that is, they are what are known as 'class functions.'

Though the theory of group representations and group characters is more general, we closely follow a portion of the work of Fulman (2009), focusing on the following three compact Lie groups:

1. $O(n, \mathbb{R})$, the orthogonal group, of all $n \times n$ real valued matrices such that $U^{\mathsf{T}}U = I$.
2. $SO(n, \mathbb{R})$, the special orthogonal group, of all elements in $O(n, \mathbb{R})$ with determinant 1.
3. $USp(2n, \mathbb{C})$, the unitary symplectic group of all $2n \times 2n$ complex matrices $U$ that satisfy

$$U^{\dagger}\Upsilon U = \Upsilon \quad \text{where } \Upsilon = \begin{bmatrix} 0 & I \\ -I & 0 \end{bmatrix},$$

with $U^{\dagger}$ is the conjugate transpose of $U$.

We will not require a precise definition of a Lie group, and refer the reader to Hall (2003).

To consider the analogs of the uniform measure in (14.1) on these groups, we select a group element with distribution according to Haar measure, that is, the unique measure $\mu$ on $G$ of total mass one that satisfies

$$\mu(gS) = \mu(S) \quad \text{for all } g \in G \text{ and all measurable subsets } S \text{ of } G.$$

Using the Haar measure, we can define an inner product on complex valued functions $\chi$ and $\psi$ on $G$ by $\langle \chi, \psi \rangle = E\chi\overline{\psi}$, that is, as

$$\langle \chi, \psi \rangle = \int_G \chi(g)\overline{\psi}(g)d\mu.$$

If $\chi$ and $\psi$ are irreducible characters, then they satisfy the orthogonality relation

$$\langle \chi, \psi \rangle = \delta_{\chi,\psi}. \tag{14.2}$$

In particular, if $W = \chi^{\tau}(g)$ where $\tau$ is a nontrivial irreducible character and $g$ is chosen according to Haar measure, this orthogonality relation implies

$$EW = 0 \quad \text{and} \quad EW^2 = 1,$$

where for the first equality we have used the orthogonality of $\chi^{\tau}$ to the character of the trivial representation 1.

We apply Stein's method to study the distribution of $W = \chi^{\tau}(g)$ for $\tau$ an irreducible character and $g$ chosen according to the Haar measure of some compact Lie group. In particular, we construct an appropriate Stein pairs $W, W'$ so that Theorem 5.5 may be applied in this context. First, to show our choices are Stein pairs, and for many subsequent calculations, we rely on Lemma 14.1 below, a result of Helgason (2000). Next, we must be able to bound the variance of a conditional expectation in order to handle the first term of Theorem 5.5. As $W$ is a function of $g$, (4.143) yields

$$\mathrm{Var}\big(E\big((W'-W)^2|W\big)\big) \le \mathrm{Var}\big(E\big((W'-W)^2|g\big)\big). \tag{14.3}$$

The conditional expectation of $(W')^2$ given $g$ is computed in Lemma 14.2 with the help of Lemma 14.1, allowing for the computation of the conditional expectation on the right hand side of (14.3), and subsequently, its variance, in Lemma 14.3. The higher order moments required for the evaluation of the second term in the bound of Theorem 5.5 are handled in Lemma 14.4; as our constructions will lead to Stein pairs, the last term of that bound will be zero.

Focusing now on the real case, let $G$ be a compact Lie group and $\chi^\tau$ a non-trivial, real valued irreducible character of $G$. To create an appropriate $W'$, let $\alpha$ be chosen independently and uniformly from some fixed self-inverse conjugacy class $C$ of $G$, that is, a conjugacy class $C$ such that $h \in C$ implies $h^{-1} \in C$. (When all the characters of $G$ are real valued, all conjugacy classes are self inverse, see Fulman 2009.) We claim that the pair

$$(W, W') = \big(\chi^\tau(g), \chi^\tau(\alpha g)\big) \tag{14.4}$$

is exchangeable. Since for all $\alpha \in G$ the product $\alpha^{-1}g$ is distributed according to Haar measure whenever $g$ is, and $\alpha^{-1} =_d \alpha$ when $\alpha$ is chosen uniformly over $C$, by the independence of $g$ and $\alpha$ we obtain

$$\big(\chi^\tau(g), \chi^\tau(\alpha g)\big) =_d \big(\chi^\tau(\alpha^{-1}g), \chi^\tau(g)\big) =_d \big(\chi^\tau(\alpha g), \chi^\tau(g)\big).$$

Furthermore, since $C$ is self-inverse and characters are constant on conjugacy classes, for $\phi$ any representation,

$$\chi^\phi(\alpha) = \chi^\phi(\alpha^{-1}) = \overline{\chi^\phi(\alpha)},$$

implying $\chi^\phi(\alpha)$ is real.

Recall that the class functions on $G$ are the ones which are constant on the conjugacy classes of $G$, so in particular they form a vector space. We have seen that the characters themselves are class functions, but more is true. The irreducible characters of a group form an orthonormal basis for the class functions. Indeed, the calculation of the bounds in the theorems that follow hinge on the expansion of given characters in terms of the irreducibles.

For the calculation of the higher order moments required for the evaluation of the second term in the bound of Theorem 5.5 we require some additional facts regarding characters and tensor products. If $\tau$ and $\rho$ are representations of groups $G$ and $H$, then we may define the tensor product representation on the product group $G \times H$ by

$$(\tau \otimes \rho)(g, h) = \tau(g) \otimes \rho(h) \quad \text{for all } g \in G, h \in H.$$

Letting $\chi, \psi$ and $\chi \otimes \psi$ be the characters associated to $\tau$, $\rho$ and $\tau \otimes \rho$, respectively, the properties of the tensor product directly imply that

$$(\chi \otimes \psi)(g,h) = \chi(g)\psi(h). \tag{14.5}$$

When $H = G$ the mapping from $g$ to $\tau(g) \otimes \tau(g)$, denoted $\tau^2$, is again a representation of $G$, and more generally we may define the $r$-fold tensor product representation $\tau^r$ for $r \in \mathbb{N}$; when $r = 0$ we let the product $\tau^0$ be the trivial representation. If $\tau$ has character $\chi$, then by (14.5) the representation $\tau^r$ has character $\chi^r$. As the irreducible characters form a basis for the class functions, the character $\chi^r$ can be so decomposed; in particular, all that is needed to specify the decomposition of $\chi^r$ in terms of irreducible characters is the multiplicity $m_\phi(\tau^r)$ of the irreducible representation $\phi$ in $\tau^r$.

The following lemma from Helgason (2000) is key in the sequel.

**Lemma 14.1** *Let $G$ be a compact Lie group and $\chi$ the character induced by the irreducible representation $\phi$ of $G$. Then*

$$\int_G \chi^\phi(h\alpha h^{-1}g)dh = \frac{\chi^\phi(\alpha)}{\dim(\phi)}\chi^\phi(g)$$

*for all $\alpha, g \in G$.*

We can now show that $W, W'$ is a Stein pair, and develop a number of its properties.

**Lemma 14.2** *On the compact Lie group $G$ let $(W, W')$ be given by* (14.4) *with $g$ chosen from Haar measure, independently of $\alpha$ having the uniform distribution over some fixed self inverse conjugacy class. Then for all $r \in \mathbb{N}_0$,*

$$W^r = \sum_\phi m_\phi(\tau^r)\chi^\phi(g) \quad \text{and} \quad E\big[(W')^r|g\big] = \sum_\phi m_\phi(\tau^r)\frac{\chi^\phi(\alpha)}{\dim(\phi)}\chi^\phi(g), \tag{14.6}$$

*where the sum is over all irreducible representations of $G$, and*

$$E(W'|W) = (1-\lambda)W \quad \text{where } \lambda = 1 - \frac{\chi^\tau(\alpha)}{\dim(\tau)} \tag{14.7}$$

*and*

$$E(W'-W)^2 = 2\left(1 - \frac{\chi^\tau(\alpha)}{\dim(\tau)}\right). \tag{14.8}$$

*Proof* The first claim is simply the decomposition of the character $W^r = \chi^\tau(g)^r$ of $\tau^r$ of the product group $G^r$ over the basis of irreducible representations of $G$. Using this decomposition on $g' = \alpha g$ we obtain

$$(W')^r = \sum_\phi m_\phi(\tau^r)\chi^\phi(g').$$

Now, since $C=\{h\alpha h^{-1}: h\in G\}$ is the conjugacy class of $\alpha$, and when $h$ is distributed according to Haar measure then $h\alpha h^{-1}$ is uniform over $C$, we have

$$E\left[\chi^{\phi}(g')|g\right]=\int_G \chi^{\phi}(h\alpha h^{-1}g)dh=\frac{\chi^{\phi}(\alpha)}{\dim(\phi)}\chi^{\phi}(g),$$

by Lemma 14.1, proving the second equality in (14.6). When $r=1$ only the summand $\phi=\tau$ is non-vanishing, yielding $E(W'|g)$ as a function of $W$, hence (14.7). The last claim is now immediate from (2.34). □

We now calculate the right hand side of (14.3) for use in bounding the first term in Theorem 5.5.

**Lemma 14.3** *With $W, W'$ and $g$ as in Lemma* 14.2,

$$\operatorname{Var}\left(E\left((W'-W)^2|g\right)\right)=\sum_{\phi}^{*} m_{\phi}(\tau^2)^2\left(1+\frac{\chi^{\phi}(\alpha)}{\dim(\phi)}-\frac{2\chi^{\tau}(\alpha)}{\dim(\tau)}\right)^2,$$

*where $*$ signifies that the sum is over all nontrivial irreducible representations of $G$.*

*Proof* Expanding the square and using the measurability of $W$ with respect to $g$, then applying (14.7), and (14.6) with $r=2$, of Lemma 14.2, we obtain

$$\begin{aligned}E\left((W'-W)^2|g\right)&=E\left((W')^2|g\right)-2WE(W'|g)+W^2\\&=E\left((W')^2|g\right)+\left(1-\frac{2\chi^{\tau}(\alpha)}{\dim(\tau)}\right)W^2\\&=\sum_{\phi} m_{\phi}(\tau^2)\left(1+\frac{\chi^{\phi}(\alpha)}{\dim(\phi)}-\frac{2\chi^{\tau}(\alpha)}{\dim(\tau)}\right)\chi^{\phi}(g).\end{aligned}$$

Now squaring and taking expectation, the orthogonality relation (14.2) for irreducible characters yields the second moment of the conditional expectation,

$$E\left(E\left((W'-W)^2|g\right)\right)^2=\sum_{\phi} m_{\phi}(\tau^2)^2\left(1+\frac{\chi^{\phi}(\alpha)}{\dim(\phi)}-\frac{2\chi^{\tau}(\alpha)}{\dim(\tau)}\right)^2. \quad (14.9)$$

We note that the square of $E(W'-W)^2$, as given in Lemma 14.2, is the summand of (14.9) corresponding to the trivial representation of multiplicity 1, and therefore subtraction of this term to yield the variance completes the proof. □

We now focus on the calculation of the fourth moment of $W'-W$ in order to bound the second term in Theorem 5.5.

**Lemma 14.4** *With $W, W'$ as in Lemma* 14.2,

$$E(W'-W)^4=\sum_{\phi} m_{\phi}(\tau^2)^2\left[8\left(1-\frac{\chi^{\tau}(\alpha)}{\dim(\tau)}\right)-6\left(1-\frac{\chi^{\phi}(\alpha)}{\dim(\phi)}\right)\right].$$

*Proof* Expanding and using the measurability of $W$ with respect to $g$, and then applying Lemma 14.2, we obtain

$$E\big[(W'-W)^4|g\big]=\sum_{r=0}^{4}(-1)^r\binom{4}{r}\chi^\tau(g)^{4-r}E\big[(W')^r|g\big]$$
$$=\sum_{r=0}^{4}(-1)^r\binom{4}{r}\chi^\tau(g)^{4-r}\sum_{\phi}m_\phi(\tau^r)\frac{\chi^\phi(\alpha)}{\dim(\alpha)}\chi^\phi(g).$$

Now, taking expectation yields

$$E(W'-W)^4=\sum_{r=0}^{4}(-1)^r\binom{4}{r}\sum_{\phi}m_\phi(\tau^r)\frac{\chi^\phi(\alpha)}{\dim(\alpha)}\int\chi^\tau(g)^{4-r}\chi^\phi(g)dg$$
$$=\sum_{r=0}^{4}(-1)^r\binom{4}{r}\sum_{\phi}m_\phi(\tau^r)m_\phi(\tau^{4-r})\frac{\chi^\phi(\alpha)}{\dim(\phi)},$$

where

$$\int\chi^\tau(g)^{4-r}\chi^\phi(g)dg=m_\phi(\tau^{4-r})$$

by the decomposition of $\chi^\tau(g)^{4-r}$ in terms of irreducible characters, and applying the orthogonality relation they satisfy.

When $\alpha$ is the identity element of $G$ then $W'=W$ and $\chi^\phi(\alpha)=\dim(\phi)$, so

$$0=\sum_{r=0}^{4}(-1)^r\binom{4}{r}\sum_{\phi}m_\phi(\tau^r)m_\phi(\tau^{4-r}),$$

so we may write, for all $\alpha$,

$$E(W'-W)^4=-\sum_{r=0}^{4}(-1)^r\binom{4}{r}\sum_{\phi}m_\phi(\tau^r)m_\phi(\tau^{4-r})\left(1-\frac{\chi^\phi(\alpha)}{\dim(\phi)}\right).$$

The $r=0,4$ terms contribute zero, since the only $\phi$ which might contribute to the sum is the trivial representation, for which the last term vanishes. For $r=2$ the contribution is

$$-6\sum_{\phi}\left(1-\frac{\chi^\phi(\alpha)}{\dim(\phi)}\right)m_\phi(\tau^2)^2.$$

For both the $r=1$ and $r=3$ terms, the only nonzero summand is the one where $\phi=\tau$, with $m_\tau(\tau)=1$, and hence these contributions sum to

$$8\left(1-\frac{\chi^\tau(\alpha)}{\dim(\tau)}\right)m_\tau(\tau^3). \tag{14.10}$$

Taking the inner product of the character of $\tau^3$ with that of $\tau$ to find $m_\tau(\tau^3)$, and then using the decomposition of $\chi^\tau(g)^2$ in terms of irreducible characters, we have

$$m_\tau(\tau^3) = \int \chi^\tau(g)^4 dg = \int \left| \sum_\phi m_\tau(\tau^2)\chi^\phi(g) \right|^2 dg = \sum_\phi m_\phi(\tau^2)^2.$$

Now substitution of this expression into (14.10) and the collection of terms yields the result. □

We now state a normal approximation theorem for general compact Lie groups with real valued characters.

**Theorem 14.1** *Let $G$ be a compact Lie group and let $\tau$ be a non-trivial irreducible representation of $G$ with real valued character $\chi^\tau$. Let $W = \chi^\tau(g)$ where $g$ is chosen from the Haar measure of $G$. Then, if $\alpha$ is any non-identity element of $G$ with the property that $\alpha$ and $\alpha^{-1}$ are conjugate,*

$$\begin{aligned}
&\sup_{z\in\mathbb{R}} \left| P(W \le z) - P(Z \le z) \right| \\
&\quad \le \frac{1}{2}\sqrt{\sum_\phi^* m_\phi(\tau^2)^2 \left[ 2 - \frac{1}{\lambda}\left(1 - \frac{\chi^\phi(\alpha)}{\dim(\phi)}\right)\right]^2} \\
&\qquad + \left[\frac{1}{\pi}\sum_\phi m_\phi(\tau^2)^2 \left(8 - \frac{6}{\lambda}\left(1 - \frac{\chi^\phi(\alpha)}{\dim(\phi)}\right)\right)\right]^{1/4},
\end{aligned}$$

*where $\lambda = 1 - \chi^\tau(\alpha)/\dim(\tau)$, the first sum is over all non-trivial representations of $G$, and the second sum over all irreducible representations of $G$.*

*Proof* Let $(W, W')$ be given by (14.4) for $g$ and an element chosen uniformly from the conjugacy class containing $\alpha$. By Lemma 14.2 we may invoke Theorem 5.5 with the given $\lambda$ and $R = 0$. The first term in the bound of Theorem 5.5 is handled using (5.4), (14.3) and Lemma 14.3, and the last term by (14.8) of Lemma 14.2, Lemma 14.4 and

$$E|W' - W|^3 \le \sqrt{E(W' - W)^2 E(W' - W)^4}.$$ □

We apply Theorem 14.1 to characters of individual Lie groups with $\tau$ their defining representation. In order to apply the bounds, for each example we need information about the decomposition of $\tau^2$ in terms of irreducible characters. We include details for the calculation of the bound for the character of $O(2n, \mathbb{R})$, and omit the similar steps for the remaining examples. For additional details see Fulman (2009).

### *14.1.1* $O(2n,\mathbb{R})$

Let $\tau$ be the $2n$-dimensional defining representation and let $x_1, x_1^{-1}, \ldots, x_n, x_n^{-1}$ be the eigenvalues of an element of $O(2n,\mathbb{R})$. The following lemma is the $k=2$ case of a result of Proctor (1990).

**Lemma 14.5** *For $n \ge 2$, the square of the defining representation of $O(2n,\mathbb{R})$ decomposes in a multiplicity free way as the sum of the following three irreducible representations*:

(i) *The trivial representation, with character* 1.
(ii) *The representation with character* $\frac{1}{2}(\sum_i x_i + \overline{x_i})^2 - \frac{1}{2}\sum_i (x_i^2 + \overline{x_i}^2)$.
(iii) *The representation with character* $\frac{1}{2}(\sum_i x_i + \overline{x_i})^2 + \frac{1}{2}\sum_i (x_i^2 + \overline{x_i}^2) - 1$.

Armed with Lemma 14.5 we may now prove the following result.

**Theorem 14.2** *Let $g$ be chosen from the Haar measure of $O(2n,\mathbb{R})$ with $n \ge 2$, and let $W$ be the trace of $g$. Then*

$$\sup_{z\in\mathbb{R}} \bigl|P(W \le z) - P(Z \le z)\bigr| \le \frac{1}{\sqrt{2}(n-1)}.$$

*Proof* We apply Theorem 14.1 with $\tau$ the defining representation and $\alpha$ a rotation by some angle $\theta$, that is, an element conjugate to the diagonal matrix with entries $\{x_1, x_1^{-1}, \ldots, x_n, x_n^{-1}\}$ where $x_1 = \cdots = x_{n-1} = 1$ and $x_n = e^{i\theta}$. Then $\alpha$ is conjugate to $\alpha^{-1}$ and

$$\lambda = 1 - \frac{\chi^{\tau}(\alpha)}{\dim(\tau)} = 1 - \frac{2(n-1) + 2\cos(\theta)}{2n} = \frac{1-\cos(\theta)}{n}.$$

To calculate the first error term in Theorem 14.1, we apply Lemma 14.5 to write the decomposition of $\tau^2$ into non-trivial irreducibles. With $\phi_1$ the non-trivial irreducible character given in (ii),

$$\chi^{\phi_1}(\alpha) = \frac{1}{2}\bigl(2(n-1) + 2\cos(\theta)\bigr)^2 - \bigl(n - 1 + \cos(2\theta)\bigr) \tag{14.11}$$

with $\dim(\phi_1) = (2n)^2/2 - 2n/2 = 2n^2 - n$, and with $\phi_2$ as in (iii)

$$\chi^{\phi_2}(\alpha) = \frac{1}{2}\bigl(2(n-1) + 2\cos(\theta)\bigr)^2 + \bigl(n - 1 + \cos(2\theta)\bigr) - 1, \tag{14.12}$$

with $\dim(\phi_2) = (2n)^2/2 + 2n/2 - 1 = 2n^2 + n - 1$.

Hence, substituting using (14.11) and (14.12), we find the first error term is

$$\frac{1}{2}\sqrt{\left[2 - \frac{1}{\lambda}\left(1 - \frac{\chi^{\phi_1}(\alpha)}{2n^2 - n}\right)\right]^2 + \left[2 - \frac{1}{\lambda}\left(1 - \frac{\chi^{\phi_2}(\alpha)}{2n^2 + n - 1}\right)\right]^2}$$
$$= \frac{1}{2}\,\frac{\sqrt{8(n^2\cos(2\theta) - 2n(n-1)\cos(\theta) + 2n^2 + 1)}}{(n+1)(2n-1)}.$$

Similarly, the second term in the error bound equals

$$\frac{1}{\pi^{1/4}}\left(8+\left[8-\frac{6}{\lambda}\left(1-\frac{\chi^{\phi_1}(\alpha)}{2n^2-n}\right)\right]+\left[8-\frac{6}{\lambda}\left(1-\frac{\chi^{\phi_2}(\alpha)}{2n^2+n-1}\right)\right]\right)^{1/4}$$
$$=\left(\frac{24n(1-\cos\theta)}{\pi(n+1)(2n-1)}\right)^{1/4}.$$

Since the bound given by the sum of these two terms holds for all $\theta$, and is continuous in $\theta$, the bound holds in the limit as $\theta\to 0$. The limiting value of the first term is $\sqrt{2}/(2n-1)<1/(\sqrt{2}(n-1))$, while the second term vanishes in the limit, thus completing the argument. □

### 14.1.2 $SO(2n+1,\mathbb{R})$

We follow the same lines of argument in Sect. 14.1.1, with corresponding notation. Let $\tau$ be the $2n+1$ dimensional defining representation of $SO(2n+1,\mathbb{R})$ and $W=\chi^{\tau}(g)$ with $g$ chosen according to Haar measure. The following result following from Sundaram (1990) gives the decomposition of $\tau^2$ into irreducibles.

**Lemma 14.6** *For $n\ge 2$, the square of the defining representation of $SO(2n+1,\mathbb{R})$ decomposes in a multiplicity free way as the sum of the following three irreducible representations*:

(i) *The trivial representation, with character* 1.
(ii) *The representation with character* $\frac{1}{2}(\sum_i x_i+x_i^{-1})^2+\frac{1}{2}\sum_i(x_i^2+x_i^{-2})+\sum_i(x_i+x_i^{-1})$.
(iii) *The representation with character* $\frac{1}{2}(\sum_i x_i+x_i^{-1})^2-\frac{1}{2}\sum_i(x_i^2+x_i^{-2})+\sum_i(x_i+x_i^{-1})$.

With the help of Lemma 14.5, we have the following result.

**Theorem 14.3** *Let $g$ be chosen from the Haar measure of $SO(2n+1,\mathbb{R})$ with $n\ge 2$, and let $W$ be the trace of $g$. Then*

$$\sup_{z\in\mathbb{R}}\left|P(W\le z)-P(Z\le z)\right|\le\frac{1}{\sqrt{2}n}.$$

*Proof* Let $\alpha$ be a rotation by some angle $\theta$, that is, an element conjugate to a diagonal matrix with entries $\{x_1,x_1^{-1},\dots,x_n,x_n^{-1},1\}$ along the diagonal, where $x_1=\cdots=x_{n-1}=1$ and $x_n=e^{i\theta}$. Then $\alpha$ is conjugate to $\alpha^{-1}$ and $\lambda=2(1-\cos(\theta))/(2n+1)$.

Using Lemma 14.6 we decompose $\tau^2$ into irreducibles. Taking the limit as $\theta\to 0$ of the first error term of Theorem 14.1 yields $1/\sqrt{2}n$. Taking the limit of second term gives 0, as in the proof of Theorem 14.2. □

### 14.1.3 $USp(2n,\mathbb{C})$

Let $\tau$ be the $2n$ dimensional defining representation of $USp(2n,\mathbb{C})$ and $W=\chi^\tau(g)$ with $g$ chosen according to Haar measure. The following result following from Sundaram (1990) gives the decomposition of $\tau^2$ into irreducibles.

**Lemma 14.7** *For $n \geq 2$, the square of the defining representation of $USp(2n,\mathbb{C})$ decomposes in a multiplicity free way as the sum of the following three irreducible representations*:

(i) *The trivial representation, with character* 1.
(ii) *The representation with character* $\frac{1}{2}(\sum_i x_i + x_i^{-1})^2 + \frac{1}{2}\sum_i (x_i^2 + x_i^{-2})$.
(iii) *The representation with character* $\frac{1}{2}(\sum_i x_i + x_i^{-1})^2 - \frac{1}{2}\sum_i (x_i^2 + x_i^{-2}) - 1$.

Using Lemma 14.7, we are able to prove the following result.

**Theorem 14.4** *Let $g$ be chosen from the Haar measure of $USp(2n,\mathbb{C})$ with $n \geq 2$, and let $W$ be the trace of $g$. Then*

$$\sup_{z\in\mathbb{R}}\left|P(W \leq z) - P(Z \leq z)\right| \leq \frac{1}{\sqrt{2n}}.$$

*Proof* Let $\alpha$ be an element of conjugate to the diagonal matrix with diagonal entries $\{x_1, x_1^{-1}, \dots, x_n, x_n^{-1}\}$ where $x_1 = \cdots = x_{n-1} = 1$ and $x_n = e^{i\theta}$. Then $\alpha$ is conjugate to $\alpha^{-1}$ and $\lambda = 2(1-\cos(\theta))/n$.

Using Lemma 14.7 we decompose $\tau^2$ into irreducibles, and obtain the limit of $\sqrt{2}/(2n+1) < 1/\sqrt{2n}$ for the first term, as $\theta \to 0$, and a limit of zero for the second term, as in the proof of Theorem 14.2. □

## 14.2 Stein's Method and Malliavin Calculus

In some real sense, the most fundamental identity underlying Stein's method for the normal, that for all absolutely continuous functions $f$ such that the expectations below exist,

$$E\big[Zf(Z)\big] = E\big[f'(Z)\big] \quad \text{if and only if} \quad Z \text{ is standard normal}, \qquad (14.13)$$

can be seen as really nothing more than integration by parts. Proceeding along these same lines in more general spaces, the fact that (14.13) is a consequence of the Malliavin calculus integration by parts formula has some profound consequences. In this section we scratch the surface of the deep connection that exists between Stein's method and the Malliavin calculus, as unveiled in Nourdin and Peccati (2009). Working exclusively below with one dimensional Brownian motion, we do not attempt to cover the generality of Nourdin and Peccati (2009), whose framework includes Gaussian fields in higher dimensions, fractional Brownian motion, and parallel results for the Gamma distribution. Neither is our presentation one which approaches a treatment with complete technical details. We refer the reader to Nourdin and Peccati (2009), the lecture notes of Peccati (2009), and the text Nourdin and Peccati (2011) for full coverage of the material here, and to the standard reference, Nualart (2006), for the needed elements of the Malliavin calculus.

Let $B(t)$ be a standard Brownian motion on $[0,1]$ on a probability space $(\Omega, \mathcal{F}, P)$, and let $L^2(B)$ be the Hilbert space of square integrable functionals of $B$, endowed with inner product $\langle X, Y\rangle_{L^2(B)} = EXY$. Starting with the definition $\int \psi dB = \int_a^b dB_t = B(b) - B(a)$ of the stochastic integral for the indicator $\psi = \mathbf{1}_{(a,b]}$ of an interval in $[0,1]$, we may extend to all $\psi \in L^2(\lambda)$, the collection of square integrable functions on $[0,1]$ with respect to Lebesgue measure $\lambda$, by taking $L^2(B)$ limits to obtain $\int \psi dB$; this integral will be denoted by $I(\psi)$. The associated map $\psi \to I(\psi)$ from $L^2(\lambda) \to L^2(B)$ satisfies the isometry property

$$\langle \psi, \phi\rangle_{L^2(\lambda)} = \big\langle I(\psi), I(\phi)\big\rangle_{L^2(B)}. \tag{14.14}$$

In addition, $I(\psi)$ is a mean zero normal random variable, which by (14.14) has variance $\|\psi\|^2_{L^2(\lambda)}$. In particular, the collection $\{I(\psi)\colon \psi \in L^2(\lambda)\}$ is a real valued Gaussian process. For $\psi = \mathbf{1}_A$, the indicator of the measurable set $A$ in $[0,1]$, we also write $B(A) = I(\mathbf{1}_A)$, and may think of $B(A)$ as measure on $[0,1]$ with values in the space of random variables. For $A = (0,t]$ the definition recovers the Brownian motion through $B((0,t]) = B(t)$.

To consider higher order integrals, for $m \ge 1$ and $L^2(\lambda^m)$, the collection of square integrable functions on $[0,1]^m$ with respect to $m$ dimensional Lebesgue measure $\lambda^m$, the higher order stochastic integrals $I_m(\psi)$ are defined as follows. Consider first elementary functions of the form

$$\psi(t_1,\ldots,t_m) = \sum_{i_1,\ldots,i_m=1}^{n} a_{i_1,\ldots,i_m} \mathbf{1}(A_{i_1} \cdots A_{i_m})(t_1,\ldots,t_m)$$

where $A_1,\ldots,A_n$ are disjoint measurable sets for all $j = 1,\ldots,n$, and the coefficients $a_{i_1,\ldots,i_m}$ are zero if any of the two indices $i_1,\ldots,i_m$ are equal. For a function $\psi$ of this form, define

$$I_m(\psi) = \sum_{i_1,\ldots,i_m=1}^{n} a_{i_1,\ldots,i_m} B(A_{i_1}) \cdots B(A_{i_m}),$$

and extend to $L^2(\lambda^m)$ by taking $L^2(B)$ limits. It is clear that $I(\psi)$ and $I_1(\psi)$ agree.

For a function $\psi : [0,1]^m \to \mathbb{R}$, define the symmetrization of $\psi$ by

$$\widetilde{\psi}(t_1,\ldots,t_m) = \frac{1}{m!}\sum_{\sigma} \psi(t_{\sigma(1)},\ldots,t_{\sigma(m)}), \tag{14.15}$$

where the sum is over all permutations $\sigma$ of $\{1,\ldots,m\}$. Letting $L^2_s(\lambda^m)$ be the closed subspace of $L^2(\lambda^m)$ of symmetric, square integrable functions on $[0,1]^m$

with respect to Lebesgue measure, we see that $\psi \in L^2(\lambda^m)$ implies $\widetilde{\psi} \in L^2_s(\lambda^m)$ by the triangle inequality,

$$\|\widetilde{\psi}\|_{L^2(\lambda)} \le \|\psi\|_{L^2(\lambda)}.$$

One can verify that the stochastic integrals $I_m(\cdot)$ have the following properties:

(i) $EI_m(\psi) = 0$ and $I_m(\psi) = I_m(\widetilde{\psi})$ for all $\psi \in L^2(\lambda^m)$.
(ii) For all $\psi \in L^2(\lambda^p)$ and $\phi \in L^2(\lambda^q)$,

$$E\big[I_p(\psi)I_q(\phi)\big] = \begin{cases} 0 & p \ne q, \\ p!\langle \widetilde{\psi}, \widetilde{\phi}\rangle_{L^2(\lambda^p)} & p = q. \end{cases} \tag{14.16}$$

(iii) The mapping $\psi \to I_m(\psi)$ from $L^2(\lambda^m)$ to $L^2(B)$ is linear.

The goal of this section, achieved in Theorem 14.6, is to obtain a bound to the normal in the total variation distance for integrals $I_q(\psi)$ for $q \ge 2$. For this purpose we require the multiplication formula, which expresses the product of the stochastic integrals of $\psi \in L^2(\lambda^p)$ and $\phi \in L^2(\lambda^q)$ in terms of sums of integrals of contractions of $\psi$ and $\phi$,

$$I_p(\psi)I_q(\phi) = \sum_{r=0}^{p\wedge q} r!\binom{p}{r}\binom{q}{r} I_{p+q-2r}(\widetilde{\psi} \otimes_r \widetilde{\phi}), \tag{14.17}$$

where, for $r = 1, \dots, p \wedge q$ and $(t_1, \dots, t_{p-r}, s_1, \dots, s_{q-r}) \in [0,1]^{p+q-2r}$, the contraction $\otimes_r$ is given by

$$(\psi \otimes_r \phi)(t_1, \dots, t_{p-r}, s_1, \dots, s_{q-r})$$
$$= \int_{[0,1]^r} \psi(z_1, \dots, z_r, t_1, \dots, t_{p-r})\phi(z_1, \dots, z_r, s_1, \dots, s_{q-r})\lambda^r(dz_1, \dots, dz_r),$$

and for $r = 0$, denoting $\otimes_0$ also by $\otimes$, by

$$(\psi \otimes \phi)(t_1, \dots, t_p, s_1, \dots, s_q) = \psi(t_1, \dots, t_p)\phi(s_1, \dots, s_q).$$

Even when $\psi$ and $\phi$ are symmetric $\psi \otimes_r \phi$ may not be, and we let $\psi \widetilde{\otimes}_r \phi$ be the symmetrization of $\psi \otimes_r \phi$ as given in (14.15).

Using the multiple stochastic integrals $I_q, q \in \mathbb{N}$, any $F \in L^2(B)$, that is, any square integrable function of the Brownian motion $B$, can be represented by the following Wiener chaos decomposition. For any such $F$, there exists a unique sequence $\{\psi_q \colon n \ge 1\}$ with $\psi_q \in L^2_s(\lambda^q)$ such that

$$F = \sum_{q=0}^{\infty} I_q(\psi_q) \tag{14.18}$$

where $I_0(\psi_0) = EF$, and the series converges in $L^2$. When all terms but one in the sum vanish so that $F = I_q(\psi_q)$ for some $q$, we say $F$ belongs to the $q$th Wiener chaos of $B$. Applying the orthogonality relation (14.16) to the symmetric 'kernels' $\psi_q$ for $F$ of the form (14.18),

$$\|F\|^2_{L^2(B)} = \sum_{q=0}^{\infty} q!\|\psi_q\|^2_{L^2(\lambda^q)}. \tag{14.19}$$

We now briefly describe two of the basic operators of the Malliavin calculus: the Malliavin derivative $D$, and the Ornstein–Uhlenbeck generator $L$. Beginning with $D$, for $g : \mathbb{R}^n \to \mathbb{R}$ a smooth function with compact support, consider a random variable of the form

$$F = g\big(I(\psi_1), \ldots, I(\psi_n)\big) \quad \text{with } \psi_1, \ldots, \psi_n \in L^2(\lambda). \tag{14.20}$$

For such an $F$, the Malliavin derivative is defined as

$$DF = \sum_{i=1}^{n} \frac{\partial}{\partial x_i} g\big(I(\psi_1), \ldots, I(\psi_n)\big)\psi_i.$$

Note in particular

$$DI(\psi) = \psi \quad \text{for every } \psi \in L^2(\lambda). \tag{14.21}$$

In general then, the $m$th derivative $D^m F$, given by

$$D^m F = \sum_{i_1,\ldots,i_m=1}^{n} \frac{\partial^m}{\partial x_{i_1} \cdots \partial x_{i_m}} g\big(I(\psi_1), \ldots, I(\psi_n)\big)\psi_{i_1} \otimes \cdots \otimes \psi_{i_m},$$

maps random variables $F$ to the Hilbert space $L^2(B, L^2(\lambda^m))$ of $L^2(\lambda^m)$ valued functionals of $B$, endowed with the inner product

$$\langle u, v\rangle_{L^2(B,L^2(\lambda^m))} = E\langle u, v\rangle_{L^2(\lambda^m)}.$$

Letting $\mathcal{S}$ denote the set of random variables of the form (14.20), for every $m \geq 1$ the domain of $D^m$ may be extended to $\mathbb{D}^{m,2}$, the closure of $\mathcal{S}$ with respect to the norm $\|\cdot\|_{m,2}$ given by

$$\|F\|^2_{m,2} = EF^2 + \sum_{i=1}^{m} E\big[\|D^i F\|^2_{L^2(\lambda^i)}\big].$$

A random variable $F \in L^2(B)$ having chaotic expansion (14.18) is an element of $\mathbb{D}^{m,2}$ if and only if the kernels $\psi_q$, $q = 1, 2, \ldots$ satisfy

$$\sum_{q=1}^{\infty} q^m q!\|\psi_q\|^2_{L^2(\lambda^q)} < \infty, \tag{14.22}$$

in which case

$$E\|D^m F\|^2_{L^2(\lambda^m)} = \sum_{q=m}^{\infty} (q)_m q!\|\psi_q\|^2_{L^2(\lambda^q)},$$

where $(q)_m$ is the falling factorial. In particular, any $F$ having a finite Wiener chaos expansion is an element of $\mathbb{D}^{m,2}$ for all $m \geq 1$.

The Malliavin derivative obeys a chain rule. If $g : \mathbb{R}^n \to \mathbb{R}$ is a continuously differentiable function with bounded derivative, and $F_i \in \mathbb{D}^{1,2}$ for $i = 1, \ldots, n$, then $g(F_1, \ldots, F_n) \in \mathbb{D}^{1,2}$ and

$$Dg(F_1, \ldots, F_n) = \sum_{i=1}^{n} \frac{\partial}{\partial x_i} g(F_1, \ldots, F_n) DF_i. \tag{14.23}$$

As we are considering Brownian motion on $[0, 1]$, and the indexed family $\{I(\psi): \psi \in L^2(\lambda)\}$ with $\lambda$ non-atomic, the derivatives of random variables $F$ of the form (14.18) can be identified with the element $L^2([0, 1] \times \Omega)$ given by

$$D_t F = \sum_{q=1}^{\infty} q I_{q-1}\big(\psi_q(\cdot, t)\big), \quad t \in [0, 1]. \tag{14.24}$$

We next introduce the Ornstein–Uhlenbeck generator $L$. For a square integrable random variable $F$ represented as in (14.18), let

$$LF = \sum_{q=0}^{\infty} -q I_q(\psi_q) \tag{14.25}$$

$$\text{and, when } EF = 0, \text{ let } L^{-1}F = \sum_{q=1}^{\infty} -\frac{1}{q} I_q(\psi_q).$$

In view of (14.19) and (14.22), we see that the operator $L^{-1}$ takes values in $\mathbb{D}^{2,2}$.

As the Malliavin derivative $D$ maps random variables to the Hilbert space $L^2(B, L^2(\lambda))$ endowed with the inner product $E\langle u, v\rangle_{L^2(\lambda)}$, by definition, the adjoint operator $\delta$ satisfies the (integration by parts) identity

$$E\big(F\delta(u)\big) = E\langle DF, u\rangle_{L^2(\lambda)}, \quad \text{for every } F \in \mathbb{D}^{1,2}, \tag{14.26}$$

when $u$ lies in the domain $\mathrm{dom}(\delta)$ of $\delta$. One of the key consequences of this identity is that for every $F \in \mathbb{D}^{1,2}$ with $EF = 0$,

$$E\big[Ff(F)\big] = E\big[\langle DF, -DL^{-1}F\rangle_{L^2(\lambda)} f'(F)\big], \tag{14.27}$$

for all real valued differentiable functions $f$ with bounded derivative; identity (14.27) also holds when $f$ is only a.e. differentiable if $F$ has an absolutely continuous law. The Stein identity (14.13) is the special case where $F = I(\psi)$ for $\psi \in L^2(\lambda)$ with $\|\psi\|_{L^2(\lambda)} = 1$. For then $F$ is standard normal, and by (14.21) and (14.25), respectively, we have,

$$DF = \psi \quad \text{and} \quad L^{-1}F = L^{-1}I(f) = -I(f) = -F,$$

so

$$\begin{aligned}&\langle DF, -DL^{-1}F\rangle_{L^2(\lambda)} \\ &\quad = \langle \psi, DF\rangle_{L^2(\lambda)} = \langle \psi, \psi\rangle_{L^2(\lambda)} = \|\psi\|^2_{L^2(\lambda)} = 1.\end{aligned} \tag{14.28}$$

Hence (14.27) implies (14.13).

Though the theory of Nourdin and Peccati (2009) supplies results in the Wasserstein, Fortet–Mourier, and the Kolmogorov distance, recalling (4.1) and (4.3), we confine ourselves to the total variation norm. In this case, by the bounds (2.12), for any random variable $F$,

$$\|\mathcal{L}(F)-\mathcal{L}(Z)\|_{\mathrm{TV}} \le \sup_{f\in\mathcal{F}_{\mathrm{TV}}} \left|E\big(f'(F)-Ff(F)\big)\right|, \tag{14.29}$$

where $\mathcal{F}_{\mathrm{TV}}$ is the collection of piecewise continuously differentiable functions that are bounded by $\sqrt{\pi/2}$ and whose derivatives are bounded by 2.

**Theorem 14.5** *Let $F \in \mathbb{D}^{1,2}$ have mean zero and an absolutely continuous law with respect to Lebesgue measure. Then*

$$\|\mathcal{L}(F)-\mathcal{L}(Z)\|_{\mathrm{TV}} \le 2E\left|\left(1-\langle DF,-DL^{-1}F\rangle_{L^2(\lambda)}\right)\right|.$$

*Proof* By (14.27),

$$E\big[f'(F)-Ff(F)\big]=E\big[f'(F)\big(1-\langle DF,-DL^{-1}F\rangle_{L^2(\lambda)}\big)\big],$$

and the proof is completed by applying (14.29). □

To make use of Theorem 14.5 it is necessary to handle the inner product appearing in the bound. We note that (14.28) gives the simplest case, and also shows the upper bound to be tight in the sense that it is zero when $F =_d Z$. Theorem 14.6 gives a much more substantial illustration of a case where computation with the inner product is possible.

We now follow Nourdin et al. (2009), which simplifies the calculations in Nourdin and Peccati (2009), as well as generalizes the results from integrals $I_2(\psi)$ to $I_q(\psi)$ for all $q \ge 2$.

**Theorem 14.6** *Let $F$ belong to the $q$th Wiener chaos of $B$ for some $q \ge 2$. Then*

$$\|\mathcal{L}(F)-\mathcal{L}(Z)\|_{\mathrm{TV}} \le 2\left|1-EF^2\right|+2\sqrt{\frac{q-1}{3q}}\sqrt{EF^4-3\big(EF^2\big)^2}.$$

The following proof shows that it is always the case that $EF^4 \ge 3(EF^2)^2$.

*Proof* Writing $F = I_q(\psi)$ with $\psi \in L_s^2(\lambda^q)$, by (14.19) we obtain

$$EF^2 = q!\|\psi\|^2_{L^2(\lambda^q)}. \tag{14.30}$$

Now, applying (14.30) for the final inequality below, by (14.24) and the multiplication formula (14.17), we have

$$\begin{aligned}\frac{1}{q}\|DF\|^2_{L^2(\lambda)} &= q\int_0^1 I_{q-1}\big(\psi(\cdot,a)\big)^2\lambda(da)\\ &= q\int_0^1\sum_{r=1}^{q-1} r!\binom{q-1}{r}^2 I_{2q-2-2r}\big(\psi(\cdot,a)\big)\,\widetilde{\otimes}_r\,\psi(\cdot,a)\lambda(da)\end{aligned}$$

$$= q\sum_{r=1}^{q-1} r!\binom{q-1}{r}^2 I_{2q-2-2r}\left(\int_0^1 \psi(\cdot,a)\,\widetilde{\otimes}_r\,\psi(\cdot,a)\lambda(da)\right)$$

$$= q\sum_{r=0}^{q-1} r!\binom{q-1}{r}^2 I_{2q-2-2r}(\psi\,\widetilde{\otimes}_{r+1}\,\psi)$$

$$= EF^2 + q\sum_{r=1}^{q-1}(r-1)!\binom{q-1}{r-1}^2 I_{2q-2r}(\psi\,\widetilde{\otimes}_r\,\psi). \tag{14.31}$$

Subtracting $EF^2$ from both sides and applying (14.16) yields

$$E\left[\left(\frac{1}{q}\|DF\|^2_{L^2(\lambda)} - EF^2\right)^2\right]$$

$$= q^2\sum_{r=1}^{q-1}(r-1)!\binom{q-1}{r-1}^4 (2q-2r)!\|\psi\,\widetilde{\otimes}_r\,\psi\|^2_{L^2(\lambda^{2q-2r})}. \tag{14.32}$$

Next, again by (14.17),

$$F^2 = \sum_{r=0}^{q} r!\binom{q}{r}^2 I_{2q-2r}(\psi\widetilde{\otimes}_r\psi). \tag{14.33}$$

Applying (14.27) and (14.25) for the second equality below, and (14.31), (14.33) and (14.16) for the third, we obtain

$$EF^4 = E\big(F\times F^3\big) = 3E\left(F^2\times\frac{1}{q}\|DF\|^2_{L^2(\lambda)}\right)$$

$$= 3E\big(F^2\big)^2 + \frac{3}{q}\sum_{r=1}^{q-1} rr!^2\binom{q}{r}^4 (2q-2r)!\|\psi\widetilde{\otimes}_r\psi\|^2_{L^2(\lambda^{2q-2r})}. \tag{14.34}$$

Comparing (14.32) and (14.34) leads to

$$E\left[\left(\frac{1}{q}\|DF\|^2_{L^2(\lambda)} - EF^2\right)^2\right] \le \frac{q-1}{3q}\big(EF^4 - 3(EF^2)^2\big).$$

Lastly, by Theorem 14.5 and (14.25),

$$\begin{aligned}\|\mathcal{L}(F)-\mathcal{L}(Z)\|_{\mathrm{TV}} &\le 2E\big|1-\langle DF,-DL^{-1}F\rangle_{L^2(\lambda)}\big|\\ &\le 2\big|1-EF^2\big| + 2E\big|EF^2 - \langle DF,-DL^{-1}F\rangle_{L^2(\lambda)}\big|\\ &\le 2\big|1-EF^2\big| + 2\sqrt{E\big(EF^2-\langle DF,-DL^{-1}F\rangle_{L^2(\lambda)}\big)^2}\\ &\le 2\big|1-EF^2\big| + 2\sqrt{\frac{q-1}{3q}}\sqrt{EF^4-3\big(EF^2\big)^2}.\end{aligned}$$

□

We note that, as one consequence of Theorem 14.6, Stein's method provides a streamlined proof of the Nualart–Peccati criterion, that is, if $F_n = I_q(\psi_n)$ for some $q \ge 2$, such that $E[F_n^2] \to \sigma^2 > 0$ as $n \to \infty$, then the following are equivalent:

(i) $\|\mathcal{L}(F_n) - \mathcal{L}(\sigma Z)\|_{\mathrm{TV}} \to 0$.
(ii) $F_n \to_d \sigma Z$.
(iii) $EF_n^4 \to 3\sigma^4$.

Though in the section we have considered the case of Brownian motion on $[0, 1]$, with corresponding Gaussian process $\{I(\psi): \psi \in L^2(\lambda)\}$, much here carries over with no essential changes when considering Gaussian processes on $\{X(\psi): \psi \in \mathcal{H}\}$ indexed by more general Hilbert spaces; see Nourdin and Peccati (2009) and Nourdin and Peccati (2011) for details.

# Appendix

## Notation

| | |
|---|---|
| $\mathbf{1}$ | indicator function |
| $=_d$ | equality in distribution |
| $\to_d$ | convergence in distribution |
| $\to_p$ | convergence in probability |
| $\mathbb{Z}$ | $\{\ldots, -1, 0, 1, \ldots\}$ |
| $\mathbb{N}$ | $\{1, 2, \ldots\}$ |
| $\mathbb{N}_0$ | $\{0, 1, \ldots\}$ |
| $\mathbb{R}$ | $(-\infty, \infty)$ |
| $\mathbb{R}^+$ | $[0, \infty)$ |
| $\mathcal{S}_n$ | the symmetric group on $n$ symbols |
| $\mathcal{N}(\mu, \sigma^2)$ | normal distribution with mean $\mu$ and variance $\sigma^2$ |
| $\mathcal{U}[a, b]$ | uniform distribution over $[a, b]$ |
| $Z$ | standard normal variable |
| $\Phi(z)$ | standard normal distribution function |
| $Nh$ | $Eh(Z)$ |
| $K_i(t)$ | $K$ function, (prototype) page 19 |
| $X^*$ | zero bias, page 26 |
| $X^s$ | size bias, page 31 |
| $X^\square$ | square bias, page 34 |
| $\|f\|$ | supremum norm of $f$ |
| $\|F - G\|_1$ | $L^1$ distance, page 64 |
| $\Gamma(\alpha)$ | Gamma function |
| $\Gamma(\alpha, \beta)$ | Gamma distribution |
| $B(\alpha, \beta)$ | Beta distribution |
| $A^{\mathrm{T}}$ | transpose of the matrix $A$ |
| $\mathrm{Tr}(A)$ | trace of the matrix $A$ |
| $\mathcal{L}(\cdot)$ | law, or distribution, of a random variable |
| $\|h\|_{L^\infty_m(\mathbb{R})}$ | page 136 |
| $\mathcal{H}_{m,\infty}$ | page 136 |

L.H.Y. Chen et al., *Normal Approximation by Stein's Method*,
Probability and Its Applications,
DOI 10.1007/978-3-642-15007-4, © Springer-Verlag Berlin Heidelberg 2011

| | |
|---|---|
| $\mathcal{H}_{m,\infty,p}$ | page 313 |
| $\Vert\mathcal{L}(\mathbf{X}) - \mathcal{L}(\mathbf{Y})\Vert_{\mathcal{H}_{m,\infty,p}}$ | page 313 |

# References


Aldous, D. (1989). *Applied mathematical sciences: Vol. 77. Probability approximations via the Poisson clumping heuristic*. New York: Springer.

Aldous, D., & Fill, J. A. (1994). *Reversible Markov chains and random walks on graphs*. Monograph in preparation. http://www.stat.berkeley.edu/aldous/RWG/book.html.

Arratia, R., Goldstein, L., & Gordon, L. (1989). Two moments suffice for Poisson approximations: the Chen–Stein method. *Annals of Probability*, *17*, 9–25.

Baldi, P., & Rinott, Y. (1989). Asymptotic normality of some graph-related statistics. *Journal of Applied Probability*, *26*, 171–175.

Baldi, P., Rinott, Y., & Stein, C. (1989). A normal approximations for the number of local maxima of a random function on a graph. In T. W. Anderson, K. B. Athreya, & D. L. Iglehart (Eds.), *Probability, statistics and mathematics, papers in honor of Samuel Karlin* (pp. 59–81). San Diego: Academic Press.

Barbour, A. D. (1990). Stein's method for diffusion approximations. *Probability Theory and Related Fields*, *84*, 297–322.

Barbour, A. D., & Chen, L. H. Y. (2005a). In A. D. Barbour & L. H. Y. Chen (Eds.), *The permutation distribution of matrix correlation statistics. Stein's method and applications*. Singapore: Singapore University Press.

Barbour, A. D., & Chen, L. H. Y. (2005b). In A. D. Barbour & L. H. Y. Chen (Eds.), *An introduction to Stein's method*. Singapore: Singapore University Press.

Barbour, A. D., & Chen, L. H. Y. (2005c). In A. D. Barbour & L. H. Y. Chen (Eds.), *Stein's method and applications*. Singapore: Singapore University Press.

Barbour, A. D., & Eagleson, G. (1986). Random association of symmetric arrays. *Stochastic Analysis and Applications*, *4*, 239–281.

Barbour, A. D., & Xia, A. (1999). Poisson perturbations. *ESAIM, P&S*, *3*, 131–150.

Barbour, A. D., Karoński, M., & Ruciński, A. (1989). A central limit theorem for decomposable random variables with applications to random graphs. *Journal of Combinatorial Theory. Series B*, *47*, 125–145.

Barbour, A. D., Holst, L., & Janson, S. (1992). *Poisson approximation*. London: Oxford University Press.

Bayer, D., & Diaconis, P. (1992). Trailing the dovetail shuffle to its lair. *The Annals of Applied Probability*, *2*, 294–313.

Bentkus, V., Götze, F., & Zitikis, R. (1994). Lower estimates of the convergence rate for $U$-statistics. *Annals of Probability*, *22*, 1707–1714.

Berry, A. (1941). The accuracy of the Gaussian approximation to the sum of independent variates. *Transactions of the American Mathematical Society*, *49*, 122–136.

Bhattacharya, R. N., & Ghosh, J. (1978). On the validity of the formal Edgeworth expansion. *Annals of Statistics*, *6*, 434–451.



L.H.Y. Chen et al., *Normal Approximation by Stein's Method*, Probability and Its Applications,
DOI 10.1007/978-3-642-15007-4, © Springer-Verlag Berlin Heidelberg 2011

Bhattacharya, R. N., & Rao, R. (1986). *Normal approximation and asymptotic expansion*. Melbourne: Krieger.
Bickel, P., & Doksum, K. (1977). *Mathematical statistics: basic ideas and selected topics*. Oakland: Holden-Day.
Biggs, N. (1993). *Algebraic graph theory*. Cambridge: Cambridge University Press.
Bikjalis, A. (1966). Estimates of the remainder term in the central limit theorem. *Lietuvos Matematikos Rinkinys*, *6*, 323–346 (in Russian).
Billingsley, P. (1968). *Convergence of probability measures*. New York: Wiley.
Bollobás, B. (1985). *Random graphs*. San Diego: Academic Press.
Bolthausen, E. (1984). An estimate of the reminder in a combinatorial central limit theorem. *Zeitschrift für Wahrscheinlichkeitstheorie und Verwandte Gebiete*, *66*, 379–386.
Borovskich, Yu. V. (1983). Asymptotics of $U$-statistics and von Mises' functionals. *Soviet Mathematics. Doklady* *27*, 303–308.
Breiman, L. (1986). *Probability*. Reading: Addison–Wesley.
Brouwer, A. E., Cohen, A. M., & Neumaier, A. (1989). *Distance-regular graphs*. Berlin: Springer.
Bulinski, A., & Suquet, C. (2002). Normal approximation for quasi-associated random fields. *Statistics & Probability Letters*, *54*, 215–226.
Cacoullos, T., & Papathanasiou, V. (1992). Lower variance bounds and a new proof of the central limit theorem. *Journal of Multivariate Analysis*, *43*, 173–184.
Chatterjee, S. (2007). Stein's method for concentration inequalities. *Probability Theory and Related Fields*, *138*, 305–321.
Chatterjee, S. (2008). A new method of normal approximation. *Annals of Probability*, *4*, 1584–1610.
Chatterjee, S., & Meckes, E. (2008). Multivariate normal approximation using exchangeable pairs. *ALEA. Latin American Journal of Probability and Mathematical Statistics*, *4*, 257–283.
Chatterjee, S., & Shao, Q. M. (2010, to appear). Non-normal approximation by Stein's method of exchangeable pairs with application to the Curie–Weiss model. *The Annals of Applied Probability*.
Chatterjee, S., Fulman, J., & Röllin, A. (2008, to appear). *Exponential approximation by Stein's method and spectral graph theory*.
Chen, L. H. Y. (1975). Poisson approximation for dependent trials. *Annals of Probability*, *3*, 534–545.
Chen, L. H. Y. (1998). Stein's method: some perspectives with applications. In L. Accardi & C. C. Heyde (Eds.), *Lecture Notes in Statistics*: *Vol. 128*. *Probability towards 2000*. Berlin: Springer.
Chen, L. H. Y., & Leong, Y. K. (2010). *From zero-bias to discretized normal approximation* (Preprint).
Chen, L. H. Y., & Röllin, A. (2010). Stein couplings for normal approximation.
Chen, L. H. Y., & Shao, Q. M. (2001). A non-uniform Berry–Esseen bound via Stein's method. *Probability Theory and Related Fields*, *120*, 236–254.
Chen, L. H. Y., & Shao, Q. M. (2004). Normal approximation under local dependence. *Annals of Probability*, *32*, 1985–2028.
Chen, L. H. Y., & Shao, Q. M. (2005). Stein's method for normal approximation. In *Lecture Notes Series, Institute for Mathematical Sciences, National University of Singapore*: *Vol. 4*. *An introduction to Stein's method* (p. 159). Singapore: Singapore University Press.
Chen, L. H. Y., & Shao, Q. M. (2007). Normal approximation for nonlinear statistics using a concentration inequality approach. *Bernoulli*, *13*, 581–599.
Chen, L. H. Y., Fang, X., & Shao, Q. M. (2009). From Stein identities to moderate deviations.
Conway, J. H., & Sloane, N. J. A. (1999). *Sphere packings, lattices and groups* (3rd ed.). New York: Springer.
Cox, D. R. (1970). The continuity correction. *Biometrika*, *57*, 217–219.
Darling, R. W. R., & Waterman, M. S. (1985). Matching rectangles in $d$ dimensions: algorithms and laws of large numbers. *Advances in Applied Mathematics*, *55*, 1–12.
Darling, R. W. R., & Waterman, M. S. (1986). Extreme value distribution for the largest cube in a random lattice. *SIAM Journal on Applied Mathematics*, *46*, 118–132.

DeGroot, M. (1986). A conversation with Charles Stein. *Statistical Science, 1*, 454–462.
Dembo, A., & Karlin, S. (1992). Poisson approximations for $r$-scan processes. *The Annals of Applied Probability, 2*, 329–357.
Dembo, A., & Rinott, Y. (1996). Some examples of normal approximations by Stein's method. In *The IMA Volumes in Mathematics and Its Applications*: *Vol. 76. Random discrete structures* (pp. 25–44) New York: Springer.
Dharmadhikari, S., & Joag-Dev, K. (1998). *Unimodality, convexity and applications*. San Diego: Academic Press.
Diaconis, P. (1977). The distribution of leading digits uniform distribution mod 1. *Annals of Probability, 5*, 72–81.
Diaconis, P., & Freedman, D. (1987). A dozen de Finetti-style results in search of a theory. *Annales de L'I.H.P. Probabilités Et Statistiques, 23*, 397–423.
Diaconis, P., & Holmes, S. (2004). *Institute of mathematical statistics lecture notes, monograph series*: *Vol. 46. Stein's method: expository lectures and applications*. Beachwood: Institute of Mathematical Statistics.
Diaconis, P., & Shahshahani, M. (1987). Time to reach stationarity in the Bernoulli–Laplace diffusion model. *SIAM Journal on Mathematical Analysis, 18*, 208–218.
Diaconis, P., & Shahshahani, M. (1994). On the eigenvalues of random matrices. *Journal of Applied Probability, 31A*, 49–62.
Donnelly, P., & Welsh, D. (1984). The antivoter problem: random 2-colourings of graphs. In B. Bollobás (Ed.), *Graph theory and combinatorics* (pp. 133–144). San Diego: Academic Press.
Efron, B., & Stein, C. (1981). The jackknife estimate of variance. *Annals of Statistics, 9*, 586–596.
Ellis, R. (1985). *Grundlehren der Mathematischen Wissenschaften. Entropy, large deviations, and statistical mechanics*. New York: Springer.
Ellis, R., & Newman, C. (1978a). The statistics of Curie–Weiss models. *Journal of Statistical Physics, 19*, 149–161.
Ellis, R., & Newman, C. (1978b). Limit theorems for sums of dependent random variables occurring in statistical mechanics. *Zeitschrift für Wahrscheinlichkeitstheorie und Verwandte Gebiete, 44*, 117–139.
Ellis, R., Newman, C., & Rosen, J. (1980). Limit theorems for sums of dependent random variables occurring in statistical mechanics. II. Conditioning, multiple phases, and metastability. *Zeitschrift für Wahrscheinlichkeitstheorie und Verwandte Gebiete, 51*, 153–169.
Erdös, P., & Rényi, A. (1959a). On the central limit theorem for samples from a finite population. *A Magyar Tudoma'nyos Akadémia Matematikai Kutató Intézetének Közleményei, 4*, 49–61.
Erdös, P., & Rényi, A. (1959b). On random graphs. *Publicationes Mathematicae Debrecen, 6*, 290–297.
Erickson, R. (1974). $L_1$ bounds for asymptotic normality of $m$-dependent sums using Stein's technique. *Annals of Probability, 2*, 522–529.
Esseen, C. (1942). On the Liapounoff limit of error in the theory of probability. *Arkiv För Matematik, Astronomi Och Fysik, 28A*, 19 pp.
Esseen, C. (1945). Fourier analysis of distribution functions. A mathematical study of the Laplace–Gaussian law. *Acta Mathematica, 77*, 1–125.
Ethier, S., & Kurtz, T. (1986). *Markov processes: characterization and convergence*. New York: Wiley.
Feller, W. (1935). Über den Zentralen Grenzwertsatz der Wahrscheinlichkeitsrechnung. *Mathematische Zeitschrift, 40*, 512–559.
Feller, W. (1968b). *An introduction to probability theory and its applications* (Vol. 2). New York: Wiley.
Feller, W. (1968a). *An introduction to probability theory and its applications* (Vol. 1). New York: Wiley.
Ferguson, T. (1996). *A course in large sample theory*. New York: Chapman & Hall.
Finkelstein, M., Kruglov, V., & Tucker, H. (1994). Convergence in law of random sums with non-random centering. *Journal of Theoretical Probability, 7*, 565–598.

Friedrich, K. (1989). A Berry–Esseen bound for functions of independent random variables. *Annals of Statistics 17*, 170–183.
Fulman, J. (2006). An inductive proof of the Berry–Esseen theorem for character ratios. *Annals of Combinatorics, 10*, 319–332.
Fulman, J. (2009). *Communications in Mathematical Physics, 288*, 1181–1201.
Fulton, W., & Harris, J. (1991). *Graduate texts in mathematics. Representation theory*. New York: Springer.
Geary, R. (1954). The continuity ratio and statistical mapping. *Incorporated Statistician, 5*, 115–145.
Ghosh, S. (2009). *$L^p$ bounds for a combinatorial central limit theorem with involutions* (Preprint).
Glaz, J., Naus, J., & Wallenstein, S. (2001). *Springer series in statistics. Scan statistics*. New York: Springer.
Goldstein, L. (2004). Normal approximation for hierarchical sequences. *The Annals of Applied Probability, 14*, 1950–1969.
Goldstein, L. (2005). Berry Esseen bounds for combinatorial central limit theorems and pattern occurrences, using zero and size biasing. *Journal of Applied Probability, 42*, 661–683.
Goldstein, L. (2007). $L^1$ bounds in normal approximation. *Annals of Probability, 35*, 1888–1930.
Goldstein, L. (2010a). Bounds on the constant in the mean central limit theorem. *Annals of Probability. 38*, 1672–1689.
Goldstein, L. (2010b). *A Berry–Esseen bound with applications to counts in the Erdös–Rényi random graph* (Preprint).
Goldstein, L., & Penrose, M. (2010). Normal approximation for coverage models over binomial point processes. *The Annals of Applied Probability, 20*, 696–721.
Goldstein, L., & Reinert, G. (1997). Stein's method and the zero bias transformation with application to simple random sampling. *The Annals of Applied Probability, 7*, 935–952.
Goldstein, L., & Reinert, G. (2005). Distributional transformations, orthogonal polynomials, and Stein characterizations. *Journal of Theoretical Probability, 18*, 237–260.
Goldstein, L., & Rinott, Y. (1996). Multivariate normal approximations by Stein's method and size bias couplings. *Journal of Applied Probability, 33*, 1–17.
Goldstein, L., & Rinott, Y. (2003). A permutation test for matching and its asymptotic distribution. *Metron, 61*, 375–388.
Goldstein, L., & Shao, Q. M. (2009). Berry–Esseen bounds for projections of coordinate symmetric random vectors. *Electronic Communications in Probability, 14*, 474–485.
Goldstein, L., & Zhang, H. (2010). *A Berry–Esseen theorem for the lightbulb problem*. Submitted for publication.
Götze, F. (1991). On the rate of convergence in the multivariate CLT. *The Annals of Applied Probability, 19*, 724–739.
Griffeath, D. (1974/1975). A maximal coupling for Markov chains. *Zeitschrift für Wahrscheinlichkeitstheorie und Verwandte Gebiete, 31*, 95–106.
Griffiths, R., & Kaufman, M. (1982). Spin systems on hierarchical lattices. Introduction and thermodynamic limit. *Physical Review. B, Solid State, 26*, 5022–5032.
Grimmett, G., & Stirzaker, D. (2001). *Probability and random processes*. London: Oxford University Press.
Haagerup, U. (1982). The best constants in the Khintchine inequality. *Studia Mathematica, 70*, 231–283.
Hájek, J. (1960). Limiting distributions in simple random sampling from a finite population. *A Magyar Tudoma'nyos Akadémia Matematikai Kutató Intézetének Közleményei, 5*, 361–374.
Hall, P. (1988). *Introduction to the theory of coverage processes*. New York: Wiley.
Hall, B. (2003). *Lie groups, Lie algebras, and representations: an elementary introduction*. Berlin: Springer.
Hall, P., & Barbour, A. D. (1984). Reversing the Berry–Esseen inequality. *Proceedings of the American Mathematical Society 90*(1), 107–110.
Helgason, S. (2000). *Groups and geometric analysis*. Providence: American Mathematical Society.

Helmers, R. (1977). The order of the normal approximation for linear combinations of order statistics with smooth weight functions. *Annals of Probability*, *5*, 940–953.

Helmers, R., & van Zwet, W. (1982). The Berry–Esseen bound for $U$-statistics. *Statistical decision theory and related topics, III, West Lafayette, Ind.* (Vol. 1, pp. 497–512). New York: Academic Press.

Helmers, R., Janssen, P., & Serfling, R. (1990). Berry–Esséen and bootstrap results for generalized $L$-statistics. *Scandinavian Journal of Statistics*, *17*, 65–77.

Henze, N. (1988). A multivariate two-sample test based on the number of nearest neighbor type coincidences. *Annals of Statistics*, *16*, 772–783.

Ho, S. T., & Chen, L. H. Y. (1978). An $L_p$ bound for the remainder in a combinatorial central limit theorem. *Annals of Probability*, *6*, 231–249.

Hoeffding, W. (1948). A class of statistics with asymptotically normal distribution. *Annals of Mathematical Statistics*, *19*, 293–325.

Hoeffding, W. (1951). A combinatorial central limit theorem. *Annals of Mathematical Statistics*, *22*, 558–566.

Horn, R., & Johnson, C. (1985). *Matrix analysis*. Cambridge: Cambridge University Press.

Huang, H. (2002). Error bounds on multivariate normal approximations for word count statistics. *Advances in Applied Probability*, *34*, 559–586.

Hubert, L. (1987). *Assignment methods in combinatorial data analysis*. New York: Dekker.

Janson, S., & Nowicki, K. (1991). The asymptotic distributions of generalized $U$-statistics with applications to random graphs. *Probability Theory and Related Fields*, *90*, 341–375.

Jing, B., & Zhou, W. (2005). A note on Edgeworth expansions for $U$-statistics under minimal conditions. *Lietuvos Matematikos Rinkinys*, *45*, 435–440; translation in *Lithuanian Math. J.*, *45*, 353–358.

Johansson, K. (1997). On random matrices from the compact classical groups. *Annals of Mathematics*, *145*, 519–545.

Jordan, J. H. (2002). Almost sure convergence for iterated functions of independent random variables. *Journal of Applied Probability*, *12*, 985–1000.

Karlub, S., & Brede, V. (1992). Chance and statistical significance in protein and DNA sequence analysis. *Science*, *257*, 39–49.

Karoński, M., & Ruciński, A. (1987). Poisson convergence of semi-induced properties of random graphs. *Mathematical Proceedings of the Cambridge Philosophical Society*, *101*, 291–300.

Klartag, B. (2007). A central limit theorem for convex sets. *Inventiones Mathematicae*, *168*, 91–131.

Klartag, B. (2009). A Berry–Esseen type inequality for convex bodies with an unconditional basis. *Probability Theory and Related Fields*, *145*, 1–33.

Knox, G. (1964). Epidemiology of childhood leukemia in Northumberland and Durham. *British Journal of Preventive & Social Medicine*, *18*, 17–24.

Kolchin, V. F., & Chistyakov, V. P. (1973). On a combinatorial limit theorem. *Theory of Probability and Its Applications*, *18*, 728–739.

Kordecki, W. (1990). Normal approximation and isolated vertices in random graphs. In M. Karoński, J. Jaworski, & A. Ruciński (Eds.), *Random graphs 1987*. New York: Wiley.

Koroljuk, V., & Borovskich, Y. (1994). *Mathematics and its applications*: *Vol. 273. Theory of U-statistics* (translated from the 1989 Russian original by P. V. Malyshev & D. V. Malyshev and revised by the authors). Dordrecht: Kluwer Academic.

LeCam, L. (1986). The central limit theorem around 1935. *Statistical Science*, *1*, 78–96.

Leech, J., & Sloane, N. (1971). Sphere packings and error-correcting codes. *Canadian Journal of Mathematics*, *23*, 718–745.

Lévy, P. (1935). Propriétes asymptotiques des sommes de variables indépendantes on enchainees. *Journal de Mathématiques Pures Et Appliquées*, *14*, 347–402.

Li, D., & Rogers, T. D. (1999). Asymptotic behavior for iterated functions of random variables. *The Annals of Applied Probability*, *9*, 1175–1201.

Liggett, T. (1985). *Interacting particle systems*. New York: Springer.

Luk, M. (1994). *Stein's method for the Gamma distribution and related statistical applications*. Ph.D. dissertation, University of Southern California, Los Angeles, USA.

Madow, W. G. (1948). On the limiting distributions of estimates based on samples from finite universes. *Annals of Mathematical Statistics, 19*, 535–545.

Mantel, N. (1967). The detection of disease cluttering and a generalized regression approach. *Cancer Research, 27*, 209–220.

Matloff, N. (1977). Ergodicity conditions for a dissonant voting model. *Annals of Probability, 5*, 371–386.

Mattner, L., & Roos, B. (2007). A shorter proof of Kanter's Bessel function concentration bound. *Probability Theory and Related Fields, 139*, 191–205.

Meckes, M., & Meckes, E. (2007). The central limit problem for random vectors with symmetries. *Journal of Theoretical Probability, 20*, 697–720.

Midzuno, H. (1951). On the sampling system with probability proportionate to sum of sizes. *Annals of the Institute of Statistical Mathematics, 2*, 99–108.

Moran, P. (1948). The interpretation of statistical maps. *Journal of the Royal Statistical Society. Series B. Methodological, 10*, 243–251.

Motoo, M. (1957). On the Hoeffding's combinatorial central limit theorem. *Annals of the Institute of Statistical Mathematics, 8*, 145–154.

Nagaev, S. (1965). Some limit theorems for large deviations. *Theory of Probability and its Applications, 10*, 214–235.

Naus, J. I. (1982). Approximations for distributions of scan statistics. *Journal of the American Statistical Association, 77*, 177–183.

Nourdin, I., & Peccati, G. (2009). Stein's method on Weiner chaos. *Probability Theory and Related Fields, 145*, 75–118.

Nourdin, I., & Peccati, G. (2011). *Normal approximations with Malliavin calculus. From Stein's method to universality.*

Nourdin, I., Peccati, G., & Reinert, G. (2009, to appear). *Invariance principles of homogeneous sums: Universality of Gaussian Wiener chaos. Annals of Probability.*

Nualart, D. (2006). *The Malliavin calculus and related topics* (2nd ed.). Berlin: Springer

Palka, Z. (1984). On the number of vertices of a given degree in a random graph. *Journal of Graph Theory, 8*, 167–170.

Papangelou, F. (1989). On the Gaussian fluctuations of the critical Curie–Weiss model in statistical mechanics. *Probability Theory and Related Fields, 83*, 265–278.

Peccati, G. (2009). Stein's method, Malliavin calculus and infinite-dimensional Gaussian analysis. Lecture notes from *Progress in Stein's method.* In http://www.ims.nus.edu.sg/Programs/stein09/files/Lecture_notes_5.pdf.

Peköz, E., & Röllin, A. (2009). *New rates for exponential approximation and the theorems of Rényi and Yaglom* (Preprint).

Penrose, M. (2003). *Random geometric graphs.* Oxford: Oxford University Press.

Petrov, V. (1995). *Oxford studies in probability: Vol. 4. Limit theorems of probability theory: sequences of independent random variables.* London: Oxford University Press.

Pickett, A. (2004). *Rates of convergence of $\chi^2$ approximations via Stein's method.* Ph.D. dissertation, Oxford.

Proctor, R. (1990). A Schensted algorithm which models tensor representations of the orthogonal group. *Canadian Journal of Mathematics, 42*, 28–49.

Rachev, S. (1984). The Monge–Kantorovich transference problem and its stochastic applications. *Theory of Probability and Its Applications, 29*, 647–676.

Raič, M. (2004). A multivariate CLT for decomposable random vectors with finite second moment. *Journal of Theoretical Probability, 17*, 573–603.

Raič, M. (2007). CLT related large deviation bounds based on Stein's method. *Advances in Applied Probability, 39*, 731–752.

Rao, R., Rao, M., & Zhang, H. (2007). One bulb? Two bulbs? How many bulbs light up? A discrete probability problem involving dermal patches. *Sankhyā. The Indian Journal of Statistics, 69*, 137–161.

Reinert, G., & Röllin, A. (2009). Multivariate normal approximation with Stein's method of exchangeable pairs under a general linearity condition. *Annals of Probability, 37*, 2150–2173.

Reinert, G., & Röllin, A. (2010). Random subgraph counts and $U$-statistics: multivariate normal approximation via exchangeable pairs and embedding. *Journal of Applied Probability*, *47*(2), 378–393.

Rinott, Y. (1994). On normal approximation rates for certain sums of dependent random variables. *Journal of Computational and Applied Mathematics*, *55*, 135–143.

Rinott, Y., & Rotar, V. (1996). A multivariate CLT for local dependence with $n^{-1/2}\log n$ rate and applications to multivariate graph related statistics. *Journal of Multivariate Analysis*, *56*, 333–350.

Rinott, Y., & Rotar, V. (1997). On coupling constructions and rates in the CLT for dependent summands with applications to the antivoter model and weighted $U$-statistics. *The Annals of Applied Probability*, *7*, 1080–1105.

Robbins, H. (1948). The asymptotic distribution of the sum of a random number of random variables. *Bulletin of the American Mathematical Society*, *54*, 1151–1161.

Ross, S., & Peköz, E. (2007). *A second course in probability*. United States: Pekozbooks.

Sagan, B. (1991). *The symmetric group*. Belmont: Wadsworth.

Sazonov, V. (1968). On the multi-dimensional central limit theorem. *Sankhyā Series A*, *30*, 181–204.

Schechtman, G., & Zinn, J. (1990). On the volume of the intersection of two $\ell_p^n$ balls. *Proceedings of the American Mathematical Society*, *110*, 217–224.

Schiffman, A., Cohen, S., Nowik, R., & Sellinger, D. (1978). Initial diagnostic hypotheses: factors which may distort physicians judgement. *Organizational Behavior and Human Performance*, *21*, 305–315.

Schilling, M. (1986). Multivariate two-sample tests based on nearest neighbors. *Journal of the American Statistical Association*, *81*, 799–806.

Schlösser, T., & Spohn, H. (1992). Sample to sample fluctuations in the conductivity of a disordered medium. *Journal of Statistical Physics*, *69*, 955–967.

Seber, G., & Lee, A. (2003). *Wiley series in probability and statistics. Linear regression analysis* (2nd ed.). New York: Wiley.

Serfling, R. (1980). *Approximation theorems of mathematical statistics*. New York: Wiley.

Serre, J.-P. (1997). *Graduate texts in mathematics. Linear representations of finite groups*. New York: Springer.

Shao, Q. M., & Su, Z. (2005). The Berry–Esseen bound for character ratios. *Proceedings of the American Mathematical Society*, *134*, 2153–2159.

Shneiberg, I. (1986). Hierarchical sequences of random variables. *Theory of Probability and Its Applications*, *31*, 137–141.

Steele, J. M. (1986). An Efron–Stein inequality for nonsymmetric statistics. *Annals of Statistics*, *14*, 753–758.

Stein, C. (1956). Inadmissibility of the usual estimator for the mean of a multivariate normal distribution. In *Proceedings of the Third Berkeley Symposium on Mathematical Statistics and Probability, 1954–1955* (Vol. I, pp. 197–206). Berkeley: University of California Press.

Stein, E. M. (1970). *Singular integrals and differentiability properties of functions*. Princeton: Princeton University Press.

Stein, C. (1972). A bound for the error in the normal approximation to the distribution of a sum of dependent random variables. In *Proceedings of the Sixth Berkeley Symposium on Mathematical Statistics and Probability* (Vol. 2, pp. 586–602). Berkeley: University of California Press.

Stein, C. (1981). Estimation of the mean of a multivariate normal distribution. *Annals of Statistics*, *9*, 1135–1151.

Stein, C. (1986). *Approximate computation of expectations*. Hayward: IMS.

Stein, C. (1995). *The accuracy of the normal approximation to the distribution of the traces of powers of random orthogonal matrices* (Technical Report no. 470). Stanford University, Statistics Department.

Stroock, D. (2000). *Probability theory, an analytic view*. Cambridge: Cambridge University Press.

Sundaram, S. (1990). Tableaux in the representation theory of compact Lie groups. In *IMA volumes in mathematics: Vol. 19. Invariant theory and tableaux* (pp. 191–225). New York: Springer.

Tyurin, I. (2010, to appear). New estimates on the convergence rate in the Lyapunov theorem.

von Bahr, B. (1976). Remainder term estimate in a combinatorial limit theorem. *Zeitschrift für Wahrscheinlichkeitstheorie und Verwandte Gebiete*, *35*, 131–139.

Wald, A., & Wolfowitz, J. (1944). Statistical tests based on permutations of the observations. *Annals of Mathematical Statistics*, *15*, 358–372.

Wehr, J. (1997). A strong law of large numbers for iterated functions of independent random variables. *Journal of Statistical Physics*, *86*, 1373–1384.

Wehr, J. (2001). Erratum on: A strong law of large numbers for iterated functions of independent random variables [Journal of Statistical Physics, 86(1997), 1373–1384]. *Journal of Statistical Physics*, *104*, 901.

Wehr, J., & Woo, J. M. (2001). Central limit theorems for nonlinear hierarchical sequences of random variables. *Journal of Statistical Physics*, *104*, 777–797.

Weyl, H. (1997). *The classical groups*. Princeton: Princeton University Press.

Zhao, L., Bai, Z., Chao, C.-C., & Liang, W.-Q. (1997). Error bound in the central limit theorem of double-indexed permtation statistics. *Annals of Statistics*, *25*, 2210–2227.

Zhou, H., & Lange, K. (2009). Composition Markov chains of multinomial type. *Advances in Applied Probability*, *41*, 270–291.

Zong, C. (1999). *Sphere packings*. New York: Springer.

# Author Index




L.H.Y. Chen et al., *Normal Approximation by Stein's Method*,
Probability and Its Applications,
DOI 10.1007/978-3-642-15007-4, © Springer-Verlag Berlin Heidelberg 2011

# Subject Index




L.H.Y. Chen et al., *Normal Approximation by Stein's Method*,
Probability and Its Applications,
DOI 10.1007/978-3-642-15007-4, © Springer-Verlag Berlin Heidelberg 2011